Pharmacology – Research, Safety Testing and Regulation Series

Biogenic Amines: Pharmacological, Neurochemical and Molecular Aspects in the CNS

PHARMACOLOGY – RESEARCH, SAFETY TESTING AND REGULATION SERIES

Antibiotic Resistance: Causes and Risk Factors, Mechanisms and Alternatives
Adriel R. Bonilla and Kaden P. Muniz (Editors)
2009. ISBN: 978-1-60741-623-4

Antibiotic Resistance: Causes and Risk Factors, Mechanisms and Alternatives
Adriel R. Bonilla and Kaden P. Muniz (Editors)
2009. ISBN: 978-1-61668-162-3 (Online Book)

Poisons: Physiologically Active Substances
S. B. Zotov and O. I. Tuzhikov
2009. ISBN: 978-1-60741-973-0

Drug Monitoring by HPLC: Recent Developments
Victoria Samanidou and Eftichia Karageorgou
2010. ISBN: 978-1-60876-183-8

Nonprescription Drugs: Considering A New Class for Behind-the-Counter Drugs
Lars P. Eliassen (Editor)
2010. ISBN: 978-1-60741-961-7

Biogenic Amines: Pharmacological, Neurochemical, and Molecular Aspects in the CNS
Tahira Farooqui and Akhlaq A. Farooqui (Editors)
2010. ISBN: 978-1-60876-625-3

PHARMACOLOGY – RESEARCH, SAFETY TESTING AND REGULATION SERIES

BIOGENIC AMINES: PHARMACOLOGICAL, NEUROCHEMICAL AND MOLECULAR ASPECTS IN THE CNS

TAHIRA FAROOQUI
AND
AKHLAQ A. FAROOQUI
EDITORS

Nova Biomedical Books
New York

For permission to use material from this book please contact us:
Telephone 631-231-7269; Fax 631-231-8175
Web Site: http://www.novapublishers.com

LIBRARY OF CONGRESS CATALOGING-IN-PUBLICATION DATA

Biogenic amines : pharmacological, neurochemical, and molecular aspects in the CNS / editors, Tahira Farooqui, Akhlaq A. Farooqui.
p. ; cm.
ISBN 978-1-60876-625-3 (hardcover)
1. Biogenic amines. 2. Neuropharmacology. I. Farooqui, Tahira. II. Farooqui, Akhlaq A.
[DNLM: 1. Biogenic Amines--pharmacology. 2. Biogenic Amines--physiology. 3. Central Nervous System--chemistry. 4. Neurotransmitter Agents--pharmacology. 5. Neurotransmitter Agents--therapeutic use. QU 61 B615 2009]
QP801.B66B565 2009
615'.78--dc22

2009041972

Published by Nova Science Publishers, Inc. ✝ New York

DEDICATION

A TRIBUTE TO PROFESSOR NORMAN J. URETSKY
(1941 – 2008)

This book is dedicated to the memory of Norman J. Uretsky, an outstanding teacher, a colleague and a dear friend who will be missed by all who knew him and his work.

CONTENTS

PREFACE

Biogenic amines are naturally occurring amines that are derived by enzymic decarboxylation of the natural amino acids. They belong to a class of neurotransmitters including catecholamines (dopamine, norepinephrine, and epinephrine), indolamine (serotonin), and imidazoleamine (histamine). In both vertebrates and invertebrates, major biogenic amines consist of five members except that in invertebrates norepinephrine and epinephrine are functionally substituted by tyramine and octopamine.

Biogenic amines have great pharmacological and physiological importance. They are synthesized in neurons, packed into vesicles, and released from vesicles at synaptic terminals into the synaptic cleft by the fusion of synaptic vesicles with the plasma membrane. Biogenic amines bind with metabotropic G protein-coupled receptors (GPCRs), leading to slower cellular responses. They are cleared from the synaptic cleft into the cytoplasm through the reuptake transporter located at the presynaptic terminal or by degrading enzymes. Biogenic amines are important neuroactive molecules in the central nervous system (CNS) of both vertebrates and invertebrates. Physiologically, biogenic amines play role as neurotransmitters, neuromodulators, or neurohormones. They are associated with functioning of the neuronal circuitry throughout the brain. Biogenic amines regulate plasticity of synaptic transmission by modulating changes in synaptic morphology, number of synapses, and receptor expression, affecting various behavioral modifications including locomotion, endocrine secretion, emotional states, cognition, motivation, reward, and learning and memory. In humans, the process of aging and the pathogenesis of various neurological disorders, such as Parkinson disease, Alzheimer disease, Huntington disease, and schizophrenia, have been linked to impaired biogenic amine signaling.

The main objective of this book is to present readers with comprehensive information on pharmacology, neurochemistry and molecular neurobiology of biogenic amine in the CNS of vertebrate and invertebrates in a single volume text. The book has been organized into chapters and sections to provide a better flow of information. This unique volume provides its readers with cutting edge information on biogenic amines. It can be used by graduate students, postdoctoral fellows, researchers, and scientists who work at the pharmaceutical industry as a handbook, which describes all aspects of biogenic amine metabolism.

In this book chapters are organized into two sections: (1) biogenic amines in vertebrate system, and (2) biogenic amines in invertebrate system. In section I, chapters 1 and 2 deal with the historical aspects of biogenic amines and life's requirements for catecholamines. Chapters 3 and 4 provide a perspective on a regulatory enzyme, aromatic L-amino acid decarboxylase, and dopamine storage and its utilization. Chapter 5 focuses on the importance of monoamine transporters in psychopharmacology. Chapter 6 discusses amine false transmitters, co-transmitters and volume transmission. Chapter 7 describes the pharmacological tools to evaluate drug-receptor interactions. Chapters 8 and 9 cover cutting-edge information on 6-hydroxydopamine animal model system and genetic basis of Parkinson disease. Chapters 10-13 are devoted to biogenic amine receptor-mediated signaling. In section II, Chapters 14-16 deal with octopamine quantification, octopamine and tyramine receptor pharmacology, and octopaminergic neurons molecular biology in invertebrate system. Finally, Chapter 17 provides readers and researchers with perspective that will be important for future research work on biogenic amines.

The publication of the current volume is made possible by the enthusiasm and personal input of all the contributors. It is a result of collaborative participation from international scientific community working on biogenic amines. The editors strongly believe that everyone who is either working or going to start their scientific career in the fascinating area of 'Biogenic Amines' would find this book as a helpful source. Editors hope that you would enjoy reading this book as much as all contributors have used their patience and efforts in carrying out their responsibilities by putting the selected and valuable information in these chapters.

Tahira Farooqui
Akhlaq A. Farooqui

FOREWORD

This book is comprised of chapters prepared by former students, postdoctoral fellows, and research colleagues of the late Dr. Norman J. Uretsky. It is a heartfelt tribute to an outstanding individual who was a wonderful teacher, superb mentor, brilliant scholar, and extraordinary colleague and friend to so many.

Norman Uretsky was born in New York City on April 29, 1941. He earned a Bachelor of Science in Pharmacy degree from Columbia University in 1962 and a Doctor of Philosophy degree in Pharmacology from the University of Chicago in 1968. Dr. Uretsky was a USPHS Postdoctoral Fellow in pharmacology at the University of Cambridge, UK, from 1968 to 1970. He was an assistant professor in pharmacology and neuropathology at Harvard Medical School, Boston, from 1971 to 1977. Professor Uretsky joined the College of Pharmacy at The Ohio State University in 1977 as associate professor of pharmacology. He taught pharmacology courses to pharmacy students and graduate students, and Professor Uretsky received numerous teaching awards during his career. He was the recipient of the annual Miriam R. Balshone Teaching Award from the College of Pharmacy in 1983, 1986, 1992, 1995, 1999, 2002, and 2003. Professor Uretsky also received the 1997 Alumni Award for Distinguished Teaching from The Ohio State University. He is remembered by his students and colleagues for his outstanding teaching skills and his devotion to pharmacology and pharmacy education.

Professor Uretsky's research and scholarship focused on neuropharmacology, dopaminergic neurotransmitters, and studies on the pharmacological actions of drugs of abuse. He published over 85 peer reviewed research publications, served as major advisor to fourteen Ph.D. students, and trained five postdoctoral fellows. His research activities were continuously supported by funds from the National Institutes of Health from 1971 through 2001. From 1978 to 1997, he served on the editorial advisory board for the Journal of Pharmacology and Experimental Therapeutics.

Professor Uretsky was an active member of the University community and served on numerous college and university committees. He served as chair of the Division of Pharmacology in the College of Pharmacy from 1995 to 2002. Dr. Uretsky retired in February 2004 as Professor Emeritus, Division of Pharmacology, College of Pharmacy. He continued to be involved in teaching pharmacology to pharmacy students on a part-time basis from 2004 through 2006. Norm Uretsky lost his courageous battle with cancer on September 20, 2008.

Dr. Uretsky was valued for his collegiality, his positive attitude, his commitment to his students, his concern for others, and his effective quiet leadership. Our lives have certainly been enriched in numerous ways through our interactions with Norm, for he was a wonderful colleague and friend.

Robert W. Brueggemeier, Ph.D.
Dean, College of Pharmacy
Professor, Medicinal Chemistry
The Ohio State University

ACKNOWLEDGMENTS

We would like to express our deepest appreciation and gratitude to all authors who took valuable time from their busy schedules and shared their expertise by contributing chapters for this book. We would also like to thank anonymous reviewers who read and commented on the first drafts of many chapters, and authors who incorporated their suggested improvements. We acknowledge with gratitude for the encouragement and advice received from Drs. Leslie L. Iversen, Michael J. Zigmond and Lane J. Wallace during the earlier stages of planning this scientific challenge. We would also like to thank Drs. Allan Burkman and Ella Uretsky for providing Dr. Norman Uretsky's photograph. This book would not have been possible without indispensable help and patience of Mr. Frank Columbus and staff at NOVA Science Publishers in completing the editorial process.

Tahira Farooqui
Akhlaq A. Farooqui

I. Biogenic Amines in Vertebrate System

In: Biogenic Amines
Editors: T. Farooqui, A. A. Farooqui
ISBN: 978-1-60876-625-3

Chapter I

BIOGENIC AMINES: A HISTORICAL PERSPECTIVE

Michael C. Gerald[*,&]
University of Connecticut, Connecticut 06268, USA

ABSTRACT

The biogenic amines include the three catecholamines, dopamine, norepinephrine, and epinephrine, and the tyrosine metabolites octopamine and tyramine. Serotonin (indoleamine) and histamine (imidazoleamine) are also biogenic amines that play a role in brain function. Pathways for the synthesis and metabolism of catecholamine are summarized. This chapter focuses on norepinephrine and dopamine, catecholamines which have neurotransmitter functions in the mammalian central nervous system. Significant events in the development of our understanding of catecholamines and a timeline are presented. Central norepinephrine is thought to play a role in the regulation of learning and memory, reinforcement, mood, arousal and sleep. Drugs used for the treatment of neuropsychiatric illnesses, such as depression and aggressive or disruptive behavior, act by interacting with adrenergic receptors in the specific region of the brain. Two types of adrenoreceptors (α-adrenoceptors and β-adrenoceptors) with numerous subtypes have been found in brain. A deficiency in norepinephrine or reduction in its receptor functional activity may be associated with depression. There are three

* Correspondence concerning this article should be addressed to: Michael C. Gerald, Ph.D., Department of Pharmacy Practice, School of Pharmacy, University of Connecticut, 69 North Eagleville Road, Storrs, CT 06269-3092, Tel 860-486-5416, E-mail Michael.gerald@uconn.edu.

& Throughout his distinguished career, our colleague and very dear friend Norman Uretsky studied the neuropharmacology of catecholamines as well as the inter-relationships of the dopaminergic and glutaminergic systems in the brain. Working with faculty and student collaborators, he studied the effects of 6-hydroxydopamine-induced chemical sympathectomy on the spontaneous motor activity of rats, the influence of dopaminergic and glutaminergic agonist and antagonists on animal behavior, 3-nitro- tyrosine-induced neurodegeneration of the striatum, and mechanisms underlying the abuse potential of amphetamine and cocaine. Working with Norm on instructional activities, committee assignments, and research projects were always both a personal delight and an enriching educational experience and ones that I will miss.

dopaminergic pathways and two families of dopamine receptors (D1-like and D2-like) in the brain that influence motor function, behavior, reward and drug dependence, attention-deficit-hyperactivity disorder, and prolactin release. A depletion of dopamine exists in the substantia nigra and striatum in Parkinson disease. One leading model of schizophrenia suggests that schizophrenia is associated with an increase in the functional activity of dopamine in the limbic system. "Typical" antipsychotic agents are potent D_2-antagonists, while "atypical" antipsychotics have dopamine and serotonin receptor blocking actions.

1. INTRODUCTION

Biogenic amines, a class of neurotransmitters, include catecholamines, indolamine, and imadazoleamine. It is an important class of neuroactive substances that are released by action potentials, long distance signaling, from neurons. After their release, these molecules bind to specific receptors that are located on the target cell membrane. Activation of biogenic amine receptors results in changes in the intracellular concentration of second messengers that are responsible for slow but long-lasting cellular responses. Pharmacological, physiological, and molecular approaches have helped in elucidating that biogenic amines regulate a multitude of cellular responses in both vertebrates and invertebrates.

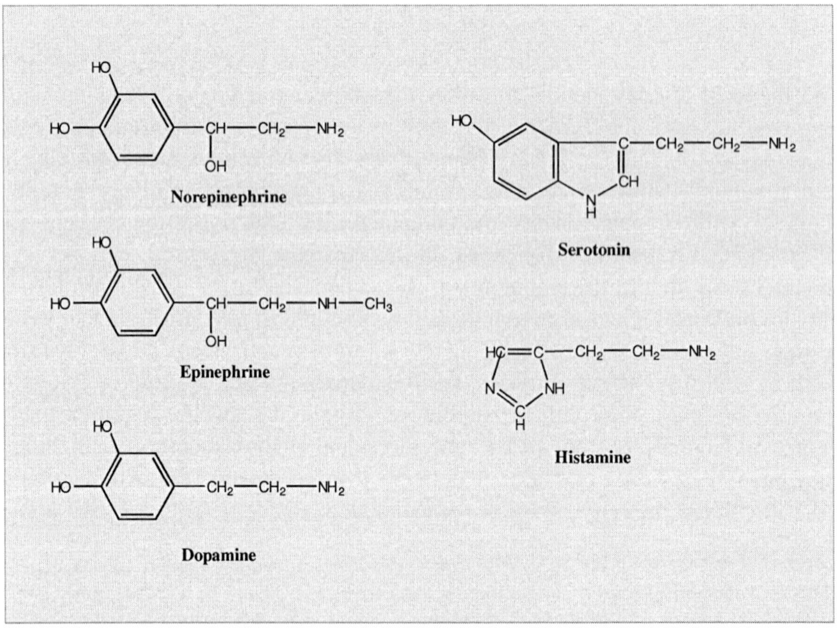

Figure. 1. Chemical structures of five classical biogenic amines.

In vertebrates, dopamine, norepinephrine, epinephrine, serotonin, and histamine have been considered to be classical biogenic amines (Figure 1). Serotonin and histamine are also present in invertebrates. Norepinephrine and epinephrine are functionally substituted by two structurally-related phenylamines, tyramine and octopamine, and both mediate diverse physiological functions in invertebrates. These amines are present at very low concentration in the mammalian brain, except in certain human disorders, suggesting that they may have some physiological significance in vertebrates.

Catecholamines include a group of important neurotransmitters that share a common ring structure, catechol ($C_6H_4(OH)_2$). Catecholamines resemble one another chemically: (1) by having an aromatic portion (catechol), and (2) a side chain, an amine or nitrogen-containing group, attached to the catechol ring. The most important catecholamines in vertebrates are dopamine, norepinephrine and epinephrine, which play key roles in neurotransmission and other signaling functions. The deficiency or excess of these amines in human brain is responsible for the symptoms of neurodegenerative, neuropsychiatric and neurological disorders (such as Parkinson's disease, Alzheimer's disease and schizophrenia). Let us next look briefly at the metabolism of catecholamines.

2. Catecholamine Biosynthesis and Inactivation

Catecholamines are biosynthesized from the common precursor tyrosine, a nonessential amino acid. Tyrosine is synthesized from the essential amino acid phenylalanine but the body's needs are more than adequately supplied by such dietary sources of tyrosine as soy products, chicken, turkey, fish, casein products, and lima beans. Tyrosine is hydroxylated by tyrosine hydroxylase to dopa (dihydroxyphenylalanine) which is decarboxylated to dopamine by aromatic L-amino acid decarboxylase (dopa decarboxylase) (Figure 2). These reactions take place in catecholamine-secreting neurons or cells of the adrenal medulla. After its movement into granular vesicles, dopamine is converted to norepinephrine, a reaction catalyzed by dopamine β-hydroxylase (DβH). Norepinephrine is N-methylated to epinephrine by phenylethanolamine-N-methyltransferase, a reaction that primarily occurs in adrenal medullary cells.

After release from their respective neurons and binding to postsynaptic and presynaptic receptors, norepinephrine and dopamine are primarily removed from these loci and from synapses into their presynaptic neurons by a reuptake process involving transporter proteins. Both catecholamines are also converted to biologically-inactive metabolites by oxidation and methylation reactions catalyzed by monoamine oxidase (MAO) and catechol-O-methyltransferase (COMT), respectively (Figure 2) [1-2].

In vertebrates, tyramine and octopamine are normally minor metabolites present in catecholamine-synthesizing tissues [3]. These "trace amines" constitute $<1\%$ of all biogenic amines in mammals. Tyrosine is converted to tyramine, a reaction catalyzed by aromatic L-amino acid decarboxylase and then to octopamine by DβH (Figure 2). The levels of these catecholamines and their biological significance are increased considerably after an MAO inhibitor (such as the antidepressant phenelzine) is taken and then ingesting a tyramine-rich diet (e.g., red wine, aged cheese, pickled herring). Tyramine builds up stimulating the massive release of norepinephrine from nerve endings resulting in a potentially fatal hypertensive crisis [2]. (For decades, pharmacology instructors have used this as a classic example of a drug-food interaction). Octopamine has been shown to play an important role as a neurotransmitter in invertebrates.

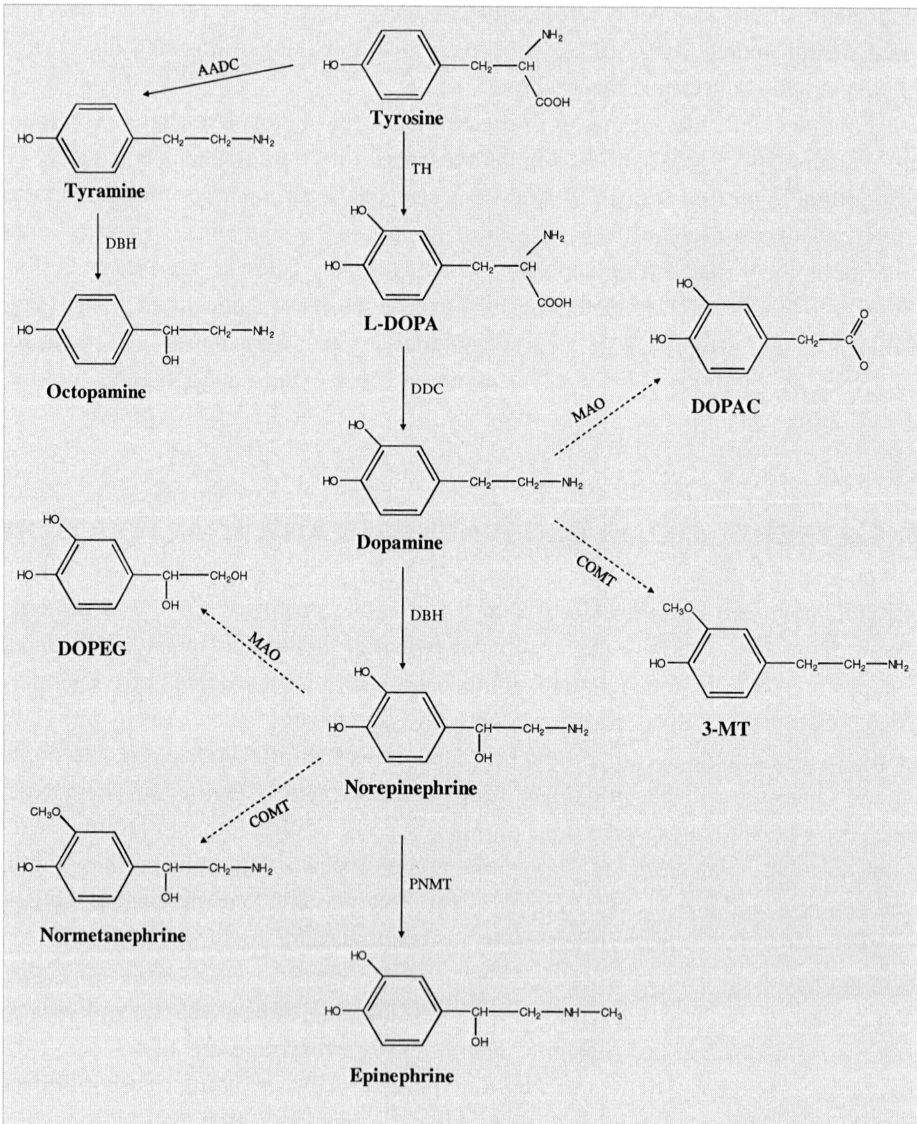

Figure. 2. Metabolic pathways of catecholamines. Enzymes involved: TH, Tyrosine hydroxylase; DDC, Dopa decarboxylase; DβH, dopamine-β-hydroxylase; PNMT, Phenylethanolamine N-methyl transferase; AADC, L-amino acid decarboxylase; MAO, Monoamine oxidase; COMT, Catechol-O-methyl transferase. L-DOPA, 3,4-dihydroxy-*L*-phenylalanine; DOPAC, Dihydroxyphenylacetic acid; 3-MT, 3-Methoxytyramine.

3. Epinephrine and Norepinephrine

Biological interest in catecholamines, in particular epinephrine, dates back to the final years of the nineteenth-century. Within one decade after Oliver and Schaefer in 1895 showed that an extract of the adrenal medulla produced a marked increase in blood pressure, its active constituent epinephrine (adrenaline) was isolated, purified, and synthesized. Epinephrine was the first hormone isolated in crystalline form. Elliott's studies in 1905 with epinephrine are of greater importance from our perspective. His results provided an important clue in the development of the nascent concept that chemicals rather than electrogenic forces were responsible for carrying messages from nerve endings to effector tissues. Direct proof of chemically-mediated neurotransmission came in 1921 with Loewi's brilliant, but simple, experiments demonstrating the release of *Vagustuff* (later shown to be acetylcholine) from the electrically-stimulated vagus nerve of a frog's heart [4].

During the 1920s and 30s, Cannon observed that the changes in blood pressure and heart rate produced by a chemical released after stimulation of sympathetic nerves, which he called "sympathin," were similar but differed somewhat from those changes produced by an injection of epinephrine. This discrepancy was clarified by von Euler in the 1946 when he discovered that norepinephrine, a biological precursor of epinephrine, was the neurotransmitter released from these nerve endings. Early research interest in norepinephrine (noradrenaline) focused upon its effects on peripheral tissues, with emphasis on the cardiovascular system, and its role as the primary sympathetic postganglionic neurotransmitter.

In the 1950s, the interest shifted to the central nervous system. Marthe Vogt found that norepinephrine was differentially distributed in different regions of the brain leading her suggestion that it might act as a central neurotransmitter. Using fluorescent techniques, Falck and Hillarp visualized noradrenergic pathways and identified the location of their cell bodies in the pons (locus ceruleus) and medulla, with their nerve terminals in the cerebral cortex, limbic system, and hypothalamus. In addition, ascending and descending noradrenergic fibers were found to pass through the medial forebrain bundle with their nerve fibers traveling to the cerebral cortex, hippocampus and cerebellum [2]. Brodie and Shore postulated that norepinephrine was the primary neurotransmitter of sympathetic centers in the brain, and that the effects of a number of psychoactive drugs, including amphetamine, ephedrine, and mescaline, might result from their interaction with norepinephrine.

If biogenic amines were associated with normal brain function, might changes in their functional activity play a role in mental disease or other disfunctions of the central nervous system? The results of neuropharmacological studies in animals supported clinical observations that drug-induced depletion of brain norepinephrine with reserpine caused depression, while elevation of its levels at sympathetic synapses alleviated clinical depression or even elevated the mood. In his 1965 "catecholamine hypothesis of affective disorders," Schildkraut proposed that depression is associated with a decrease and mania an excess in the functional activity of norepinephrine at these synapses. This model provided a unified framework to explain the mechanisms of action of antidepressant and antimanic drugs. While elements of this hypothesis remain appealing four decades later, and norepinephrine is still thought to play a role in some depressive illnesses, significant inconsistencies and new research findings now place greater emphasis on serotoninergic involvement in depression and the use of selective serotonin reuptake inhibitors for its treatment [5].

In addition to mood disorders, norepinephrine has been implicated in mechanisms involved with arousal and sleep [6]. Electrodes implanted in the locus ceruleus in animals show increased activity during periods of arousal, while the electrical activity in this area is absent during sleep. Sleep is not a uniform state but rather cycles at 90-minute intervals between periods of rapid-eye movements (REM), which is associated with dreaming, and non-REM sleep. Changes in the release of norepinephrine, serotonin, and acetylcholine are observed during and between these sleep phases. During deep sleep, norepinephrine levels decline and serotonin levels rise, while REM sleep is thought to be induced by the release of norepinephrine and acetylcholine [7].

4. Dopamine

Dopamine, once an inconspicuous intermediate in the biosynthesis of norepinephrine, is now the most widely researched catecholamine. During the late 1950s, Carlsson utilized a variety of experimental approaches that demonstrated the presence of dopamine in the brain and its function as a neurotransmitter.

Three dopaminergic pathways are present in the brain: The *nigrostriatal pathway* contains three-quarters of brain dopamine, has cell bodies originating in the substantia nigra and terminating in the corpus striatum, and is associated with motor control. The *mesolimbic-mesocortical pathway*, with cell bodies in the ventral tegmental area and projections to the limbic system, is involved with behavior, more specifically, reward and addiction; defects in this system may account for some of the symptoms of schizophrenia. The cell bodies of the *tuberoinfundibular pathway* originate in the ventral hypothalamus and terminate in the anterior pituitary gland, and its activation is responsible for inhibition of the release of prolactin from the anterior pituitary [1,2,8].

Based on their pharmacological and structural properties, five subtypes of dopamine receptors have been identified, and these have been grouped into two families: D1-like, which includes D_1 and D_5, and D2-like that consists of D_2, D_3, and D_4 subtypes. The two families are present in common (striatal) and distinct regions of the brain. Each of these subtypes is a G-protein transmembrane receptor. Activation of D_1 receptors stimulate adenylyl cyclase leading to the formation of cAMP by activating the stimulatory G protein Gs, while the D_2 receptors inhibit adenylyl cyclase and cAMP levels by activating the inhibitory G protein Gi [1-2].

Carlsson detected high levels of dopamine in the basal ganglia. Depletion of dopamine in the substantia nigra with reserpine caused Parkinsonian symptoms, which were alleviated by administration of the dopamine precursor levodopa. (This is a classic example of replacement therapy). Unlike dopamine, levodopa crosses the blood-brain barrier, making it useful for the treatment of this neurological disorder. Other drugs used for the treatment of Parkinson's disease include those that are dopamine agonists (bromocriptine, pergolide) and inhibitors of dopa decarboxylase (carbidopa), MAO-B (selegiline) or COMT (entacapone), enzymes that catalyze the inactivation of levodopa or dopamine [9-10].

The development of antiparkinson drugs was based on a rational approach, namely, determination of the neurochemical defect in the disorder followed by treatment with specific drugs that corrected the defect. By contrast, drugs possessing antipsychotic effects were first found to be clinically effective and, based on their effects, neurobiological theories were subsequently formulated that sought to explain the cause(s) of schizophrenia.

The first "typical" antipsychotic drugs, chlorpromazine and haloperidol, appeared in clinics in the mid-1950s and 1960s. These prototypes were followed by some dozen drugs that were more similar than different in their wide spectrum of receptor interactions and adverse neurological effects [11]. In 1976 Seeman noted a striking correlation between the clinical potency of antipsychotic drugs (expressed as the average daily dose used to treat schizophrenia) and their activity in binding to and blocking the D_2- receptor in the striatum *in vitro* [12]. These findings, linking an increased functional activity of dopamine and schizophrenia, were reinforced by the observations that the dopamine-releaser amphetamine elicits stereotyped behavior in animals analogous to that seen in some psychotic patients, and high doses of amphetamine can cause a psychotic-like state in humans. Moreover, the D_2-receptor agonists apomorphine and bromocriptine have been shown to worsen the symptoms of schizophrenic patients. When coupled with other clinical findings, it appears that D_2-receptor antagonist antipsychotics are more effective in treating the positive symptoms of schizophrenia than the negative or cognitive symptoms.

Clozapine is an effective antipsychotic drug that has a different spectrum of pharmacological effects and is devoid of the extrapyramidal side effects that plague the chlorpromazine (phenothiazine) and haloperidol (butyrophenone)-like drugs. With the appearance of clozapine, the prototype of a series of "atypical" antipsychotic drugs introduced in the United States in 1989, this weak D_2- receptor blocker and potent serotonin 5-HT_{2A} antagonist, has undermined the earlier focus on dopamine as the exclusive cause of schizophrenia [11,13].

One of our greatest societal challenges deals with a better understanding of the nature of drug dependence and more effective approaches to its treatment. Drugs capable of producing a dependent state and their compulsive use, such as opioids, cocaine, amphetamines, nicotine, and alcohol, produce different behavioral effects. However, they appear to share a common link, namely, all these drugs activate the mesolimbic dopaminergic pathway leading to a release of dopamine in the nucleus accumbens. Activation of this pathway is thought to be responsible for the rewarding or positive reinforcing effects of these drugs and their continued use. (Compulsive gambling and sexual behavior may be added to this list). Chemical or surgical interference with this dopamine pathway reduces the rewarding properties of such drugs [14]. As noted above for schizophrenia, serotonin may also play a role in mechanisms of reward.

Attention-deficit-hyperactivity disorder (ADHD), makes its appearance in childhood, and is characterized by increased motor activity and impulsive behavior and difficulty in maintaining attention and focus. Recent evidence suggests that ADHD may result from reduced levels of dopamine in the hippocampus and amygdala. Amphetamines and methylphenidate increase the postsynaptic levels of dopamine and have long been used to effectively treat ADHD [15]. Dopamine inhibits prolactin release, and D_2-antagonist antipsychotic drugs stimulate lactation, a side effect seen even in male patients. There are many physical and chemical causes of nausea and vomiting. Drugs that enhance dopaminergic activity, such as those used to treat Parkinson's disease, cause emetic effects, while dopamine antagonists, including some antipsychotics and metoclopramide, have antiemetic properties.

5. Timeline for Significant Catecholamine-Related Events

Year	Scientist(s)	Event
1895	G. Oliver & E Schaefer	Adrenal medullary extract increases blood pressure
1897	JJ Abel & AC Crawford	Isolated & purified adrenal medullary extract ("epinephrin"), 1st hormone
1905	TR Elliott	Similarity in response between sympathetic nerve stimulation & epinephrine injection
1905	J Langley	Postulates "receptive substance" in tissues
1910	G Barger & J Hill	Synthesis of dopamine
1921	O Loewi	Proof of neurochemical transmission with *Vagusstuff* (acetylcholine)
1920s-1930s	W Cannon	Release of "sympathin" after sympathetic nerve stimulation
1932		Amphetamine marketed
1939	H Blaschko	Postulation of catecholamine biosynthetic pathway
1946	U von Euler	Sympathin E = norepinephrine, adrenergic neurotransmitter
1948	R Ahlquist	α & β- adrenergic receptors
1948	V Erspamer	Octopamine discovered in salivary gland of octopus
1952	J Delay & P Deniker	Clinical trials of chlorpromazine ("typical") antipsychotic; 1954: FDA approved
1953		Reserpine FDA approved as antipsychotic; 1931: Sen & Bose demonstrate clinical effectiveness of rauwolfia in mania & hypertension
1954	M Vogt	Differential distribution of norepinephrine in brain; possible neurotransmitter
1957	BB Brodie & P Shore	Norepinephrine primary neurotransmitter in sympathetic nerves in brain; psychoactive drugs interact with norepinephrine
1957-1960	A Carlsson & N-A Hillarp	Dopamine is neurotransmitter in brain; dopaminergic pathways in brain identified; reserpine depletion of brain catecholamines, with levodopa restoration of brain dopamine
1959	P Janssen	Clinical introduction of haloperidol as antipsychotic in Europe; 1967: FDA approval
1960	O Hornykiewicz	Low dopamine in substantia nigra & striatum in Parkinsonian patients
1965	JJ Schildkraut	Catecholamine hypothesis of affective disorders postulated
1967	GC Cotzias	Established effectiveness of levodopa for Parkinson's disease; 1961: clinical trials initiated
1975		Clozapine used in Europe as "atypical" antipsychotic; withdrawn after agranulocytosis; 1989: FDA approved
1976	P Seeman	Correlation between clinical potency of antipsychotic agents and blockade D_2 receptors in striatum
1979	JW Kebabian & DB Calne	Classification of dopamine receptors into D_1 & D_2

6. Conclusion

In the peripheral autonomic nervous system, neurochemical transmission is relatively simple. Acetylcholine is the preganglionic neurotransmitter, while acetylcholine and norepinephrine are the neurotransmitter substances released from postganglionic cholinergic and adrenergic nerves, respectively. The situation is far more complicated and fascinating in the central nervous system. While there is a clear association between norepinephrine and depression and sleep, and between dopamine and Parkinson disease, schizophrenia, and drug dependence, these associations are no longer believed to be exclusive ones. Mounting evidence suggests that there are multiple etiologies of these conditions that involve some of the 40-50 neurotransmitters/ neuromodulators/ neurotropic factors present in the brain. Biogenic amines regulate a plethora of biologic responses. Further research into the molecular events associated with the actions of the various biogenic amines on specific subtype of receptors may be beneficial for the development of specific agonists and antagonists to be used for the treatment of various brain disorders.

References

[1] Bloom FE. (2006). Neurotransmission and the central nervous system. In: L.L Brunton, J.S. Lazo, K.L. Parker (Eds.), *Goodman and Gilman's The Pharmacological Basis of Therapeutics* (11th edition, pp. 317-339). New York, NY: McGraw-Hill.

[2] Rang HP, Dale MM, Ritter JM, Moore, PK. (2003). *Pharmacology* (5th edition). New York: Churchill Livingstone.

[3] Usdin E, Sandler M., Eds. (1976). *Trace Amines and the Brain.* New York: Dekker.

[4] Rubin RP. A brief history of great discoveries in pharmacology: in celebration of the centennial anniversary of the founding of the American Society of Pharmacology and Experimental Therapeutics. *Pharmacological Reviews*, 2007;59: 289-359.

[5] Frazer A. Pharmacology of antidepressants. *Journal of Clinical Psychopharmacology*, 1997;17:2S-18S.

[6] Ordway GA, Schwartz MA, Frazer A., Eds. (2009). *Brain Norepinephrine: Neurobiology and Therapeutics.* New York: Cambridge University Press.

[7] Vadim R. REM sleep function and brain monoamine regulation: an application of the search activity concept. Cited 2009 June 9 from http://www.bio-epidermis.org/4r.html.

[8] Marsden CA. Dopamine: the rewarding years. *British Journal of Pharmacology*, 2006;147(S1):136-144.

[9] Lang AE, Lozano AM. Parkinson's disease. *New England Journal of Medicine*, 1998;339:1044-1053.

[10] Triarhou LC. Dopamine and Parkinson's disease. *Advances in Experimental Medicine and Biology,*2002;517:1-14.

[11] Waddington J, Casey D. (2000). Comparative pharmacology of classical and novel (second generation) antipsychotics. In P.F. Buckley, J.L. Waddington (Eds.), *Schizophrenia and Mood Disorders: The new drug therapies in clinical practice.* Butterworth-Heinemann, Boston, pp. 1-13.

[12] Seeman P. Dopamine receptors and the dopamine hypothesis of schizophrenia. *Synapse.*1987;1:133-152.

[13] Jones HM, Pilowsky LS. Dopamine and antipsychotic drug action revisited. *British Journal of Psychiatry,*2002;181:271-275.

[14] Spanagel R, Weiss F. The dopamine hypothesis of reward: past and current status. *Trends in Neuroscience.*1999;22:521-527.

[15] Elia J, Ambrosini PJ, Rapoport JL. Treatment of attention-deficit-hyperactivity disorder. *New England Journal of Medicine*, 1999;340:780-788.

In: Biogenic Amines
Editors: T. Farooqui, A. A. Farooqui ISBN: 978-1-60876-625-3

Chapter II

LIFE'S REQUIREMENTS FOR CATECHOLAMINES

Suzanne Roffler-Tarlov [*]
Tufts University School of Medicine, Boston, MA 02111, USA.

ABSTRACT

Strategies developed to determine the biological functions of catecholamines have, to a great extent, depended on destroying them. This has been accomplished by surgical lesions and by chemical lesions as Norman Uretsky pioneered using 6 hydroxy-dopamine injected into the ventricular system of the brain of experimental animals. The results of these and later studies made it clear that catecholamines and especially dopamine and norepinephrine are involved in myriad tasks for the conduct of life. The creation of genetically engineered mice in which catecholamines are eliminated, even specific catecholamines, norepinephrine, epinephrine, and dopamine have widened the view to emphasize that catecholamines are not just involved, they are essential from nearly the beginning of life: between midgestion, as soon as the rate limiting enzyme tyrosine hydroxylase is expressed until the end. These mice have taught us that life's stages have special catecholamine requirements.

INTRODUCTION

It was sometime in 1968, daylight hours. We were postdoctoral fellows sitting, listening to a terrific presentation at a meeting of the British Pharmacologists in Cambridge. The Swedish speaker described the effects of administration of a compound known as 6-hydroxydopamine on the sympathetic nervous system. I heard excited rustling and urgent whispering just behind me where Norman Uretsky and his mentor Leslie Iversen were sitting

[*] Correspondence concerning this article should be addressed to: Suzanne Roffler-Tarlov, Department of Neuroscience, Tufts University School of Medicine, 136 Harrison Ave., Boston, MA, 02111. Suzanne.Tarlov@Tufts.edu.

together. I looked around and heard their words. I saw their big grins too. The following day, Norm and Les began studies that opened new views into brain function through the use of that Trojan horse-like poison, 6 hydroxy-dopamine, by destroying catecholamine containing neurons in the brains of experimental rats [1,2]. Thus, the use of stealthy poisons, e.g. 6 hydroxydopamine and 1-methyl-4-phenyl-1,2,3,6-tetrahydropyridine (MPTP), replaced surgical lesions to assess the effects of lack of catecholamines and thus to infer function and to develop models of neurodegenerative diseases.

Much later, gene "knockout" strategies joined the force for the study of catecholamine function. Through what is known as homologous recombination in embryonic stem cells, it is possible to replace a normal gene in an embryonic stem cell with a defective copy of the same gene. And then after the implantation of blastocysts that harbor these altered stem cells, chimeric mice are created from which are bred mice that lack the normal gene to be studied -- -- if luck prevails [3,4]. In this way, genes that code for the three enzymes that act in sequence and that are specific for catecholamine synthesis have been eliminated one by one: tyrosine hydroxylase [5-7], dopamine β-hydroxylase [8], and phenylethanolamine N-methyltransferase [9]. Another very cleverly engineered mouse lacks dopamine but does synthesize norepinephrine [10]. The "knockout" mice have added much new information, some of it surprising.

Genetically Engineered Mice That Lack All Catecholamines

Disruption of functional tyrosine hydroxylase (TH) causes loss of all the catecholamines as this enzyme is required as the first step for the neural synthesis of catecholamines. Tyrosine hydroxylase catalyzes the conversion of tyrosine to L-Dopa (3,4,dihydroxy phenylalanine) which is the substrate for the next downstream enzyme, aromatic amino dehydroxylase. Thus, elimination of TH activity would be expected to prevent synthesis of dopamine, norepinephrine, and epinephrine formed in neurons and in the adrenal medulla in addition to L-Dopa.

All three genetically engineered mouse lines that lack TH, from three different laboratories, point to the absolute requirement for catecholamines before birth and before the development of the nervous systems [5-7]. The TH-null mutants in each of these lines die *in utero* though the time of death varies a bit between lines. Certainly genetic background is a factor. In the TH-null line that we made initially, fetuses without TH begin to die at embryonic day (E) 9.5, shortly after expression of TH occurs in normal mice. No mutant fetuses survived birth [5].These mice, on a C57Bl6 background, were crossed onto a CD-1 background and produced TH-nulls that live longer *in utero* and 5 – 10% percent survive birth [5]. At birth, the TH-null pups are indistinguishable by eye from their heterozygous and wild-type littermates though TH-null mutants lack all three catecholamines: dopamine, norepinephrine, and epinephrine.

NOREPINEPHRINE IS REQUIRED FOR FETAL LIFE

Of the three catecholamine transmitters, it is norepinephrine alone that is required during fetal life. Fetuses with TH but which lack dopamine β hydroxylase, the enzyme that catalyzes the conversion of dopamine to norepinephrine, also die midgestation [8]. This is their fate in spite of their accumulation of abnormally large quantities of dopamine. Elimination of epinephrine synthesis, on the other hand, has no effect on viability of mouse fetuses [9]. Mutants that lack dopamine but synthesize norepinephrine also survive fetal life [10].

Consistent with the specificity of need for norepinephrine during fetal life is the fact that most TH-null fetuses can be rescued by replacement of some of the missing norepinephrine by administration of the catecholamine precursor L-Dopa to the pregnant heterozygote dams in their the drinking water [5,6,8]. The exogenous L-Dopa is converted first to dopamine and dopamine is subsequently converted to norepinephrine providing two catecholamines. Administration of L-threo-3,4 dihydroxyphenylserine (DOPS) is also effective in the rescue of the TH-null pups to birth without producing dopamine. DOPS is converted directly to norepinephrine by aromatic amino acid decarboxylase. An extremely effective rescue is provided by an even more specifically targeted drug. This is isoproterenol, the adrenergic drug that acts as an agonist at β receptors. Isoproterenol dissolved in the drinking water of the pregnant dams rescues nearly 100 per cent of expected TH-null mice in our experience [11,12].

An agonist that is specific for the ß1 subtype adrenergic receptor rescued effectively also: 74% of expected nulls [11,12]. A key finding from the Chikaraishi laboratory is that β1 receptors are present on the fetal mouse heart and exclusively on the heart at E12.5 [11]. Thus the action of norepinephrine at the β1 receptor on heart is the critical need for catecholamines before birth.

The sympathetic system has not yet innervated the heart at the time the TH-null fetuses are vulnerable. What's more, the sympathetic innervation of the heart is not functional until postnatal stages. As the fetal heart is not controlled by sympathetic innervation, why does the fetal heart fail without a supply of norepinephrine?

The results of histological examination of hearts of TH-null fetuses do point to heart failure as a probable cause of prenatal death. In the middle of gestation the TH-null fetus is typically found with blood congestion in the heart and major blood vessels [5,6,13-15]. Pools of blood are sometimes seen in liver and lungs. The mutants' hearts may have dilated atria, with thinning of the atrial wall, and reduced cell density in the ventricles [5,6,13-15]. That the normal fetal heart is responsive to norepinephrine has been known for a long time. This fact indicates that interaction of norepinephrine with the fetal heart normally precedes mature sympathetic function. The finding that the fetal heart has already made functional β1 receptors explains its early responsiveness to norepinephrine [11].

Without norepinephrine, the fetal TH-null resting heart rate is slow compared to wild-type monitored *in vivo* [12] and *ex vivo* [6-8], but it has a steady rhythm which is sufficient to maintain sufficient oxygenation to keep the mutant alive unless stress that requires an increased heart rate occurs. Stress does occur.

The results of studies conducted by Chikaraishi and her colleagues explain the need for norepinephrine and the heart failure that results from its absence *in utero* [11,12,14,15]. The conclusion of these careful and imaginative studies is that a supply of norepinephrine is essential for fetal heart function during the periodic bouts of fetal hypoxia that occur during gestation [15]. Episodes of increased hypoxia are normal at this time. Fetuses live in a hypoxic environment (compared to postnatal animals) normally as many developmental events that take place *in utero* require low oxygen tension. The "normal" hypoxic condition is exacerbated from time to time due to periodic spontaneous uterine contractions. Such contractions are also normal and they are quite frequent, reported to occur hourly on average and to reduce fetal vascular PO2 by 10 – 25%. This data was reported after studies of fetal sheep [16]. Normal fetuses though, avoid the hyper- hypoxia created periodically during gestation. This happens because hypoxia signals release of norepinephrine which counters the hypoxic condition by stimulating the heart rate [15]. Hypoxia in the normal mouse fetus is followed by a 13-fold increase in plasma norepinephrine [15]. The periodic extreme hypoxia induced by uterine contractions is fatal for the TH-nulls because they have no norepinephrine to maintain their heart rate and oxygenation during these episodes. Stimulation of adrenergic β receptors on the heart is essential during fetal life: thus the TH-null mutants are rescued by administration of NE precursors such as L-Dopa and DOPS or β adrenergic agonists to the pregnant mother.

It follows that fetal hypoxia is the root cause of the heart failure and subsequent death of fetuses that lack norepinephrine and Chikaraishi and collaborators proved that. They showed that prevention of hypoxia *in utero* rescues the otherwise untreated TH-null fetuses [15]. For these experiments pregnant females were placed in an elevated oxygen environment (33% and 63% rather than the normal ambient 21%) from E11.5 to term. The increased oxygen eliminated bradycardia in TH-null fetuses and rescued them in spite of the absence norepinephrine and of stimulation of the heart's β1 receptors. The 63% oxygen condition rescued almost 80% of TH-null fetuses. These investigators concluded that adrenergic stimulation, specifically norepinephrine activation of β1 receptors on fetal heart, reverses hypoxia caused by periodic uterine contractions [15]. Norepinephrine is essential for this reason; without it the uterine contractions result in death of most of the fetuses because oxygen homeostasis cannot be maintained.

After birth, pups are freed of the perils of uterine contractions. Are the TH-null mutants home free?

The Crises of the Newborn: Norepinephrine is Still a Necessity

Rescued TH-null mutants and the few that survive birth without supplementation are born with normal weight, length, appearance, and blood glucose levels. However they've no sooner survived the hypoxic crises of their life *in utero*, but they encounter a new crisis: life *ex utero*. The untreated TH-null pups, at first indistinguishable from their littermates, live days to weeks but not beyond weaning unless they are specially treated.

Life of the newborn is similar to that of the fetus in that norepinephrine is the single essential catecholamine for life between birth and weaning. Fetuses appear to have a single problem that norepinephrine conquers; that is avoiding hypoxia. Newborns, on the other hand, are confronted with a multitude of problems if they are born without norepinephrine. Eating is not one of these. Newborns have no need for dopamine for motivation to eat. Like the fetuses, the heart rate of the neonatal pups is abnormally slow. They would not be able to mount a "fight or flight" response but their laboratory-bound life style does not permit such emergencies.

The TH-null mutants are so very hungry; they suck avidly and their mothers take good care of them. However, in spite of stomachs full of milk throughout their lives, the mutants fail to thrive. Mutant neonates do not gain weight after their second or third day; they weigh between 3 and 4 grams throughout their lives. The TH-nulls maintain the weight of a normal 4 day-old pup which would be expected to weigh between 8 and 10 grams at two weeks of age. The mutants are deficient in length, crown to tail, as if they had been starved (Figure 1). But unlike unfed normal pups with equivalent weight and length, the TH-null mutants become severely hypoglycemic though they always have milk in their stomachs and their livers store glycogen. The TH-null mutants display normal insulin responses though they do not experience the normal rise in blood insulin levels that occur with age. After a glucose challenge, blood glucose levels rise to the levels that are produced in controls after the same challenge. After a glucose challenge, the controls return to the pre-challenge glucose levels and so do the mutants. The mutants return to the same low glucose values as before the challenge. These low values apparently are sensed by the TH-nulls as normal glucose levels. That the TH-null mutants respond to glucose normally shows that the pancreas has developed in that it secretes insulin in response to glucose.

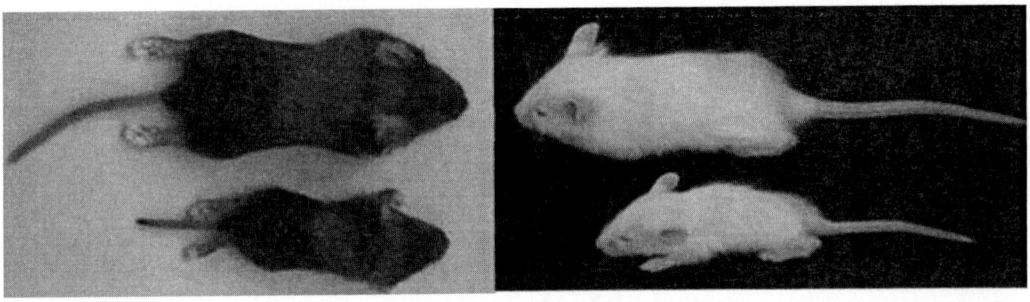

Figure 1. Black mice and white mice. TH-null mutants with wild-type littermates on postnatal day 18. The pair on the left are on a C57Bl6 background. The mutant makes small amount of catecholamines in spite of lack of tyrosine hydroxylase. The pair on the right lack pigment because they carry a mutation that disables the gene that codes for tyrosinase in the melanin pathway. The small white mouse lacks both tyrosine hydroxylase and tyrosinase.

The failure-to-grow phenotype of mice that lack catecholamines is illustrated in Figure 1, which pictures two TH-null pups and two wild-type mice. One pair is black, on a C57 Bl /6J background and the other is albino. The albino mice were created after the pigmented heterozygous carriers were backcrossed onto albino CD1 mice. The albino pair are homozygous tyrosinase-deficient due to a naturally occurring mutation in the gene that codes for tyrosinase. In this way we created mice that lack activity of both TH and tyrosinase [5]. Each pair of pups pictured were 18 day-old littermates produced by heterozygote carriers of the TH-null mutation. At this age, the weight disparity between normal and mutant is very great: TH-null pups weigh only about one-third of their littermates. The mutant pups fail to grow length as well. There are no such differences between the wild-type and the TH-null heterozygote which indicates that a single copy of TH produces sufficient catecholamine and in fact we've noted that the heterozygotes production of catecholamine is equivalent to that of the wild-type.

The TH-null pups are developmentally behind their normal littermates as well as being very small. For example, eye opening in normal mice occurs by postnatal day 14, the eyes of TH-null mutants remain closed at least a week longer than is normal.

The Contribution of Tyrosinase Activity to Stores of Catecholamines

Tyrosine hydroxylase is not the only enzyme that converts tyrosine to L-dopa. Tyrosinase, the enzyme that is present in melanosomes does the same job for the synthesis of pigments (for discussion of this pathway see [17]. The first TH knockout made in our lab was on a black background, the C57Bl6. As expected, the TH-null fetuses produced no catecholamines [5]. The assays were both biochemical using HPLC separation followed by electrochemical detection, and histological by glyoxylic acid induced histofluorescence. The latter can pinpoint the locations of small quantities of catecholamines present in neuronal processes, for example.

Initially, we were perplexed to find that a week after the pigmented TH-null pups were born, they did make catecholamines and quantities were not insignificant during the first weeks of life. Levels of norepinephrine were especially high in heart and skin where quantities found in two week-old TH-nulls averaged 17 percent of wild-type littermates in heart and 23 percent of wild-type in skin. In the adrenal gland norepinephrine and epinephrine levels of the pigmented nulls amounted to less than 5 percent of wild-type littermates but considering the great quantity of these substances in the postnatal adrenal --- that is a lot. And in the brain, there was measurable norepinephrine and dopamine in the pigmented TH-null, less than 10 percent of wild type values (5). These catecholamines were in normal locations; glyoxylic acid-induced histofluorescence showed catecholamines present in sympathetically innervated structures all over the body and even in the dopamine-containing pathways of the brain.

We discovered that the catecholamines found in the postnatal pigmented TH-null pups were due to the activity of tyrosinase, the pigment pathway enzyme which in skin becomes active postnatally. The tyrosinase-based origin of these catecholamines was proven after we crossed our pigmented line of heterozygous TH-null carriers onto a line of CD1 albino mice. Catecholamines were absent in two week-old albino mice that were also TH-null [5].

Does the small amount of norepinephrine and dopamine synthesized from L-Dopa generated in the melanin pathway help the TH-null pups? No and yes. It does not help in the most fundamental way: i.e. reversing their failure to gain weight and to thrive during neonatal life. The TH nulls that are pigmented i.e. have normal tyrosinase activity, do not live longer on average than do the albino TH nulls, nor do they gain weight more readily. The disparity in weight, growth, and gross development is the same whether the mouse is pigmented and has some catecholamine neonatally or albino, without the contribution made by tyrosinase activity (Figure 1).

And yes, the L-dopa from tyrosinase activity that is a substrate for dopamine and then norepinephrine formation is functional. We tested the function of tyrosinase-derived norepinephrine at the sympathetically-innervated sweat gland. The sympathetic innervation of the sweat glands is a well known anomaly; the transmitter at the sweat glands is acetylcholine, not norepinephrine. Stimulation of the innervation in the mature mouse releases acetylcholine rather than norepinephrine to produce sweating. However, early on as the sympathetic innervation of the sweat glands first contacts the glands, the neurons have an adrenergic phenotype. The initial transmitter produced by the neonate's sympathetic innervation is norepinephrine. The transmitter type changes to acetylcholine as postnatal development proceeds. Sweating occurs in response to cholinergic stimulation after the disappearance of adrenergic characteristics.

Story Landis and her colleagues constructed an hypothesis to explain the transmitter switch based on a great deal of data some of it acquired *in vivo* and some *in vitro* [18]. By this hypothesis the early action of norepinephrine is essential for the transmitter switch to take place. Sweating would not occur if the nerve endings did not switch transmitter from norepinephrine to acetylcholine. The prediction then for a test of the TH-null mice, would be that sweating would not occur in the mice that lacked both TH and tyrosinase because their sympathetic innervation had never produced norepinephrine. And secondly the double mutants would not have "switched" to the cholinergic transmitter, as they had not produced the original transmitter, norepinephrine. The result did follow part of the prediction: the double mutant did not produce sweat in response to the muscarinic agonist pilocarpine. The surprise result was that the double mutant did display the cholinergic phenotype in spite of not having produced norepinephrine earlier. Thus in spite of having developed the cholinergic phenotype, no sweating occurred. That was contrary to the prediction [19].

The pigmented TH-nulls showed another result. Unlike their albino counterparts, they did sweat and they did also acquire the cholinergic phenotype. Thus the fraction of the normal quantity of norepinephrine (that was about 20% of wild-type levels) made by the mutant's sympathetic innervation of the sweat gland functioned to permit sweating in the mouse in response to pilocarpine [19]. However, as the double mutant also developed the cholinergic phenotype, the norepinephrine is unnecessary to produce the switch. The early adrenergic phenotype is, however, essential for sweating to occur. Thus the sympathetic innervation to the sweat glands develops the cholinergic phenotype without a "switch". That is not enough: ability to sweat needs a noradrenergic introduction [19].

After we found the functional importance of the norepinephrine that was produced through the activity of tyrosinase, we used the double mutant for study to avoid this complication.

The Failure to Thrive Phenotype

Another system that goes awry in both pigmented and albino TH-nulls is the gut. It seems likely that their failure to grow is connected with problems with gut function. An abnormal phenotype is seen histologically in the gut, most prominent in the small intestine (Figure 2). All cell types have been generated for the gut and are in the appropriate positions in the mutants.

More prominent in the TH-null pup than in wild-type are large vacuoles contained in cells in all of the villae of the mutant (examples shown with black arrows in B). These are rarely present in the wild-type. In the TH-null, most of the cells contain such vacuoles. The exceptions are the newly generated cells at the base of the villae. A yellow arrow in B points to a newly generated goblet cell without large vacuole at the base of a villus. The vacuolated cells have been found in all the TH-null pups examined; ages P 9.5 up to P 20.5. The condition does not worsen with postnatal age. The vacuoles are present in all regions of the mutant's small intestine.

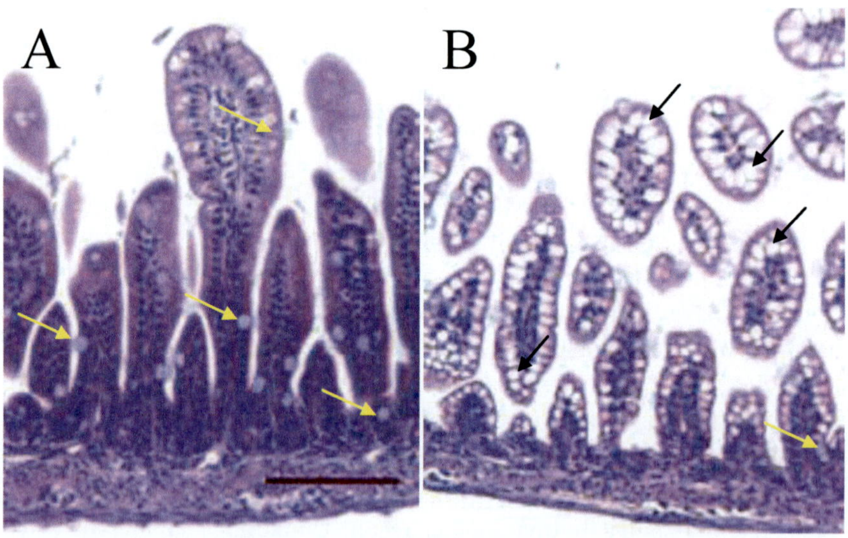

Figure 2. Photomicrographs of histological sections of ileum stained with hematoxylin and eosin. The tissue sections were taken from a P9.5 wild-type albino pup (A) and a TH-null albino pup (B). The pups were littermates. The yellow arrows in A and B point to goblet cells in the wild-type and in the TH-null tissue. Some of the cells are secreting muscin into the lumen of the intestine.

We found pathology in most organs of the TH-null pups that were examined. Another example of damaged tissue, one that has been published, is the adrenal gland [20]. Chromaffin cell structure is altered in the TH-null as was seen at the ultrastructural level. Chromaffin cell vesicles are depleted and an increase in rough endoplasmic reticulum is present in the TH-nulls. The few remaining chromaffin vesicles in the mutant pup line up proximally to the cell membranes. The expression of mRNA for proenkephalin and neuropeptide were increased.

Signs of reduced cortical function appear in the mutant's adrenal cortex. Cells contain increased numbers of liposomes. Tubular mitochondria have a reduced mass of internal membranes. Without catecholamines in the adrenal medulla, the adrenal cortex is malformed and corticosterone levels in blood are reduced to 20% of wild-type values. ACTH was not affected [20].

The weight and glucose deficient phenotypes of the newborn TH-nulls are partially prevented by treatment with drugs that supply NE or β adrenergic stimulation.

Both L-Dopa and DOPS fortify the TH-null pups, as does isoproterenol. As in the case of the fetuses, administration of L-Dopa, DOPS, and isoproterenol effectively help the TH-nulls to gain weight during the first weeks of postnatal life. Administration of these compounds also rescue the TH-null pups until about 20 days of age. Without treatment many of the TH-nulls die before reaching this age. Regardless of the agent used for their rescue, the mutant pups do gain weight, unlike the untreated nulls. However, the weight gain of the rescued mutants is much less than that of their wild-type and heterozygous littermates.

At the end of the third postnatal week, treated and untreated TH-null mutants fail to muster further gains though they do continue to nurse and the mothers permit that. After this event the treated mutant pups lose weight and die, unless they have been receiving L-Dopa. The other rescue strategies allow the mutant pups to live only until weaning. Then life drops its requirement for NE.

Adolescence and Maturity: Norepinephrine is Dispensable. Life Demands Dopamine

After weaning, a switch takes place. Life moves beyond a requirement for norepinephrine to the need for dopamine.

Sophia Savage working in my lab administered L-Dopa daily by intraperitoneal injection to both male and female TH-null mice from the time of weaning until the mice were nearly a year old. These mutants were small throughout their lives. Before their dose of L-Dopa each day they were inactive and looked forlorn. The administration of L-Dopa by intraperitoneal injection caused nearly instant activity and frantic eating on the part of the TH-null mice. The nulls, both male and female were able to breed. The females successfully carried their litters and birthed them. They took care of them as well. A mating between a male TH-null and a female TH-null produced an entire litter of TH-nulls.

Conclusions

Life without catecholamines is nasty, brutish, and short.[&]

Life has evolved no substitute for norepinephrine during critical periods of demand for an enhanced supply of oxygen or for developmental needs. The mice without TH emphasize also, the critical importance of that amino acid L-Dopa and the synthetic pathway for pigment. It seems that the nervous system can borrow from the production of L-Dopa from the pigment pathway but the pigment pathway cannot benefit from L-Dopa generated in neurons.

[&] after Thomas Hobbes.

Acknowledgments

The published and unpublished experiments from the SR-T lab were done with support from NIH RO1 NS 31673 and GRASP at Tufts University School of Medicine.

References

[1] Uretsky NJ, Iversen LL. Effects of 6-hydroxydopamine on noradrenaline-containing neurones in the rat brain. *Nature*. 1969;221:557-559.

[2] Uretsky NJ, Iversen LL. Effects of 6-hydroxydopamine on catecholamine containing neurones in the rat brain. *Journal of Neurochemistry*. 1970; 17:269-278.

[3] Capecchi MR. Targeted gene replacement. *Scientific American*. 1994;207:52-59.

[4] Papaioannou V, and Johnson, R. Production of chimeras and genetically defined offspring from targeted ES cells. In: Joyner A, editor. Gene Targeting A Practical Approach. New York: Oxford University Press; 1993.

[5] Rios M, Habecker B, Sasaoka T, Eisenhofer G, Tian H, Landis S, Chikaraishi D, Roffler-Tarlov S. Catecholamine synthesis is mediated by tyrosinase in the absence of tyrosine hydroxylase. *Journal of Neuroscience*. 1999;19:3519-3526.

[6] Zhou QY, Quaife CJ, Palmiter RD. Targeted disruption of the tyrosine hydroxylase gene reveals that catecholamines are required for mouse fetal development. *Nature*. 1995;374:640-643.

[7] Kobayashi K, Morita S, Sawada H, Mizuguchi T, Yamada K, Nagatsu I, Hata T, Watanabe Y, Fujita K, Nagatsu T. Targeted disruption of the tyrosine hydroxylase locus results in severe catecholamine depletion and perinatal lethality in mice. *Journal of Biological Chemistry*. 1995;270:27235-27243.

[8] Thomas SA, Matsumoto AM, Palmiter RD. Noradrenaline is essential for mouse fetal development. *Nature*. 1995 Apr 13;374(6523):643-6.

[9] Bao X, Lu CM, Liu F, Gu Y, Dalton ND, Zhu BQ, Foster E, Chen J, Karliner JS, Ross J Jr, Simpson PC, Ziegler MG. Epinephrine is required for normal cardiovascular responses to stress in the phenylethanolamine N-methyltransferase knockout mouse. *Circulation*. 2007;116:1024-1031.

[10] Zhou QY, Palmiter RD. Dopamine-deficient mice are severely hypoactive, adipsic, and aphagic. *Cell*. 1995; 83:1197-1209.

[11] Chandra R, Portbury AL, Ray A, Ream M, Groelle M, Chikaraishi DM. Beta1-adrenergic receptors maintain fetal heart rate and survival. *Biology of the Neonate*. 2006;89:147-158.

[12] Portbury AL, Chandra R, Groelle M, McMillian MK, Elias A, Herlong JR, Rios M, Roffler-Tarlov S, Chikaraishi DM. Catecholamines act via a β-adrenergic receptor to maintain fetal heart rate and survival. *American Journal of Physiology- Heart and Circulatory Physiology*. 2003;284:H2069-H2077.

[13] Roffler-Tarlov S. and Rios, Catecholamines Before Nervous Activity. Kalverboer AF, Gramsbergen AA. Handbook of brain and behaviour in human development. Dordrecht ; Boston: Kluwer; p. 165-183, 2001.

[14] Ream M, Ray AM, Chandra R, Chikaraishi DM. Early fetal hypoxia leads to growth restriction and myocardial thinning. *American Journal of Physiology - Regulatory, Integrative and Comparative Physiology*. 2008;295:R583-R595.

[15] Ream MA, Chandra R, Peavey M, Ray AM, Roffler-Tarlov S, Kim HG, Westel WC, Rockman HA, Chikaraishi DM. High oxygen prevents fetal lethality due to lack of catecholamines. *American Journal of Physiology - Regulatory, Integrative and Comparative Physiology*. 2008;295:R942-R953.

[16] Jansen CA, Krane EJ, Thomas AL, Beck NF, Lowe KC, Joyce P, Parr M, Nathanielsz PW. Continuous variability of fetal PO2 in the chronically catheterized fetal sheep. *American Journal of Obstetrics and Gynecology*. 1979;134:776-783.

[17] Riley PA. The great DOPA mystery: the source and significance of DOPA in phase I melanogenesis. *Cellular and Molecular Biology* (Noisy-le-grand). 1999; 45:951-960.

[18] Habecker BA, Landis SC. Noradrenergic regulation of cholinergic differentiation. *Science*. 1994;264:1602-1604.

[19] Tian H, Habecker B, Guidry G, Gurtan A, Rios M, Roffler-Tarlov S, Landis SC. Catecholamines are required for the acquisition of secretory responsiveness by sweat glands. *Journal of Neuroscience*. 2000;20:7362-7369.

[20] Bornstein SR, Tian H, Haidan A, Bottner A, Hiroi N, Eisenhofer G, McCann SM, Chrousos GP, Roffler-Tarlov S. Deletion of tyrosine hydroxylase gene reveals functional interdependence of adrenocortical and chromaffin cell system in vivo. *Proceedings of the National Academy of Sciences of the United States of America*. 2000;97:14742-14747.

In: Biogenic Amines
Editors: T. Farooqui, A. A. Farooqui
ISBN: 978-1-60876-625-3

Chapter III

AROMATIC L-AMINO ACID DECARBOXYLASE

Maria Hadjiconstantinou*, Anne-Marie Duchemin, Abul Azad and Norton H. Neff
Ohio State University, Columbus, Ohio 43210, USA.

ABSTRACT

Aromatic L-amino acid decarboxylase (AAAD) is a required enzyme for the formation of catecholamines, indolamines and trace amines. AAAD gene and protein analyses have provided evidence for enzyme activity regulation at the transcriptional and post-translational levels. Extensive studies in neuronal tissues have demonstrated that AAAD activity is regulated by activation and induction, protein phosphorylation and gene transcription. A number of neurotransmitter receptors including dopamine (D1-4), glutamate (NMDA), serotonin (5-HT1A, 5-HT2A) and nicotinic acetylcholine receptors regulate AAAD activity in the rodent brain. Generally, receptor antagonists enhance AAAD activity; while, agonists have no effect or diminish it. Second messenger systems, particularly cAMP, participate in the short- and long-term enzyme regulation in the brain. Following a lesion of nigrostriatal dopaminergic neurons, AAAD in striatum responds more robustly to pharmacological manipulations indicating ability for compensatory responses. We provide a historical perspective of the current knowledge and review evidence supporting the notion that AAAD is a regulated enzyme. Because of the importance of AAAD for L-DOPA treatment in Parkinson disease, we propose that this knowledge can be exploited for the benefit of Parkinson patients.

Keywords: Aromatic L-amino acid decarboxylase, regulation, neurotransmitter receptors, L-DOPA, Parkinson disease.

---

* Correspondence concerning this article should be addressed to: Maria Hadjiconstantinou, MD, Department of Psychiatry, Division of Molecular Neuropsychopharmacology, College of Medicine, Ohio State University, 333 West Tenth Avenue, Columbus, Ohio 43210. neff.6@osu.edu; Telephone: 614-292-6168; Fax: 614-292-7232.

In 1969 a seminal paper by Uretsky and Iversen [1] showed that an intracerebroventricular (icv) injection of 6-hydroxydopamine reduced the norepinephrine content in rat brain. The authors then went on to provide evidence that 6-hydroxydopamine was a selective neurotoxin for the catecholaminergic neurons, and that the long-lasting cetecholamine depletion was most likely due to neuronal destruction [2]. Furthermore, they demonstrated that the catecholamine synthetic enzymes tyrosine hydroxylase and aromatic L-amino acid decarboxylase (AAAD) were diminished by the neurotoxin. These initial reports were the basis for the thousands of studies that used selective neurotoxins to investigate the physiological function of norepinephrine and dopamine of brain. Over the years, the neurotoxins changed but the concept of revealing the role of brain neurotransmitters after administration of a selective neurotoxin has remained a mainstay of neuroscience research. Many of the studies on Parkinson disease (PD) have utilized the neurotoxins 6-hydroxydopamine or MPTP (1-methyl-4-phenyl-1,2,3,6-tetrahydropyridine) to destroy nigrostriatal dopaminergic neurons to mimic the neuropathology of PD. In tribute to Dr. Norman J. Uretsky, we review studies on the regulation of AAAD, much of which is based on the original concept of using selective neurotoxins to unmask the function of brain catecholamine systems.

Historical Perspective

In 1938 Holtz et al. [3] were the first to report on the enzymatic activity of AAAD (EC 4.1.1.28, aromatic L-amino acid decarboxylase [AAAD; AADC; DDC], also known as, 5-hydroxytryptophan decarboxylase, DOPA decarboxylase, hydroxytryptophan decarboxylase, L-DOPA decarboxylase, tryptophan decarboxylase) in mammalian tissue. The enzyme is found in both neuronal (monoaminergic neurons in the CNS, sympathetic neurons, and adrenal medulla) and non-neuronal cells (liver, intestine, and kidney) and this distribution is ascribed to tissue-specific expression of the AAAD gene as a result of alternative promoter usage and differential splicing [4]. Originally it was thought that decarboxylation of 3,4-dihydroxyphenylalanine (DOPA) to form dopamine and 5-hydroxytryptophan to form serotonin occurred by two separate enzymes [5], but now it is well established that the same enzyme is responsible for the decarboxylation of both substrates. There is evidence that m-tyrosine, o-tyrosine and other amino acids are substrates as well, and hence the general term aromatic L-amino acid decarboxylase [6]. AAAD is recognized as the required enzyme for the synthesis of catecholamines, serotonin and trace amines [7]. The observation that AAAD is also found in non-monoaminergic neurons in brain (AAAD+) [8,9], peripheral tissues [10] as well as in glial cells [11] supports the idea of additional, yet to be determined, roles for the enzyme. Dopamine and serotonin are necessary for normal nervous system function, and changes in their levels contribute to various pathologies. AAAD deficiency is a very rare autosomal recessive genetic disorder characterized by developmental delay, intellectual disability, abnormal movements and autonomic dysfunction [12].

In 1939 Blaschko postulated the sequence of biochemical events leading to the synthesis of catecholamines after finding that L-DOPA was decarboxylated by preparations from guinea pig kidney and liver to oxytyramine (dopamine) [13]. Brain AAAD gained a renewed interest in 1951 when Raab and Gigee showed that administered DOPA enhanced a catecholamine like substance in rat brain [14], and subsequent studies revealed higher concentrations of dopamine in the brain than in other tissues [15]. At the same time, Carlsson et al. [16] demonstrated that DOPA, but not 5-hydroxytryphan, administration restored reserpine-induced motor deficits in experimental animals, and reported that dopamine was the major catecholamine in brain highly concentrated in the caudate nucleus [17]. Based on these observations, Carlsson postulated that parkinsonism was associated with loss of brain dopamine [17].

Perhaps the study that generated the most interest in brain dopamine was the observation by Ehringer and Hornykiewicz [18] that there was a severe dopamine loss in the caudate and putamen of PD patients. This was followed in the 1960s by reports from several research groups [19-21] that administered L- DOPA, the endogenous substrate for AAAD, alleviated the clinical symptoms of PD by replacing lost dopamine from nigrostriatal neurons. Strong support for replenishing dopamine in the nigrostriatal neurons of PD patients was provided by Lloyd et al. [22] when they demonstrated that L-DOPA administration to PD patients resulted in as much as a fifteen-fold higher level of dopamine in the striatum compared with non-treated patients. Indeed, administration of L-DOPA is now the standard treatment for PD [23]. Protracted therapy with L-DOPA, unfortunately, is associated with diminishing effectiveness and motor complications. Although many theories have been put forward to explain the loss of effectiveness with time, the notion that loss of AAAD activity as dopaminergic nigrostriatal neurons degenerate [24], might be partly responsible for the diminished effectiveness of L-DOPA therapy has not been seriously addressed. Decarboxylation of L-DOPA by AAAD is the controlling step for the formation of dopamine in PD patients, and diminished enzyme activity could contribute to the decreasing therapeutic response of L-DOPA [25].

Despite its clinical importance for L-DOPA treatment, AAAD is a relatively ignored enzyme. Perhaps, this is because to many investigators AAAD is an uninteresting enzyme; it is relatively nonspecific, it is rather widely distributed, and it is normally not rate-limiting for dopamine synthesis [26]. Unfortunately, this dogmatic and erroneous viewpoint still persists despite the newer developments in AAAD neurochemistry. There is now substantial evidence that the activity of AAAD is regulated. Since decarboxylation of L-DOPA by AAAD is the controlling step for the formation of dopamine in PD, altered enzyme activity and regulation might be critical for the therapeutic response of L-DOPA. This Chapter is a comprehensive review of the current knowledge about the molecular biology, chemistry and pharmacology of brain AAAD, which supports the concept that AAAD is a potential target for pharmacological intervention in PD.

Molecular Structure of AAAD

AAAD cDNA has been cloned from rat [27], human [28], and bovine [29] tissues, and the mouse brain (Neff et al.; Genebank accession number AF 071068). The neuronal and non-neuronal forms of AAAD are coded by a single gene [29], and originate from the alternative use of two distinct promoters followed by alternative splicing [30-33].

Transcription from the downstream neuronal promoter gives rise to the expression of an AAAD mRNA with a specific 5' untranslated region located primarily in neuroectodermal tissue. Transcription from the upstream non-neuronal promoter gives rise to the expression of an AAAD mRNA with a different 5' untranslated region located primarily in kidney and liver. Rat, human and mouse genes are highly homologous, with more than 80 kb, containing 16 exons with the first and second exons encoding the non-neuronal and neuronal specific 5' untranslated regions, respectively. The mouse gene has been mapped to chromosome 11A [34], while the human gene has been mapped to chromosome 7p11 [35,36].

Analysis of the neuronal promoter of the human, rat and mouse AAAD gene identified several potential positive and negative cis-acting elements [37] within the first 2.4 kb, which are highly conserved. The binding sites for transcription factors include, AP2, octamer-binding factor, Oct-1, N-Oct-3, SP1, NFY, GATA-1, HNF-1, HNF-3, CRE-BP, CP-1, Ptx3, NF-kB, SRF, EGR1, and Pet-1. Cooperation between some of these transcription factors has been found necessary for activation of the human AAAD neuronal promoter, which requires interaction between the NFY factor and the cell specific N-Oct-3 POU domain protein [38]. Several transcription factor binding sites detected in the AAAD promoter are also present in the tyrosine hydroxylase promoter, such as CRE-BP1, SP1, AP1, Ptx3, NF-kB, SRF, EGR1 and several POU/octamers [39], suggesting that the regulation of the transcription of the two enzymes can be coordinated. Of interest is the transcription factor Ptx3, which is restricted to the dopaminergic neurons of the substantia nigra, and has been found to regulate the transcription of tyrosine hydroxylase [40]. Although, in some experimental models, the expression of mRNA for tyrosine hydroxylase is enhanced in parallel with the expression of mRNA for AAAD [41], conjoint expression is not always the case, suggesting that other factors participate in the regulation of the two enzymes [42]. Pet-1, an ETS domain factor specific for serotonergic neurons [43,44] with binding sites present on many serotonergic genes including tryptophan hydroxylase and the serotonin transporter, is also present in the AAAD gene indicating that transcriptional expression of AAAD could be regulated in parallel with the expression of other serotonergic proteins. Mice with disrupted Pet-1 gene have decreased setotonin and serotonergic neurons; while, the catecholaminetgic neurons are normal [45].

There are several protein kinase phosphorylation recognition motifs on mouse brain AAAD sequence (Table 1) pointing to phosphorylation as a mechanism for post-translational regulation. Taken together, these observations suggest that brain AAAD can be regulated at the transcriptional and posttranslational levels, and there is the potential for coordinated and neuron phenotype specific enzyme regulation.

The substrate binding site of AAAD has been hypothesized from comparison of several catechol modifying enzymes. The partial structural homology within a stretch of 58 amino acids (residues 58-115 of AAAD) between AAAD, tyrosine hydroxylase, dopamine β-hydroxylase and phenylethanolamine-N-methyltransferase, all needed to synthesize catecholamines, suggested that this part of the enzyme may represent the catechol binding site [27]. Indeed, crystallization of the AAAD molecule in the presence of carbiDOPA identified Ile101 and Phe103, conserved between species, as two active residues of the catechol binding pocket [46]. Mutation of Cys111 of AAAD leads to reduced enzyme

Table 1. Phosphorylation Recognition Motifs for Protein Kinases Present in the Mouse Brain AAAD Sequence

Phosphorylation Site	PDPK	PKA	PKG	PKC	CaMKII
S^{73}	+				
S^{108}	+				
S^{171}	+				
S^{216}				+	
S^{220}		+	+	+	
T^{320}		+	+		
S^{336}			+		
S^{359}		+	+	+	+
S^{429}		+	+	+	

Proline-Directed Kinase = PDPK; cAMP-Dependent Kinase = PKA; cGMP-Dependent Kinase = PKG; Protein Kinase C = PKC; Calmodulin-Dependent Kinase II + CaMKII

Table 2. Mutational Analysis of Mouse Brain AAAD

Description	AAAD Activity
1) mbAAAD:Ser73→Ala	ND
2) mbAAAD:Ser108→Ala	Yes
3) mbAAAD:Ser220→Ala	Yes
4) mbAAAD:Ser359→Ala	Yes
5) mbAAAD:His269→Ala, Arg, Lys	ND
6) mbAAAD:Asp271→Val, Arg, Lys	ND
7) mbAAAD:His269Asp271→AlaVal	ND

Constructs were expressed in the *E. coli* BL21(DE3)pLysS and activity measured *in vitro*. ND = not detectable.

Table 3. Kinetic Analysis of Mouse Brain AAAD Mutants Retaining Catalytic Activity (Table 2)

	Wild Type	Ser108→Ala	Ser220→Ala	Ser359→Ala
L-DOPA				
Vmax	103±0.1	68±0.4*	140±1.2*	146±3.4*
Km	37±1	106±2*	30±1	58±5*
Pyridoxal 5' phosphate				
Vmax	92±0.8	58±1.2*	140±3.2*	117±1.6*
Km	0.36±0.01	0.66±0.04*	0.37±0.03	0.434±0.02
5-HTP				
Vmax	24±0.4	14±0.15*	30±0.5*	31±0.4*
Km	16±1	28±1*	17±1	29±1*

Estimated Vmax = pmol DA/µg AAAD prot/ min ± SEM; and estimated Km = µM ± SEM. *P<0.05 compared to corresponding Wild Type.

activity and cofactor binding [47], and Cys111 is believed to be at or near the pyridoxal 5'-phosphate binding site. Based on amino acid sequence analysis, AAAD belongs to group II in the classification of Sandmeier et al. [48]. Early studies with selective reagents pointed to arginine [49], histidine [50] and cysteine [51] residues as part of the enzyme structure. Ishii et al. [52] performed mutational analysis on 13 group II amino acid decarboxylases and showed that replacement of His192, Asp253, Asp271, Ser296, Lys303, Tyr332 and Arg355 essentially abolished AAAD activity. Mutational analysis of Asp271 and His269, two residues conserved across several species, in the mouse brain AAAD demonstrated that replacement of Asp271 and His269 abolished AAAD activity, supporting the importance of these amino acids for enzyme catalytic activity in brain (Table 2). Although mutants with substitution of the putative phosphoracceptor residues Ser108, Ser220 and Ser359 retained AAAD activity, the apparent Vmax and Km for the substrates L-DOPA and 5-HTP and the cofactor pyridoxal 5' phosphate were altered (Tables 2 and 3), indicating that the mutations might affect enzyme kinetics in the expression system. The optimal pH of about 7.0 for the decarboxylation of L-DOPA appeared unaltered. The Ser73 mutant displayed no catalytic activity (Table 2), suggesting that the mutation induces unfavorable conformational changes at the active site or substrate binding site. Taken together, mutational analysis of mouse brain AAAD gene has provided evidence that Ser73, His269 and Asp271 are essential for AAAD catalytic activity.

REGULATION OF AAAD ACTIVITY

There is now substantial evidence that the activity of AAAD, the second enzyme in the pathway leading to the biosynthesis of dopamine, is regulated and L-DOPA decarboxylation mirrors enzyme regulation. Modulation of AAAD activity *in vivo* and *in vitro,* involves both enzyme activation and induction. Activation *in vivo* occurs in response to physiological stimuli, drugs that act at neurotransmitter receptors, or modulation of the activity of endogenous kinases/phosphatases [53-63]. In striatum, kinetic activation of AAAD is rapid, short-lasting and characterized by changes in the apparent Vmax for both the substrate and the cofactor pyridoxal-5'-phosphate, suggesting partial phosphorylation of the enzyme [54]. Indeed, phosphorylation appears to play a role in the activation of the enzyme *in vivo.* Injection of either forskolin [59] or phorbol-12,13-myristic acid (PMA) [58] icv increases the activity of AAAD in striatum, a response that can be blocked by selective inhibitors of PKA or PKC. In addition, okadaic acid, a protein phosphatase PP1 and PP2A inhibitor, exaggerates the response to PMA [58]. Notably, the diminished serine phosphorylation and AAAD activity in dopaminergic cells that overexpress α-synuclein has been attributed to enhanced PP2A activity [64]. *In vitro* studies with recombinant and brain (striatum and midbrain) AAAD demonstrated that the enzyme is phosphorylated and activated by PKA, but not PKC [65], indicating that the effect of PKC activators *in vivo* is apparently indirect. Additionally, *in vitro* evidence suggests that recombinant and brain AAAD can be phosphorylated and activated by PKG (Figure 1), but its biological role remains to be elucidated.

The early activation of AAAD is followed by a late, longer lasting (hours) response, which is accompanied by an increase in AAAD mRNA and protein [54,60,66]. Prolonged induction of AAAD (days) also occurs after chronic administration of neuroactive drugs [42]. Although the transcription of AAAD in neuronal cells is not fully explored, the presence of putative regulatory elements on the promoter [31,37,67, Genbank AY37370] and the observation that pharmacological agents, hormones, and trophic factors change the abundance of AAAD mRNA [41,60,66,68-74] imply that transcriptional regulation of the enzyme is possible. The finding that icv forskolin induces a late increase of AAAD activity and mRNA in midbrain neurons suggests that cyclic-AMP might regulate the transcription of AAAD *in vivo* [57].

Even if AAAD is a regulated enzyme, there is no evidence supporting a rate limiting action for the synthesis of dopamine under normal circumstances. Despite commonalities in regulatory mechanisms and responses [41,57,66,71,75] often the pharmacological regulation of AAAD and tyrosine hydroxylase, the rate-limiting enzyme, in the striatum is discordant [42,60,74] suggesting 1) that AAAD might serve additional function(s); and/or 2) that AAAD regulation might, in part, occur in extra-dopaminergic sites in striatum. Indeed, AAAD is the rate-limiting enzyme for the synthesis of trace amines [26]; it is present in serotonergic and noradrenergic neurons and in a subset of $AAAD^+$ neurons in striatum [76-78]; and AAAD mRNA increases in the locus coeruleus and raphe nuclei, in addition to substantia nigra pars compacta and ventral tegmental area, after pharmacological manipulations [66].

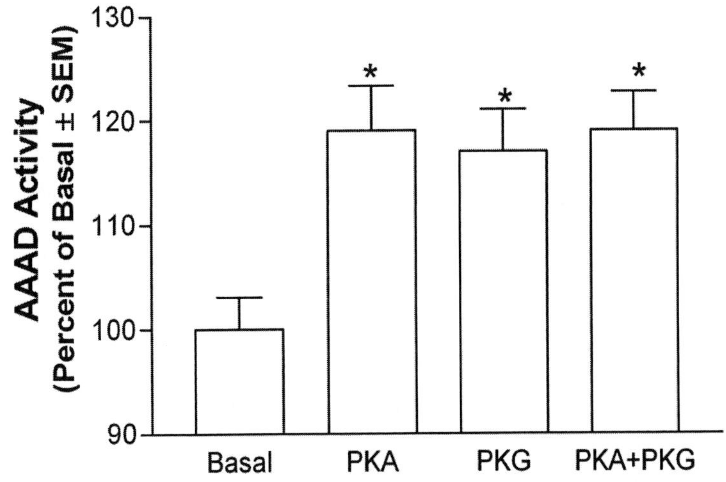

Figure 1. Exogenous PKA or PKG increases AAAD activity in striatal homogenates *in vitro*. *P<0.05 compared to Basal.

NEUROTRANSMITTER RECEPTOR REGULATION OF AAAD

Cumulative evidence suggests that acute or chronic blockade of D1- and D2-like receptors increase the activity of AAAD in rodent striatum (Table 4). Selective or mixed D1, D2, D3 or D4 antagonists administered acutely, all elevate striatal AAAD activity, and a similar effect has been observed with multireceptor drugs displaying affinity for D2-like receptors, such as spiperone and clozapine [54,56,60,66,79]. In general, there is an early transient activation and a late induction, and the latter is accompanied by a rise in enzyme mRNA in midbrain and protein content in the striatum [54,60]. Like-wise, chronic blockade of D1- and D2-like receptors results in a prolonged increase of AAAD activity in striatum and mRNA in brain and midbrain nigrostriatal neurons [42,80], and similar results are observed after icv administration of a D2 receptor antisense oligodeoxynucleotide [72]. In agreement with the enzyme studies, enhancement of AAAD activity following flupenthixol has been observed by following [^{3}H]DOPA conversion to [^{3}H]dopamine in the rat striatum *in vivo* [81]. In the same line, PET studies have documented an increase in AAAD activity in the human and porcine striatum after haloperidol [82,83]. In schizophrenic patients, PET studies measuring ^{18}F-L-DOPA uptake have shown that suppressed AAAD activity in the ventral striatum of drug-free schizophrenics is increased with typical or atypical antipsychotic treatment [84]. Finally, depletion of dopamine by reserpine increases AAAD activity in striatum [54,85,86]. In humans, AAAD activity appears to be genetically determined, as healthy subjects carrying the A1 allele of the human D2 receptor gene have increased enzyme activity, presumably due to decreased D2 autoreceptor function [87].

Studies with agonists suggest that dopamine receptor activation decreases enzyme activity, but the responses are variable and modest. In rodents, acute or chronic administration of D2-like agonists, particularly bromocryptine, suppresses the activity of AAAD in striatum; while, the effect of D1-like agonists SKF 38393 is inconsistent [42,54,60,63,79]. Correspondingly, acute or chronic administration of L-DOPA [54,63] and inhibition of MAO A, which raises synaptic dopamine, also decreases AAAD activity in striatum [74,88]. Deprenyl attenuates AAAD activity in striatum at doses that inhibit both MAO B and MAO A [74] and elevates mRNA in the substantia nigra pars compacta and ventral tegmental area [70]. Diminution of AAAD activity by apomorphine has been observed using conversion of [^{3}H]DOPA into [^{3}H]dopamine in rat striatum in vivo [81], and PET studies have shown a decrease in AAAD activity in the macaque striatum after MAO B inhibition [89].

Even though the majority of the studies have focused on the regulation of AAAD by dopamine receptors, a notable body of evidence suggests that the enzyme is subjected to regulation by other neurotransmitter receptors as well (Table 4). The emerging picture supports the idea that serotonergic, glutamatergic and nicotinic receptors (Table 4) modulate AAAD activity; whereas, involvement of alpha adrenergic, muscarinic and GABA receptors appears less likely in the brain. For example, the 5-HT2A/5-HT2C antagonist ketanserin, the 5-HT1A/5-HT2A antagonist metergoline and the 5-HT1A selective antagonist Way 100635 all increase AAAD activity in striatum after acute administration, while selective antagonists of 5-HT2C and 5-HT3 receptors do not influence enzyme activity [66]. Acute blockade of the glutamate NMDA ionotropic receptors by the non-competitive antagonist MK-801 enhances the activity of AAAD in striatum [73], and similar observations have been made for other NMDA channel blockers [61-63]. Unlike NMDA competitive antagonists, antagonists acting at the glycine or polyamine site, and AMPA non-competitive antagonists have little effect [61,62]. AAAD in dopaminergic and non-dopaminergic neurons might be regulated by

Table 4. Effect of Neurotransmitter Receptor Agonists or Antagonists on AAAD Activity in Mouse Striatum *in vivo* [91]

Receptor Type	AAAD Activity
D1-Like	
Agonists	↔
Antagonists	↑
D2-Like	
Agonists	↓,↔
Antagonists	↑
Alpha Adrenergic	
Alpha 1 Agonists	↔
Alpha 1 Antagonists	↔
Alpha 2 Agonists	↔
Alpha 2 Antagonists	↔
Cholinergic	
nAChR Agonists	↔
nAChR Antagonists	↑
mAChR Agonists	↔
mAChR Antagonists	↔
Serotonergic	
5-HT1A Antagonists	↑
5-HT2A Antagonists	↑
5-HT2C Antagonists	↔
5-HT3 Antagonists	↔
GABAergic	
GABAA Antagonists	↔
GABAB Antagonists	↔
Glutamatergic	
NMDA Channel Blockers	↑
NMDA Polyamine Site Antagonists	↔
NMDA Glycine Site Antagonists	↔
NMDA Competitive Antagonists	↔
AMPA Noncompetitive Antagonists	↔

NMDA receptors, as acute administration of MK-801 increases the expression of AAAD mRNA in midbrain [73]; whereas, chronic administration of phencyclidine has been shown to augment AAAD mRNA in striatum [90] indicating possible regulation in $AAAD^+$ neurons of the region. Cholinergic nicotine, but not muscarinic, receptors may be involved in the regulation of striatal AAAD; as mecamylamine, a non-competitive nicotine receptor antagonist, increases enzyme activity [66,91]. Alpha1- and alpha2-adrenergic receptors do not appear to play a role in the acute regulation of AAAD in striatum [66], and we have made similar observations for GABAA and GABAB receptors [91]. In the retina in contrast to the striatum, blockade of alpha2-adrenergic receptors increases, while activation of the receptor decreases enzyme activity [55]. Perhaps, in different regions of the nervous system the biochemical mechanisms and receptors for regulating AAAD activity vary.

Taken together, the dopaminergic, glutamatergic and setotonergic system have been identified as candidate targets for pharmacological regulation of AAAD via selective receptors (Table 4). Apparently in nigrostriatal neurons AAAD is under dopaminergic control, and dopamine receptors regulate enzyme activity via pre-synaptic (D2, D3) and post-synaptic (D1, D2) mechanisms [60]. Collectively, the studies reviewed so far imply that under normal conditions, dopamine exerts tonic inhibition of AAAD activity in dopaminergic neurons via autoreceptors. Glutamate, NMDA, and serotonin, 5-HT1A and 5-HT2A, receptors modulate dopaminergic neuron firing, dopamine synthesis and release [92-95] and their respective antagonists could affect AAAD activity, at least in part, by disinhibiting dopaminergic control. Although studies exploring the regulation of AAAD in serotonergic neurons are lacking, we speculate that in analogy with dopaminergic neurons, AAAD in serotonergic neurons is under the control of serotonin and 5-HT1A, antagonists could regulate AAAD expression in raphe nuclei [66] via somatodendritic autoreceptors [94].

AAAD Activity in PD Models

AAAD activity in nigrostriatal neurons appears to be up-regulated soon after a MPTP lesion [96] implying the involvement of compensatory or altered regulatory mechanisms. In MPTP-treated mice, AAAD in the denervated striatum responds more robustly to D1- and D2-like antagonists [52], and the same appears to be true for clozapine [66]. We have termed this exaggerated responsiveness of AAAD after a lesion of the nigrostriatal neurons "pre-synaptic supersensitivity", and it is manifested by a shift of the time- and dose-response curves to the left and upward; i.e. AAAD activity is more pronounced and observed earlier and with lower antagonist doses. Altered post-translational modification of AAAD after MPTP and/or altered receptor/second messenger signaling cascades secondary to the dopaminergic lesion might underlie this phenomenon (Figure 2). Indeed, the magnitude of exogenous PKA-induced AAAD phosphorylation is greater in homogenates prepared from striatum and midbrain of MPTP-lesioned mice, with the response increasing with time following the lesion (Figure 2A), and icv administration of forskolin in MPTP-mice amplifies the activation of the enzyme (Figure 2B), and causes an earlier induction of AAAD mRNA [57,59].

Imaging studies using ^{18}F-DOPA [97,98] have shown that the PD process differentially affects AAAD in various brain regions and within the striatum over time [24,99-102] and have estimated a relative upregulation of striatal AAAD by comparison to other neurochemical, neuroanatomical and imaging dopaminergic markers [103-106]. Although this has been attributed to a compensatory increase of AAAD activity in the remaining dopaminergic neurons, serotonergic, noradrenergic and intrinsic striatal neurons contribute to the AAAD measurements and enzyme activity in striatum [100,102] and might add to the calculated AAAD "upregulation" [106] as well. Regardless, radiotracer imaging studies have strengthened our original proposal that AAAD is a regulated enzyme [53,54] and provided clues supporting the idea that AAAD can be subjected to pharmacological manipulation in PD. For example, therapeutic infusion of L-DOPA or apomorphine caused a small decrease in L-[^{11}C]DOPA influx in the striatum with early but not advanced PD [107,108], and decreases in putamen AAAD activity following a 2 year treatment with L-DOPA, -20.3%, or a D2 agonist, -13.4% have been reported [109].

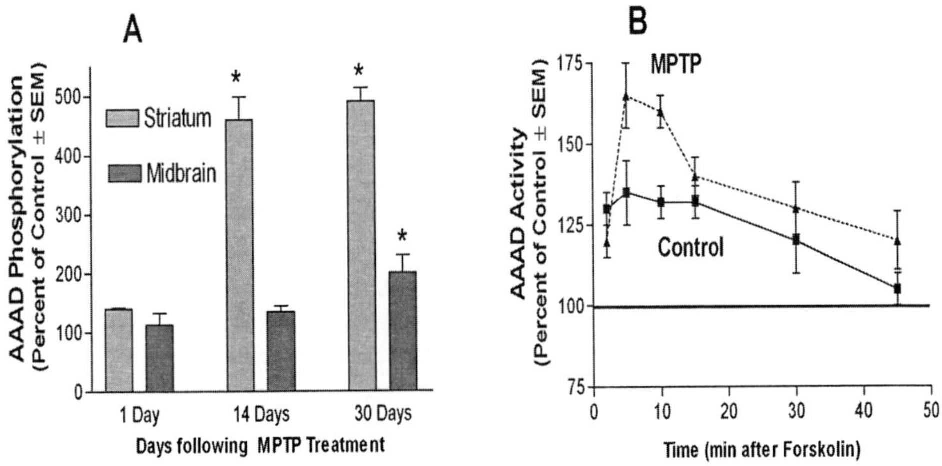

Figure 2. Exogenous PKA-induced phosphorylation of AAAD in the striatum and midbrain is enhanced in MPTP-treaded mice over time (Figure 2.A). AAAD activity in the striatum of MPTP-treated mice is enhanced following icv administration of forskolin (Figure 2.B). *P<0.05 compared with Control.

Clinical Implications

When treating PD patients with exogenous L-DOPA, AAAD becomes the rate limiting step for the formation of dopamine and the potential to modulate the activity of the enzyme by pharmacological means should not be disregarded. We propose that enhancing and stabilizing AAAD activity might be an alternative approach to achieve and maintain optimal dopaminergic transmission with L-DOPA treatment, and AAAD inducers may prove to be useful L-DOPA adjuncts. Drugs that increase AAAD activity in the striatum enhance the formation of dopamine from L-DOPA, and are expected to decrease the dose of L-DOPA, smooth the rise and fall of striatal dopamine following intermittent administration of L-DOPA, and maintain relatively constant levels of intrasynaptic dopamine resulting, thus, to a more physiological receptor stimulation. Hence, L-DOPA decarboxylation enhancers could meet two important tenets for successful L-DOPA therapy: 1) to use the least amount of drug that improves disability and 2) to achieve a relatively continuous stimulation of dopamine receptors [110-112]. High doses of L-DOPA are associated with an increased risk of motor complications [113,114] and continuous dopaminergic stimulation in early PD is believed to reduce this risk [111,112,115].

Clinical studies provide implicit support for the translational inferences we made afore. The L-DOPA adjuvants amantadine, budipine and memantine, increase AAAD activity and L-DOPA decarboxylation in experimental animals (61-63), improve parkinsonian symptoms

[116-121] and reduce motor complications [121,122]. In L-DOPA treated patients, the atypical antipsychotic clozapine moderately improves motor performance [123-130] and reduces motor fluctuations and dyskinesias [122,130-136]; while, in a PD model, it augments AAAD activity and L-DOPA decarboxylation [66]. Finally, the antidepressants mianserin, nefazodone and ritanserin improve parkinsonian symptoms [137-139] and the blockade of 5-HT2A receptors enhances AAAD activity in the striatum of mice [66]. These observations call for further studies to understand the regulation of AAAD and its role in L-DOPA treatment and adjuvant L-DOPA therapies.

REFERENCES

[1] Uretsky NJ, Iversen LL. Effects of 6-hydroxydopamine on noradrenaline-containing neurones in the rat brain. *Nature*. 1969;8:221:557-559.

[2] Uretsky NJ, Iversen LL. Effects of 6-hydroxydopamine on catecholamine containing neurons in the rat brain. *Journal of Neurochemistry*. 1970;17:269-278.

[3] Holtz P, Heise R, Ludtke K. Fermentativer Abbau von L-Dioxyphenylalanin (Dopa) durch Niere. *Naunyn-Schmiedeberg's Archives of Pharmacology*.1938;191:87-118.

[4] Jahng JW, Wessel TC, Houpt TA, Son JH, Joh TH. Alternate promoters in the rat aromatic L-amino acid decarboxylase gene for neuronal and nonneuronal expression: an in situ hybridization study. *Journal of Neurochemistry*. 1996;66:14-9.

[5] Clark CT, Weissbach H, Udenfriend S. 5-Hydroxytryptophan decarboxylase: preparation and properties. *Journal of Biological Chemistry*. 1954;210:139-48.

[6] Sourkes TL. DOPA Decarboxylase (aromatic amino acid decarboxylase). In: Horn AS, Korf J, Westerink BHC, editors. The Neurobiology of Dopamine. London: Academic Press;1979:123-132.

[7] Berry MD. Mammalian central nervous system trace amines. Pharmacologic amphetamines, physiologic neuromodulators. *Journal of Neurochemistry*. 2004;90:257-271.

[8] Jaeger CB, Teitelman G, Joh TH, Albert VR, Park DH, Reis DJ. Some neurons of the rat central nervous system contain aromatic-L-amino-acid decarboxylase but not monoamines. *Science*.1983;219:1233-1235.

[9] Nagatsu I, Sakai M, Yoshida M, Nagatsu T. Aromatic L-amino acid decarboxylase-immunoreactive neurons in and around the cerebrospinal fluid-contacting neurons of the central canal do not contain dopamine or serotonin in the mouse and rat spinal cord. *Brain Research*.1988;475:91-102.

[10] Lauweryns JM, Van Ranst L. Immunocytochemical localization of aromatic L-amino acid decarboxylase in human, rat, and mouse bronchopulmonary and gastrointestinal endocrine cells. *Jouranl of Histochemistry and Cytochemistry*.1988;36:1181-1186.

[11] Juorio AV, Li XM, Walz W, Paterson IA. Decarboxylation of L-dopa by cultured mouse astrocytes. *Brain Research*. 1993;626:306-309.

[12] Abdenur JE, Abeling N, Specola N, Jorge L, Schenone AB, van Cruchten AC, et al. Aromatic l-aminoacid decarboxylase deficiency: unusual neonatal presentation and

additional findings in organic acid analysis. *Molecular Genetics and Metabolism*.2006;87:48-53.

[13] Blaschko H. The specific actions of l-DOPA decarboxylase. *Journal of Physiology*. 1939;96:50-51.

[14] Raab W, Gigee W. Concentration and distribution of "encephalin" in the brain of humans and animals. *Proceedings of the Society for Experimental Biology and Medicine*.1951;76:97-100.

[15] Weil-Malherbe H, Bone AD. Intracellular distribution of catecholamines in the brain. *Nature*.1957;180:1050-1051.

[16] Carlsson A, Lindqvist M, Magnusson T. 3,4-Dihydroxyphenylalanine and 5-hydroxytryptophan as reserpine antagonists. *Nature*.1957;180:1200.

[17] Carlsson A. The occurrence, distribution and physiological role of catecholamines in the nervous system. *Pharmacological Reviews*.1959;11:490-493.

[18] Ehringer H, Hornykiewicz O. [Distribution of noradrenaline and dopamine (3-hydroxytyramine) in the human brain and their behavior in diseases of the extrapyramidal system. *Journal Klin Wochenschr*.1960;38:1236-1239.

[19] Birkmayer W, Hornykiewicz O. [The L-3,4-dioxyphenylalanine (DOPA)-effect in Parkinson-akinesia.]. *Wien Klin Wochenschr*.1961;73:787-788.

[20] Cotzias GC, Van Woert MH, Schiffer LM. Aromatic amino acids and modification of parkinsonism. *New England Journal of Medicine*.1967;276:374-379.

[21] Yahr MD, Duvoisin RC, Schear MJ, Barrett RE, Hoehn MM. Treatment of parkinsonism with levodopa. *Archives of Neurology*.1969t;21:343-354.

[22] Lloyd KG, Davidson L, Hornykiewicz O. The neurochemistry of Parkinson's disease: effect of L-dopa therapy. *Journal of Pharmacology Experimental Therapeutics*.1975;195:453-464.

[23] Fahn S. The history of dopamine and levodopa in the treatment of Parkinson's disease. *Movement Disorders*. 2008;23 Suppl 3:S497-508.

[24] Kuwabara H, Cumming P, Yasuhara Y, Leger GC, Guttman M, Diksic M, et al. Regional striatal DOPA transport and decarboxylase activity in Parkinson's disease. *Journal of Nuclear Medicine*. 1995;36:1226-1231.

[25] Gjedde A, Leger GC, Cumming P, Yasuhara Y, Evans AC, Guttman M, et al. Striatal L-dopa decarboxylase activity in Parkinson's disease in vivo: implications for the regulation of dopamine synthesis. *Journal of Neurochemistry*. 1993;61:1538-1541.

[26] Berry MD, Juorio AV, Li XM, Boulton AA. Aromatic L-amino acid decarboxylase: a neglected and misunderstood enzyme. *Neurochemical Research*. 1996;21:1075-87.

[27] Tanaka T, Horio Y, Taketoshi M, Imamura I, Ando-Yamamoto M, Kangawa K, et al. Molecular cloning and sequencing of a cDNA of rat dopa decarboxylase: partial amino acid homologies with other enzymes synthesizing catecholamines. *Proceedings of the National Academy of Science* U S A. 1989;86:8142-8146.

[28] Ichinose H, Kurosawa Y, Titani K, Fujita K, Nagatsu T. Isolation and characterization of a cDNA clone encoding human aromatic L-amino acid decarboxylase. *Biochemical and Biophysical Research Communications*. 1989;164:1024-1030.

[29] Albert VR, Allen JM, Joh TH. A single gene codes for aromatic L-amino acid decarboxylase in both neuronal and non-neuronal tissues. *Journal of Biological Chemistry*. 1987;262:9404-9411.

[30] Albert VR, Lee MR, Bolden AH, Wurzburger RJ, Aguanno A. Distinct promoters direct neuronal and nonneuronal expression of rat aromatic L-amino acid decarboxylase. *Proceedings of the National Academy of Science* U S A. 1992;89:12053-12057.

[31] Hahn SL, Hahn M, Kang UJ, Joh TH. Structure of the rat aromatic L-amino acid decarboxylase gene: evidence for an alternative promoter usage. *Journal of Neurochemistry*. 1993;60:1058-1064.

[32] Sumi-Ichinose C, Hasegawa S, Ohtsuki M, Nomura H, Nomura T, Hagino Y, et al. Analysis of an alternative promoter that regulates tissue-specific expression of the human aromatic L-amino acid decarboxylase gene in cultured cell lines. *Journal of Neural Transmission*. 1996;103:1-15.

[33] Le Van Thai A, Coste E, Allen JM, Palmiter RD, Weber MJ. Identification of a neuron-specific promoter of human aromatic L-amino acid decarboxylase gene. *Brain Research Molecular Brain Research*. 1993;17:227-238.

[34] Bruneau G, Thibault J, Gros F, Mattei MG. Mapping of the dopa decarboxylase gene to the 11A band of the murine genome. *Biochemical Biophysical Research Communications*. 1992;186:926-930.

[35] Bruneau G, Gross MS, Krieger M, Bernheim A, Thibault J, Nguyen VC. Preparation of a human DOPA decarboxylase cDNA probe by PCR and its assignment to chromosome 7. *Annals of Genetics*. 1990;33:208-213.

[36] Scherer LJ, McPherson JD, Wasmuth JJ, Marsh JL. Human dopa decarboxylase: Localization to human chromosome 7p11 and characterization of hepatic cDNAs. *Genomics*. 1992;13:469-471.

[37] Aguanno A, Lee MR, Marden CM, Rattray M, Gault A, Albert VR. Analysis of the neuronal promoter of the rat aromatic L-amino acid decarboxylase gene. *Journal of Neurochemistry*. 1995;65:1944-1954.

[38] Dugast C, Weber MJ. NF-Y binding is required for transactivation of neuronal aromatic L-amino acid decarboxylase gene promoter by the POU-domain protein Brn-2. *Brain Research Molecular Brain Research*. 2001;89:58-70.

[39] Schimmel JJ, Crews L, Roffler-Tarlov S, Chikaraishi DM. 4.5 kb of the rat tyrosine hydroxylase 5' flanking sequence directs tissue specific expression during development and contains consensus sites for multiple transcription factors. *Brain Research Molecular Brain Research*. 1999;74:1-14.

[40] Cazorla P, Smidt MP, O'Malley KL, Burbach JP. A response element for the homeodomain transcription factor Ptx3 in the tyrosine hydroxylase gene promoter. *Journal of Neurochemistry*. 2000;74:1829-1837.

[41] Wessel TC, Joh TH. Parallel upregulation of catecholamine-synthesizing enzymes in rat brain and adrenal gland: effects of reserpine and correlation with immediate early gene expression. *Brain Research Molecular Brain Research*. 1992;15:349-360.

[42] Cho S, Duchemin AM, Neff NH, Hadjiconstantinou M. Tyrosine hydroxylase, aromatic L-amino acid decarboxylase and dopamine metabolism after chronic treatment with dopaminergic drugs. *Brain Research*. 1999;830:237-245.

[43] Hendricks T, Francis N, Fyodorov D, Deneris ES. The ETS domain factor Pet-1 is an early and precise marker of central serotonin neurons and interacts with a conserved element in serotonergic genes. *Journal of Neuroscience*. 1999;19:10348-10356.

[44] Krueger KC, Deneris ES. Serotonergic transcription of human FEV reveals direct GATA factor interactions and fate of Pet-1-deficient serotonin neuron precursors. *Journal of Neuroscience*. 2008;28:12748-12758.

[45] Hendricks TJ, Fyodorov DV, Wegman LJ, Lelutiu NB, Pehek EA, Yamamoto B, et al. Pet-1 ETS gene plays a critical role in 5-HT neuron development and is required for normal anxiety-like and aggressive behavior. *Neuron*. 2003;37:233-247.

[46] Burkhard P, Dominici P, Borri-Voltattorni C, Jansonius JN, Malashkevich VN. Structural insight into Parkinson's disease treatment from drug-inhibited DOPA decarboxylase. *Nature Structural & Molecular Biology*. 2001;8:963-967.

[47] Dominici P, Moore PS, Castellani S, Bertoldi M, Voltattorni CB. Mutation of cysteine 111 in Dopa decarboxylase leads to active site perturbation. *Protein Science*. 1997;6:2007-2015.

[48] Sandmeier E, Hale TI, Christen P. Multiple evolutionary origin of pyridoxal-5'-phosphate-dependent amino acid decarboxylases. *European Journal of Biochemistry*. 1994;221:997-1002.

[49] Tancini B, Dominici P, Barra D, Voltattorni CB. An essential arginine residue at the binding site of pig kidney 3,4-dihydroxyphenylalanine decarboxylase. *Archives of Biochemistry and Biophysics*. 1985;238:565-573.

[50] Dominici P, Tancini B, Borri Voltattorni C. Chemical modification of pig kidney 3,4-dihydroxyphenylalanine decarboxylase with diethyl pyrocarbonate. Evidence for an essential histidyl residue. *Journal of Biological Chemistry*. 1985;260:10583-10589.

[51] Dominici P, Maras B, Mei G, Borri Voltattorni C. Affinity labeling of pig kidney 3,4-dihydroxyphenylalanine (Dopa) decarboxylase with N-(bromoacetyl)pyridoxamine 5'-phosphate. Modification of an active-site cysteine. *European Journal of Biochemistry*. 1991;201:393-397.

[52] Ishii S, Mizuguchi H, Nishino J, Hayashi H, Kagamiyama H. Functionally important residues of aromatic L-amino acid decarboxylase probed by sequence alignment and site-directed mutagenesis. *Journal of Biochemistry*. 1996;120:369-376.

[53] Hadjiconstantinou M, Rossetti Z, Silvia C, Krajnc D, Neff NH. Aromatic L-amino acid decarboxylase activity of the rat retina is modulated in vivo by environmental light. *Journal of Neurochemistry*. 1988;51:1560-1564.

[54] Hadjiconstantinou M, Wemlinger TA, Sylvia CP, Hubble JP, Neff NH. Aromatic L-amino acid decarboxylase activity of mouse striatum is modulated via dopamine receptors. *Journal of Neurochemistry*. 1993;60:2175-2180.

[55] Rossetti Z, Krajnc D, Neff NH, Hadjiconstantinou M. Modulation of retinal aromatic L-amino acid decarboxylase via alpha 2 adrenoceptors. *Journal of Neurochemistry*. 1989;52:647-652.

[56] Zhu MY, Juorio AV, Paterson IA, Boulton AA. Regulation of aromatic L-amino acid decarboxylase by dopamine receptors in the rat brain. *Journal of Neurochemistry.* 1992;58:636-641.

[57] Young EA, Duchemin AM, Neff NH, Hadjiconstantinou M. Parallel modulation of striatal dopamine synthetic enzymes by second messenger pathways. *European Journal of Pharmacology.* 1998;357:15-23.

[58] Young EA, Neff NH, Hadjiconstantinou M. Phorbol ester administration transiently increases aromatic L-amino acid decarboxylase activity of the mouse striatum and midbrain. *Journal of Neurochemistry.* 1994;63:694-697.

[59] Young EA, Neff NH, Hadjiconstantinou M. Evidence for cyclic AMP-mediated increase of aromatic L-amino acid decarboxylase activity in the striatum and midbrain. *Journal of Neurochemistry.* 1993;60:2331-2333.

[60] Cho S, Neff NH, Hadjiconstantinou M. Regulation of tyrosine hydroxylase and aromatic L-amino acid decarboxylase by dopaminergic drugs. *European Journal of Pharmacology.* 1997;323:149-157.

[61] Fisher A, Biggs CS, Starr MS. Differential effects of NMDA and non-NMDA antagonists on the activity of aromatic L-amino acid decarboxylase activity in the nigrostriatal dopamine pathway of the rat. *Brain Research.* 1998;792:126-132.

[62] Fisher A, Biggs CS, Starr MS. Effects of glutamate antagonists on the activity of aromatic L-amino acid decarboxylase. *Amino Acids.* 1998;14:43-49.

[63] Fisher A, Starr MS. Opposite effects of glutamate antagonists and antiparkinsonian drugs on the activities of DOPA decarboxylase and 5-HTP decarboxylase in the rat brain. *Brain Research.* 2000;868:268-274.

[64] Tehranian R, Montoya SE, Van Laar AD, Hastings TG, Perez RG. Alpha-synuclein inhibits aromatic amino acid decarboxylase activity in dopaminergic cells. *Journal of Neurochemistry.* 2006;99:1188-1196.

[65] Duchemin AM, Berry MD, Neff NH, Hadjiconstantinou M. Phosphorylation and activation of brain aromatic L-amino acid decarboxylase by cyclic AMP-dependent protein kinase. *Journal of Neurochemistry.* 2000;75:725-731.

[66] Neff NH, Wemlinger TA, Duchemin AM, Hadjiconstantinou M. Clozapine modulates aromatic L-amino acid decarboxylase activity in mouse striatum. *Journal of Pharmacology and Experimental Therapeutics.* 2006;317:480-487.

[67] O'Malley KL, Harmon S, Moffat M, Uhland-Smith A, Wong S. The human aromatic L-amino acid decarboxylase gene can be alternatively spliced to generate unique protein isoforms. *Journal of Neurochemistry.* 1995;65:2409-2416.

[68] Li XM, Juorio AV, Boulton AA. NSD-1015 alters the gene expression of aromatic L-amino acid decarboxylase in rat PC12 pheochromocytoma cells. *Neurochemical Research.* 1993;18:915-919.

[69] Li XM, Juorio AV, Boulton AA. Induction of aromatic L-amino acid decarboxylase mRNA by interleukin-1 beta and prostaglandin E2 in PC12 cells. *Neurochemical Research.* 1994;19:591-595.

[70] Li XM, Juorio AV, Paterson IA, Zhu MY, Boulton AA. Specific irreversible monoamine oxidase B inhibitors stimulate gene expression of aromatic L-amino acid decarboxylase in PC12 cells. *Journal of Neurochemistry.* 1992;59:2324-2327.

[71] Kim KT, Park DH, Joh TH. Parallel up-regulation of catecholamine biosynthetic enzymes by dexamethasone in PC12 cells. *Journal of Neurochemistry.* 1993;60:946-951.

[72] Hadjiconstantinou M, Neff NH, Zhou LW, Weiss B. D2 dopamine receptor antisense increases the activity and mRNA of tyrosine hydroxylase and aromatic L-amino acid decarboxylase in mouse brain. *Neuroscience Letters.* 1996;217:105-108.

[73] Hadjiconstantinou M, Rossetti ZL, Wemlinger TA, Neff NH. Dizocilpine enhances striatal tyrosine hydroxylase and aromatic L-amino acid decarboxylase activity. *European Journal of Pharmacology.* 1995;289:97-101.

[74] Cho S, Duchemin AM, Neff NH, Hadjiconstantinou M. Modulation of tyrosine hydroxylase and aromatic L-amino acid decarboxylase after inhibiting monoamine oxidase-A. *European Journal of Pharmacology.* 1996;314:51-59.

[75] Baptista MJ, O'Farrell C, Daya S, Ahmad R, Miller DW, Hardy J, et al. Co-ordinate transcriptional regulation of dopamine synthesis genes by alpha-synuclein in human neuroblastoma cell lines. *Journal of Neurochemistry.* 2003;85:957-968.

[76] Tashiro Y, Kaneko T, Sugimoto T, Nagatsu I, Kikuchi H, Mizuno N. Striatal neurons with aromatic L-amino acid decarboxylase-like immunoreactivity in the rat. *Neuroscience Letters.* 1989;100:29-34.

[77] Eaton MJ, Gudehithlu KP, Quach T, Silvia CP, Hadjiconstantinou M, Neff NH. Distribution of aromatic L-amino acid decarboxylase mRNA in mouse brain by in situ hybridization histology. *Journal of Comparative Neurology.* 1993;337:640-654.

[78] Meredith GE, Farrell T, Kellaghan P, Tan Y, Zahm DS, Totterdell S. Immunocytochemical characterization of catecholaminergic neurons in the rat striatum following dopamine-depleting lesions. *European Journal of Neuroscience.* 1999;11:3585-3596.

[79] Zhu MY, Juorio AV, Paterson IA, Boulton AA. Regulation of striatal aromatic L-amino acid decarboxylase: effects of blockade or activation of dopamine receptors. *European Journal of Pharmacology.* 1993;238:157-164.

[80] Buckland PR, O'Donovan MC, McGuffin P. Changes in dopa decarboxylase mRNA but not tyrosine hydroxylase mRNA levels in rat brain following antipsychotic treatment. *Psychopharmacology* (Berl). 1992;108:98-102.

[81] Cumming P, Ase A, Laliberte C, Kuwabara H, Gjedde A. In vivo regulation of DOPA decarboxylase by dopamine receptors in rat brain. *Journal of Cerebral Blood Flow and Metabolism.* 1997;17:1254-1260.

[82] Vernaleken I, Kumakura Y, Cumming P, Buchholz HG, Siessmeier T, Stoeter P, et al. Modulation of [18F]fluorodopa (FDOPA) kinetics in the brain of healthy volunteers after acute haloperidol challenge. *Neuroimage.* 2006;30:1332-1339.

[83] Danielsen EH, Smith D, Hermansen F, Gjedde A, Cumming P. Acute neuroleptic stimulates DOPA decarboxylase in porcine brain in vivo. *Synapse.* 2001;41:172-175.

[84] Elkashef AM, Doudet D, Bryant t, Cohen RM, Li SH Wyatt, RJ. 6-(18)F-DOPA PET study in patients with schizophrenia. Positiron emission tomography. *Psychiatry Research* 2000;100:1-11.

[85] Biggs CS, Fisher A, Starr MS. The antiparkinsonian drug budipine stimulates the activity of aromatic L-amino acid decarboxylase and enhances L-DOPA-induced dopamine release in rat substantia nigra. *Synapse.* 1998;30:309-17.
[86] Fisher A, Biggs CS, Eradiri O, Starr MS. Dual effects of L-3,4-dihydroxyphenylalanine on aromatic L-amino acid decarboxylase, dopamine release and motor stimulation in the reserpine-treated rat: evidence that behaviour is dopamine independent. *Neuroscience.* 2000;95:97-111.
[87] Laakso A, Pohjalainen T, Bergman J, Kajander J, Haaparanta M, Solin O, et al. The A1 allele of the human D2 dopamine receptor gene is associated with increased activity of striatal L-amino acid decarboxylase in healthy subjects. *Pharmacogenetics and Genomics.* 2005;15:387-391.
[88] Cumming P, Kuwabara H, Ase A, Gjedde A. Regulation of DOPA decarboxylase activity in brain of living rat. *Journal of Neurochemistry.* 1995;65:1381-1390.
[89] DeJesus OT, Flores LG, Roberts AD, Dick DW, Bartlett RM, Murali D, et al. Aromatic L-amino acid decarboxylase (AAAD) activity in rhesus macaque striatum after MAO-B inhibition by Ro 16-6491. *Synapse.* 2005;56:54-56.
[90] Buckland PR, Marshall R, Watkins P, McGuffin P. Does phenylethylamine have a role in schizophrenia?: LSD and PCP up-regulate aromatic L-amino acid decarboxylase mRNA levels. *Brain Research Molecular Brain Research.* 1997;49:266-270.
[91] Hadjiconstantinou M, Neff NH. Enhancing aromatic L-amino acid decarboxylase activity: implications for L-DOPA treatment in Parkinson's disease. *CNS Neuroscience and Therapeutics.* 2008:14:340-351.
[92] Zigmond MJ, Castro SL, Keefe KA, Abercrombie ED, Sved AF. Role of excitatory amino acids in the regulation of dopamine synthesis and release in the neostriatum. *Amino Acids.* 1998;14:57-62.
[93] David HN, Ansseau M, Abraini JH. Dopamine-glutamate reciprocal modulation of release and motor responses in the rat caudate-putamen and nucleus accumbens of "intact" animals. *Brain Research Brain Research Reviews.* 2005; 0:336-360.
[94] Alex KD, Pehek EA. Pharmacologic mechanisms of serotonergic regulation of dopamine neurotransmission. *Pharmacology and Therapeutics.* 2007;113:296-320.
[95] Bustos G, Abarca J, Campusano J, Bustos V, Noriega V, Aliaga E. Functional interactions between somatodendritic dopamine release, glutamate receptors and brain-derived neurotrophic factor expression in mesencephalic structures of the brain. *Brain Research Brain Research Reviews.* 2004;47:126-144.
[96] Hadjiconstantinou M, Neff NH. Differential recovery of dopamine synthetic enzymes following MPTP and the consequences of GM1 ganglioside treatment. *European Journal of Pharmacology.* 1990;181:137-139.
[97] Heiss WD, Hilker R. The sensitivity of 18-fluorodopa positron emission tomography and magnetic resonance imaging in Parkinson's disease. *European Journal of Neurology.* 2004;11:5-12.
[98] Ravina B, Eidelberg D, Ahlskog JE, Albin RL, Brooks DJ, Carbon M, et al. The role of radiotracer imaging in Parkinson disease. *Neurology.* 2005;64:208-215.

[99] Morrish PK, Sawle GV, Brooks DJ. Clinical and [18F] dopa PET findings in early Parkinson's disease. *Journal of Neurology, Neurosurgery & Psychiatry.* 1995;59:597-600.

[100] Rakshi JS, Uema T, Ito K, Bailey DL, Morrish PK, Ashburner J, et al. Frontal, midbrain and striatal dopaminergic function in early and advanced Parkinson's disease A 3D [(18)F]dopa-PET study. *Brain.* 1999;122:1637-1650.

[101] Bruck A, Aalto S, Nurmi E, Vahlberg T, Bergman J, Rinne JO. Striatal subregional 6-[18F]fluoro-L-dopa uptake in early Parkinson's disease: a two-year follow-up study. *Movement Disorders.* 2006;21:958-963.

[102] Moore RY, Whone AL, Brooks DJ. Extrastriatal monoamine neuron function in Parkinson's disease: an 18F-dopa PET study. *Neurobiology of Disease.* 2008;29:381-90.

[103] Lee CS, Samii A, Sossi V, Ruth TJ, Schulzer M, Holden JE, et al. In vivo positron emission tomographic evidence for compensatory changes in presynaptic dopaminergic nerve terminals in Parkinson's disease. *Annals Neurology.* 2000;47:493-503.

[104] Tedroff J, Ekesbo A, Rydin E, Langstrom B, Hagberg G. Regulation of dopaminergic activity in early Parkinson's disease. *Annals Neurology.* 1999;46:359-365.

[105] Whone AL, Moore RY, Piccini PP, Brooks DJ. Plasticity of the nigropallidal pathway in Parkinson's disease. *Annals Neurology.* 2003;53:206-213.

[106] Brooks DJ, Piccini P. Imaging in Parkinson's disease: the role of monoamines in behavior. *Biological Psychiatry.* 2006;59:908-918.

[107] Torstenson R, Hartvig P, Langstrom B, Westerberg G, Tedroff J. Differential effects of levodopa on dopaminergic function in early and advanced Parkinson's disease. *Annals Neurology.* 1997;41:334-340.

[108] Ekesbo A, Rydin E, Torstenson R, Sydow O, Laengstrom B, Tedroff J. Dopamine autoreceptor function is lost in advanced Parkinson's disease. *Neurology.* 1999;52:120-125.

[109] Whone AL, Watts RL, Stoessl AJ, Davis M, Reske S, Nahmias C, et al. Slower progression of Parkinson's disease with ropinirole versus levodopa: The REAL-PET study. *Annals Neurology.* 2003;54:93-101.

[110] Koller WC. Treatment of early Parkinson's disease. *Neurology.* 2002;58(4 Suppl):S79-86.

[111] Olanow W, Schapira AH, Rascol O. Continuous dopamine-receptor stimulation in early Parkinson's disease. *Trends Neuroscience.* 2000;23(10 Suppl):S117-26.

[112] Chase TN. The significance of continuous dopaminergic stimulation in the treatment of Parkinson's disease. *Drugs.* 1998;55 (Suppl) 1:1-9.

[113] Schrag A, Quinn N. Dyskinesias and motor fluctuations in Parkinson's disease. A community-based study. *Brain.* 2000;123:2297-305.

[114] Fahn S, Oakes D, Shoulson I, Kieburtz K, Rudolph A, Lang A, et al. Levodopa and the progression of Parkinson's disease. *New England Journal of Medicine.* 2004;351:2498-2508.

[115] Stocchi F, Olanow CW. Continuous dopaminergic stimulation in early and advanced Parkinson's disease. *Neurology.* 2004;62(Suppl 1):S56-63.

[116] Jellinger K, Bliesath H. Adjuvant treatment of Parksinson's disease with budipine: a double-blind trial versus placebo. *Journal of Neurology*. 1987;234:280-282.

[117] Merello M, Nouzeilles MI, Cammarota A, Leiguarda R. Effect of memantine (NMDA antagonist) on Parkinson's disease: a double-blind crossover randomized study. *Clinical Neuropharmacology*. 1999;22:273-276.

[118] Rabey JM, Nissipeanu P, Korczyn AD. Efficacy of memantine, an NMDA receptor antagonist, in the treatment of Parkinson's disease. *J Neural Transmission Parkinson Disease Dementia Section*. 1992;4:277-282.

[119] Przuntek H, Bittkau S, Bliesath H, Buttner U, Fuchs G, Glass J, et al. Budipine provides additional benefit in patients with Parkinson disease receiving a stable optimum dopaminergic drug regimen. *Archives of Neurology*. 2002;59:803-806.

[120] Crosby N, Deane KH, Clarke CE. Amantadine in Parkinson's disease. *Cochrane database of systematic reviews*. 2003:CD003468.

[121] Goetz CG, Poewe W, Rascol O, Sampaio C. Evidence-based medical review update: pharmacological and surgical treatments of Parkinson's disease: 2001 to 2004. *Movement Disorders*. 2005;20:523-39.

[122] Fabbrini G, Brotchie JM, Grandas F, Nomoto M, Goetz CG. Levodopa-induced dyskinesias. *Movement Disorders*. 2007;22:1379-1389.

[123] Factor SA, Brown D. Clozapine prevents recurrence of psychosis in Parkinson's disease. *Movement Disorders*. 1992;7:125-131.

[124] Linazasoro G, Marti Masso JF, Suarez JA. Nocturnal akathisia in Parkinson's disease: treatment with clozapine. *Movement Disorders*. 1993;8:171-174.

[125] Bennett JP, Jr., Landow ER, Dietrich S, Schuh LA. Suppression of dyskinesias in advanced Parkinson's disease: moderate daily clozapine doses provide long-term dyskinesia reduction. *Movement Disorders*. 1994;9:409-414.

[126] Wagner ML, Defilippi JL, Menza MA, Sage JI. Clozapine for the treatment of psychosis in Parkinson's disease: chart review of 49 patients. *Journal of Neuropsychiatry and Clinical Neuroscience*. 1996;8:276-280.

[127] Trosch RM, Friedman JH, Lannon MC, Pahwa R, Smith D, Seeberger LC, et al. Clozapine use in Parkinson's disease: a retrospective analysis of a large multicentered clinical experience. *Movement Disorders*. 1998;13:377-382.

[128] Ellis T, Cudkowicz ME, Sexton PM, Growdon JH. Clozapine and risperidone treatment of psychosis in Parkinson's disease. *Journal of Neuropsychiatry and Clinical Neuroscience*. 2000;12:364-369.

[129] Goetz CG, Blasucci LM, Leurgans S, Pappert EJ. Olanzapine and clozapine: comparative effects on motor function in hallucinating PD patients. *Neurology*. 2000;55:789-794.

[130] Morgante L, Epifanio A, Spina E, Di Rosa AE, Zappia M, Basile G, et al. Quetiapine versus clozapine: a preliminary report of comparative effects on dopaminergic psychosis in patients with Parkinson's disease. *Neurological Science*. 2002;23 Suppl 2:S89-90.

[131] Arevalo GJ, Gershanik OS. Modulatory effect of clozapine on levodopa response in Parkinson's disease: a preliminary study. *Movement Disorders*. 19931;8:349-354.

[132] Bennett JP, Jr., Landow ER, Schuh LA. Suppression of dyskinesias in advanced Parkinson's disease. II. Increasing daily clozapine doses suppress dyskinesias and improve parkinsonism symptoms. *Neurology*. 1993;43:1551-1555.

[133] Durif F, Vidailhet M, Assal F, Roche C, Bonnet AM, Agid Y. Low-dose clozapine improves dyskinesias in Parkinson's disease. *Neurology*. 1997;48:658-662.

[134] Pierelli F, Adipietro A, Soldati G, Fattapposta F, Pozzessere G, Scoppetta C. Low dosage clozapine effects on L-dopa induced dyskinesias in parkinsonian patients. *Acta* ***Neurologica*** *Scandinavica*. 1998;97:295-299.

[135] Morgante L, Epifanio A, Spina E, Zappia M, Di Rosa AE, Marconi R, et al. Quetiapine and clozapine in parkinsonian patients with dopaminergic psychosis. *Clinical Neuropharmacology*. 2004;27:153-156.

[136] Waters C. Other pharmacological treatments for motor complications and dyskinesias. *Movement Disorders*. 2005;20 Suppl 11:S38-44.

[137] Henderson J, Yiannikas C, Graham JS. Effect of ritanserin, a highly selective 5-HT2 receptor antagonist, on Parkinson's disease. *Clinical and Experimental Neurology*. 1992;29:277-282.

[138] Ikeguchi K, Kuroda A. Mianserin treatment of patients with psychosis induced by antiparkinsonian drugs. *European Archives of Psychiatry and Clinical Neuroscience*. 1995;244:320-324.

[139] Avila A, Cardona X, Martin-Baranera M, Maho P, Sastre F, Bello J. Does nefazodone improve both depression and Parkinson disease? A pilot randomized trial. *Journal of Clinical Psychopharmacology*. 2003;23:509-513.

In: Biogenic Amines
Editors: T. Farooqui, A. A. Farooqui

ISBN: 978-1-60876-625-3

Chapter IV

STRIATAL DOPAMINE STORAGE AND UTILIZATION

Lane J. Wallace *
The Ohio State University, Columbus, Ohio, USA.

ABSTRACT

Studies on striatal dopamine turnover indicate a large excess of available dopamine relative to dopamine used in signaling. In addition, electron micrographs show a small percentage of storage vesicles close to the membrane and a substantial percentage of storage vesicles not associated with the membrane. These data suggest the concept of an available or readily used population of dopamine storage vesicles and a reserve or storage population of dopamine storage vesicles. A variety of kinetic experiments have provided data interpreted as supporting two populations of dopamine. These include dopamine and metabolite kinetics after inhibition of dopamine synthesis, dopamine and metabolite specific activity after administration of radiolabeled tyrosine, and amount of newly synthesized dopamine in extracellular space. Other studies probed effects of inhibition of dopamine synthesis and of inhibition of dopamine uptake when dopaminergic neurons are experimentally stimulated at supra-physiological rates. Computer simulation studies of a variety of these experiments show that data from many can be explained using a single compartment of stored dopamine when all factors involved in regulation of dopamine secretion and reuptake are taken into account. Some of the experiments can not be explained by either a one compartment or a two compartment model of dopamine storage but can be explained if a second source of DOPAC – direct metabolism of newly synthesized dopamine – is present. Both physiological data and modeling simulation data support the concept of dopamine secretion where vesicles only partially empty and then are reused ("kiss and run"). Recent biochemical data suggest that a membrane associated dopamine storage pool has distinct characteristics from a cytosol-associated storage pool. A reasonable

* Correspondence concerning this article should be addressed to: Lane J. Wallace, 500 West 12th Avenue, Columbus, OH 43210. 614-292-9917; wallace.8@osu.edu.

interpretation of the total set of data reviewed is that most signaling is done by membrane-associated storage that uses a "kiss and run" mechanism and rapid recycling of used dopamine. The signaling population is small enough that it is difficult to distinguish it from the larger storage/reserve pool. This leaves unanswered the reason/function for a large store/reserve of dopamine.

INTRODUCTION

Dopaminergic neurons have a major role in controlling goal directed movements, determining saliency of reward, and comparing actual to expected reward [1-3]. In addition, all addicting drugs increase extracellular dopamine, a measure of signaling dopamine, in the nucleus accumbens region of the brain [4]. Most dopamine neurons in the brain have cell bodies in either the substantia nigra or the ventral tegmental area and project to the striatum, nucleus accumbens, and prefrontal cortex. The striatum has the highest concentration of dopamine in the brain.

Morphological studies show that dopaminergic neurons have only a few long dendrites, one major axon that travels between the region of the cell bodies and the target areas, and a very large arborization of axons in the terminal areas [2,5,6]. Dopamine release sites appear to be distributed along the axonal branches rather than being limited to the axon terminals. The density of terminals in the dorsal striatum is estimated to be as high as 10^8 per mm^3 [7,8]. Thus, dopamine is very highly concentrated in the terminal area.

Turnover studies provide information on rate of loss and replenishment of a neurotransmitter. Turnover can be estimated from rate of loss of a neurotransmitter after inhibition of its synthesis or from rate of accumulation after inhibition of its metabolism. In the case of dopamine, α-methyl-*p*-tyrosine is an effective inhibitor of tyrosine hydroxylase [9], the rate limiting enzyme in dopamine synthesis. Pargyline and deprenyl are inhibitors of monoamine oxidase [10,11], the major enzyme involved in dopamine metabolism. While turnover studies provide solid information on the average lifetime of a neurotransmitter molecule, they usually only provide clues to the subcellular travels of a molecule during its lifetime. Portions of this chapter will focus on inferences regarding subcellular travels.

A dopamine molecule can move through several subcellular compartments during its lifetime. Dopamine is synthesized from the amino acid tyrosine by the action of the enzymes tyrosine hydroxylase (rate limiting) and aromatic amino acid decarboxylase (fast). The newly synthesized dopamine is thought to be deposited in the cytosolic compartment of a varicosity. From there, a dopamine molecule is moved into a storage vesicle by the action of vesicular monoamine transporter protein. Dopamine leaves vesicles either by diffusion into the cytosolic compartment or by exocytosis into the extracellular compartment. Dopamine is moved from extracellular compartment to cytosolic compartment by action of dopamine transporter protein. Some dopamine in extracellular compartment is metabolized, and some may be taken up into glial cells. In either case, dopamine molecules removed from extracellular compartment by actions other than movement by dopamine transporter are lost from the varicosity. Some dopamine molecules in cytosolic compartment are metabolized by monoamine oxidase (rate limiting) followed by aldehyde dehydrogenase (fast) to make DOPAC.

The major metabolite made from cytosolic dopamine is DOPAC. DOPAC is found in both cytosolic compartments and extracellular compartments. DOPAC can be removed from either compartment by metabolism. It is thought that most DOPAC in extracellular compartment was originally made in cytosolic compartment. However, the mechanism of movement of this acidic compound between compartments is not known.

The vast majority of dopamine in a varicosity is stored in vesicles. Vesicles are approximately 50 nm in outside diameter. There may be 100 to 300 vesicles associated with each varicosity. Clearly, some of these vesicles are close to the membrane release sites, while others are somewhat removed from the membrane. This suggests that there is a higher probability that vesicles closer to the membrane would be used in a signaling event compared to those that are more distant from the membrane. In order to maintain a relatively constant level of signal, vesicles that are distant from the membrane must be moved to the membrane to replace vesicles that have been involved in a release event. Thus, vesicle trafficking would appear to be crucial to maintain effective synaptic signaling.

For some neurotransmitters in some types of neurons, it has been clearly established that there is a pool of vesicles actively involved in exocytotic events (called available pool) and a pool of vesicles providing a reserve or storage that is only called upon when the active pool depletes. The size of these pools and the ability to move between them varies substantially between types of neurons. Sometimes, the exchange is sufficiently rapid to accommodate very rapid firing rates [12-14]. Occasionally, exchange between pools is not sufficiently rapid to maintain the available pool when firing rates are very high [15-17]

Regarding dopaminergic neurons, some studies have provided data that have been interpreted as supporting the concept of a pool of vesicles involved in signaling and another pool held in reserve or storage. Two very divergent types of experimental paradigms have provided data that have been interpreted as supporting this hypothesis. One paradigm involves turnover studies of dopamine and DOPAC. The other paradigm involves interaction of drugs that inhibit dopamine transporter with one of two other manipulations. One of these manipulations is the stimulation of firing rate of dopaminergic neurons at supra-physiological rates. The other is addition of antipsychotic drugs.

We have been using computer simulations models of dopaminergic varicosities in studies of various quantitative questions regarding dopamine and DOPAC [18-20]. Through these studies, we became interested in the hypothesis of reserve and available compartments, potential sizes of compartments, and relation of size of compartments to size of signaling events. In this chapter, we review these concepts, relying heavily on data from our simulations.

Turnover Studies of Dopamine and Dopac

An early suggestion of the hypothesis of two storage compartments for dopamine came from studies on the rate of disappearance of dopamine from the striatum after inhibition of dopamine synthesis by α-methyl-*p*-tyrosine. The kinetics of disappearance reported by Javoy and Glowinski [21] showed an initial rapid decrease in tissue dopamine, followed by an increase in tissue dopamine, followed by a gradual decline of dopamine over a long period of time. These data were interpreted as indicative of two compartments of dopamine, and turnover rate constants were determined for each compartment. As we analyzed these results, we noted that an increase in amount of dopamine cannot be accomplished by shifting dopamine between compartments. The increase can only be explained by some decrement in the extent of inhibition of dopamine synthesis or addition of a new source of dopamine synthesis. To illustrate, the Javoy data are compared with two simulations (Figure 1). Both simulations use the beginning and ending points of the Javoy data, and one shows rate of dopamine disappearance from a one-compartment model and the other the rate of dopamine disappearance from a two compartment model. Clearly, neither solution matches the published data.

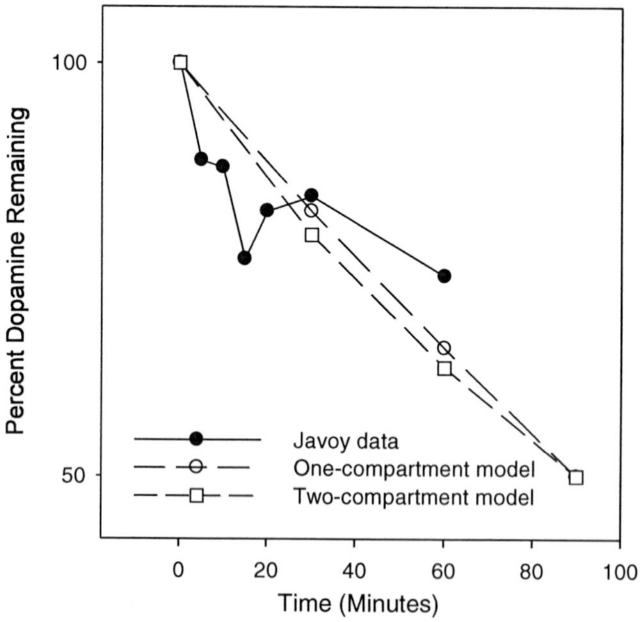

Figure 1. Dopamine disappearance after inhibition of dopamine synthesis comparing the data from Javoy and Glowinski [21] to a one-compartment and a two-compartment simulation model.

We considered the possibility that the lack of a uniform rate of disappearance of dopamine in the data presented by Javoy might result from a "bad" data point. We compared rates of disappearance of dopamine after inhibition of dopamine synthesis presented in 9 different publications (Figure 2). While in general all studies have very similar results, the few studies reporting on early time points show a rapid rate of loss of dopamine occurring in the first 30 minutes after administration of α-methyl-*p*-tyrosine followed by a short period with either an increase in dopamine or no change in dopamine, followed by a sustained gradual loss of dopamine. Doteuchi *et al* [22] and Paden [23], both of whom confirmed an early rapid disappearance of dopamine after inhibition of dopamine synthesis, also reported on specific activities of newly synthesized dopamine after administration of labeled tyrosine, finding no evidence that newly synthesized dopamine selectively entered one of two compartments. Doteuchi *et al* did note that some α-methyl-*p*-tyrosine is converted to *p*-hydroxyamphetamine, which might release some dopamine from storage. However, this does not resolve the problem of increasing levels of dopamine shortly after inhibition of dopamine synthesis. While the rapid early decline in of dopamine after inhibition of dopamine synthesis could possibly suggest the presence of a small available dopamine compartment dependent on dopamine synthesis, we conclude that the data are not sufficiently rigorous to support analyses of compartment size and kinetic constants.

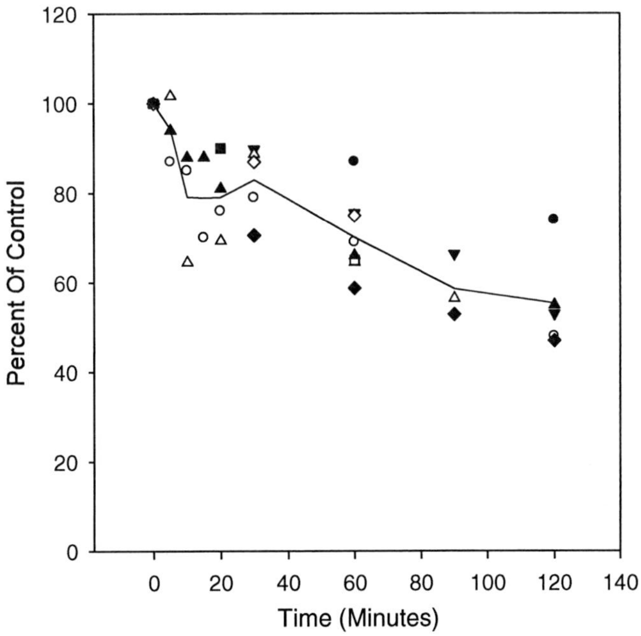

Figure 2. Rate of decline of tissue dopamine after inhibition of dopamine synthesis. Data are derived from: filled circle [26], open circle [21], down filled triangle [72], up open triangle [23], filled square [73], open squares [74], filled diamond [75], and up filled triangle [22].

An argument for the hypothesis of two storage compartments for dopamine comes from analysis of the specific activity of dopamine and its metabolites DOPAC and HVA after injection of radioactive tyrosine. In a study reported by Gropetti *et al* [24], the specific activity of DOPAC was approximately 4 times higher than that of dopamine at five minutes after the injection of labeled tyrosine and three times higher at 20 minutes. Gropetti *et al* argued that this could only occur if dopamine was distributed amongst more than one compartment. As we explored models that might simulate these data, it became apparent that the only way a metabolite can have a higher specific activity than its parent is if the metabolite is made from a compartment with a higher specific activity than the overall specific activity of the parent. This argues that there must be two dopamine compartments and that newly synthesized dopamine preferentially enters one of them. Shortly after addition of radioactive precursor, this would produce a dopamine compartment with a higher specific activity than the average of the two compartments. A significant fraction of DOPAC synthesis must originate from the compartment with the higher specific activity. DOPAC is synthesized from dopamine by the enzyme monoamine oxidase. As monoamine oxidase enzyme is known to be localized on mitochondria, it seemed unlikely that mitochondria would be associated with a pool of newly synthesized dopamine. Therefore, for our simulations, we chose to have two sources of DOPAC formation: one from mitochondria distributed generally throughout the cytoplasmic pool and a second associated with a hypothesized complex where a fraction of newly synthesized dopamine is immediately converted to DOPAC. This model provided an excellent fit to the data from Gropetti *et al* (Figure 3) when 20% of newly synthesized dopamine is immediately converted to DOPAC.

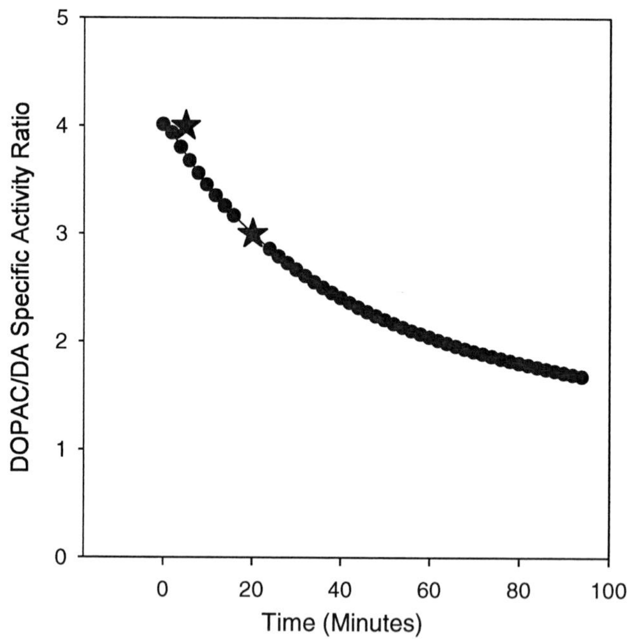

Figure 3. Ratio of the specific activity of DOPAC divided by the specific activity of dopamine after administration of labeled tyrosine. Stars show data derived from Gropetti [24] . Circles show simulation output from the model where 20% of DOPAC is made from newly synthesized dopamine.

We were interested in the size of the compartment containing the newly synthesized dopamine. Our simulation output shown in Figure 3 has a steady state level of 8.6 dopamine molecules in this compartment compared to 470,000 molecules in the total varicosity. Using simulations with a more populated compartment for newly synthesized dopamine produce simulation output with DOPAC/dopamine specific activity ratios much higher than those observed by Gropetti et al. We therefore conclude that the compartment containing newly synthesized dopamine is very small and is likely part of a synthesis complex. Our simulations indicate that this compartment is so small that it would be difficult to detect in experiments examining kinetics of the total dopamine content of a varicosity.

An exploration was done to determine whether there are other evidences for a source of DOPAC by metabolism of newly synthesized dopamine. A comparison of rates of decline of dopamine and DOPAC after inhibition of dopamine synthesis shows that DOPAC disappears substantially faster than dopamine [25,26]. In the study by Soares-da-Silva and Garrett, rate constants are reported of 0.2 per hour for loss of dopamine and 0.5 per hour for loss of DOPAC in the striatum. These investigators argue that this condition can only occur if a substantial amount of DOPAC in brain derives from a newly formed pool of dopamine. We confirmed this conclusion using simulations of a dopaminergic varicosity. Figure 4 shows results of simulations comparing a one-compartment model with the two compartment model with 20% of newly synthesized dopamine being immediately converted to DOPAC. These results show that after inhibition of dopamine synthesis, DOPAC and dopamine disappear at the same rate in the one-compartment model. DOPAC disappears faster than dopamine in the model in which 20% of newly synthesized dopamine is immediately converted to DOPAC.

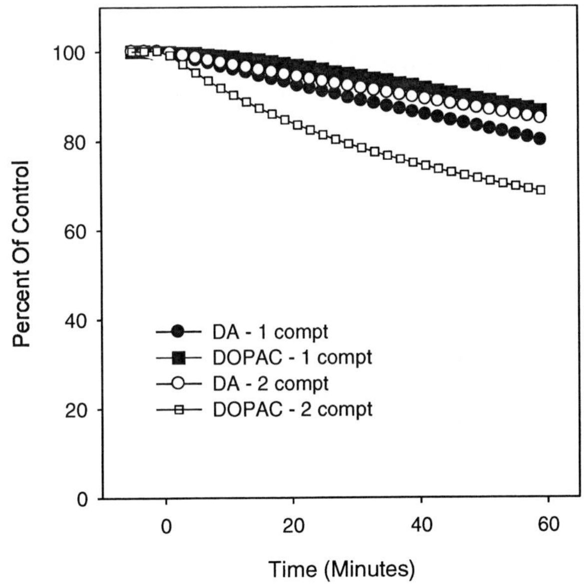

Figure 4. Rate of decline of dopamine and DOPAC after inhibition of dopamine synthesis. Simulation outputs show that DOPAC declines somewhat more slowly than dopamine when using a one-compartment model. However, DOPAC declines more rapidly than dopamine when using a model in which 20% of DOPAC is made from newly synthesized dopamine.

Another hypothesis relative to compartments is that newly synthesized dopamine enters a compartment from which it is preferentially secreted into the extracellular space. One early experiment supporting this hypothesis involved removing cortical tissue above the striatum in an anesthetized cat, placing a perfusion cup apparatus on top of the striatal tissue, and perfusing fluid across the surface of the striatal tissue [27]. When radioactive tyrosine was added to the perfusion, radioactive dopamine recovered in the perfusate increased to a steady state. When α-methyl-*p*-tyrosine was added to the perfusate, recovery of labeled dopamine declined to zero. This was interpreted as evidence that newly synthesized dopamine is secreted into extracellular compartment. Additional evidence for this concept can be drawn from data showing that rate of decline of extracellular dopamine is much faster than rate of decline of tissue dopamine after inhibition of dopamine synthesis (Figure 5). It is important to be aware that the two measurements shown in figure 5 are not made under similar experimental conditions. Tissue dopamine measurements are from animals that do not have dialysis probes in the striatum, while extracellular dopamine measurements are from animals that have probes in the striatum. We have done extensive simulations of one and two compartments models with newly synthesized dopamine being preferentially secreted. The simulations have led us to identify a condition that must be met in order for extracellular dopamine levels to decline at a rate faster than tissue dopamine levels when dopamine synthesis is inhibited. The rate of dopamine movement from the storage compartment to the compartment from which dopamine is being secreted must be very slow relative to rate of dopamine use. As dopamine depletes from the extracellular space over a time period of nearly an hour in the cat cortical cup experiment and over a time period of hours in rodent dialysis experiments, this requirement means that movement between compartments must be close to zero. However, our simulations show an alternate interpretation of the data. If we include extraction of dopamine molecules into the perfusion fluid in our simulations, extracellular dopamine decreases after inhibition of dopamine synthesis. Figure 6 shows a comparison of the data from the cortical cup experiment described above to a simulation output when 5% of the secreted dopamine molecules are lost in the perfusate and 95% are recycled via dopamine transporter. There is an excellent match between simulation output and experimental data. The simulation output also shows an important piece of information not shown in the plot. Total tissue dopamine begins declining when the perfusion starts and reaches a value of zero at the same time that the labeled extracellular dopamine simulation reaches zero. Unfortunately, we are not able to find published data where measurements of tissue dopamine have been made under conditions where dialysis has been done after inhibition of dopamine synthesis. There is one caveat that should be mentioned here. While our simulations suggest that experiments measuring extracellular dopamine after inhibition of dopamine synthesis should not be interpreted as evidence for preferential secretion of newly synthesized dopamine, our simulations do not rule out this hypothesis.

Data from simulations have been used to determine the potential contribution immediate secretion of newly synthesized dopamine might make to a signaling event. With average firing rates for dopaminergic neurons of 5 per second [28,29], a secretory event occurs once every 0.2 seconds. Our simulation models are using a synthetic rate of six dopamine molecules per varicosity per 0.2 second. It is difficult to determine the number of dopamine molecules secreted. However, various models we have used to explain a variety of different kinds of experimental manipulations used from 180 to 1000 molecules per signaling event. In our most recent models, the number of molecules per signaling event for baseline conditions is 470. At this rate, newly synthesized dopamine could contribute only 1.3% of the signal. This suggests that under normal conditions, newly synthesized dopamine has little contribution to overall dopamine signaling.

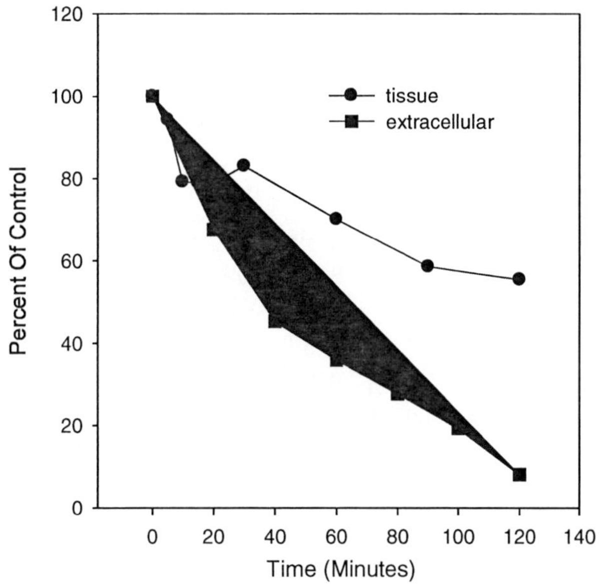

Figure 5. Comparison of rate of decline of tissue dopamine and extracellular dopamine after inhibition of dopamine synthesis. Tissue dopamine is the average value from the references listed in the legend for Figure 2. Extracellular dopamine is the average value from the three different experiments published by Butcher [76].

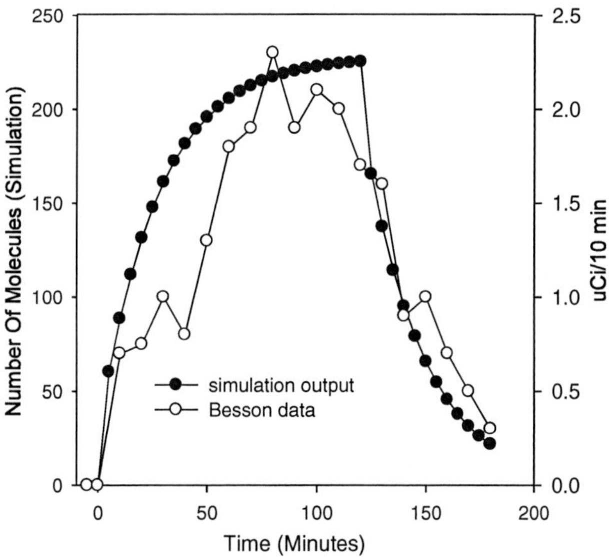

Figure 6. Labeled dopamine in extracellular fluid. Open circles show data derived from Besson [27]. Labeled tyrosine was added to the perfusion fluid at time zero, and α-methyl-*p*-tyrosine was administered to the cat at time 120 minutes. The simulation model had 95% of extracellular dopamine recaptured by dopamine transporter and 5% extracted into the perfusate.

Quantitative information relative to number of tyrosine hydroxylase molecules, potential synthetic capacity, and actual synthetic capacity suggest that dopamine synthetic capacity far exceeds actual baseline synthetic rate. Dopamine is synthesized from the amino acid tyrosine by the action of the enzymes tyrosine hydroxylase (rate limiting) and aromatic amino acid decarboxylase (fast). Striatal density of tyrosine hydroxylase is estimated to be 1.6 microgram per mg protein [30-35]. This calculates to 2.4×10^4 tyrosine hydroxylase molecules per varicosity (assuming 10^8 varicosities per mm^3 [7,8], and a density of brain tissue of 1). The turnover number for this enzyme as determined in in vitro enzyme activity studies is 1.2 DOPA molecules produced per second per molecule [36-39]. Even assuming that in vitro synthetic rate is roughly 10% of optimum rate under in vivo conditions, this suggests a synthetic capacity in vivo of thousands of dopamine molecules per second. DOPA, the product of the tyrosine hydroxylase enzyme is normally present in very low amounts. In the presence of inhibitors of aromatic amino acid decarboxylase, levels of DOPA accumulate linearly. The rate of this accumulation is commonly used as an estimate of the rate of synthesis of dopamine. The average of various studies is 5×10^{-9} molar/second [40-45]. This rate calculates to approximately 30 molecules per second in each varicosity, which is the value used in our simulations. This value is far below the 1000 molecules per second potential maximum rate calculated above. Another evidence for a slow basal rate of dopamine synthesis compared to synthetic capacity comes from studies of animals lacking dopamine transporter. Based on rates of DOPA accumulation after inhibition of aromatic amino acid decarboxylase, these knockout animals have a baseline dopamine synthetic rate approximately 2-fold higher than wildtype animals [46]. However, they have less than 10% the number of tyrosine hydroxylase molecules as the wildtype animals. Thus, the knock-out animals are producing far more dopamine with much less enzyme than are wildtype animals. On average, the mice lacking dopamine transporter have 20-fold more production from each tyrosine hydroxylase enzyme molecule. This suggests the presence of a large reserve synthetic capacity in the wildtype animals.

Studies of the Effects of Inhibitors of Dopamine Transporter when Exocytosis is Stimulated at Supra-Physiological Rates

Several studies on the effects of inhibitors of dopamine transporter have been interpreted as supporting the hypothesis that an effect of these inhibitors is to move dopamine from a storage to an available pool [47-49]. In some of these studies, amperometric techniques have been used to measure the increase in extracellular dopamine in the striatum elicited by stimulating dopaminergic neurons at supra-physiological rates. After pharmacological inhibition of dopamine synthesis, the amount of dopamine elicited by such stimulation declines much more rapidly than the decline in total tissue dopamine. If an inhibitor of dopamine transporter is then added, the increase in extracellular dopamine elicited by stimulation at supra-physiological rates increases to near control levels. One interpretation of these results is that inhibition of the transporter causes movement of dopamine from a storage pool to an available pool.

We have recently published an alternate interpretation of these results derived from simulations of the studies [20]. A kinetic analysis of dopamine movement by dopamine transporter is necessary for understanding these new interpretations. When the number of dopamine molecules in the extracellular space is less than the number of dopamine transporters, the rate limiting step for transport is the rate of binding of dopamine to the transporter. When the number of dopamine molecules in extracellular space is greater than the number of dopamine transporters, the rate limiting step in transport is the turnover rate of the transporter. The rate of binding is faster than the duration of the transport cycle. Therefore, at smaller numbers of dopamine molecules in extracellular space, clearance rate of dopamine is faster than at higher numbers of dopamine molecules in extracellular space.

Under baseline conditions, the number of dopamine molecules involved in each secretory event is much smaller than the number of dopamine transporters, and binding of dopamine molecules to dopamine transporters is the rate limiting step in dopamine clearance. When stimulating dopamine neuron firing rate at supra-physiological rates, the amount of extracellular dopamine increases roughly in proportion to stimulation rate until the number of dopamine molecules in extracellular space matches the number of dopamine transporters. As stimulation rate increases beyond this point, the rate limiting step becomes the slower turnover rate of the dopamine transporters and stimulated levels of extracellular dopamine increase to values much greater than proportional to firing rate increase (Figure 7). In this experimental paradigm, inhibition of dopamine transporter is functionally equivalent to decreasing the number of available transporter molecules. In the presence of an inhibitor of dopamine transporter, the curve relating increase in extracellular dopamine to stimulation frequency lies above that for the control condition (Figure 7).

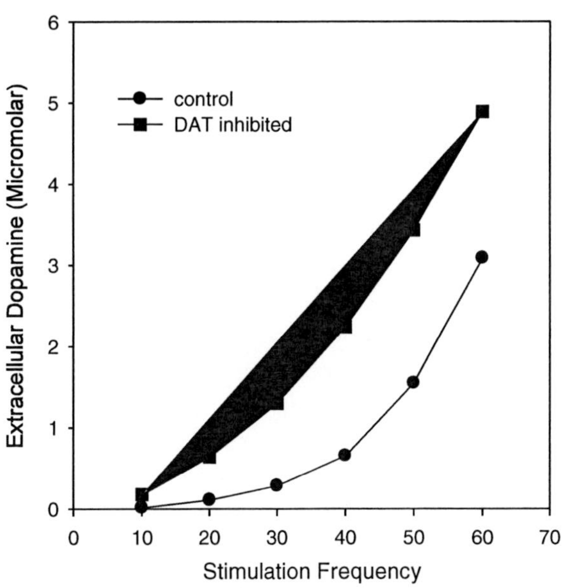

Figure 7. Effects of dopamine neuron stimulation frequency on elicited increase in extracellular dopamine. The circles show data from simulations [20] that provide an output that is a good match to data published by Garris and Wightman [77]. The squares show simulation output for a model in which a competitive inhibitor of dopamine transporter is added to extracellular compartment.

After inhibition of dopamine synthesis, dopamine stores are decreased. This results in fewer dopamine molecules secreted to extracellular space at each stimulation pulse. The functional effect is a decrease in rate of dopamine secretion, which can be visualized as moving to the left along the control curve in figure 7. For example, under the control condition, stimulation at the 60 Hz rate will increase extracellular dopamine to about 3 μM. When dopamine stores are depleted by 33%, the same stimulation rate is now effectively the same as a stimulation rate of 40 Hz under control conditons, and the increase in extracellular dopamine is about 0.5 μM. When an inhibitor of dopamine transporter is now added to the system, the increase in extracellular dopamine as a function of stimulation frequency is given by the upper curve of Figure 7. For the 40 Hz stimulation (actual effect of 60 Hz with 33% dopamine depletion), the increase in extracellular dopamine now jumps to more than 2 μM. Thus, the transporter kinetic explanation can explain the experimental results without need for inhibitors of dopamine transporter moving dopamine between storage compartments.

Studies of the Effects of Combination of Inhibitors of Dopamine Transporter and Antipsychotic Drugs

Other studies interpreted as inhibition of dopamine transporter moving dopamine from storage to available pools examine the rate of decrease of dopamine after the combination of an antipsychotic drug and an inhibitor of the dopamine transporter. Neither drug by itself causes a loss of tissue dopamine. However, a combination causes a substantial loss in tissue dopamine, reaching values of 50% by 90 minutes after drug administration [50,51]. Such studies have been interpreted as moving dopamine from a storage pool where it is protected from metabolism to an available pool where dopamine can be metabolized.

Results from our simulation studies suggest that the non-protected compartment into which dopamine is moved by the combination of inhibition of dopamine transporter and antipsychotic drug is the extracellular compartment. While most of the dopamine entering the extracellular compartment is captured by dopamine transporter and recycled, a small fraction is lost either through metabolism or uptake into glial cells. Under normal conditions, this small fraction that is lost is replaced by ongoing dopamine synthesis. However, in the presence of the combination of inhibition of dopamine transporter and an antipsychotic drug, this fraction of dopamine lost from extracellular space exceeds rate of dopamine synthesis, resulting in the loss of total dopamine in the varicosity. The following three paragraphs will explain how this occurs.

Typical antipsychotic drugs such as haloperidol induce substantial changes in striatal dopamine turnover (Table 1). These drugs increase the rate of dopamine synthesis and increase the rate of dopamine neuron firing. In addition, our simulation data suggest that the majority of newly synthesized dopamine stimulated by these drugs is immediately converted to DOPAC and that rate of DOPAC metabolism is decreased. Extracellular dopamine is increased as a result of the increase in firing rate. As there is excess dopamine transporter capacity, the excess dopamine is cleared rapidly, which minimizes increases in extracellular dopamine. However, the relatively modest increase in extracellular dopamine has an impact on dopamine turnover in the varicosity as the rate of loss of extracellular dopamine is increased in proportion to the increase in extracellular dopamine. In addition, a significant fraction of cytosolic dopamine arrives into cytosolic compartment via movement from extracellular compartment mediated by dopamine transporter. Rate of metabolism of cytosolic dopamine also increases proportionally to increases in extracellular dopamine. In the case of antipsychotic drugs, the increase in dopamine synthesis producing dopamine for storage and signaling balances the increase in metabolism of dopamine. Therefore, there is no net change in total dopamine in the varicosity.

Table 1. Effects of haloperidol and amfonelic acid on various measures of dopamine, DOPAC, and their synthesis and metabolism. Conclusions were derived from the following: [50,51,67,78,79].

parameter	*haloperidol*	*amfonelic acid*	*combo*
firing rate	↑	↓	?
DAT	=	↓	↓
TH	↑	=	↑
DOPAC metabolism	↓	=	↓
DAext	↑	↑	↑
DAves	=	=	↓
DOPAC	↑	↑	↑

The major effect of inhibition of dopamine transporter is to increase extracellular dopamine. This occurs as the drug competes with dopamine for occupancy of the transporter. Another effect of drugs that inhibit dopamine transporters is to slow dopamine neuron firing rate. This action decreases rate of dopamine secretion into extracellular space. This lessens the extent to which extracellular dopamine is increased by inhibition of dopamine transporter. Although the increase in extracellular dopamine does increase rate of loss of dopamine (as discussed in the previous paragraph), the increased loss over the duration of action of a drug would be a small fraction of total dopamine in the varicosity and might not be detected in biochemical measurements.

Our strategy for simulating effects of the combination of inhibition of dopamine transporter and antipsychotic drug was to preserve the inhibition of dopamine transporter (effect of dopamine transporter inhibitor) and the increased rate of dopamine synthesis with the majority of this increase converted immediately to DOPAC (effect of antipsychotic drug). We then varied the firing rate until our simulation output matched published experimental data. The simulation output showed a large (6-fold over control) increase in extracellular dopamine (Figure 8). Analysis of rate of dopamine entering and exiting the extracellular compartment during the initial five seconds after the combination of drugs was imposed on the system (Table 2) shows that the combination of increased firing rate (10 Hz as compared to a control of 5 Hz) coupled with inhibition of dopamine transporter accounts for the large increase in extracellular dopamine. We then looked at rates of addition of dopamine to and removal of dopamine from the system (Table 3). We note that, with the combination of drugs, the amount of dopamine disappearing from extracellular space via mechanisms other than captured by the dopamine transporter substantially exceeds the rate of synthesis of new dopamine. This accounts for the net loss of dopamine from the system. In summary, the redistribution of dopamine by the combination of inhibitor of dopamine transporter and antipsychotic drug from a protected to a non-protected compartment appears to be movement from storage vesicles to extracellular space.

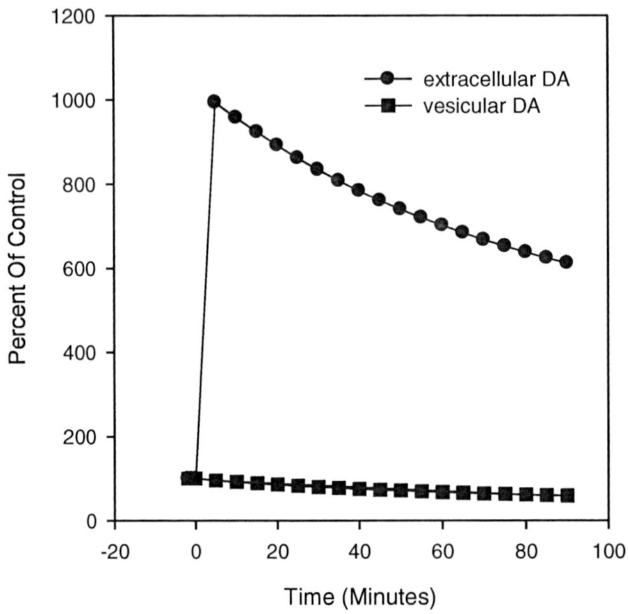

Figure 8. Simulations of effects of the combination of inhibition of dopamine transporter (amfonelic acid) and antipsychotic drug (haloperidol) on amount of dopamine in vesicle and extracellular compartments.

Table 2. Movement of dopamine into and out of extracellular compartment. Data show total number of molecules moving over a period of 5 seconds after imposition of effects of drugs into the simulation. Dopamine enters extracellular compartment by exocytosis and is removed by action of dopamine transporter (DAT) or either metabolism or up take into glial cells (Non_DAT). Balance is the difference between number of molecules entering and leaving extracellular compartment

Condition	*Exocytosis*	*DAT*	*Non_DAT*	*Balance*
Control	11755	11716	39	0
Amfonelic Acid	4703	4335	60	+308
Haloperidol	38909	38791	60	+58
Amfonelic Acid and Haloperidol	23302	18268	322	+4712

Table 3. Addition and loss of dopamine from a varicosity. Data show total number of molecules added or removed over a period of 5 seconds after imposition of effects of drugs into the simulation. TH_DA indicates newly synthesized dopamine entering storage and signaling compartments and excludes newly synthesized dopamine immediately made into DOPAC. DAmetab is the cytosolic dopamine metabolized to DOPAC, and Non_DAT is dopamine lost from extracellular compartment

Condition	*TH_DA*	*DAmetab*	*Non_DAT*	*Balance*
Control	120	81	39	0
Amfonelic Acid	171	58	65	+48
Haloperidol	209	173	60	+24
Amfonelic Acid and Haloperidol	218	106	361	-249

POTENTIAL SIZE OF AVIALABLE POOL

In thinking about the concepts of a storage and an available pool, the question arises as to the size of pool needed to supply normal signaling via dopamine neurons. The number of molecules used in each signaling event has not been rigorously determined. In our models, we have used a range of values. The low end of the range occurs in the pulse paradigm in which we are simulating effects of stimulating neuronal firing at a supra-physiological rate. For these simulations, we used a value of 180 molecules per pulse. In our models using unstimulated firing rates (5 Hz), we are currently using a value of 470 molecules per signaling event. If we assume that 470 molecules per signaling event is a reasonable estimate, this calculates to 0.1% of the total dopamine in the varicosity being used in each signaling event. This is a very small fraction of the stored dopamine. In fact, at this rate signaling could be maintained for just over 16 minutes in the absence of any new dopamine synthesis or any recycling of dopamine. These data suggest that an available pool of dopamine need not be large.

Another line of reasoning suggests that the total pool of dopamine is very large compared to that needed for signaling. In a previous publication [20], we provided an estimate of the effects of stimulating dopaminergic neurons 900 times at 60 Hz [49], which is the most intense stimulation paradigm we have found published. Our estimates show that this paradigm would only release 65% of the total dopamine in the varicosity. If we assume that this 15-second train of pulses occurs much faster than an available pool could be replenished by a storage pool, this would put an upper limit on the requirement for an available pool. However, the size of the available pool would be much smaller because dopamine is recycled via mechanisms involving dopamine transporter and vesicular monoamine transporter. Capacity of dopamine transporters is sufficient to capture 80% of the released dopamine within the time of a 15-second stimulus train, which would reduce the needed available dopamine by 80%. This results in a total of 61,000 dopamine molecules, or 13% of the total dopamine in the varicosity. Realizing that a 900 pulse 60 Hz stimulus is far greater than the 5-11 spikes of up to 15 Hz reported for dopaminergic neuron burst firing [52], these calculations suggest that a small pool consisting of less than 10% of the total dopamine in the varicosity could supply all signaling needs of dopamine neurons under physiological conditions.

The values we have estimated for numbers of molecules used in each dopamine signaling event suggests that the kiss and run mechanism of neurotransmitter secretion is involved in the striatum. A reasonable estimate of the number of vesicles in a varicosity is 250, based on electron micrographs of varicosities [53] and on volume of the varicosity related to total volume of a vesicle. This would result in about 1880 dopamine molecules per vesicles. Our current value of 470 molecules per dopamine release event is 25% of the number of molecules in the vesicle. This number is close to an estimate determined from a study of stimulated dopamine release from cultured ventral midbrain neurons [54]. The small fraction is best explained by a kiss and run hypothesis rather than by full exocytosis.

The conclusion in the previous paragraph is based on a very rough estimate of the number of dopamine molecules per release event. One can ask whether this estimate might be too low. We have attempted to validate the value for number of molecules per signaling event by comparing simulation output with estimated extracellular concentrations of dopamine measured by no net flux dialysis. Using a value of 5 nM [46,55] for baseline extracellular dopamine coupled with an extracellular compartment size associated with each varicosity of $2x10^{-15}$ liters (10^8 varicosities per mm^3 [7,8] and 20% extracellular space [56]) leads to an average number of 6 molecules in extracellular compartment. For our current baseline value of 470 molecules per signaling event, this value could be achieved if all of the molecules in a signaling event were cleared from the extracellular space within 3 ms. This is much faster than the estimated half-rate for clearance of 50 ms [57]. This suggests that either our estimate of dopamine secretion rate is too high or that much of the secreted dopamine does not diffuse rapidly enough to be detected by the dialysis measurement.

One question arises from the analysis in which we demonstrate that signaling can be accomplished with a small fraction of the total dopamine in the varicosity both under normal conditions and conditions where the dopamine system is stimulated substantially beyond baseline levels. Because dopamine recycling is highly efficient, signaling can be done for a long time in the absence of synthesis of new dopamine. For example, data shown in Figures 1 and 2 indicate that approximately 40% of dopamine is lost over a period of one hour after inhibition of dopamine synthesis. At a signaling rate of 5 events per second, 18,000 signaling events occur during that hour. This translates into a rate of loss of dopamine of about 0.002% per signaling event. This brings up the question as to why so much dopamine is needed. This question becomes more perplexing when one considers that storage vesicles are somewhat leaky [18]. A small percentage of their contents is continually being lost and must be replaced. This process requires energy. It seems highly inefficient that an energy storing process would be utilized for storing dopamine that does not need to be there for dopamine signaling. This brings into question whether this dopamine might be used in a way other than the normal signaling processes that have been studied to date. While at the current time this proposal is entirely speculation, we suggest that local activation of presynaptic glutamate or nicotinic acetylcholinergic receptors might locally release substantial amounts of this stored dopamine.

Possible Locations of Avialable Dopamine

Information discussed in this chapter presents the concept that most studies purporting to support two compartments of striatal dopamine can be explained using other mechanisms and that signaling could be accomplished using a small fraction of the total dopamine existing in a varicosity. We now offer some speculation that signaling dopamine might not be associated with a cytosolic vesicular compartment. Recently, a dopamine storage mechanism associated with the plasma membrane has been identified [58]. Storage in this compartment requires vesicular monoamine transporter, but kinetics and regulation of the size of the compartment by some drugs are different from those associated with the cytosolic vesicular compartment [58]. Studies on neurosecretion of GABA have also identified a plasma membrane storage that can be released by conditions known to promote synaptic signaling [58,59]. These new ideas suggest the possibility that the dopamine used in signaling may not originate from storage or available pools of cytosolic vesicles.

Conclusion

Several experiments that have been interpreted as providing evidence for two compartments for dopamine storage can also be adequately explained using a one storage compartment model. This conclusion eliminates the inconsistency of having two-compartments between which dopamine is not exchanged in kinetic studies occurring over hours of duration and two-compartments between which dopamine can be rapidly exchanged within minutes or seconds but only when dopamine transporters are inhibited.

A very large storage pool of dopamine exists that does not need to be used for normal rates of signaling. Our simulations suggest that each signaling event uses only 0.1% of dopamine located in a varicosity and that the dopamine used in a signaling event is very efficiently recycled. There is also a very large reserve dopamine synthetic capacity. These observations suggest that dopamine might be used for signaling other than that associated with dopaminergic neuron firing events.

Our computer simulations of two very divergent experimental paradigms require that a percentage of the DOPAC in a varicosity be synthesized from newly synthesized dopamine rather than from dopamine distributed throughout the cytosolic pool. We hypothesize that a synthetic complex contains a monoamine oxidase enzyme in association with tyrosine hydroxylase and aromatic amino acid decarboxylase enzymes. Our simulations suggest that under normal conditions, about 20% of the newly synthesized dopamine is directly converted to DOPAC and that the percent going to DOPAC is markedly increased by antipsychotic drugs. Other data suggesting that DOPAC production might be regulated include observations that a variety of drug treatments such as amphetamines and antipsychotic drugs markedly change amounts of extracellular DOPAC [60-67]. Furthermore, extracellular DOPAC levels are several hundred fold higher than extracellular dopamine levels [68-71]. We suggest that these observations indicate that DOPAC has a regulatory role that has not yet been determined.

Both physiological data and modeling simulation data support the concept of dopamine secretion where vesicles only partially empty and then are reused ("kiss and run"). Recent biochemical data suggest that a membrane associated dopamine storage pool has distinct characteristics from a cytosol-associated storage pool. A reasonable interpretation of the total set of data reviewed is that most signaling is done by membrane-associated storage that uses a "kiss and run" mechanism and rapid recycling of used dopamine.

Acknowledgments

Contributions of work done by Rachel M. Hughes and Celia E. Fenell on undergraduate student honors research projects are gratefully acknowledged.

References

[1] Bjorklund A, Dunnett SB. Dopamine neuron systems in the brain: An update. Trends in Neurosciences. 2007;30:194-202.

[2] Joel D, Weiner I. The connections of the dopaminergic system with the striatum in rats and primates: An analysis with respect to the functional and compartmental organization of the striatum. Neuroscience. 2000;96:451-474.

[3] Schultz W. Predictive reward signal of dopamine neurons. Journal of Neurophysiology. 1998;80:1-27.

[4] Di Chiara G, Imperato A. Drugs abused by humans preferentially increase synaptic dopamine concentrations in the mesolimbic system of freely moving rats. Procedings of the National Academy of Sciences, USA. 1988;85:5274-7278.

[5] Matsuda W, Furuta T, Nakamura KC, Hioki H, Fujiyama F, Arai R, Kaneko T. Single nigrostriatal dopaminergic neurons form widely spread and highly dense axonal arborizations in the neostriatum. Journal of Neuroscience. 2009;29:444-453.

[6] Prensa L, Parent A. The nigrostriatal pathway in the rat: A single-axon study of the relationship between dorsal and ventral tier nigral neurons and the striosome/matrix striatal compartments. Journal of Neuroscience. 2001;21:7247-7260.

[7] Anden NE, Fuxe K, Hamberger B, Hokfelt T. A quantitative study on the nigro-neostriatal dopamine neuron system in the rat. Acta Physiologica Scandinavica. 1966;67:306-312.

[8] Doucet G, Descarries L, Garcia S. Quantification of the dopamine innervation in adult-rat neostriatum. Neuroscience. 1986;19:427-445.

[9] Udenfriend S, Zaltzman-Nirenberg P, Nagatsu T. Inhibitors of purified beef adrenal tyrosine hydroxylase. Biochemical Pharmacology. 1965;14:837-845.

[10] Johnston J. Some observations upon a new inhibitor of monoamine oxidase in brain. Biochemical Pharmacology. 1968;17:1285-1297.

[11] Knoll J, Magyar K. Some puzzling pharmacological effects of monoamine oxidase inhibitors. Advances in Biochemical Psychopharmacology. 1972;5:393-408.

[12] Crowley JJ, Carter AG, Regehr WG. Fast vesicle replenishment and rapid recovery from desensitization at a single synaptic release site. Journal of Neuroscience. 2007;27:5448-5460.

[13] Griesinger CB, Richards CD, Ashmore JF. Fast vesicle replenishment allows indefatigable signalling at the first auditory synapse. Nature. 2005;435:212-215.

[14] Saviane C, Silver RA. Fast vesicle reloading and a large pool sustain high bandwidth transmission at a central synapse. Nature. 2006;439:983-987.

[15] Foster KA, Regehr WG. Variance-mean analysis in the presence of a rapid antagonist indicates vesicle depletion underlies depression at the climbing fiber synapse. Neuron. 2004;43:119-131.

[16] Schneggenburger R, Sakaba T, Neher E. Vesicle pools and short-term synaptic depression: Lessons from a large synapse. Trends in Neurosciences. 2002;25:206-212.

[17] von Gersdorff H, Matthews G. Depletion and replenishment of vesicle pools at a ribbon-type synaptic terminal. Journal of Neuroscience. 1997;17:1919-1927.

[18] Wallace LJ. A small dopamine permeability of storage vesicle membranes and end product inhibition of tyrosine hydroxylase are sufficient to explalin changes occurring in dopamine synthesis and storage after inhibitionof neuron firing. Synapse. 2007;61:715-723.

[19] Wallace LJ, Connell LE. Mechanisms by which amphetamine redistributes dopamine out of vesicles: A computational study. Synapse. 2008;62:370-378.

[20] Wallace LJ, Hughes RM. Computational analysis of stimulated dopaminergic synapses suggests release largely occurs from a single pool of vesicles. Synapse. 2008;62:909-919.

[21] Javoy F, Glowinski J. Dynamic characteristics of the 'functional compartment' of dopamine in dopaminergic terminals of the rat striatum. Journal of Neurochemistry. 1971;18:1305-1311.

[22] Doteuchi M, Wang C, Costa E. Compartmentation of dopamine in rat striatum. Molecular Pharmacology. 1974;10:225-234.

[23] Paden CM. Disappearance of newly synthesized and total dopamine from the striatum of the rat after inhibition of synthesis: Evidence for a homogeneous kinetic compartment. Journal of Neurochemistry. 1979;33:471-479.

[24] Groppetti A, S. A, Cattabeni F, Di Giulio AM, Galli CL, Ponzio F, Spano PF. Changes in specific activity of dopamine metabolites as evidence of multiple compartmentation of dopamine in striatal neurons. Journal of Neurochemistry. 1977;28:193-197.

[25] Di Giulio AM, Groppetti A, Algeri S, Ponzio F, Cattabeni F, Galli CL. Measurement of 3,4-dihydroxyphenylacetic acid and 3-methoxytyramine specific activity in rat striatum. Analytical Biochemistry. 1979;92:82-90.

[26] Soares da Silva P, Garrett MC. A kinetic-study of the rate of formation of dopamine, 3,4-dihydroxyphenylacetic acid (DOPAC) and homovanillic-acid (HVA) in the brain of the rat - implications for the origin of DOPAC. Neuropharmacology. 1990;29:869-874.

[27] Besson MJ, Cheramy A, Glowinski J, Gauchy C. In vivo release of 3H-DA from the cat caudate nucleus. In: Usdin E, Snyder SH, eds. Frontiers in catecholamine research. New York: Pergamon Press Inc.; 1973: 557-559.

[28] Bunney BS, Aghajanian GK, Roth RH. Comparison of effects of l-dopa, amphetamine and apomorphine on firing rate of rat dopaminergic neurons. Nature New Biology. 1973;245:123-125.

[29] Koeltzow TE, Xu M, Cooper DC, Hu XT, Tonegawa S, Wolf ME, White FJ. Alterations in dopamine release but not dopamine autoreceptor function in dopamine D3 receptor mutant mice. Journal of Neuroscience. 1998;18:2231-2238.

[30] Bowyer JF, Frame LT, Clausing P, Nagamoto-Combs K, Osterhout CA, Sterling CR, Tank AW. Long-term effects of amphetamine neurotoxicity on tyrosine hydroxylase mrna and protein in aged rats. Journal of Pharmacology and Experimental Therapeutics. 1998;286:1074-1085.

[31] Bowyer JF, Holson RR, Miller DB, O'Callaghan JP. Phenobarbital and dizocilpine can block methamphetamine-induced neurotoxicity in mice by mechanisms that are independent of thermoregulation. Brain Research. 2001;919:179-183.

[32] Haycock JW. Quantitation of tyrosine-hydroxylase protein-levels - spot immunolabeling with an affinity-purified antibody. Analytical Biochemistry. 1989;181:259-266.

[33] O'Callaghan JP, Miller DB, Reinhard JF, Jr. Characterization of the origins of astrocyte response to injury using the dopaminergic neurotoxicant 1-methyl-4-phenyl-1,2,3,6-tetrahydropyridine. Brain Research. 1990;521:73-80.

[34] Reinhard JF, Ocallaghan JP. Measurement of tyrosine-hydroxylase apoenzyme protein by enzyme-linked-immunosorbent-assay (elisa) - effects of 1-methyl-4-phenyl-1,2,3,6-tetrahydropyridine (MPTP) on striatal tyrosine-hydroxylase activity and content. Analytical Biochemistry. 1991;196:296-301.

[35] Salvatore MF, Zhang JL, Large DM, Wilson PE, Gash CR, Thomas TC, Haycock JW, Bing GY, Stanford JA, Gash DM, Gerhardt GA. Striatal GDNF administration increases tyrosine hydroxylase phosphorylation in the rat striatum and substantia nigra. Journal of Neurochemistry. 2004;90:245-254.

[36] Daubner SC, Lauriano C, Haycock JW, Fitzpatrick PF. Site-directed mutagenesis of serine-40 of rat tyrosine-hydroxylase - effects of dopamine and camp-dependent phosphorylation on enzyme-activity. Journal of Biological Chemistry. 1992;267:12639-12646.

[37] Daubner SC, Lohse DL, Fitzpatrick PF. Expression and characterization of catalytic and regulatory domains of rat tyrosine-hydroxylase. Protein Science. 1993;2:1452-1460.

[38] Fitzpatrick PF, Chlumsky LJ, Daubner SC, Omalley KL. Expression of rat tyrosine-hydroxylase in insect tissue-culture cells and purification and characterization of the cloned enzyme. Journal of Biological Chemistry. 1990;265:2042-2047.

[39] Richtand NM, Inagami T, Misono K, Kuczenski R. Purification and characterization of rat striatal tyrosine-hydroxylase - comparison of the activation by cyclic amp-dependent phosphorylation and by other effectors. Journal of Biological Chemistry. 1985;260:8465-8473.

[40] Bannon MJ, Michaud RL, Roth RH. Mesocortical dopamine neurons. Lack of autoreceptors modulating dopamine synthesis. Molecular Pharmacology. 1981;19:270-275.

[41] Demarest KT, Lawson-Wendling KL, Moore KE. D-Amphetamine and gamma-butyrolactone alteration of dopamine synthesis in the terminals of nigrostriatal and mesolimbic neurons. Possible role of various autoreceptor sensitivities. Biochemical Pharmacology. 1983;32:691-697.
[42] Demarest KT, Moore KE. Comparison of dopamine synthesis regulation in the terminals of nigrostriatal, mesolimbic, tuberoinfundibular and tuberohypophyseal neurons. Journal of Neural Transmission. 1979;46:263-277.
[43] Kilts CD, Anderson CM, Ely TD, Nishita JK. Absence of synthesis-modulating nerve-terminal autoreceptors on mesoamygdaloid and other mesolimbic dopamine neuronal populations. Journal of Neuroscience. 1987;7:3961-3975.
[44] Liljequist S, Engel J. Attenuation of the gamma-butyrolactone-induced increase in dopa accumulation by chronic ethanol administration. Journal of Neural Transmission. 1979;46:195-204.
[45] Walters JR, Roth RH. Dopaminergic neurons: Drug-induced antagonism of the increase in tyrosine hydroxylase activity produced by cessation of impulse flow. Journal of Pharmacology and Experimental Therapeutics. 1974;191:82-91.
[46] Jones SR, Gainetdinov RR, Jaber M, Giros B, Wightman RM, Caron MG. Profound neuronal plasticity in response to inactivation of the dopamine transporter. Proceedings of the National Academy of Sciences of the United States of America. 1998;95:4029-4034.
[47] Ewing AG, Bigelow JC, Wightman RM. Direct invivo monitoring of dopamine released from 2 striatal compartments in the rat. Science. 1983;221:169-171.
[48] Kruk ZL, Stamford JA. Nomifensine increases stimulated dopamine release in vivo by an action on the dopamine storage pool: Voltammetric evidence. British Journal of Pharmacology. 1985;86:635P.
[49] Venton BJ, Seipel AT, Phillips PEM, Wetsel WC, Gitler D, Greengard P, Augustine GJ, Wightman RM. Cocaine increases dopamine release by mobilization of a synapsin-dependent reserve pool. Journal of Neuroscience. 2006;26:3206-3209.
[50] Shore PA. Actions of amfonelic acid and other non-amphetamine stimulants on dopamine neuron. Journal of Pharmacy and Pharmacology. 1976;28:855-857.
[51] Waldmeier PC, Huber H, Heinrich M, Stoecklin K. Discrimination of neuroleptics by means of their interaction with amfonelic acid - an attempt to characterize the test. Biochemical Pharmacology. 1985;34:39-44.
[52] Grace AA, Bunney BS. The control of firing pattern in nigral dopamine neurons - burst firing. Journal of Neuroscience. 1984;4:2877-2890.
[53] Pothos EN, Davila V, Sulzer D. Presynaptic recording of quanta from midbrain dopamine neurons and modulation of the quantal size. Journal of Neuroscience. 1998;18:4106-4118.
[54] Staal RGW, Hogan KA, Liang CL, German DC, Sonsalla PK. In vitro studies of striatal vesicles containing the vesicular monoamine transporter (VMAT2): Rat versus mouse differences in sequestration of 1-methyl-4-phenylpyridinium. Journal of Pharmacology and Experimental Therapeutics. 2000;293:329-335.

[55] Kalivas PW, Duffy P. Time course of extracellular dopamine and behavioral sensitization to cocaine. I. Dopamine axon terminals. Journal of Neuroscience. 1993;13:266-275.

[56] Rice ME, Nicholson C. Diffusion characteristics and extracellular volume fraction during normoxia and hypoxia in slices of rat neostriatum. Journal of Neurophysiology. 1991;65:264-272.

[57] Garris PA, Ciolkowski EL, Pastore P, Wightman RM. Efflux of dopamine from the synaptic cleft in the nucleus accumbens of the rat brain. Journal of Neuroscience. 1994;14:6084-6093.

[58] Volz TJ, Farnsworth SJ, King JL, Riddle EL, Hanson GR, Fleckenstein AE. Methylphenidate administration alters vesicular monoamine transporter-2 function in cytoplasmic and membrane-associated vesicles. Journal of Pharmacology and Experimental Therapeutics. 2007;323:738-745.

[59] Vautrin J, Maric D, Sukhareva M, Schaffner AE, Barker JL. Surface-accessible GABA supports tonic and quantal synaptic transmission. Synapse. 2000;37:38-55.

[60] Anderzhanova E, Rayevsky KS, Saransaari P, Oja SS. Effect of sulpiride on the amphetamine-induced changes in extracellular dopamine, DOPAC, and hydroxyl radical generation in the rat striatum. Neurochemical Research. 2003;28:1241-1248.

[61] Brown EE, Damsma G, Cumming P, Fibiger HC. Interstitial 3-methoxytyramine reflects striatal dopamine release - an invivo microdialysis study. Journal of Neurochemistry. 1991;57:701-707.

[62] Heeringa MJ, Abercrombie ED. Biochemistry of somatodendritic dopamine release in substantia-nigra - an in-vivo comparison with striatal dopamine release. Journal of Neurochemistry. 1995;65:192-200.

[63] Hurd YL, Ungerstedt U. In vivo neurochemical profile of dopamine uptake inhibitors and releasers in rat caudate-putamen. European Journal of Pharmacology. 1989;166:251-260.

[64] Imperato A, Dichiara G. Trans-striatal dialysis coupled to reverse phase high-performance liquid-chromatography with electrochemical detection - a new method for the study of the invivo release of endogenous dopamine and metabolites. Journal of Neuroscience. 1984;4:966-977.

[65] Nishijima K, Kashiwa A, Hashimoto A, Iwama H, Umino A, Nishikawa T. Differential effects of phencyclidine and methamphetamine on dopamine metabolism in rat frontal cortex and striatum as revealed by in vivo dialysis. Synapse. 1996;22:304-312.

[66] O'Dell SJ, Weihmuller FB, Marshall JF. Methamphetamine-induced dopamine overflow and injury to striatal dopamine terminals: Attenuation by dopamine D1 or D2 antagonists. Journal of Neurochemistry. 1993;60:1792-1799.

[67] Rivest R, Marsden CA. Differential-effects of amfonelic acid on the haloperidol-induced and clozapine-induced increase in extracellular DOPAC in the nucleus-accumbens and the striatum. Synapse. 1992;10:71-78.

[68] Golembiowska K, Dziubina A. Effect of the adenosine A(2a) receptor antagonist 8-(3-chlorostyryl)caffeine on l-dopa biotransformation in rat striatum. Brain Research. 2004;998:208-217.

[69] Marburger A, Sohr R, Reum T, Morgenstern R. Comparison by microdialysis of striatal l-dopa after its systemic administration in rats with probes implanted acutely or through a guide cannula. Journal of Neuroscience Methods. 2000;102:127-132.

[70] Pereira FC, Lourenco ES, Borges F, Morgadinho T, Ribeiro CF, Macedo TR, Ali SF. Single or multiple injections of methamphetamine increased dopamine turnover but did not decrease tyrosine hydroxylase levels or cleave caspase-3 in caudate-putamen. Synapse. 2006;60:185-193.

[71] Rodriguez M, Morales I, Gonzalez-Mora JL, Gomez I, Sabate M, Dopico JG, Rodriguez-Oroz MC, Obeso JA. Different levodopa actions on the extracellular dopamine pools in the rat striatum. Synapse. 2007;61:61-71.

[72] Broxterman HJ, Noach EL, Vanvalkenburg CFM. Differential-effects of acute and sub-acute HA-966 treatment on storage and release of striatal dopamine. European Journal of Pharmacology. 1979;60:153-161.

[73] Elverfors A, Nissbrandt H. Effects of d-amphetamine on dopaminergic neurotransmission - a comparison between the substantia-nigra and the striatum. Neuropharmacology. 1992;31:661-670.

[74] Westerink BHC, Vanes TP, Spaan SJ. Effects of drugs interfering with dopamine and noradrenaline biosynthesis on the endogenous 3,4-dihydroxyphenylalanine levels in rat-brain. Journal of Neurochemistry. 1982;39:44-51.

[75] Westerink BHC, Spaan SJ. On the significance of endogenous 3-methoxytyramine for the effects of centrally acting drugs on dopamine release in the rat-brain. Journal of Neurochemistry. 1982;38:680-686.

[76] Butcher SP, Fairbrother IS, Kelly JS, Arbuthnott GW. Amphetamine-induced dopamine release in the rat striatum - an invivo microdialysis study. Journal of Neurochemistry. 1988;50:346-355.

[77] Garris PA, Wightman RM. Different kinetics govern dopaminergic transmission in the amygdala, prefrontal cortex, and striatum: An in vivo voltammetric study. Journal of Neuroscience. 1994;14:442-450.

[78] Brougham LR, Conway PG, Ellis DB. Effect of ritanserin on the interaction of amfonelic acid and neuroleptic-induced striatal dopamine metabolism. Neuropharmacology. 1991;30:1137-1140.

[79] Gudelsky GA, Nwajei EE, Defife K, Nash JF. Interaction of amfonelic acid with antipsychotic-drugs on dopaminergic-neurons. Synapse. 1992;12:304-311.

In: Biogenic Amines
Editors: T. Farooqui, A. A. Farooqui
ISBN: 978-1-60876-625-3

Chapter V

THE IMPORTANCE OF MONOAMINE TRANSPORTERS IN THE DEVELOPMENT OF PSYCHOPHARMACOLOGY

Leslie Iversen*
University of Oxford, UK

ABSTRACT

The discovery of norepinephrine uptake in sympathetic nerves by Julius Axelrod and colleagues in the 1960's is an example of how a new scientific discovery can have a profound impact on the development of new therapies. The understanding that antidepressants acted as inhibitors of monoamine reuptake led to the development of many new drugs of this type. The older tricyclic antidepressants suffered from serious toxicology due to secondary pharmacological effects, notably on the heart They were followed by the much safer serotonin selective reuptake inhibitors (SSRIs) (e.g. fluoxetine – "Prozac"), and subsequently by a new generation of mixed norepinephrine/serotonin reuptake inhibitors with improved safety profiles.

In addition monoamine uptake inhibitors are used in the treatment of obesity (sibutramine); smoking cessation (bupropion); and attention deficit hyperactivity disorder (ADHD) (atomoxetine). The amphetamines exert their psychostimulant effects by acting as selective ligands for norepinephrine and dopamine uptake, subsequently displacing dopamine and norepinephrine. Amphetamine and the analog methyl phenidate ("Ritalin") are widely used in the treatment of ADHD. The psychostimulant drug cocaine also acts by virtue of its ability to inhibit the dopamine transporter.

Finally, the selective neurotoxin 6-hydroxydopamine – whose use in brain was pioneered by Norman Uretsky – also acts by virtue of its recognition and uptake into monaminergic neurons by the norepinephrine and dopamine transporters.

* Correspondence concerning this article should be addressed to: les.iversen@pharm.ox.ac.uk

Introduction

The discovery of monoamine transporters by Julius Axelrod and colleagues and their importance in the inactivation of monoamine neurotransmitters is a classic example of the value of basic research and serendipity. It led to the development of many important drugs used in the treatment of neuropsychiatric disorders. The concept of neurotransmitter reuptake is only some forty years old. Previously it was assumed that the inactivation of neurotransmitters after their release from nerves involved rapid enzymatic breakdown, as seen with acetylcholinesterase. Indeed the degradation of monoamines by the enzyme monoamine oxidase (MAO) was known early on and MAO inhibitors were the first effective antidepressant drugs [1]. In the 1950's a second enzyme catechol-O-methyl transferase (COMT) was discovered by Julius Axelrod and was also thought to play a key role in inactivating norepinephrine and other catecholamines [2]. Experiments performed in the Axelrod laboratory at the National Institutes of Health in the late 1950's with [^{3}H]-epinephrine and later with [^{3}H]-norepinephrine were designed to evaluate further the importance of COMT vis a vis MAO in the inactivation of the catecholamines in vivo. But the experiments yielded an unexpected result. Although in laboratory animals most of the injected dose of labelled catecholamine was rapidly metabolized (mainly by COMT), a substantial proportion of the injected monoamine (30-40%) was removed from the circulation by a rapid uptake into tissues, where it remained for some time unchanged [3]. A key observation was that the uptake of [^{3}H]-norepinephrine into the heart was virtually eliminated in animals in which the sympathetic innervation had been destroyed by surgical removal of the superior cervical ganglion [4].This led Hertting and Axelrod [5] to propose that the reuptake of norepinephrine by the same nerves from which it had been released might represent a novel mechanism for inactivating this neurotransmitter

The discovery of epinephrine and norepinephrine uptake was followed by the finding that similar but distinct transporters were involved in the inactivation of serotonin and dopamine, and that similar mechanisms existed for the inactivation of the amino acid neurotransmitters [6,7].

Monoamine Transporters

The norepinephrine transporter (NET) was cloned [8] and this soon led to the discovery of other related members of the monoamine transporter gene family. Separate transporters exist for serotonin (SERT) and dopamine (DAT) [7]. The monoamine transporters use the electrochemical gradient of sodium between the outside and inside surfaces of the cell membrane to provide the thermodynamic energy required to pump neurotransmitters from low concentrations outside the cell to the much higher concentrations inside the cell. Chloride ions accompany the entry of neurotransmitter and sodium, and there is a net movement of positively charged ions into the cell, although not in sufficient amounts to appreciably alter the resting membrane potential.

The vesicular neurotransmitter transporters represent another family whose function is to maintain the very high concentrations of monoamine and amino acid neurotransmitters in storage vesicles. They use the proton gradient that exists across the vesicular membrane as the motive force. The vesicular monoamine transporters (VMAT-1 & VMAT-2) recognize serotonin, dopamine, norepinephrine, epinephrine and histamine. VMAT-2 is the predominant form found in monaminergic neurons in CNS. It is also expressed in the histamine-containing cells of the stomach, and in the adrenal medulla and in blood cells. The Na^+/Cl^+ -dependent transporters and the vesicular transporters are membrane proteins consisting of a single polypeptide chain of 5-600 amino acid residues, with 12 α-helical membrane-spanning domains [9].

Drugs as Inhibitors of Monoamine Transporters

The most important CNS drugs that target the norepinephrine and serotonin neurotransmitter transporters (NET and SERT) are the tricyclic antidepressants and their modern counterparts. The discovery that imipramine potently inhibited NET in sympathetic nerves [10] and the finding that this also applied in the brain [11] led to the first understanding of the mechanism of action of the tricyclic antidepressants. Following the discovery of the serotonin uptake system in brain it soon became apparent that the classical tricyclic drugs imipramine and amitriptyline were also potent inhibitors of SERT. This reinforced the "monoamine hypothesis of depression" as a monoamine deficiency state, and stimulated much further research in the pharmaceutical industry to discover new inhibitors of monoamine uptake. An effort to improve the selectivity of antidepressants was made in the 1970's by scientists at the CIBA-GEIGY Company in Switzerland (now Novartis), who developed the selective norepinephrine uptake inhibitor maprotiline [12]. This proved to be clinically effective as an antidepressant but it was not a great success commercially and had few clear advantages over the classical TCAs. The serotonin selective reuptake inhibitors (SSRIs) developed in the 1990's proved to be far more important. The first SSRI zimelidine was launched by Astra in Europe in the 1980's, but it had to be withdrawn because of serious adverse side effects [13]. The first widely used SSRI was fluoxetine ("Prozac") [14]. It was followed by a number of other SSRI's, several of which met with considerable commercial success. Although the SSRI's were no more efficacious than the first generation tricyclic antidepressants, and did not act any faster, they were considerably safer in over-dose and were used more widely. The first generation of tricyclic antidepressants were very effective in the treatment of major depression but suffered from a number of adverse side effects caused by secondary actions on a variety of monoamine receptors and conduction mechanisms in the heart. The latter effects could prove lethal if the drugs were used in too high a dose – and many depressed patients committed suicide in this way in the 1960' & 70's. The greatly improved safety profile of the SSRI's allowed an expansion of the approved indications for the use of these drugs to include a variety of other psychiatric indications (Table 1). Apart from the use of these drugs in various phobic conditions, a major change in recent years has been the recognition of their effectiveness in the treatment of generalized anxiety disorder, and the first FDA approval of antidepressant compounds for this indication (venlafaxine, duloxetine, and escitalopram). In addition, agents in this class in controlled trials have shown usefulness in treating pre-menstrual dysphoria, borderline personality disorder, obesity, smoking cessation, and alcoholism.

Table 1. FDA-Approved Indications for SSRI and mixed NE/5-HT Antidepressants in addition to major depression

Bulimia nervosa
Panic disorder
Obsessive compulsive disorder (OCD)
Generalized anxiety disorder
Social anxiety disorder
Post-traumatic stress syndrome

Table 2. Antidepressants: Inhibition of Human Serotonin (SERT), Norepinephrine (NET), and Dopamine (DAT) Transporters

Generic Name	Human SERT, K_d (n*M*)	Human NET, K_d (n*M*)	Human DAT, K_d (n*M*)	Selectivity: SERT *vs* NET
Amitriptyline	4.3	35	3250	8
Amoxepine	58	16	4310	0.3
Bupropion	9100	52,000	520	5.7
Citalopram	1.2	4070	28,100	**3500**
Clomipramine	0.3	38	2190	130
Desipramine	17.6	0.8	3190	0.05
Dothiepin	8.6	46	5310	5.3
Doxepine	68	29.5	12,100	0.4
Duloxetine	0.8	7.5	240	9.4
Escitalopram	2.5	2514	>100,000	>1000
Fluoxetine	0.8	240	3600	300
Fluvoxamine	2.2	1300	9200	580
Imipramine	1.4	37	8500	27
Lofepramine	70	5.4	18,000	0.08
Maprotiline	5800	11.1	1000	0.002
Mirtazapine	>100,000	4600	>100,000	—
Nefazodone	200	360	360	1.8
Nortriptyline	18	4.4	1140	0.24
Paroxetine	0.13	40	490	300
Protriptyline	19.6	1.4	2100	0.07
Reboxetine	129	1.1	—	0.008
Sertraline	0.29	420	25	1400
Trazodone	160	8500	7400	53
Trimipramine	149	2450	780	16
Venlafaxine	8.9	1060	9300	120

(From Glennon and Iversen, [20])

Some of the most recently introduced antidepressants hark back to the less-selective compounds of the earlier era. Thus duloxetine [15] and venlafaxine [16] are drugs that combine both norepinephrine and serotonin re-uptake inhibition although *in vitro* binding data show that venlafaxine binds with more than 100 times higher affinity to human SERT than to NET, (Table 2). Reboxetine is the first antidepressant drug since maprotiline in a new class of NET-selective inhibitors [17,18]. Reboxetine is reported to be as effective as the SSRI's or older tricylics, but is not associated with sexual dysfunction, a common side effect of the SSRI's. It is claimed to be more effective than fluoxetine in improving the social adjustment of depressed patients. The SSRI's and the more recently developed mixed NET/SERT antidepressants have proved hugely successful commercially, with worldwide sales well in excess of $10 billion. Table 2 summarizes the affinities of currently used antidepressants on cloned human monoamine transporters expressed in tissue culture cell lines [19].

Some Unanswered Questions

Although the monoamine uptake inhibitors have proved very successful in the treatment of depression and anxiety states many questions remain unanswered. How can drugs that are selective norepinephrine reuptake inhibitors be equally effective as those that selectively target serotonin reuptake? In reality none of the antidepressants is completely selective for NET or SERT. In some cases the formation of active metabolites alters the drug selectivity profile. Thus the non-selective compound imipramine and the partially NET-selective compound lofepramine are extensively metabolized to desipramine, a highly potent and selective NE reuptake inhibitor. Similarly whereas amitriptyline has little selectivity for NET or SERT, the metabolite nortriptyline is a selective NET inhibitor. Alternatively some have suggested that the SSRI's act indirectly to modulate noradrenergic function [21,22].

The long term changes in the brain that are triggered by the antidepressants, remain obscure. Although inhibition of monoamine uptake is an immediate effect, all antidepressant drugs require a period of several weeks before they become fully effective. Many theories have been proposed to explain this delayed action, including alterations in the expression of monoamine receptors, changes in transcription factors and neurotrophic factors, and morphological alterations in the connectivity of monoaminergic nerves and the promotion of new nerve cell formation [23].

The mechanisms underlying the prominent placebo effect seen in all clinical trials of antidepressant drugs also remain obscure. It is not uncommon for the placebo effect to be so large as to obscure any significant effect attributable to the active drug in clinical trials. An analysis of 74 Phase III clinical trials for 12 antidepressants submitted to the Food and Drug Administration USA between 1987 and 2004 revealed that only half of these trials succeeded in demonstrating a statistically significant benefit from drug treatment [24].

Other Medical Uses of Inhibitors of Monoamine Uptake

Amphetamines represent another important group of compounds that act on monoamine transporters. They act by promoting the release of dopamine in the brain by virtue of their high affinity for the dopamine transporter and the vesicular monoamine transporter. They displace endogenous dopamine by a combination of a depletion of vesicular stores and counter transport of dopamine outwards via the transporter [25]. Amphetamine itself and the related drug methylphenidate ("Ritalin®") are widely used in the treatment of children with attention deficit hyperactivity disorder (ADHD). A norepinephrine-selective NET inhibitor, atomoxetine is also used for the treatment of ADHD.

The older antidepressant bupropion, acts as a weak inhibitor of norepinephrine and dopamine uptake, with little effect on serotonin uptake. Bupropion had little success as an antidepressant, but is approved in the USA and Europe as an aid to smoking cessation [26].

The anti-obesity drug sibutramine acts as a non-specific inhibitor of NET, SERT and DAT [27].

Inhibitors of Monoamine Uptake as Drugs of Abuse

The monoamine transporters are also important targets for drugs of abuse. The dopamine transporter (DAT) is the key site of action for the psychostimulant amphetamines (*d*-amphetamine (dexedrine), and methamphetamine ("crystal meth") and for cocaine. Mice that are genetically engineered to knock out the expression of the DAT gene are profoundly hyperactive and fail to show any further stimulation of activity in response to cocaine or *d*-amphetamine [28]. Such animals, nevertheless, will continue to self administer cocaine [29], suggesting that the rewarding properties of the drug cannot be explained entirely by its ability to inhibit DAT. Cocaine is also a potent inhibitor of both serotonin and norepinephrine uptake.

Concern has been expressed that amphetamine and the congener methyl phenidate used to treat ADHD in children might have abuse and dependence liabilities. In practice, however, there is very little evidence for this, probably because when used medically the drugs are administered orally, leading to slow absorption and brain penetration. In contrast, when abused the amphetamines are administered intravenously or by smoking (methamphetamine), routes that lead to the rapid absorption and brain exposure that addicts crave.

Future Developments?

Pharmaceutical companies have virtually exhausted the permutations of different types of monoamine transporter. Some have resorted to developing single enantiomer forms of existing drugs – for example, escitalopram – the S-enantiomer of the SSRI citalopram, or an active metabolite- for example, desvenlefaxine – the desmethyl metabolite of venlafaxine. These can extend the commercial patent life of a compound but do little to provide real medical advances. Although inhibitors of DAT tend to be psychostimulant in character, several companies are \attempting to develop "triple" monoamine uptake inhibitors that target NET, SERT and DAT as potential antidepressants. It is too early to know whether this will prove successful.

Uptake 2

Another monoamine transporter, known originally as Uptake2 [6,30]. is present in several peripheral tissues and in brain. It is not dependent on Na^+ or Cl^-, has a low affinity for substrates and a high capacity. It is sensitive to inhibition by O-methylated catecholamine metabolites and by steroids [6]. Uptake2 has been cloned in animals, where it is termed "organic cation transporter 3" and in man where it is named "extraneuronal monoamine transporter" [31]. This uptake system may represent a second line of defense that inactivates monoamines which have escaped neuronal uptake, and thus prevents uncontrolled spread of the signal. It has been suggested that inhibitors of this transporter might represent faster acting antidepressants [32], but this idea has not been pursued.

6-Hydroxydopamine – A Neurotoxic Substrate for Monoamine Transport

The 6-hydroxy analogue of dopamine acts as a uniquely selective neurotoxin for noradrenergic and dopaminergic neurons. After systemic administration it causes a long-lasting depletion of norepinephrine from sympathetic nerves, and Thoenen and Tranzer [33] were the first to show that this resulted from drug-induced degeneration and disappearance of the sympathetic nerve endings. This is due to the selective accumulation of 6-OHDA by NET in sympathetic nerve terminals, followed by the formation of a toxic quinone metabolite. Although 6-OHDA does not penetrate the blood brain barrier, it has similar effects on catecholamine-containing neurons in brain if administered directly into CSF or by local micro-infusion. Norman Uretsky, working in my laboratory in Cambridge in the 1960's, was one of the first to demonstrate a long-lasting (> 30 days) depletion of both norepinephrine and dopamine from rat brain after injection of 6-OHDA into the lateral ventricle [34,35]. This was accompanied by a marked reduction in the uptake of ^{3}H-norepinephrine after icv injection or in slices taken from 6-OHDA treated rats. Uretsky went on to show [36] that the motor stimulant effects of amphetamine were significantly reduced in animals pre-treated with 6-OHDA. The effects of 6-OHDA on central monoaminergic neurons were also reported at about the same time by other laboratories [37,38], and 6-OHDA became widely used as a valuable tool for causing selective lesions of specific catecholamine-containing pathways in brain and examining the effects of such lesions on behavior.

Conclusion

The accidental discovery of monoaminergic uptake had a major influence on the development of effective drugs for the treatment of depression, phobic disorders, and anxiety, as well as explaining the molecular basis of the psychostimulant effects of the amphetamines used in the treatment of ADHD. Drugs of abuse and the selective neurotoxin 6-OHDA also target these mechanisms. Overall, this demonstrates the value of "blue skies" basic research which can be translated into clinically valuable treatments.

References

[1] Pletscher A. Monoamine oxidase inhibitors. Pharmacol Rev. 1966. 18, 121-129.

[2] Axelrod, J., Senoh, S. & Witkop, B. O-Methylation of catecholamines in vivo. J. Biol.Chem. 233:697-701.

[3] Axelrod, J. The fate of adrenaline and noradrenaline. In: CIBA Foundation Symposium on Adrenergic Mechanisms, 1960.Vane, J.R., Wolstenholme, G.E.W. & O' Connor, M (eds), pp 28-39, J.A.Churchill Ltd, London.

[4] Hertting, G., Axelrod,J., Kopin, I.J. & Whitby, L. Lack of uptake of catecholamines after chronic denervation of sympathetic nerves. Nature, 1961, 189, 66.

[5] Hertting, G. & Axelrod, J Fate of tritiated noradrenaline at the sympathetic nerve endings. Nature, 1961. 192, 172-173.

[6] Iversen, L.L. Role of transmitter uptake mechanisms in synaptic neurotransmission. Br J Pharmacol., 1971. 41, 571-591.

[7] Masson J., et al. Neurotransmitter transporters in the central nervous system. Pharmacol Rev, 1999. 51,439-464.

[8] Pacholczyk, T., Blakeley, R.D. & Amara, S.G. Expression cloning of an antidepressant-sensitive noradrenaline transporter. Nature, 1991. 350, 350-353

[9] Kavanaugh, M.P. Neurotransmitter transport: models in flux. Proc Natl Acad Sci US, 1998. 95, 12737-12738.

[10] Axelrod, J., Whitby, L. & Hertting, G. Effect of psychotropic drugs on the uptake of 3H-norepinephrine by tissues. Science, 1961. 133, 383.

[11] Glowinski,J. & Axelrod, J. Inhibition of uptake of tritiated-noradrenaline in the intact rat brain by imipramine and structurally related compounds. Nature, 1964. 204, 1318-1319.

[12] Waldmeier, P. From mental illness to neurodegeneration. In "The Psychopharmacologists", 1996. ed D.Healy, Chapman and Hall, London, pp565-586.

[13] Carlsson A. A paradigm shift in brain research. Science, 2001. 294, 1021-1024.

[14] Wong, D.T., Bymaster, F.P. & Engelman, E.A. Prozac (fluoxetine, Lilly 110140), the first serotonin uptake inhibitor and an antidepressant drug. Twenty years since its first publication. Life Sci., 1995. 57, 411-441.

[15] Kirwin, J.L. & Goren, J.L.(Duloxetine: a dual serotonin-norepinephrine reuptake inhibitor for the treatment of major depressive disorder. Pharmacotherapy,2004. 25,396-401.

[16] Mendlewicz, J. Pharmacologic profile and efficacy of venlafaxine. Int Clin Psychopharmacol., 1995. Suppl 2, 5-13.

[17] Hajos, M., Fleishaker, J.C., Filipiak-Reisner, J.K., Brown, M.T. & Wong, E.H. The selective norepinephrine reuptake inhibitor antidepressant reboxetine: pharmacological and clinical profile. CNS Drug Rev., 2004. 10, 23-44.

[18] Wong, E.H.F., et al. Reboxetine: a pharmacologically potent, selective, and specific norepinephrine reuptake inhibitor. Biol Psychiat, 2000. 47, 818-829.

[19] Tatsumi, M., Groshan, K., Blakeley, R.D. &.Richelson, E. Pharmacological profile of antidepressants and related compounds at human monoamine transporters. Eur J Pharmacol., 1997. 340, 249-258.

[20] Glennon, R.A. and Iversen, L. Antidepressants. In Burger's Medicinal Chemistry and Drug Discovery.2008. 7th Edtn, Vol 5.

[21] Svensson, T.H. Brain noradrenaline and the mechanisms of action of antidepressant drugs. Acta Psychiatr Scand,2000. 101(Suppl 402), 18-27.

[22] Gorman, J.M. & Sullivan, G. Noradrenergic approaches to antidepressant therapy. J Clin Psychiatry, 2000. 61, (Suppl 1), 13-16.

[23] Banasr M, Duman RS. Regulation of neurogenesis and gliogenesis by stress and antidepressant treatment. CNS Neurol Disord Drug Targets. 2007.6, 311-20.

[24] Turner, E.H. et al Selective publication of antidepressant trials and its influence on apparent efficacy. N.Engl.J. Med. 2008, 358, 252-260.

[25] Rothman, R.B. & Baumann, M.H. Monoamine transporters and psychostimulant drugs. European Journal of Pharmacology, 2003. 479, 23-40.

[26] Hurt, J.R., Goldstein, M.G., Hurt,R.D. & Shiffman, S. Recent advances in the pharmacotherapy of smoking. JAMA., 1999. 281, 72-76.

[27] Arterburn DE, Crane PK, Veenstra DL. The efficacy and safety of sibutramine for weight loss: a systematic review. Archives of Internal Medicine. 2004. 164: 994-1003.

[28] Giros,B., Jaber,M., Jones, S.R., Wightman, R.M. & Caron,M.G. Hyperlocomotion and indifference to cocaine and amphetamine in mice lacking the dopamine transporter. Nature, 1996. 379, 606-612.

[29] Rocha, B.A., Fumagalli, F., Gainetdinov, R.R., Jones, S.R., Ator,R., Giros, B.,Miller, G.W. & Caron, M.G. Cocaine self-administration in dopamine transporter knockout mice. Nature Neurosci. 1998. 1, 132-137.

[30] Iversen, L. L. The uptake of catecholamines at high perfusion concentrations in the isolated rat heart: a novel catecholamine uptake process. Br J Pharmacol., 1965. 25, 18-33.

[31] Martel, F. & Azevedo, I. An update on the extraneuronal monoamine Transporter (EMT):characteristics, distribution and regulation. Curr Drug Metab., 2003. 4, 313-8.

[32] Schildkraut, J.J. & Mooney, J.J. Toward a rapidly acting antidepressant: the normetanephrine and extraneuronal monoamine transporter (uptake 2) hypothesis. Am J Psychiatry, 2004. 161, 909-11.

[33] Thoenen H & Tranzer P. Chemical sympathectomy by selective destruction of adrenergic nerve endings with 6-hydoxydopamine. N.S.Archiv Pharmakol. 1968. 262,271-288.

[34] Uretsky N J, Iversen L L Effects of 6-hydroxydopamine on noradrenaline containing neurones in the rat brain. Nature 1969. 221,557-559.

[35] Uretsky N J, Iversen L L Effects of 6-hydroxydopamine on catecholamine containing neurones in the rat brain J Neurochem 1970. 17,269-278.

[36] Evetts, K.D., Uretsky, N.J., Iversen,L.L., Iversen, S.D. Effects of 6-hydroxydopamine on CNS monoamines, spontaneous motor activity and amphetamine-induced hyperactivity in rats. Nature, 1970. 225, 961-962.

[37] Bloom F.E., Algeri, S., Groppetti, A.,Revuelta, A., & Costa,E. Lesions of central norepinephrine terminals with 6-hydroxydopamine. Biochemistry and fine structure. Science, 1969. 166, 1284-1289.

[38] Ungerstedt, U. 6-hydroxydopamine-induced degeneration of central monoamine neurons. Eur.J.Pharmacol.1968. 5, 107-110.

In: Biogenic Amines
Editors: T. Farooqui, A. A. Farooqui

ISBN: 978-1-60876-625-3

Chapter VI

Amine False Transmitters, Co-transmitters and Volume Transmission: Fifty Years of Exceptions to the Standard Model of Chemical Neurotransmission

S. Robert Snodgrass *

UCLA Geffen School of Medicine and Harbor-UCLA Medical Center, Torrance, USA.

'In honor of my friend Norman Uretsky: humble, thoughtful and mindful of others'

Abstract

The standard model of chemical neurotransmission began with the converging ideas of Waldeyer, Cajal and Sherrington in the 1890s. Early data came primarily from studies of the effects of adrenal extracts, sympathetic nerve stimulation, and parallel studies of the neuromuscular junction. The accessibility of peripheral nervous tissues and the special importance of autonomic drugs in human cardiovascular pharmacology gave the neuromuscular junction and the autonomic nervous system disproportionate influence in pharmacology and neuroscience. The discovery of miniature endplate or postsynaptic potentials by intracellular recording, and then synaptic vesicles in the 1950s seemed to complete the theory. I selectively review three exceptions to the standard model of chemical neurotransmission, emphasizing the historical themes of thesis, antithesis, synthesis and reconsideration. The limited specificity of vesicular monoamine transporters permits other amines to enter storage vesicles and to be released by

* Correspondence concerning this article should be addressed to: S. Robert Snodgrass, Pediatric Neurology, Bldg. N-25, Harbor-UCLA Medical Center, 1000 West Carson St., Torrance, CA 90509. Tel (310)-222-4168; FAX (310)-320-4006; email: srsnodgrass@labiomed.org.

depolarization. Some actions of amines such as tyramine and octopamine may be explained by these less potent agonists displacing the natural transmitter. However, the so-called trace amines, including those derived from the diet, may interact with other receptors including trace amine-associated receptors (TAARs).

Co-transmitters pose more complex problems and emphasize how much the nervous system changes during its development. The false transmitter phenomenon assumes stability of "transmitter machinery" [transporters, storage, synthesis, receptors]. First we learned that other chemicals might displace "natural transmitters", and then we found that the transmitter machinery itself may change. I review the ability of sympathetic neurons to change from adrenergic to cholinergic phenotype and three of the many known co-transmitters (ATP, substance P, and neuropeptide Y). They are found in adrenal chromaffin cells and have special links to noradrenaline and the sympathetic nervous system. Purines are probably the most common co-transmitters and were the first recognized co-transmitters. Slow synaptic potentials and transmitter effects focus attention on neuropeptides, which may diffuse relatively long distances to reach receptors. Endocannabinoids introduce more complexity; the endocannabinoid 2-AG (2-arachidonoylglycerol) is the best studied and understood CNS retrograde messenger. Glutamate, the prototypical CNS fast excitatory transmitter, also modulates the effects of other transmitters. It is released from cholinergic motor neurons and modulates cholinergic effects at the neuromuscular junction.

Messages cross the synapse in either direction, and may act at extrasynaptic sites.

The existence of extrasynaptic receptors with different functional effects from synaptic receptors for the same transmitter is now well established. Volume transmission, the discharge of signaling molecules into the extracellular space, is established, but its quantitative significance remains uncertain. Receptor mismatches, mismatched synapses, and silent synapses remain poorly understood. These exceptions to the standard model, along with submicroscopic entities such as gap junctions and tunneling nanotubes may be more important in early development and pathological states, than in the hypothetical healthy young adult. Increasing attention to high frequency discharges and network oscillations suggest that gap junctions may become more important in human physiology and pharmacology. These exceptions increase the complexity of nervous function, but they do not invalidate the basic models of Cajal or Dale, which provide useful first approximations.

Keywords: neurotransmission, cotransmission, extracellular space, vesicular transport, synapse, receptor specificity, receptor interactions

List of Abbreviations

5-HT	5-hydroxytryptamine or serotonin;
A	adrenaline;
ACh	acetylcholine;
A1 Receptor	One of the adenosine receptor family of GCPRs, which preferentially recognizes adenosine;
AMPA	α-amino-3-hydroxy-5-methyl-4-isoxazolepropionic acid, an artificial amino acid and glutamate analog, AMPA receptors are a

	non-NMDA-type ionotropic glutamate receptor that mediates fast synaptic transmission in the CNS;
ATP	adenosine 5' triphosphate;
CB1 receptor	receptor for endogenous cannabinoids;
CGRP	calcitonin gene related peptide, a member of the calcitonin family
CNS	central nervous system;
D2 receptor	the dopamine receptor most related to neuroleptic function [173];
DA	dopamine;
DOPA	the amino acid dihydroxyphenylalanine, immediate precursor of dopamine;
Epsp	excitatory postsynaptic potential;
G protein	Guanine nucleotide binding proteins;
GCPR	G protein coupled receptor;
GDNF	glial derived neurotrophic factor;
GR	glucocorticoid receptor;
NA	noradrenaline (norepinephrine in the US);
NMDA	N-methyl-D-aspartate, a ligand which binds to a major class of glutamate receptors, which are called NMDA receptors;
NPY	neuropeptide Y;
P2X	refers to a division of purinergic receptors begun by Burnstock, P1 receptors prefer adenosine, P2 receptors respond best to ATP or ADP; P2x receptors are ionotropic, P2Y are GCPRs;
PHI	peptide histidine isoleucine, a peptide included in Prepro-vasoactive intestinal peptide (prepro-VIP) along with VIP and peptide histidine valine (PHV)
PD	Parkinson's Disease;
PET	Positron emission tomography;
TAAR	trace amine-associated receptor;
VIP	vasoactive intestinal peptide
VMAT	vesicular monoamine transporter.

1. Background History

Cell theory profoundly changed biology and medicine in the mid 19th century. Cells and the germ theory of disease gradually displaced the old humoral theories of health and disease. However, many authorities, including ES Russell and AN Whitehead in the 1920s, feared that a parliament of separate cells meant chaos. Nobody imagined the thousands of macromolecules and signaling molecules that we know today.

Early microscopists examined unstained brain tissue and reported "globules", often uniform in size. Cell theory had to wait for better optics. Achromatic microscopes became available in about 1830 and brought major changes. Ehrenberg reported the existence of cerebral ganglion cells in 1833 and Purkynê visualized neurons and dendrites (Figure 1). He presented drawings of neurons to scientific meetings [1].

Theodor Schwann proclaimed that the animal body was an organized collection of cells [2]. This implied mechanisms of intercellular communication and coordination, which were unknown at the time. Neural and endocrine mechanisms of communication and coordination began to emerge in the late 19th century. The neuromuscular junction was identified and studied before Sherrington developed the concept of synapses. Vulpian theorized in 1866 that curare acted on an intermediate zone between the nerve and the muscle [3]. DuBois-Reymond noted in 1877 that nerves might use chemical or electrical messages to communicate with effector elements [4]. The first systematic neuron theory and the word neuron came from von Waldeyer-Hartz in 1891 [5]. Waldeyer organized the observations of others into a coherent scheme without any new data of his own. His contemporaries Santiago Ramon y Cajal and Charles Sherrington played major roles in promoting and expanding the neuron concept. Cajal stressed the idea of neuronal polarization: axons carried outgoing messages, while dendrites processed incoming messages. Messages crossed the synapse in one direction only. His drawings often included arrows to stress this polarization. However, Cajal knew that retinal amacrine cells lacked axons and that conduction might go in the "wrong direction" in pathological states [6].

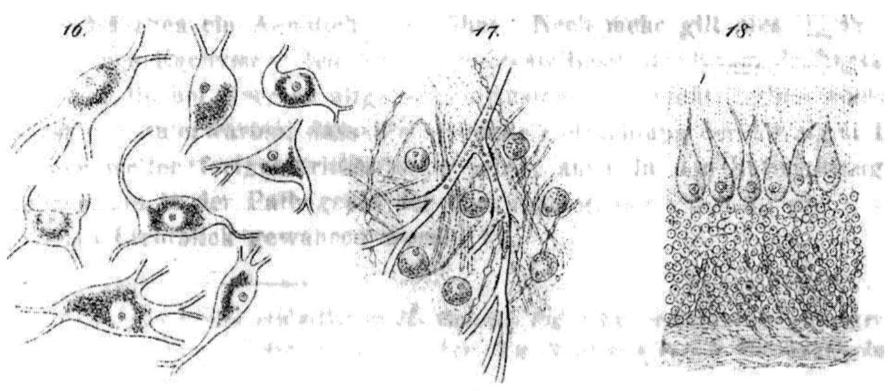

Figure 1. These drawings of nerve cells were used by Purkyne in his 1837 lecture to the Prague Congress. Purkyne outlined a cell theory (Kornchen) in advance of Schwann's famous book. This drawing includes cells from the brain stem (#16), thalamus (#17) and cerebellum containing what we now call Purkinje cells (#18). It is reproduced from reference [1] with the permission and assistance of author Larry Swanson who reproduced it from an original.

Neurophysiologist Charles Sherrington first wrote of synapses in 1897. His influential book, *The Integrative Action of the Nervous System* appeared in 1906 [7]. Sherrington's functional and physiological emphasis offset the limitations of neuroanatomy; light microscopy could not visualize the synaptic cleft. Sherrington and his students thought primarily in terms of electrical communication between cells.

Adrenaline[1] was identified in 1899; it reproduced the features of some forms of endocrine and neural communication. The autonomic nervous system became a critical focus of monoamine pharmacology and theoretical development. Immunologist Paul Ehrlich and

[1] I use adrenaline, noradrenaline, A, NA in this paper instead of the U.S. nomenclature (epinephrine) because of the historical emphasis of the paper and the importance of European pharmacologists in identifying these exceptions.

Cambridge physiologist John Langley were the first receptor theorists. They knew each other's work and were influenced by the Emil Fischer's well known studies of enzymes. Fischer suggested that enzyme-substrate specificity depended on something like a lock-and-key mechanism. Langley first mentioned "receptive substances" in 1878 and "the autonomic nervous system" in 1898. He showed that nicotine and curare acted on the same site or mechanism at the neuromuscular junction, with opposite actions [8]. Langley said in his 1906 Croonian lecture to the Royal Society:

"...stimuli passing the nerve can only affect the contractile molecules by the radical which combines with nicotine and curare. And this seems in its turn to require that the nervous impulse should not pass from nerve to muscle by an electrical discharge, but by the secretion of a special substance at the end of the nerve" [9].

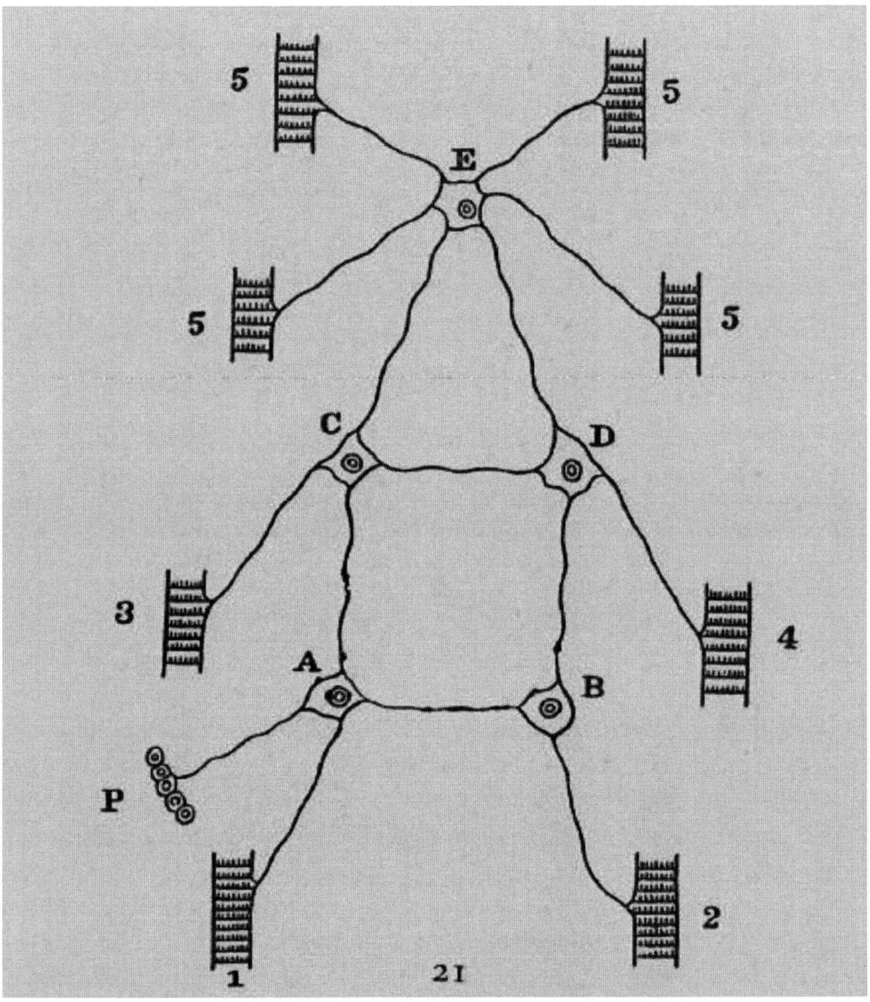

Figure 2. This drawing of a hypothetical reticular network is reproduced from reference [1] with the permission of author Larry Swanson. There are no synapses; nerve cells are shown in direct continuity with muscle cells as well as other nerve cells. He reproduced it from an original in Landois and Sterling's *Textbook of Human Physiology* (1891).

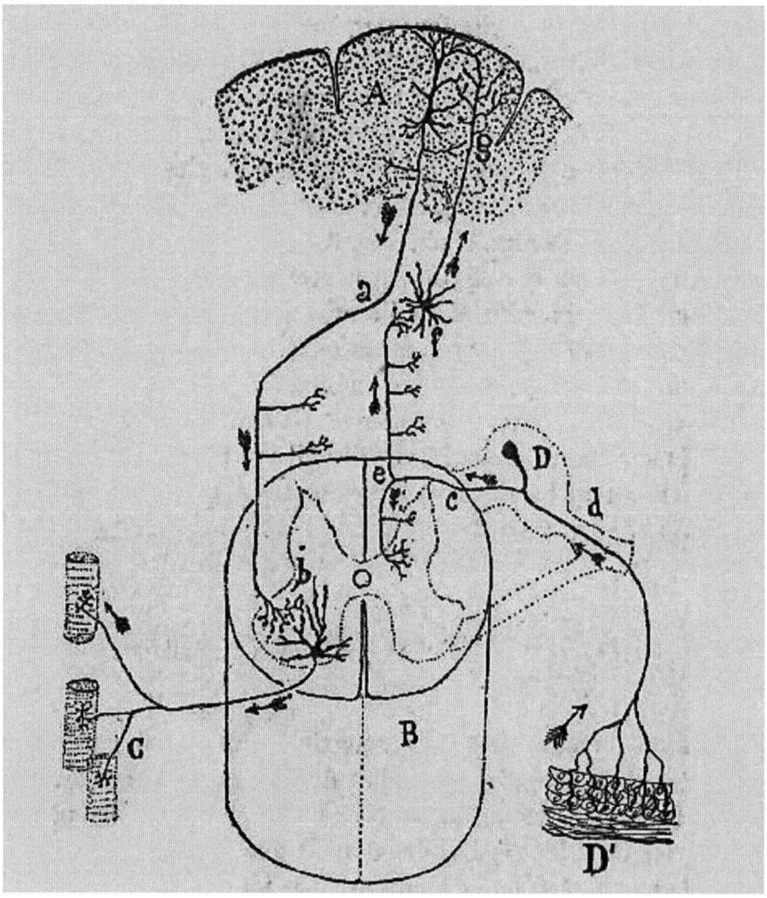

Figure 3. This 1894 drawing of a spinal reflex arc shows Cajal's functional emphasis and his use of arrows to illustrate the flow of information. Compare it to Figure 2, from the same era. It was figure 9 in reference [1] and is reproduced with permission and assistance of Larry Swanson, who reproduced it from an original.

Langley had a broader theory of chemical neurotransmission than did his student, Thomas Elliott, who suggested in an abstract that sympathetic nerve impulses might liberate adrenaline [10]. Walter Dixon proposed in 1907 that muscarine might be the vagus nerve messenger. Henry Dale reported in 1914 that adrenaline and acetylcholine duplicated the effects of sympathetic and parasympathetic nerve stimulation [11]. Dale and Otto Loewi constructed an outline of autonomic pharmacology between 1915-1945, which assumed that each cell used the same message at all its branches. Loewi showed that vagus nerve stimulation caused either an inhibitory substance or an excitatory one, depending upon the experimental details, to appear in the Ringer's solution perfusing a frog heart, as judged by the effect on a second heart. Loewi repeated this experiment at the 1926 International Congress of Physiology in Stockholm, but this was not enough to establish acetylcholine as "vagustoff", the vagus nerve neurotransmitter. Many authorities thought that transmission was too fast at ganglia and the neuromuscular junction to be chemically mediated. Acetylcholine wasn't demonstrated in animal tissues until 1929. Feldberg and Krayer used the leech bioassay to demonstrate that vagal stimulation produced acetylcholine-like material in the coronary venous blood of dogs and cats in 1933 [12]. Acetylcholine effects on muscle were nicotine-sensitive, its cardiovascular effects were not [13]. Dale proposed a division of autonomic pharmacology into cholinergic and adrenergic domains (he invented these words). John Eccles elevated Dale's principle to a formal theory of neurotransmission later, in the 1950s, after intracellular recording demonstrated a chemical basis for CNS neurotransmission. Dale himself never ruled out multiple transmitters. After observing atropine-resistant effects of vagus nerve stimulation, he said:

"We may suppose that vagus effects not paralysed by atropine are not humorally transmitted at all, or that the transmitter is not a choline ester, but in the latter case, we shall have to postulate not one, but several other transmitters with different degrees of liability to the antagonism of atropine." [14].

After World War II, intracellular recording demonstrated quantal transmitter release [15] and several different electron microscopic laboratories discovered synaptic vesicles. Now the skeptics were convinced. The standard model was summarized in classical publications of Bernard Katz and John Eccles. Excitatory transmission produced currents that depolarized the postsynaptic membrane, while inhibitory transmission hyperpolarized it, changing the membrane potential in the opposite direction [16,17]. These studies and theories considered only rapid responses, on a millisecond time scale. They didn't study stability of transmitter properties over periods longer than a day or the effects of disease and development.

The first paper in the new 1949 Pharmacological Reviews was Bacq's synthesis of adrenergic transmitter metabolism, hinting at the coming era of transmitter uptake and storage [18]. Reserpine was used by psychiatrists (it was approved by the US Food and Drug administration for clinical use in 1953) before its mode of action was understood. Bernard Brodie and his colleagues reported that reserpine released monoamines [19] and ascribed reserpine effects to serotonin depletion, which was only partly correct. Arvid Carlsson, who had studied with Brodie, showed that DOPA antagonized reserpine effects in rodents, whereas the serotonin precursor 5-hydroxytryptophan did not. [20]. This suggested the presence of a previously unknown brain amine and led to Carlsson's discovery that dopamine was a CNS neurotransmitter [21].

Reserpine could deplete all stored monoamine transmitters [22]. The first specific study of the storage granule transporter in 1962 [23] used adrenal chromaffin cells. Transporter specificity was limited; uptake was coupled to ATP hydrolysis. Carlsson found that reserpine was the most potent inhibitor of vesicular amine transport [22]. Tetrabenazine and ketanserin are also potent inhibitors. Tetrabenazine has little effect on the peripheral vesicular monoamine transporter; its depletion of CNS vesicular amine content lasts hours rather than days, unlike reserpine effects [24]. Once used extensively for treatment of psychosis and hypertension, reserpine is little used today because of many side effects, including depression and even suicide. Tetrabenazine is used today for chorea and other hyperkinetic movement disorders. It was never used for hypertension. The ketanserin derivative, (C^{11}dihydrotetrabenazine [^{11}C-CIDTBZ) has been used as a ligand for positron emission tomography (PET scanning) for in vivo studies of the type 2 vesicular monoamine transporter (VMAT2) in the human brain. Studies with DTBZ indicated fewer VMAT2 binding sites in the striatum of Parkinsonian patients than in controls [25]. Binding decreased during normal aging, less so than in Parkinson's disease. However, DTBZ binding is altered by dopamine depletion; it provides only an approximate measure of vesicle number [26].

Parkinson's disease [PD] is generally sporadic, but some patients have documented single gene mutations. PET studies of two kindreds with LRRK2 gene mutations revealed heterogeneous forms of presynaptic DA dysfunction, similar to the findings in sporadic Parkinson's disease. A small number of asymptomatic relatives with the mutation showed presynaptic abnormalities, often related to the plasma membrane DA transporter [27]. Only fifteen family members were studied, enough to suggest that multiple different presynaptic factors may contribute to PD. Follow-up studies showed progression of these presynaptic abnormalities as symptoms progressed [28]. No patients had evidence of postsynaptic D2 receptor dysfunction. There is a mouse model of PD produced by mutations of the vesicular monoamine transporter [29], but no analogous human mutations have been reported.

Two different genes encode vesicular monoamine transporters, VMAT1 and VMAT2 [30]. Very little VMAT1 is found in the adult CNS, and VMAT1 is inhibited only very weakly by tetrabenazine. Vesicular transporter expression is one factor regulating the neurotransmitter content of secretory vesicles and influences the amount of transmitter released [31]. Pothos et al showed that more DA molecules were released per quantum in cultured neurons treated with glial-derived neurotrophic factor (GDNF) and with DOPA [32]. This was the first demonstration that presynaptic factors could modulate quantal transmitter size of CNS neurons. Quantal size (the number of neurotransmitter molecules released by a single synaptic vesicle during exocytosis) was assumed to be invariant in the classical model. Quoting from Del Castillo and Katz,

"transmission at a nerve-muscle junction takes place in all-or-none quanta whose sizes are indicated by the spontaneously occurring miniature discharges...The average 'quantum content' of the e.p.p. depends on the probability of response of the individual units, and this varies with the external Ca and Mg concentrations...." [33]

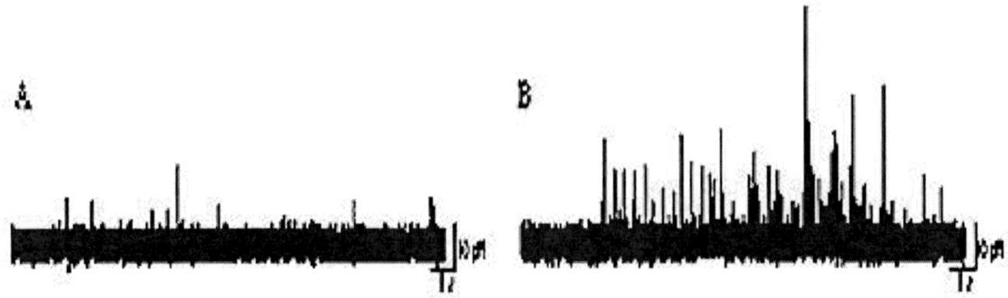

Figure 4. These are amperometric recordings of quantal DA release from cultured rat neurons [34]. Control (A) and experimental neurons overexpressing the VMAT2 gene (E) were stimulated with 40 mM potassium to increase quantal release. The experimental neurons have increased amplitude, mean of 7800 molecules © and 11800 molecules (E), and frequency (increased about 10 fold) both of which were statistically significant at the P<.01 level. Permission for reproduction obtained from Elsevier Limited, publisher of Behavioral Brain Research.

However, GDNF and DOPA have complex actions and the role of VMAT2 was not clear in the Pothos experiment. Pathos then used VAMT2 transfections into cultured midbrain and hippocampal neurons to show that more transporter protein caused increased quantal size [34]. Transporter protein reduction had the opposite effect, as shown in Figure 4.

Hippocampal nondopaminergic neurons became able to store and release DA when VMAT2 was expressed. Vesicular monoamine storage is also regulated by G proteins [35,36].

Mice with deletions of the G protein $G_{o2\alpha}$ have a decreased motor response to cocaine and less stored cerebral DA [37]. G proteins appear to interact with calcium dependent activator proteins CAPS I and II, which regulate monoamine uptake and storage, perhaps by determining when filling is complete [36]. Cocaine is a well known inhibitor of the plasma membrane DA transporter [38]; many drugs act at both plasma and vesicular monoamine transporters.

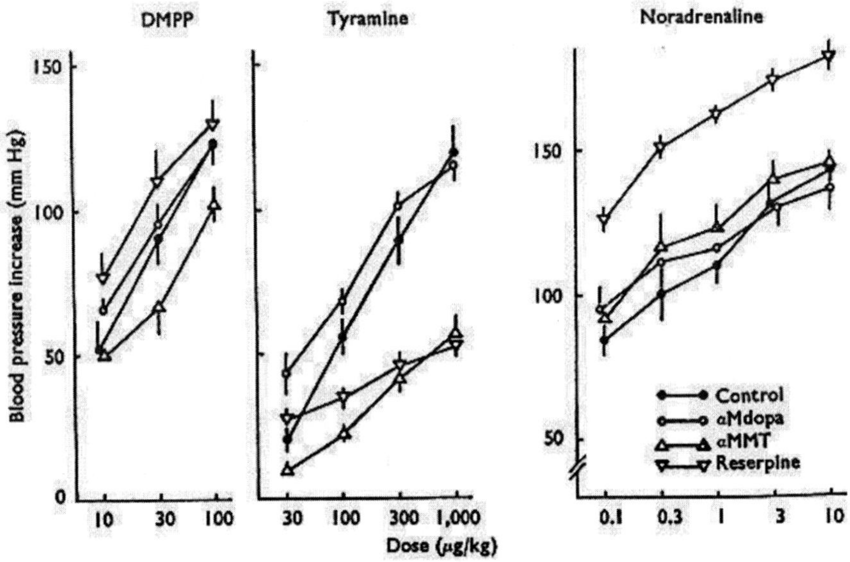

Figure 5. This figure, modified from Haefely [42] and reproduced with permission, shows blood pressure responses to three different treatments in cats treated with two false transmitter precursors (α-methyl dopa and α-methyl meta tyrosine) and reserpine. DMPP (dimethyl phenylpiperazinium), a nicotinic ganglionic stimulant; tyramine, an indirect acting "sympathomimetic drug"; and NA, the natural transmitter, were studied. Reserpinized cats had the greatest BP response to DMPP & NA; responses to the false transmitters were similar with NA and divergent with tyramine- false transmitters explain only part of these results.

2. False Transmitters

Sourkes reported that α-methylDOPA inhibited DOPA decarboxylase in 1954. Its antihypertensive actions were initially ascribed to this effect. However, decarboxylase inhibition did not explain its antihypertensive action. Drug administration led to α-methyldopamine storage in the brain and sympathetic nerves. Carlsson and Lindqvist contrasted reserpine-induced monoamine depletion with that caused by the DOPA analogs, α-methyldopa and α-methyl-m-tyrosine [39]. Carlsson and others proposed that α-methyldopa's antihypertensive action was due to displacement of the natural transmitter [40]. Stimulation of cardiac sympathetic nerves liberated α-methylnoradrenaline from the heart of α-methyldopa treated rabbits [41]. Haefely et al. showed that the ratio of α-methylnoradrenaline to noradrenaline released by splenic nerve stimulation correlated with that found in the spleen [42]. However the relationship between storage of the false transmitter α-methyl NA and the effects of α-methyl dopa on sympathetic nerve function was complex. Figure 5 illustrates the complexity of false transmitter actions. The response to intravenous noradrenaline was little changed by pretreatment with either of two false transmitters, α-methyl dopa or α-methyl-meta-tyrosine, but tyramine injections had different effects. Haefely and colleagues also found that contractions of the nictitating membrane evoked by sympathetic stimulation were not consistently decreased in spite of marked displacement of NA by the two false transmitters (their figure 2, not shown here).

Reserpine eliminates all stored amines, while the decarboxylated DOPA analogs displace the natural transmitter from storage vesicles. They are released from the storage sites but are less potent than the natural transmitter. Decarboxylation and subsequent β-hydroxylation of α-methyl-meta-tyrosine produces metaraminol. It is much less potent than noradrenaline, so that replacement of the physiological transmitter with metaraminol should produce a greater sympathetic deficit than replacement with a-methylnoradrenaline. Udenfriend and Zaltzman-Nirenberg showed that α-methyl-meta-tyrosine (α-MMT) must be decarboxylated in order to release NA from the guinea pig heart (Figure 6). Metaraminol and other metabolic products of α-MMT were potent NA releasers [43]. Crout and coworkers showed that sympathetic nerve stimulation released metaraminol from the perfused cat heart after injections of metaraminol [44].

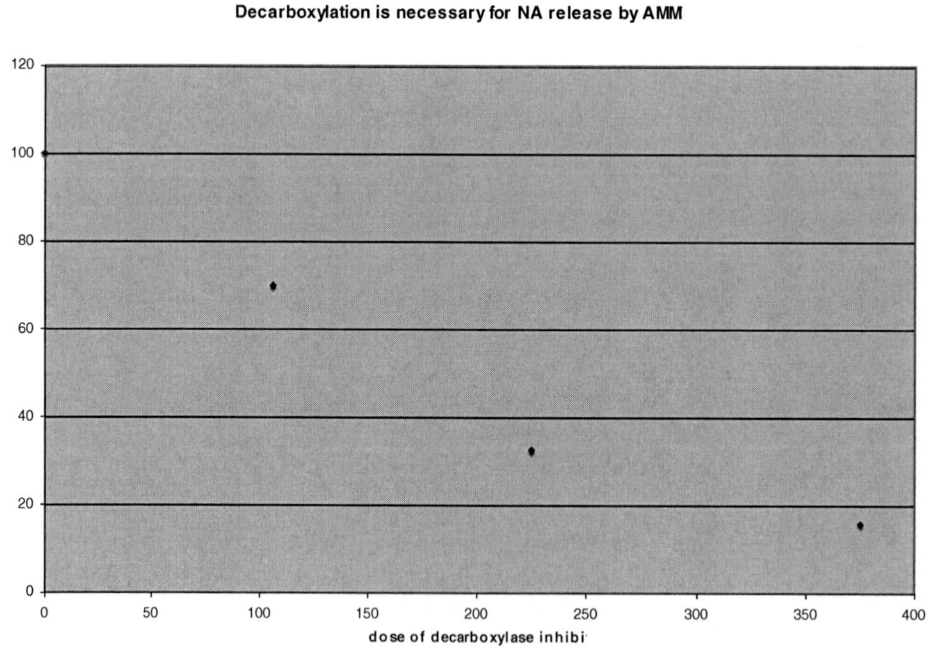

Figure 6. This figure 6 uses data from reference 43. Guinea pigs received one injection of α, β-methylene α-MMT and variable doses of the decarboxylase inhibitor α-methyl dopa hydrazine and were killed 16 hrs later. As decarboxylation is progressively inhibited, α-MMT released progressively less NA from the heart.

Monoamine oxidase inhibition reduces blood pressure and causes octopamine accumulation in sympathetic nerves [45]. Blood and tissue content of tyramine and many other amines increase. Octopamine is synthesized from tyramine and acts as a false transmitter [46]. However, octopamine is a primary transmitter in other circumstances, with its own receptors. Octopamine was identified as a lobster neurotransmitter in the early 1970s [47] and was soon found in many more invertebrates. Peter Evans suggested in 1978 that it functioned in invertebrates much as noradrenaline does in vertebrates, as the primary sympathetic nervous system messenger [48]. Octopamine receptors were abundant, and their drug responses generally resembled those of mammalian α- adrenergic receptors.

Octopamine appears to be a stress hormone in many insects [49]. Its uptake, receptors and second messengers are analogous to those for catecholamines; it regulates learning and complex behavior in many invertebrates [49,50]. Small amounts of tyramine and octopamine are found in all mammals. This is easily understood from the reactions shown below (Figure 7). Larger amounts of octopamine accumulate in hepatic encephalopathy [51]; certain fermented foods are rich in tyramine and octopamine [52].

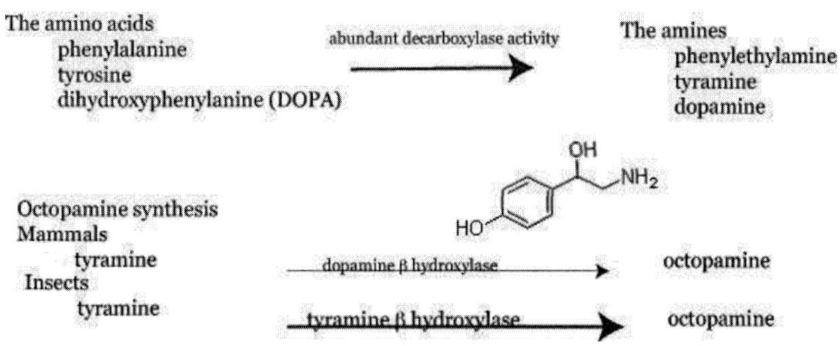

Figure 7. Octopamine is produced by hydroxylation of tyramine. Tyramine hydroxylation by dopamine β hydroxylase is a minor pathway in mammals. Tyramine synthesis by a separate but related enzyme, tyramine β hydroxylase, is a major pathway in insects.

The classical biogenic amines (adrenaline, noradrenaline, dopamine, serotonin and histamine) interact with specific families of G protein-coupled receptors (GPCRs). Borowsky and coworkers discovered a new GCPR family responding to tyramine and β-phenylethyl-amine but not to classical biogenic amines in 2001 [53]. Several closely related genes were identified and designated as trace amine associated receptors [TAARs]. Many of these TAAR subtypes respond poorly to tyramine, β-phenylethylamine, tryptamine or octopamine, suggesting the existence of additional endogenous ligands [54]. However, TAAR-1 modulates dopaminergic and motor function [55] and some trace amines probably have direct vascular effects [56]. These receptors may be useful therapeutic targets in human disease including psychiatric diseases [57].

Limited specificity of the transport, synthesis, storage and release machinery meant that many false transmitters could be created, such as octopamine from tyramine. DOPA replacement therapy for Parkinson's disease produces dopamine accumulation in nondopaminergic nerve endings; part of its effect is a false transmitter effect. The NA precursor L-threo-3, 4-dihydroxyphenylserine (DOPS) and the serotonin precursor 5-hydroxytryptophan (5-HTP) have similar effects, but they are used much less frequently than

DOPA is. The clinical use of a-methyl Dopa, like reserpine, is mostly of historical interest because better antihypertensive drugs are now available. Metaraminol and other sympathomimetic amines are used less freely for hypotension and shock than they were in 1970 because many other treatments are now available [58]. Metaraminol's false transmitter action means that its initial pressor effect, due to NA release, may be followed by hypotension if infusion continues, as metaraminol displaces stored NA [59].

Owman and others reported in the 1960s that pineal adrenergic nerves appeared to contain two different signaling amine molecules: 5-HT and octopamine [60,61]. Soon thereafter, Zweig and Axelrod reported a reciprocal relationship between pineal content of NA and 5-HT. Depletion of one amine produced an increase in the other [62]. This was a stable relationship unlike false transmitters related to drug therapy. The pineal situation became even more complex as octopamine appeared to follow the same pattern, increasing to fill storage space when 5-HT was depleted [63]. Pineal octopamine content increased when rats received tyramine and a MAO inhibitor; it fell when the rats received reserpine [64]. This illustrates multiple stable transmitters and competition between related amines for storage space.

The arylalkylguanidine meta-iodobenzylguanidine (MIBG) is structurally similar to noradrenaline [65]. It is a false transmitter that is concentrated within secretory granules of catecholamine-producing cells [66,67]. Iodinated MIBG is used for localization of neural crest tumors such as neuroblastoma and pheochromocytoma. The first use of ^{131}I-MIBG to localize a pheochromocytoma was in 1981 [68]. MIBG does not cross the blood-brain barrier. A recent study used a fluorescent false transmitter to monitor dopamine release from living mouse striatal slices [69]. This study showed that the fraction of synaptic vesicles releasing transmitter with each stimulus was a function of stimulus frequency. False transmitters remain a useful research tool; they have a small place in clinical medicine.

Table 1. Synopsis of false transmitters

Thesis	A neuron releases the same transmitter at all branches; transmitter identity never changes.
Antithesis	Endogenous or exogenous amines may displace natural amine transmitters and be released by nerve stimulation
Synthesis	Transmitter uptake, storage and synthesis are only partially specific. Drug treatment, disease and unusual diets may produce false transmitters, with or without physiological consequences
Reconsideration	False transmitters have multiple effects. Some classical false transmitters, such as octopamine, act at trace amine-activated receptors in addition to their false transmitter actions

3. Co- Transmitters

Co-transmitters are more important than false transmitters. They are very common in nature, even in the most primitive animals. Most neurons release more than one transmitter

[70]. The transmitter phenotype of individual neurons correlates with their peripheral targets. Increasing evidence indicates that transmitter phenotype (the mix of transmitters utilized) changes during development because of changes in transmitter synthetic enzymes and transporters. Burn and Rand reported in 1959 that adrenergic nerves contained and released both acetylcholine (ACh) and noradrenaline (NA) [71]. Their work could not be reproduced, but later studies of Furshpan and Patterson showed that sympathetic neurons, cultured from young rat cervical ganglia, produced and released both transmitters [72,73]. Further study of these dual function neurons revealed unexpected plasticity of sympathetic transmitter phenotype [74]. Sympathetic neurons are adrenergic at birth. Most remain adrenergic, but those innervating sweat glands become cholinergic; injury to the cell bodies or treatment with various cytokines such as leukemia inhibitory factor causes many to assume a cholinergic phenotype [75]. We are just beginning to understand the molecular determinants of transmitter phenotype [76]. Knowing that some adult sympathetic neurons are cholinergic, we can reconsider the failed Burn and Rand theory and suspect that it may have been partially valid [77,78].

Blaschko and colleagues found large amounts of ATP in the adrenal gland in the 1950s [79]. Release of ATP, proteins and adrenaline was demonstrated a decade later [80]. Although many laboratories reported multiple transmitters in invertebrate neurons [81], purines were the first well-established vertebrate co-transmitters. Geoffrey Burnstock suggested that some neurons released more than one transmitter in a pioneering 1976 essay [82]. By 1979, the new journal Trends in Neurosciences asked "Is Dale's principle valid?" [83].

Burnstock noted that both American workers [84] and Day and Rand from Oxford [71] had suggested corelease of ACh with another transmitter (oxytocin, NA) in the 1950s. Both groups thought that acetylcholine might facilitate transmitter release rather than acting at postsynaptic receptors. Pamela Holton of St. Mary's Hospital Medical School in London had reported in 1959 that ATP was liberated from rabbit sensory nerves by electrical stimulation [85]. She suggested two roles for ATP in nerves, as an energy source and also as a signal, but probably not the only signal, released from certain nerves. Burnstock and colleagues were able to demonstrate ATP release from guinea pig enteric nerves in the 1970s [86,87].

ATP and other purines are difficult to study because they are metabolically labile and found in all tissues. In general, purines may have presynaptic [typically inhibitory] effects or postsynaptic [mostly excitatory] effects [88,89]. ATP is known to mediate fast synaptic transmission at certain CNS [90] and peripheral synapses [91,92]. ATP, noradrenaline [NA] and neuropeptide Y [NPY] are important transmitters in perivascular sympathetic nerves; ATP acts at postjunctional P2X receptors to evoke vascular smooth muscle contraction [93]. It is a neurotransmitter in perivascular non-adrenergic non-cholinergic nerves. ATP is rapidly metabolized to adenosine which can also regulate the release of neurotransmitters via inhibitory prejunctional A1 adenosine receptors. The relative contributions of ATP and NA as functional co-transmitters varies with species, age, type and size of blood vessel, frequency and duration of stimulation, the tone/pressure of the blood vessel, and in disease.

ATP produces excitatory junctional potentials (EJPs) in response to single impulses in sympathetic nerves, and the initial twitch responses of the smooth muscle; NA produces the longer-lasting tonic contractions. The proportions of ATP to NA vary between different sympathetic nerves; they also change during development and in some pathological conditions, including hypertension [94]. Release of the two co-transmitters appears to involve independent mechanisms and is frequency dependent; they may be stored in separate vesicles but this isn't proven [95].

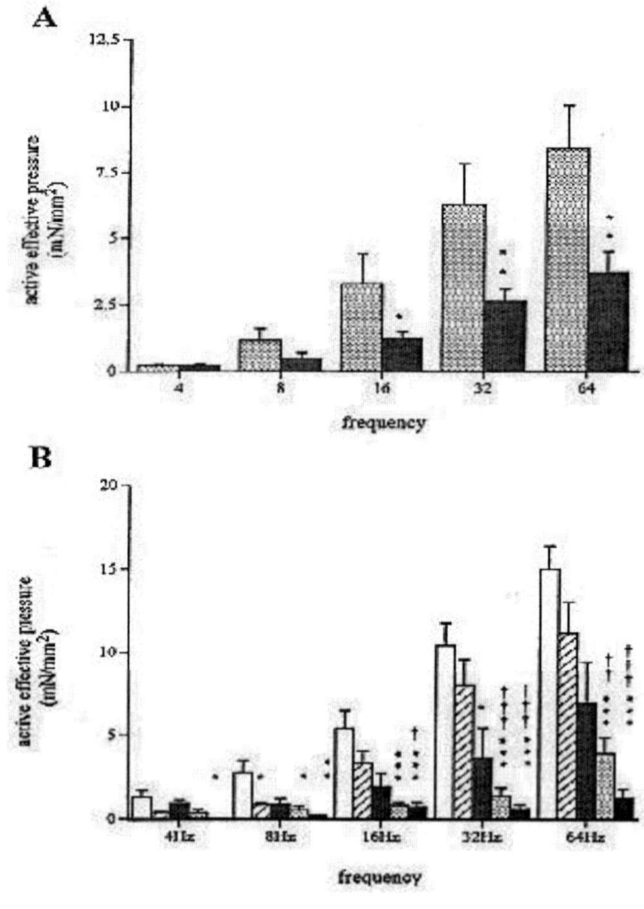

Figure 8. This figure from Smith [96] shows the relative contributions of α-adrenergic and purinergic receptors to the pressure response of the rabbit cutaneous resistance artery. Panel A shows that the pressor response increased with stimulation frequency, and that the response was progressively diminished by the α-adrenergic receptor blocker prazosin. (light bars control, dark bars with prazosin). Panel B compares the effects of purinergic blockade with α, β-methylene ATP to those of increasing doses of prazosin. The purinergic blocker had a large effect at 4 Hz stimulation, while the reduction due to maximal doses of prazosin was much greater at 64 Hz, shown by the difference between the left most and right most bars for frequencies 8 - 64 Hz in panel B. Modified and reproduced with permission from Oxford University press, reference [96].

Adenosine is an important neuromodulator in its own right. It generally modulates ongoing synaptic transmission rather than carrying direct messages. Adenosine modulation is most often seen in excitatory synapses [97]. The best known adenosine CNS effects operate through inhibitory A_1 receptors, which are one of the most abundant GCPRs in the brain [97]. A_1 receptors are mainly found in synapses; adenosine receptor heteromers regulate striatal glutamatergic input [98] and are targets for caffeine actions [99]. Presynaptic A1 receptors inhibit glutamate release [as well as that of other neurotransmitters]; post-synaptically they inhibit calcium influx through voltage-sensitive calcium channels and NMDA receptors and also inhibit potassium currents, leading to membrane hyperpolarization, as reviewed in [100]. Activation of A_1 receptors may be neuroprotective [101], probably because of extra-synaptic effects, in particular their ability to decrease brain metabolism at times of stress [102]. Human astrocytes are larger and more complex than rodent astrocytes [103]. Human neurons are probably more dependent on astrocytic signals and support than rodent or invertebrate neurons. Purines released from astrocytes appear to be important signals in complex human glioneural networks [103,104].

Substance P (for powder) was discovered in 1931 by von Euler and Gaddum as they surveyed horse tissues for acetylcholine following the 1929 report of Dale and Dudley who found histamine and acetylcholine in the spleens of horses and oxen [105,106]. Von Euler and Gaddum found something that was neither acetylcholine nor histamine and that caused vasodilatation and intestinal contraction. It was most abundant in intestine and brain. Hellauer and Umrath reported in 1948 that dorsal roots contained much more vasodilator activity than spinal ventral roots [107]. Angelucci demonstrated the release of something like Substance P [SP] from frog spinal cord by cutaneous stimuli. [108]. Many laboratories gathered evidence supporting a neurotransmitter role for SP during the 1970s [109-115]. Otsuka and Konishi showed that SP evoked slow spinal epsps [115]. Calcium-dependent release of spinal Substance P was shown and SP - like immunoreactivity found in synaptosomes and dense core vesicles [116]. The association of Substance P with spinal afferents suggested by Hellauer, Umrath and Angelucci was a reproducible finding. Jessell and Iversen suggested that Substance P might be the factor or a factor that opened the hypothetical "pain gate" in the dorsal horn of the spinal cord previously postulated by Melzack and Wall [113,117]. Their paper suggested central roles for two peptides at the "pain gate"- the old peptide, SP, known for decades, as pain messenger, and the new enkephalin peptides, which inhibited secretion of the pain messenger and partly explained the effects of opioid analgesia.

The "pain gate" and pain processing seem much more complex today, but SP and its neurokinin 1 receptor in the dorsal horn of the spinal cord retain a central role [118,119]. Substance P has analgesic effects in supraspinal regions; neurokinin 1 receptors are prominent in brain regions that process nociceptive information, such as the periaqueductal gray [120]. Pain management remains a complex, partly solved problem, for modern medicine.

Neuropeptide Y (NPY) has a shorter story, because it was discovered in 1982. However, it may be the most abundant neuropeptide in mammalian brain [121] and it has a strong association with NA, stress and the sympathetic nervous system. Tatemoto and colleagues at Karolinska Institute isolated large quantities of a peptide amide from pig brain [122]. This was neuropeptide Y, a 36 amino acid peptide that is sometimes called neuropeptide tyrosine. Hökfelt and co-workers soon demonstrated NPY within peripheral and medullary catecholamine neurons [123,124]. The highest concentration of NPY in the rat brain was found in the paraventricular nuclei of the hypothalamus [125]. Further study showed that NPY was often found in sympathetic neurons and was stored in large dense core vesicles.

Mild sympathetic stimulation released NA but little NPY. More severe or sustained stimulation released both NA and NPY [126].

NPY was found to be a very potent appetite stimulant in 1984 [127]. Elevations in hypothalamic NPY were reported after food deprivation and in genetically obese rodents. This stimulated a vigorous search for NPY antagonists for therapeutic use against obesity and metabolic syndrome. However, genetic deletions of the NPY gene surprised everyone- there was no consistent effect on weight and feeding behavior [128]. Today, we recognize multiple transmitters and factors that influence eating. NPY is still an important player. It acts through six different receptors, Y1-Y6; our hopes for simple solutions to the modern epidemics of obesity and addiction have been chastened [129-131]. I can't go into all the complex and interesting NPY effects here. The roles of adenosine in pain processing [132] and purines in obesity and eating [133] underline the complexity of CNS function. Figure 9 reminds us how many different peptide co-transmitters are found in the sympathetic nervous system [134].

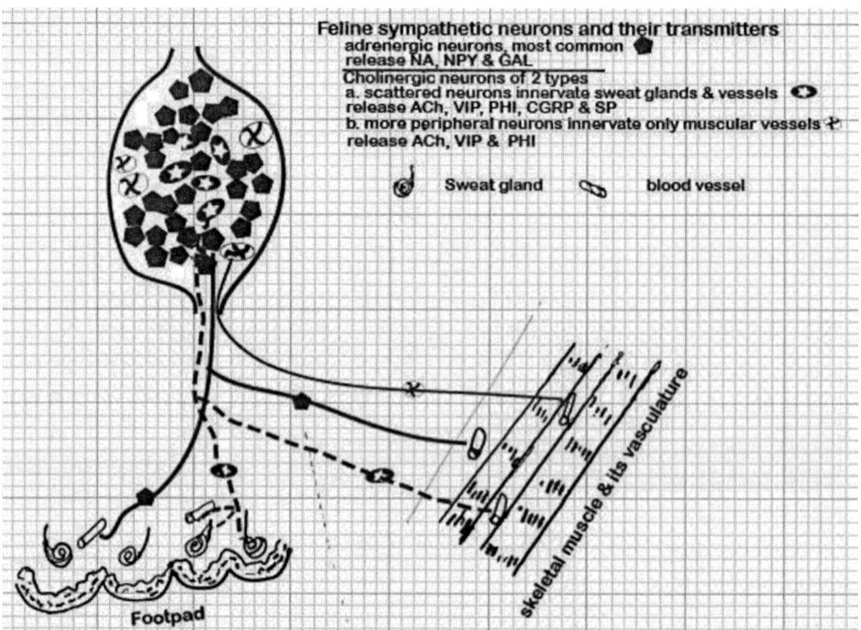

Figure 9. This figure is based on references [134,135,136] and shows that different populations of neurons with different peptide cotransmitters occupy different spaces within the cat lumbar sympathetic ganglion, and project to different regions and structures of the hind paw. Recent studies of Samano et al suggest that peptides and cholinergic transmission machinery may occupy different terminals of the same neuron [136]. The peptide abbreviations may be found in the List of Abbreviations.

GABA and glycine were the first fast acting classical neurotransmitters proven to be co-released from the mammalian CNS. This was initially shown in the neonatal rat spinal cord [137]. Later studies showed that the proportion of GABA and glycine released was a function of developmental age [138]. The demonstration that some GABA/glycine neurons in the medial nucleus of the trapezoid body [MTNB] transiently released glutamate at certain developmental stages was very surprising [139]. The release of both fast acting excitatory and inhibitory transmitters seemed paradoxical. Further study showed that this transient phenomenon, the first recorded glutamate release at a GABA/glycine synapse, coincided with major developmental changes. The synapses changed from depolarizing to hyperpolarizing because of developmental changes in chloride transporters [119], the transmitter phenotype of the MNTB neurons changed from predominantly GABAergic to mostly glycinergic and numerous synapses were eliminated [140].

We also briefly review recently discovered glutamate effects at the cholinergic neuromuscular junction. Claude Bernard studied the action of curare, which was used by South American peoples to paralyze prey. Dale and other early pharmacologists established the neuromuscular junction as a nicotinic cholinergic synapse [13]. However, the neuromuscular junction [NMJ] contains all the apparatus for glutamatergic signaling: glutamate transporters (GLAST and GLT) [141], vesicular glutamate transporter [142], ionotropic and metabotropic [143] glutamate receptors and glutamate itself [144]. This glutamatergic machinery is functional and modulates synaptic transmission [143]. Motor neurons also release glutamate at their central synapses [145]. Both NMDA and AMPA receptors are found post-synaptically at the mouse NMJ. They are not associated with the acetylcholine receptor, and are colocalized with DIg and with Kir2 potassium channels [146].

What is Glutamate's Function at the Neuromuscular Junction?

First, recall that many invertebrates have dual innervation of the neuromuscular junction by inhibitory and excitatory neurons. Next consider the effects of deletions of AMPA and NMDA receptors on NMJ function. Mutations of AMPA and NMDA receptor subunits produce weakness or death, but haven't been studied in detail [146]. There is a cost to producing and precisely targeting these glutamatergic synaptic elements; assuming that they convey functional benefit, but the nature of that benefit is unknown. The general utility of cotransmission remains somewhat ambiguous, in spite of its ubiquity. Authorities often suggest that a co-transmitter may act presynaptically to decrease the release of the primary transmitter, while acting postsynaptically at the same time to increase the efficacy or effect of the primary transmitter [89]. Trudeau and Gutierrez have reviewed the common coexistence of two primary neurotransmitters in the same mammalian neuron, often glutamate and a monoamine, and the ability of this transmitter phenotype to change with injury or disease, even in adult life [70].

Table 2. Synopsis of multiple transmitters and co-transmitters

Thesis	Neurons have one transmitter that never changes
Antithesis	Most neurons have more than one transmitter, this may change during development, disease or neurotrophin treatment
Synthesis	Most neurons have multiple transmitters, it is often possible to say that one is more functionally important than the others
Reconsideration	No change from above

4. VOLUME TRANSMISSION

Descarries reported "naked" CNS axon terminals (lacking adjacent post-synaptic elements) labeled by 3H- noradrenaline and in later studies with 3H- 5-hydroxytryptamine [147]. Only 5 % of the varicosities formed with 5-HT were associated with identifiable synapses [148]. The suggestion of transmitter release from non-synaptic sites led Agnati and Fuxe to construct a theory of volume transmission [149,150]. Jansson and colleagues at Karolinska Institute studied both transmitter releasing 5-HT axon varicosities and 5-HT2A receptor elements identified by electron microscopic immunocytochemistry [151]. They found few close contacts between 5-HT terminal-like varicosities and 5-HT2A-IR neuronal structures, suggesting that 5-HT2A receptor mediated 5-HT transmission in the rat forebrain is mostly volume transmission mediated by short distance diffusion in the extra-cellular space.

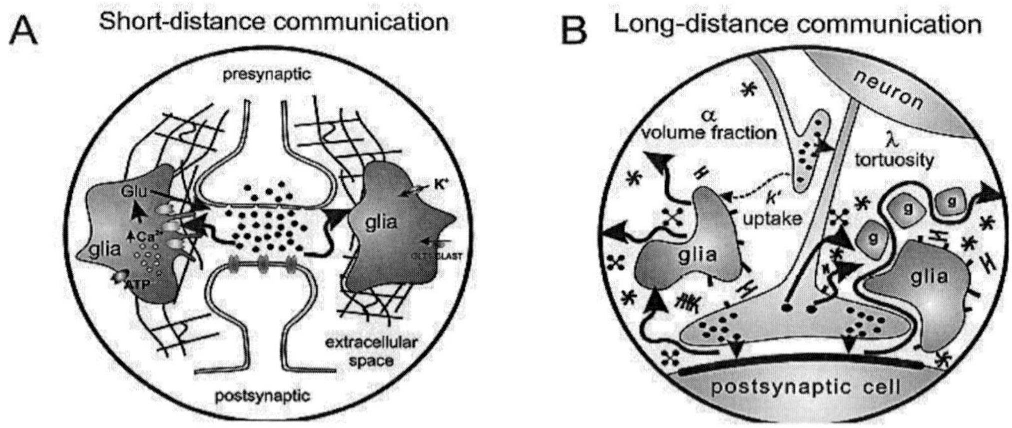

Figure 10. This figure is from reference [164]. Panel A represents a "tripartite synapse" with pre- and postsynaptic elements tightly ensheathed by glial cells and extracellular space shown containing a grid or extracellular matrix. Panel B shows an open synapse, which facilitates communication at a distance via diffusion. The rate of diffusion varies, depending on the ECS diffusion parameters α for volume fraction, λ for tortuosity, and nonspecific uptake κ'. Reproduced with permission from Physiological Research.

This theory of volume transmission attracted supporters [152,153] and detractors [154,155]. Similar nonsynaptic nerve endings have been found in peripheral autonomic ganglia [136] and probably function in some circumstances. Another group of workers focused on the complex and highly organized CNS extracellular space [156] and the extracellular proteoglycan matrix that envelops some synapses [157,158]. These extracellular proteoglycans provide an extracellular anionic cap around certain synapses; manipulations of the proteoglycan matrix alter their function [159]. The extracellular matrix restricts diffusion of synaptic membrane receptors [160,161]. Astrocytes are important barriers to diffusion, retarding "spillover" to extrasynaptic receptors [162]. Glial reactions may change diffusion properties near synapses. The magnitude and significance of volume transmission remain uncertain. Volume transmission may be important in certain regions at certain stages of development and unimportant in others. Exceptions to the standard model continue to be found. Guillery wrote of "a Doctrine of the Neuroscience Church" and said, "today there has to be serious doubt about this" [163] (the centrality of the standard neuron doctrine). Complex CNS networks use fast wiring transmission, including gap junctions, slow synaptic messages and slow volume transmission using extracellular messages possibly influenced by the proteoglycan nets ensheathing certain classes of neurons.

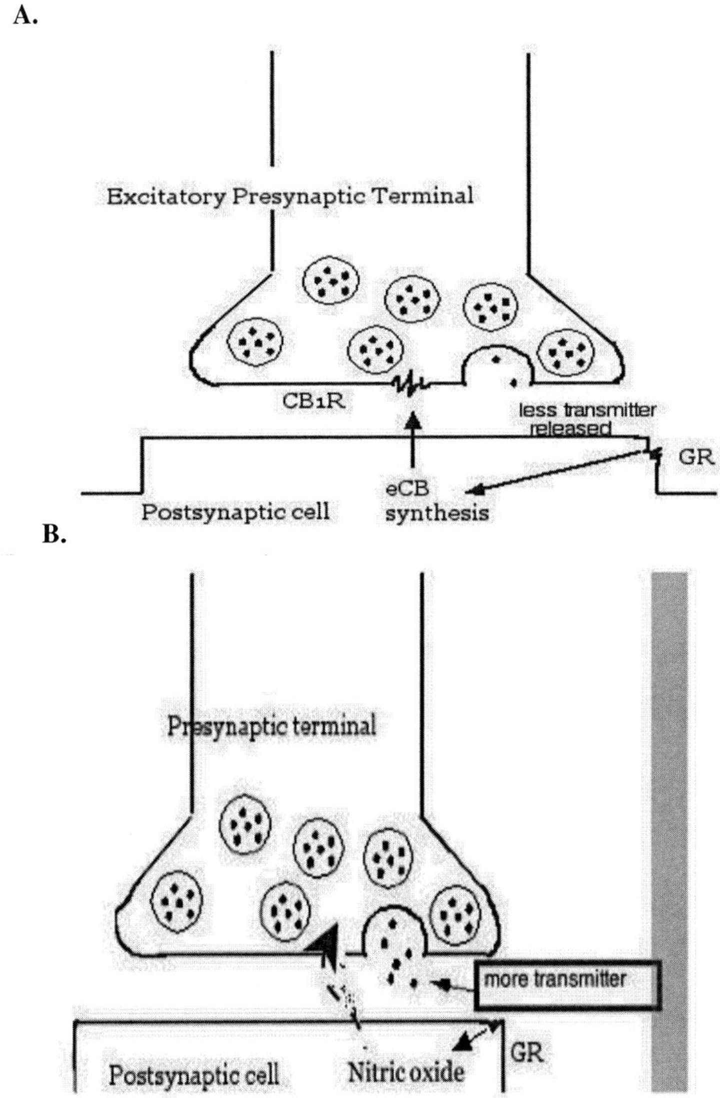

Figure 11. (A) This drawing illustrates mechanisms of glucocorticoid modulation of excitatory hypothalamic synapses, as reported in reference [167]. Glucocorticoids act at GR (glucocorticoid receptor) to stimulate endocannabinoid (eCB) synthesis in the postsynaptic cell. Cannabinoid acts at its receptors (CB1 receptors) on the presynaptic membrane. This reduces the amount of transmitter (glutamate) released. (B) Inhibitory hypothalamic synapse (GABA synapses): GCs act by membrane receptors to activate nNOS in postsynaptic elements, producing nitric oxide which diffuses to presynaptic GABA terminal and stimulates release by as yet undefined pathways.

Retrograde synaptic transmission is now an acknowledged fact. Retrograde transmission and its complexities are sobering. Endocannabinoids modulate both electrical and chemical transmission; 2-AG (2- arachidonoylglycerol) is a well studied retrograde synaptic messenger [165,166] Consider the recently described regulatory effects of glucocorticoids on retrograde transmission in fast amino acid synapses [167] shown in Figure 11.

Fast wiring transmission is dominated by GABA and glutamate neurons. Monoamines and endocannabinoids are important slow transmitters, often acting indirectly to modulate fast synapses. Wiring becomes preponderant over volume transmission in the mature brain, but disease may change this.

Gap junctions are one class of electrical synapses, not readily visualized by electron microscopy [168,169] and are prominent in the developing nervous system. They provide cytoplasmic continuity between cells, often between groups of astrocytes [170]. They often use purinergic messages. Hemichannels (one gap junction includes two channels) discharge messages into the extracellular space [171].

Table 3. Change is the critical variable.

1.	The mammalian body, especially the nervous and immune systems, is far more changeable than foreseen in 1900
2.	Immature cells release messages and respond to external messages, but are poorly differentiated.
3.	Synapses form in excessive numbers, and are continuously remodeled and eliminated throughout life [177]
4.	Learning involves processes similar to long term potentiation. They include changing receptor nature and location, recycling of vesicular machinery and changes in dendritic number and shape. [178, 179]
5.	CNS information processing is influenced by physical injuries, which alter brain extracellular spaces and produce increased numbers of glial cells, as well as influx of bone marrow-derived inflammatory cells.
6.	The nervous system uses both chemical and electrical signals. Each can influence the other- synaptic currents alter chemical transmission and vice versa [180, 181]

How may one distinguish transmitters and neuromodulators? We may be tempted to say that a neurotransmitter is a substance that acts to change the function of ion channels and modulators act at other kinds of receptors. However, the same molecule may act at many different kinds of receptors. Some signals may change macromolecules rather than being recognized by receptors- nitric oxide is a classical example. It can change tyrosine residues to nitrotyrosine and alter the function of important macromolecules [172]. We often use speed of signaling to categorize synaptic transmission [173]. Adenosine, nitric oxide, prostaglandins and CO2 may be considered modulators because they are released at non-synaptic sites and their effects develop more slowly than those of classical transmitters [174]. Slow synaptic potentials, such as the M-current, are well known [175]. Drugs like retigabine, which modulate M currents, may have important roles in treatment of epilepsy, pain and psychiatric disease [176].

5. Discussion

Distinctions between paracrine signals and volume transmission are arbitrary. We can't limit volume transmission to neuronal activities because of the now overwhelming evidence for important signaling functions of glia and other non-neuronal cells found with the nervous system. The list of exceptions to the classical model will surely grow as technology advances.

Have these exceptions and complexities invalidated the neuronal model as described by Cajal and Dale? I think not. Einstein's theory of relativity did not invalidate the use of Newtonian mechanics in many situations. Newton's theory produced a flawed but useful estimate for Mercury's orbit, neglecting precession seen near to the perihelion. Once 19th century astronomers detected this discrepancy, they postulated that this deviation from the expected orbit must reflect the presence of an undetected planet, which they called Vulcan. This followed the reasoning which had led to the discovery of Neptune, because of its effect on the orbit of Uranus. Einstein's theory of general relativity correctly predicted the perihelion shift of Mercury's orbit, because it included the curvature of space-time caused by the sun [182]. The standard model is deficient in certain developmental and pathological situations, but it is usually a good approximation to the findings in normal adult tissues [154]. We need not disparage the work of Cajal and Dale. Both were progressive and functionally oriented. They followed Occam's razor in assuming that neurotransmission should be thought of in the simplest terms possible. However, we have learned in the last fifty years that Occam's razor is a poor guide to synaptic and nervous function.

Table 4. Grand synopsis

1.	Chemical transmission is the primary means of mammalian neuronal communication
2.	It uses many different mechanisms, some poorly understood
3.	Gap junctions transmit both chemical and electrical signals
4.	New technology will continue to reveal new communication modalities, such as tunneling nanotubes [183] and synaptic exosomal transfer of proteins and RNA [184]

References

[1] Swanson, L. W. Quest for the basic plan of nervous system circuitry. *Brain Research Reviews.* 2007;55:356-72.

[2] Schwann T. Mikroskopische *Untersuchungen über die Übereinstimmung in der Struktur und dem Wachstum der Thiere und Pflanzen.* Berlin. Sander'schen Buchhandlung, 1839.

[3] Vulpian EF. *Lecons sur la Physiologie Generale et Comparee du Systeme Nerveux.* Bailliere, Paris, 1866.

[4] DuBois-Reymond, E. (1875-1877), *Gesammette Abhandtungen fur allegememen Muskel--und Nervenphysik* Veil v. Comp., Leipzig, 2 vols.,

[5] von Waldeyer-Hartz HWG. (1891) Uber einigeneuere Forschungen im Gebiete der Anatomie des Centralnervensysystems. *Deutsche Med. Wochenschr 17*,1213–1218, 1244–1246, 1267–1269, 1287–1289, 1331–1332, 1352–1356.

[6] Bennett MVL. Neoreticularism and neuronal polarization. *Progress in Brain Research*. 2002;136: 189-201.

[7] Sherrington CS. *The Integrative Action of the Nervous System*, New Haven, Yale University Press; 1906.

[8] Langley JN. On the reaction of cells and of nerve endings to certain poisons, chiefly as regards the reaction of striated muscle to nicotine and to curari. *Journal of Physiology*. 1905;33: 374–413.

[9] Langley JN. Croonian Lecture: On Nerve Endings and on Special Excitable Substances in Cells. Proceedings of The Royal Society of London, series B, 1906; 78: 170-194.

[10] Elliott T R. On the action of adrenaline. *Journal of Physiology*, 1904; 31:xx–xxi.

[11] Dale HH. Some chemical factors in the control of the circulation. Lancet, 1929; I: 1285-1290.

[12] Feldberg W, Krayer O. Das Auftreten eines Azetylcholinartigen Stoffes im Herzenblut von Warmblüttern bei Reizung der Nervi vagi. *Naunyn-Schmiedeberg's Archiv der experimentelle Pathologie und Pharmakologie*. 1933; 172:170–193.

[13] Dale HH. Pharmacology and Nerve-endings (Walter Ernest Dixon Memorial Lecture). *Proceedings of the Royal Society of Medicine*. 1935;28:319-332.

[14] Dale HH, Gaddum JH. Reactions of denervated voluntary muscle and their bearing on the mode of action of parasympathetic and related nerves. *Journal of Physiology*. 1930;70: 109-144.

[15] Fatt P, Katz B. Spontaneous subthreshold activity at motor nerve endings, *Journal of Physiology*. 1952;115: 320-370.

[16] Katz B. *Nerve, muscle and synapse*. New York, McGraw-Hill; 1966.

[17] Eccles JC. The mechanism of synaptic transmission. *Ergebnisse der Physiologie, biologischen Chemie und experimentellen Pharmakologie*. 1961;31: 299-430.

[18] Bacq ZM. The metabolism of adrenaline. *Pharmacological Reviews*. 1949;1: 1-26.

[19] Pletscher A, Shore PA, Brodie BB. Serotonin release as a possible mechanism of reserpine action. *Science*. 1955;122: 374-375.

[20] Carlsson A, Lindqvist M, Magnusson T. 3,4-Dihydroxyphenylalanine and 5-hydroxytryptophan as reserpine antagonists. *Nature*. 1957;180:1200..

[21] Carlsson A, Lindqvist M, Magnusson T, Waldeck B. On the presence of 3-hydroxytyramine in brain. *Science*. 1958;127:471.

[22] Carlsson A. Drugs which block the storage of 5-hydroxytryptamine and related amines. *Handbook of Experimental Pharmacology*. Berlin, Springer-Verlag. 1965;19:529-592.

[23] Kirshner N. Uptake of catecholamines by a particulate fraction of the adrenal medulla. *Journal of Biological Chemistry*. 1962;237: 2311-2317.

[24] Parsons SM. Transport mechanisms in acetylcholine and monoamine storage. *FASEB Journal*. 2000;14:2423-2434

[25] Frey KA, Koeppe RA, Kilbourn MR, Vander Borght TM, Albin RL, Gilman S, Kuhl D E. Presynaptic monoaminergic vesicles in Parkinson's disease and normal aging. *Annals of Neurology*. 1996;40:873-84.

[26] Tong J, Wilson AA, Boileau I, Houle S, Kish SJ. Dopamine modulating drugs influence striatal (+)-[11C]DTBZ binding in rats: VMAT2 binding is sensitive to changes in vesicular dopamine concentration. *Synapse*. 2008;62:873-6.

[27] Adams JR, van Netten H, Schulzer M, Mak E, Mckenzie J, Strongosky A., Sossi V, Ruth TJ, Lee CS, Farrer M, Gasser T, Uitti RJ, Calne DB, Wszolek ZK, Stoessl AJ. PET in LRRK2 mutations: comparison to sporadic Parkinson's disease and evidence for presymptomatic compensation. *Brain*. 2005;128: 2777-2785.

[28] Nandhagopal R, Mak E, Schulzer M, McKenzie J, McCormick S, Sossi V, Ruth T J, Strongosky A, Farrer M J, Wszolek ZK, Stoessl AJ. Progression of dopaminergic dysfunction in a LRRK2 kindred: a multitracer PET study. *Neurology*. 2008;71:1790-5.

[29] Taylor TN, Caudle WM, Shepherd KR, Noorian A, Jackson CR, Iuvone PM, Weinshenker D, Greene JG, Miller GW. Nonmotor symptoms of Parkinson's disease revealed in an animal model with reduced monoamine storage capacity. *Journal of Neuroscience*. 2009;29: 8103-13.

[30] Henry JP, Sagné C, Bedet C, Gasnier B. The vesicular monoamine transporter: from chromaffin granule to brain. *Neurochemistry International*.1998;32:227-46.

[31] Edwards RH. The neurotransmitter cycle and quantal size. Neuron. 2007;55:835-858.

[32] Pothos EN, Davila V, Sulzer D. Presynaptic recording of quanta from midbrain dopamine neurons and modulation of quantal size. *Journal of Neuroscience*. 1998;18:4106–4118.

[33] Del Castillo J, Katz B. Quantal components of the end-plate potential. *Journal of. Physiology*. 1954;124: 560-573.

[34] Pothos EN. Regulation of dopamine quantal size in midbrain and hippocampal neurons. *Behavioral Brain Research*. 2002;130:203-207

[35] Brunk I, Höltje M, von Jagow B, Winter S, Sternberg J, Blex C, Pahner I, Ahnert-Hilger G. Regulation of vesicular monoamine and glutamate transporters by vesicle-associated trimeric G proteins: new jobs for long-known signal transduction molecules. *Handbook of Experimental Pharmacology*. Berlin, Springer-Verlag. 2006; 175:305-325.

[36] Brunk I, Blex C, Speidel D, Brose N, Ahnert-Hilger G. Ca^{2+}-dependent activator proteins of secretion promote vesicular monoamine uptake. *Journal of Biological Chemistry*. 2009; 284:1050-1056

[37] Brunk I, Blex C, Sanchis-Segura C, Sternberg J, Perreau-Lenz S, Bilbao A, Hörtnagl H, Baron J, Juranek J, Laube G, Birnbaumer L, Spanagel R, Ahnert-Hilger G. Deletion of Go2alpha abolishes cocaine-induced behavioral sensitization by disturbing the striatal dopamine system. *FASEB Journal*. 2008; 22:3736-46.

[38] Iversen LL. Neurotransmitter transporters and their impact on the development of psychopharmacology. *British Journal of Pharmacology*. 2006;147 Suppl 1:S82-88.

[39] Carlsson A, Lindqvist M. In vivo decarboxylation of α-methyl-Dopa and α-methyl-meta-tyrosine. *Acta Physiologica Scandanavica*. 1962;54:87-94..

[40] Day MD, Rand MJ. A hypothesis for the mode of action of a-methyl-dopa in relieving hypertension. *Journal of Pharmacy and. Pharmacology*. 1963;15:221-224.

[41] Muscholl E, Maitre L. Release by sympathetic stimulation of a-methylnoradrenaline stored in the heart after administration of a-methyldopa. *Experientia*. 1963;19:658-659.

[42] Haefely W, Hürlimann A, Thoenen H. Adrenergic transmitter changes and response to sympathetic nerve stimulation after differing pretreatment with alpha-methyldopa. *British Journal of Pharmacology and Chemotherapy*. 1967;31:105-19

[43] Udenfriend S, Zaltzman-Nirenberg P. On the mechanism of the norepinephrine release produced by a-methyl-meta-tyrosine. *Journal of Pharmacology and Experimental Therapeutics*. 1962;138:194-199.

[44] Crout JR, Alpers HS, Tatum EL, Shore PA. Release of metaraminol (Aramine) from the heart by sympathetic nerve stimulation. *Science*. 1964;145: 828-829.

[45] Kopin IJ, Fischer JE, Musacchio J, Horst WD. Evidence for a false neurochemical transmitter as a mechanism for the hypotensive effect of monoamine oxidase inhibitors. *Proceedings of the National Academy of Sciences, U. S. A*. 1964;52: 716-721.

[46] Kopin IJ. False adrenergic transmitters. *Annual Review of Pharmacology*. 1968; 8:377-94.

[47] Evans PD, Kravitz EA, Talamo BR, Wallace BG. The association of octopamine with specific neurones along lobster nerve trunks. *Journal of Physiology*. 1976; 262:51-70

[48] Evans PD. Octopamine: from metabolic mistake to modulator. *Trends in the Neurosciences*. 1978;1: 154-157.

[49] Roeder T. Octopamine in invertebrates. *Progress in Neurobiology*. 1999; 59:533-61.

[50] Farooqui, T. Octopamine-mediated neuronal plasticity in honeybees: implications for olfactory dysfunction in humans. *Neuroscientist*. 2007;13:304-22

[51] Mousseau DD, Butterworth RF. Trace amines in hepatic encephalopathy. *Progress in Brain Research*. 1995;106:277-84.

[52] Bover-Cid S, Izquierdo-Pulido M, Vidal-Carou MC. Influence of hygienic quality of raw materials on biogenic amine production during ripening and storage of dry fermented sausages. *Journal of Food Protection*. 2000;63:1544-50.

[53] Borowsky B, Adham N, Jones KA, Raddatz R, Artymyshyn R, Ogozalek KL, Durkin MM, Lakhlani PP, Bonini JA, Pathirana S, Boyle N, Pu X, Kouranova E, Lichtblau H, Ochoa FY, Branchek TA, Gerald, C. Trace amines: identification of a family of mammalian G protein-coupled receptors. *Proceedings of the National Academy of Science U. S. A*. 2001;98:8966-8971.

[54] Zucchi R, Chiellini G, Scanlan TS, Grandy DK. Trace amine associated receptors and their ligands. *British Journal of Pharmacology*. 2006;149:967-78

[55] Lindemann, L, Meyer CA, Jeanneau K, Bradaia A, Ozmen L, Bluethmann H, Bettler B, Wettstein JG, Borroni E, Moreau JL, Hoener MC. Trace amine-associated receptor 1 modulates dopaminergic activity. *Journal of Pharmacology and Experimental Therapeutics*. 2008;324:948-56.

[56] Herbert AA, Kidd EJ, Broadley KJ. Dietary trace amine-dependent vasoconstriction in porcine coronary artery. *British Journal of Pharmacology*. 2008;155:525-34.

[57] Sotnikova TD, Caron MG, Gainetdinov RR. Trace amine associated receptors as emerging therapeutic targets. *Molecular Pharmacology*. 2009; In press.

[58] Ellender TJ, Skinner JC. The use of vasopressors and inotropes in the emergency medical treatment of shock. *Emergency Medicine Clinics of North America*. 2008;26:759-86

[59] Smith NT, Corbascio AN. The use and misuse of pressor agents. *Anesthesiology*. 1970;33:58-101.

[60] Owman C. Sympathetic nerves probably storing two types of monoamines in the rat pineal gland. *International Journal of Neuropharmacology*. 1964; 2:105-112.

[61] Bertler A, Falck B, Owman C. Studies on 5-hydroxytryptamine stores in pineal gland of rat. *Acta Physiologica Scandanavica Supplementum*. 1964;Suppl. 239: 1-18.

[62] Zweig M, Axelrod J. Relationship between catecholamines and serotonin in sympathetic nerves of the rat pineal gland. *Journal of Neurobiology*. 1969;1:87-97.

[63] Jaim-Etcheverry G, Zieher LM. Octopamine probably coexists with noradrenaline and serotonin vesicles of pineal adrenergic nerves. *Journal of Neurochemistry*. 1975;25:915-917.

[64] Buck SHH, Murphy RC, Molinoff PB. The normal occurrence of octopamine in the central nervous system of the rat. *Brain Research*. 1977;122: 281-297.

[65] Pashankar FD, O'Dorisio MS, Menda Y. MIBG and somatostatin receptor analogs in children: current concepts on diagnostic and therapeutic use. *Journal of Nuclear Medicine*. 2005;46 Suppl 1:55S-61S

[66] Jaques S Jr, Tobes MC, Sisson JC, Baker JA, Wieland DM. Comparison of the sodium dependency of uptake of meta-iodobenzylguanidine and norepinephrine into cultured bovine adrenomedullary cells. *Molecular Pharmacology*. 1984;26:539 –546.

[67] Smets LA, Loesberg C, Janssen M, Metwally EA, Huiskamp R. Active uptake and extravesicular storage of m-iodobenzylguanidine in human neuroblastoma cells. *Biochemical Pharmacology*. 1989;39:1959-1964.

[68] Sisson JC, Frager MS, Valk TW, Gross MD, Swanson DP, Wieland DM, Tobes MC, Beierwaltes WH, Thompson NW. Scintigraphic localization of pheochromocytoma. *New England Journal of Medicine*. 1981;305:12–17.

[69] Gubernator NG, Zhang H, Staal RG, Mosharov EV, Pereira DB, Yue M, Balsanek V, Vadola PA, Mukherjee B, Edwards RH, Sulzer D, Sames D. Fluorescent false neurotransmitters visualize dopamine release from individual presynaptic terminals. *Science*. 2009;324:1441-4.

[70] Trudeau LE, Gutiérrez R. On cotransmission and transmitter phenotype plasticity. *Molecular Interventions*. 2007;7:138-146.

[71] Bum JH, Rand M. J. Sympathetic postganglionic mechanisms. *Nature*. 1959; 184:163-65

[72] Furshpan E, MacLeish PR, O'Lague PH, Potter DD. Chemical transmission between rat sympathetic neurons and cardiac myocytes developing in microcultures: evidence for cholinergic, adrenergic, and dual-function neurons. *Proceedings of the National Academy of Science U S A*. 1976;73:4225-4229.

[73] Furshpan EJ, Potter DD, Landis SC. On the transmitter repertoire of sympathetic neurons in culture. *Harvey Lectures*. 1982;76:149-91

[74] Landis SC, Keefe D. Evidence for neurotransmitter plasticity in vivo: developmental changes in properties of cholinergic sympathetic neurons. *Developmental. Biology*. 1983;98:349-72

[75] Rao MS, Landis SC. Cell interactions that determine sympathetic neuron transmitter phenotype and the neurokines that mediate them. *Journal of Neurobiology*. 1993;24:215-32.

[76] Nakatani T, Minaki Y, Kumai M, Ono Y. Helt determines GABAergic over glutamatergic neuronal fate by repressing Ngn genes in the developing mesencephalon. *Development*. 2007;134:2783-93.

[77] Burn JH. The development of the adrenergic fibre, *British Journal of. Pharmacology*. 1978;32: 575–582.

[78] Ferry CB. Cholinergic link hypothesis in adrenergic neuroeffector transmission. *Physiological. Reviews*. 1966;46:420-56

[79] Blaschko H, Born GVR, D'iorio A, Eade NR. Observations on the distribution of catecholamines and adenosinetriphosphate in the bovine adrenal medulla. *Journal of Physiology*. 1956;133:548-557

[80] Douglas WW, Poisner AM. On the relation between ATP splitting and secretion in the adrenal chromaffin cell: extrusion of ATP (unhydrolysed) during release of catecholamines. *Journal of Physiology*. 1966;183:249-56

[81] Brownstein MJ, Saavedra J.M, Axelrod J, Zeman GH, Carpenter DO. Coexistence of several putative neurotransmitters in single identified neurons of Aplysia. *Proceedings of the National. Academy of Science. U. S. A*. 1974;71:4662-4665.

[82] Burnstock G. Do some nerve cells release more than one transmitter? *Neuroscience*, 1976;1: 239-248.

[83] Osborne NN. Is Dale's principle valid? *Trends in Neurosciences*. 1979;2: 73-75.

[84] Abrahams VC, Koelle GB, Smart P. Histochemical demonstration of cholinesterases in the hypothalamus of the dog. *Journal of. Physiology*. 1957; 139:137-144.

[85] Holton P. The liberation of adenosine triphosphate on antidromic stimulation of sensory nerves. *Journal of Physiology*. 1959;145: 494-504.

[86] Burnstock G, Campbell G, Satchell D, Smythe A. Evidence that adenosine triphosphate or a related nucleotide is the transmitter substance released by non-adrenergic inhibitory nerves in the gut. *British Journal of Pharmacology*. 1970; 40: 668-688.

[87] Su C, Bevan JA, Burnstock G. [^{3}H]adenosine triphosphate: release during stimulation of enteric nerves. *Science*. 1971;173: 336-338.

[88] Stone TW. Physiological roles for adenosine and adenosine 5'- triphosphate in the nervous system. *Neuroscience*. 1981;6: 523-555,

[89] Kupfermann I. Functional studies of cotransmission. *Physiological Reviews*. 1991;71:683-732.

[90] Edwards FA, Gibb AJ. ATP – a fast neurotransmitter. *FEBS Letters*. 1993;325: 86-89.

[91] Evans RJ, Derkach V, Surprenant A. ATP mediates fast synaptic transmission in mammalian neurons. *Nature*. 1992;357 503-505.

[92] Silinsky EM, Gerzanich V. On the excitatory effects of ATP and its role as a neurotransmitter in coeliac neurons of the guinea-pig. *Journal of Physiology*. 1993;464: 197-212.

[93] Wier WG, Zang WJ, Lamont C, Raina H. Sympathetic neurogenic Ca2+ signalling in rat arteries: ATP, noradrenaline and neuropeptide Y. *Experimental Physiology*. 2009;94:31-7.

[94] Ralevic R. Purines as neurotransmitters and neuromodulators in blood vessels. *Current Vascular Pharmacology*. 2009;7:3-14.

[95] Burnstock G. Noradrenaline and ATP: cotransmitters and neuromodulators. *Journal of Physiology and Pharmacology*. 1995;46 365-84

[96] Smith KM, Macmillan JB, McCulloch KM, McGrath JC. NOS inhibition potentiates norepinephrine but not sympathetic nerve-mediated co-transmission in resistance arteries. *Cardiovascular Research*. 1999;43:762-71.

[97] Dunwiddie TV, Masino SA. The role and regulation of adenosine in the central nervous system. *Annual Review of Neuroscience*. 2001;24: 31-55.

[98] Ciruela F, Casadó V, Rodrigues RJ, Luján R, Burgueño J, Canals M, Borycz J, Rebola N, Goldberg SR, Mallol J, Cortés A, Canela EI, López-Giménez JF, Milligan G, Lluis C, Cunha RA, Ferré S, Franco R. Presynaptic control of striatal glutamatergic neurotransmission by adenosine A1-A2A receptor heteromers. *Journal of Neuroscience*. 2006;26: 2080-7.

[99] Ferré S. An update on the mechanisms of the psychostimulant effects of caffeine. *Journal of Neurochemistry*. 2008;105:1067-79.

[100] Schotanus SM, Fredholm BB, Chergui K. NMDA depresses glutamatergic synaptic transmission in the striatum through the activation of adenosine A1 receptors: evidence from knockout mice. *Neuropharmacology*. 2006;51:272-82.

[101] Fredholm, B. B., Chen, J.F., Cunha, R.A., Svenningsson, P., Vaugeois, J.M.. Adenosine and brain function. *International Review of Neurobiology*. 2005;63: 191–270.

[102] Chen, J. F., Sonsalla, P. K., Pedata, F., Melani, A., Domenici, M. R., Popoli, P., Geiger, J., Lopes, L., de Mendonça, A. Adenosine A2A receptors and brain injury: broad spectrum of neuroprotection, multifaceted actions and "fine tuning" modulation. *Progress in Neurobiology*. 2007;83:310-31.

[103] Oberheim NA, Wang X, Goldman S, Nedergaard M. Astrocytic complexity distinguishes the human brain. *Trends in Neuroscience*. 2006;29:547–553.

[104] Takano T, Oberheim N, Cotrina ML, Nedergaard M. Astrocytes and ischemic injury. *Stroke*. 2009;40 (3 Suppl):S8-12.

[105] von Euler US, Gaddum JH. An unidentified depressor substance in certain tissue extracts. *Journal of Physiology*. 1931;72:74-87

[106] Dale HH, Dudley HW. The presence of histamine and acetylcholine in the spleen of the ox and the horse. *Journal of Physiology*. 1929;68:97-123

[107] Hellauer HF, Umrath K. Uber die Aktionssubstanz der sensiblen Nerven. *Pflüger's Archiv für die gesamte Physiologie des Menschen und der Tiere*. 1948; 249:619-30

[108] Angelucci L. Experiments with perfused frog's spinal cord. *British Journal of Pharmacology*. 1956;11:161-170.

[109] Lembeck F, Zetler G. Substance P: A polypeptide of possible physiological significance, especially within the nervous system. *International Review of. Neurobiology*. 1962;4:159- 215

[110] Krnjevic K, Morris ME. An excitatory action of substance P on cuneate neurones. *Canadian Journal Physiology and Pharmacology*. 1974;52:736-44

[111] Hökfelt T, Kellerth JO, Nilsson G, Pernow B. Substance P: localization in the central nervous system and in some primary sensory neurons. *Science*. 1976;190: 889-890.

[112] Iversen LL, Jessell T, Kanazawa I. Release and metabolism of substance P in rat hypothalamus. Nature. 1976;264:8 1-83.

[113] Jessell TM, Iversen LL. Opiate analgesics inhibit substance P release from rat trigeminal nucleus. *Nature*. 1977;268:549-51

[114] Ljungdahl A, Hökfelt T, Nilsson G, Goldstein M. Distribution of substance P-like immunoreactivity in the central nervous system of the rat. II. Light microscopic localization in relation to catecholamine-containing neurons. *Neuroscience*. 1978;3:945-976

[115] Otsuka M, Konishi S. Substance P and excitatory transmitter of primary sensory neurons. *Cold Spring Harbor Symposium on Quantitative Biology*. 1976;40:135-143.

[116] Cuello AC, Jessell TM, Kanazawa I, Iversen LL. Substance P: Localization in synaptic vesicles in rat central nervous system. *Journal of Neurochemistry*. 1977; 29:747-751.

[117] Melzack R, Wall PD. Pain mechanisms: a new theory. *Science*. 1965;150:971-979.

[118] Todd AJ. Anatomy of primary afferents and projection neurones in the rat spinal dorsal horn with particular emphasis on substance P and the neurokinin 1 receptor. *Experimental Physiology*. 2002;87:245–249.

[119] Price TJ, Cervero F, Gold MS, Hammond DL, Prescott SA. Chloride regulation in the pain pathway. *Brain Research Reviews*. 2009;60:149-70.

[120] Drew GM, Lau BK, Vaughan CW. Substance P drives endocannabinoid-mediated disinhibition in a midbrain descending analgesic pathway. *Journal of Neuroscience*. 2009;29:7220-7229.

[121] Eaton K, Sallee FR, Ash R. Relevance of neuropeptide Y (NPY) in psychiatry. *Current Topics in Medicinal Chemistry*. 2007;7:1645-1659.

[122] Tatemoto K, Carlquist M, Mutt V. Neuropeptide Y: a novel brain peptide with structural similarities to peptide YY and pancreatic polypeptide. *Nature*. 1982; 296:659-60

[123] Lundberg JM, Terenius L, Hökfelt T, Goldstein M. High levels of neuropeptide Y in peripheral noradrenergic neurons in various mammals including man. *Neuroscience Letters*. 1983;42:167-172.

[124] Hökfelt T, Lundberg JM, Lagercrantz H, Tatemoto K, Mutt V, Lindberg J, Terenius L, Everitt BJ, Fuxe K, Agnati L, Goldstein M. Occurrence of neuropeptide Y (NPY)-like immunoreactivity in catecholamine neurons in the human medulla oblongata. *Neuroscience Letters*. 1983;36: 217-22.

[125] Allen YS, Adrian TE, Allen JM, Tatemoto K, Crow TJ, Bloom SR, Polak JM. Neuropeptide Y distribution in the rat brain. *Science*. 1983;221: 877–879.

[126] Zukowska-Grojec Z. Neuropeptide Y. A novel sympathetic stress hormone and more. *Annals of the New York Academy of Sciences*. 1995;771:219-33.

[127] Clark JT, Kalra PS, Crowley WR, Kalra SP. Neuropeptide Y and human pancreatic polypeptide stimulate feeding behavior in rats. *Endocrinology*. 1984; 115:427-9.

[128] Erickson JC, Hollopeter G, Palmiter RD. Attenuation of the obesity syndrome of ob/ob mice by the loss of neuropeptide Y. *Science*. 1996;274: 1704–1707.

[129] Kalra SP, Kalra PS. To subjugate NPY is to improve the quality of life and live longer. *Peptides*. 2007;28:413-418.

[130] Volkoff H. The role of neuropeptide Y, orexins, cocaine and amphetamine-related transcript, cholecystokinin, amylin and leptin in the regulation of feeding in fish. *Comparative Biochemistry and Physiology A Molecular and Integrative Physiology*. 2006;144: 325-31.

[131] Myslobodsky M. Multitasking of neuropeptide Y through the lens of motifs. *Recent Patterns of CNS Drug Discovery*. 2009;4:52-60.

[132] Zahn PK, Straub H, Wenk M, Pogatzki-Zahn EM. Adenosine A1 but not A2a receptor agonist reduces hyperalgesia caused by a surgical incision in rats: a pertussis toxin-sensitive G protein-dependent process. *Anesthesiology*. 2007; 107:797-806.

[133] Peters A, Pellerin L, Dallman MF, Oltmanns KM, Schweiger U, Born J, Fehm H. L. Causes of obesity, looking beyond the hypothalamus. *Progress in Neurobiology*. 2007;81:61-88.

[134] Elfvin LG, Lindh B, Hökfelt T. The chemical neuroanatomy of sympathetic ganglia. *Annual Review of Neuroscience*. 1993;16:471-507

[135] Sámano C, Zetina ME, Marín MA, Cifuentes F, Morales MA. Choline acetyl transferase and neuropeptide immunoreactivities are colocalized in somata, but preferentially localized in distinct axon fibers and boutons of cat sympathetic preganglionic neurons. *Synapse*. 2006;60: 295-306.

[136] Lindh BH, Lundberg JM, Hökfelt T. NPY-, Galanin-, VIP/PHI-, CGRP- And Substance P-Immunoreactive Neuronal Subpopulations In Cat Autonomic And Sensory Ganglia And Their Projections. *Cell and Tissue Research*. 1989;256:259-73.

[137] Jonas P, Bischofberger J, Sandkühler J. Corelease of two fast transmitters at a central synapse. *Science*. 1998;281: 419-24.

[138] Nabekura J, Katsurabayashi S, Kakazu Y, Shibata S, Matsubara A, Jinno S, Mizoguchi Y, Sasaki A, Ishibashi H. Developmental switch from GABA to glycine release in single central synaptic terminals. *Nature Neuroscience*. 2004; 7:17-23.

[139] Gillespie DC, Kim G, Kandler K. Inhibitory synapses in the developing auditory system are glutamatergic. *Nature Neuroscience*. 2005;8:332-8.

[140] Nishimaki T, Jang IS, Ishibashi H,Yamaguchi J, Nabekura J. Reduction of metabotropic glutamate receptor-mediated heterosynaptic inhibition of developing MNTB-LSO inhibitory synapses. *European Journal of Neuroscience*. 2007; 26:323-30.

[141] Rinholm JE, Slettalokken G, Marcaggi P, Skare O, Storm-Mathisen J, Bergersen LH. Subcellular localization of the glutamate transporters GLAST and GLT at the neuromuscular junction in rodents. *Neuroscience*. 2007; 145:579-591.

[142] Boulland JL, Qureshi T, Seal RP, Rafiki A, Gundersen V, Bergersen LH, Fremeau RT Jr, Edwards RH, Storm-Mathisen, J, Chaudhry FA. Expression of the vesicular glutamate transporters during development indicates the widespread corelease of multiple neurotransmitters. *Journal of. Comparative Neurology*. 2004;480:264-280.

[143] Pinard A, Lévesque S, Vallée J, Robitaille R. Glutamatergic modulation of synaptic plasticity at a PNS vertebrate cholinergic synapse. *European Journal of Neuroscience*. 2003;18:3241-50.

[144] Waerhaug O, Ottersen OP. Demonstration of glutamate-like immunoreactivity at rat neuromuscular junctions by quantitative electron microscopic immunocytochemistry. *Anatomy and Embryology*. (1993) 188:501-13.

[145] Nishimaru H, Restrepo CE, Ryge J, Yanagawa Y, Kiehn O. Mammalian motor neurons corelease glutamate and acetylcholine at central synapses. *Proceedings of the National Academy of. Science*. U.S.A. 2005;102: 5245–5249.

[146] Mays TA, Sanford JL, Hanada T, Chishti AH, Rafael-Fortney JA. Glutamate receptors localize postsynaptically at neuromuscular junctions in mice. *Muscle and Nerve*. 2009;39:343-9

[147] Beaudet A, Descarries L. The monoamine innervation of rat cerebral cortex: synaptic and nonsynaptic axon terminals. *Neuroscience*. 1978;3: 851-860.

[148] Descarries L, Beaudet A, Watkins KC. Serotonin nerve terminals in adult rat neocortex. *Brain Research*. 1975; 100: 563-58.

[149] Agnati LF, Fuxe K. Volume transmission as a key feature of information handling in the central nervous system possible new interpretative value of the Turing's B-type machine. *Progress in Brain Research*. 2000;125:3-19.

[150] Zoli M, Torri C, Ferrari R, Jansson A, Zini I, Fuxe K, Agnati LF. The emergence of the volume transmission concept. *Brain Research Reviews*. 1998;26:136-47.

[151] Jansson A, Tinner B, Bancila M, Vergé D, Steinbusch HW, Agnati LF, Fuxe K. Relationships of 5-hydroxytryptamine immunoreactive terminal-like varicosities to 5-hydroxytryptamine-2A receptor-immunoreactive neuronal processes in the rat forebrain. *Journal of Chemical Neuroanatomy*. 2001;22:185-203.

[152] Freeman WJ. Ndn, volume transmission, and self-organization in brain dynamics. *Journal of Integrative Neuroscience*. 2005;4:407-21.

[153] Sørensen AT, Kanter-Schlifke I, Lin E J, During MJ, Kokaia M. Activity-dependent volume transmission by transgene NPY attenuates glutamate release and LTP in the subiculum. *Molecular and Cellular Neuroscience*. 2008;39:229-237.

[154] Bloom FE. Integration of wiring transmission and volume transmission. *Progress in Brain Research*. 2000;125: 21-26

[155] Sarter M, Parikh V, Howe WM. Phasic acetylcholine release and the volume transmission hypothesis: time to move on. *Nature Reviews Neuroscience*. 2009; 10:383-90.

[156] Sykova E, Nicholson C. Diffusion in brain extracellular space. *Physiological Reviews*. 2008;88:1277-1340.

[157] Brückner G, Szeöke S, Pavlica S, Grosche J, Kacza J. Axon initial segment ensheathed by extracellular matrix in perineuronal nets. *Neuroscience*. 2006;138: 365-375

[158] Koeppe G, Bruckner G, Brauer K, Hartig W, Bigl V. Developmental patterns of proteoglycan-containing extracellular matrix in perineuronal nets and neuropil of the postnatal rat brain. *Cell and Tissue Research*. 1997;288:33-41

[159] Brückner G, Brauer K, Härtig W, Wolff JR, Rickmann MJ, Derouiche A, Delpech B, Girard N, Oertel WH, Reichenbach A. Perineuronal nets provide a polyanionic, glia-associated form of microenvironment around certain neurons in many parts of the rat brain. *Glia*. 1993;8:183-200.

[160] Viggiano D, Ibrahim M, Celio MR. Relationship between glia and the perineuronal nets of extracellular matrix in the rat cerebral cortex: importance for volume transmission in the brain. *Progress in Brain Research*. 2000;125:193-198.

[161] Frischknecht R, Heine M, Perrais M, Seidenbecher CI, Choquet D. Gundelfinger ED. Brain extracellular matrix affects AMPA receptor lateral mobility and short-term synaptic plasticity. *Nature Neuroscience*. 2009;12:897-904.

[162] Piet R, Vargová L, Syková E, Poulain DA, Oliet SH. Physiological contribution of the astrocytic environment of neurons to intersynaptic crosstalk. *Proceedings of the National Academy of Science U S A*. 2004;101:2151-2155.

[163] Guilery RW. Observations of synaptic structures: origins of the neuron doctrine and its current status. *Philosophical Transactions of the Royal Society of London, series B Biological Sciences*. 2005;360:1281-307.

[164] Vargova L, Sykova E. Extracellular space diffusion and extrasynaptic transmission. *Physiological Research*. 2008;57 (suppl. 3): S89-S99.

[165] Cachope R, Mackie K, Triller A, O'Brien J, Pereda AE. Potentiation of electrical and chemical synaptic transmission mediated by endocannabinoids. *Neuron*. 2007;56:1034-47.

[166] Kano M, Ohno-Shosaku T, Hashimotodani Y, Uchigashima M, Watanabe M. Endocannabinoid-mediated control of synaptic transmission. *Physiological Reviews*. 2009;89:309-380.

[167] Di S, Maxson MM, Franco A, Tasker JG. Glucocorticoids regulate glutamate and GABA synapse-specific retrograde transmission via divergent nongenomic signaling pathways. *Journal of Neuroscience*. 2009;29393-401.

[168] Fukuda T. Structural organization of the gap junction network in the cerebral cortex. *Neuroscientist*. 2007;13:199-207.

[169] Bruzzone R, Dermietzel R. Structure and function of gap junctions in the developing brain. *Cell and Tissue Research*. 2006;326:239-248.

[170] Orthmann-Murphy JL, Abrams CK, Scherer SS. Gap junctions couple astrocytes and oligodendrocytes. *Journal of Molecular Neuroscience*. 2008;35:101-116.

[171] Rana S, Dringen R. Gap junction hemichannel-mediated release of glutathione from cultured rat astrocytes. *Neuroscience Letters*. 2007;415:45-8.

[172] Nott A, Watson PM, Robinson JD, Crepaldi L, Riccio A. S-Nitrosylation of histone deacetylase 2 induces chromatin remodelling in neurons. *Nature*. 2008;455: 411-415.

[173] Greengard P. The Neurobiology of Slow Synaptic Transmission. *Science*. 2001; 294:1024-1030.

[174] Iversen LL, Iversen SD, Bloom FE, Roth RH. *Introduction to Neuropsychopharmacology*. Oxford, New York, 2009.

[175] Schroeder BC, Hechenberger M, Weinreich F, Kubisch C, Jentsch TJ. KCNQ5, a novel potassium channel broadly expressed in brain, mediates M-type currents. *Journal of Biological Chemistry*. 2000;275:24089-24095

[176] Gribkoff VK. The therapeutic potential of neuronal KCNQ channel modulators. *Expert Opinion on Therapeutic Targets*. 2003;7:737-748.

[177] De Roo M, Klauser P, Garcia PM, Poglia L, Muller D. Spine dynamics and synapse remodeling during LTP and memory processes. *Progress in Brain Research.* 2008;169:199-207.

[178] Muller D, Nikonenko I, Jourdain P, Alberi S. LTP, Memory and structural plasticity. *Current Molecular Medicine.* 2002;2:605-11.

[179] Dalton GL, Wang YT, Floresco SB, Phillips, AG. Disruption of AMPA receptor endocytosis impairs the extinction, but not acquisition of learned fear. *Neuropsychopharmacology.* 2008;33: 2416-26

[180] Li Q, Burrell BD. CNQX and AMPA inhibit electrical synaptic transmission: a potential interaction between electrical and glutamatergic synapses. *Brain Research,* 2008;1228:43-57.

[181] Sylantyev S, Savtchenko LP, Niu YP, Ivanov AI, Jensen TP, Kullmann DM, Xiao MY, Rusakov DA. Electric fields due to synaptic currents sharpen excitatory transmission. *Science,* 2008;319:1845–1849.

[182] Kuhn KF, Koupelis T. *In Quest of the Universe, 5th edition,* Jones & Bartlett Publishers, Sudbury, MA; 2007.

[183] Gousset K, Schiff E, Langevin C., Marijanovic Z, Caputo A, Browman DT, Chenouard N, de Chaumont F, Martino A, Enninga J, Olivo-Marin JC, Männel D, Zurzolo C. Prions hijack tunnelling nanotubes for intercellular spread. *Nature Cell Biology.* 2009;11:328-36.

[184] Smalheiser NR. Exosomal transfer of proteins and RNAs at synapses in the nervous system. *Biology Direct.* 2007;2:35.

In: Biogenic Amines
Editors: T. Farooqui, A. A. Farooqui

ISBN: 978-1-60876-625-3

Chapter VII

PHARMACOLOGICAL APPROACHES FOR THE EVALUATION OF DRUG-RECEPTOR RELATED EVENTS*

Norman J. Uretsky[1,&], *Margarita Salazar-Bookaman*[2] *and Popat N. Patil*[1]

[1]The Ohio State University, Columbus, OH 43210, USA;
[2]Department of Pharmacology, Universidad Central de Venezuela, Caracas, Venezuela.

ABSTRACT

During the last one hundred years there has been a unprecedented growth and understanding of receptor pharmacology. Isolation of natural products as well as elucidation of chemical structures provided a synergistic development based on the receptor therapeutics. In preclinical pharmacology receptor mediated drug quantitation remained at the cutting edge of discovery. The receptor related efficacy of the drug, the receptor reserve, the functional reserve, dose-response relationships, ED_{50} and the maximal effects, Emax are the most useful parameters in research. The classical contribution of well-known pharmacologists, A.J. Clark, E.J. Ariens, R.P. Stephenson, Sir J.H. Gaddum, H.O. Schild, Robert F. Furchgott established the foundation of quantitative parameters for evaluation of agonists and antagonists. Since the time of Henry Dale, R.P. Ahlquist, A.M. Lands and others, relative potency of agonists and dissociation constants of antagonists were used for the classification of receptors. Isolation and structural characterization of receptors added new dimensions for understanding the functionally relevant interactions of ligand at molecular sites of the

* "The review on the topic was enthusiastically initiated by the professor Norman J. Uretsky in the early 1990s for a book. After completion of the chapter the project was tabled. The current manuscript is a tribute to a scholar, a passionate teacher and a wonderful colleague."

& Correspondence concerning this article should be addressed to: Margarita Salazar-Bookaman, Department of Pharmacology, College of Pharmacy, University of Venezuela, Caracas, Venezuela. Email: mmsalazarb@gmail.com.

receptors in the central or the peripheral nervous system. Thus, specific and potent therapeutic agents can de developed with precision.

1. INTRODUCTION

Medicinal chemists and pharmacologists are constantly in the pursuit of discovering new chemical entities that are useful in treating pathological conditions. Based on rational therapeutics, new drugs are either chemically synthesized or isolated from natural products and evaluated on biological systems. In addition to an understanding of classical pharmacological methods, the evaluation of new drugs requires knowledge of anatomy, biochemistry, physiology, and pathology. In the past, diseases were treated with preparations derived from natural resources. Around the end of the 15th century, the "Doctrine of Signatures," endorsed by Giambattista Porta, promoted the concept that the ability of a plant preparation to cure a diseased body part was linked to the physical shape of the plant. For example, heart shaped bulbs were supposed to cure diseases of the heart. Leaves or flowers that were similar in shape and size to the lobes of the liver were supposed to cure liver disease. Eventually, physicians realized that the chemical constituents of a plant rather than its general appearance determined the effect of the plant preparation. An important step in the refinement of disease treatment was the use of active constituents which were extracted and isolated from natural products. Beginning in 1803, the isolation of alkaloids such as atropine, emetine, ergotoxine, curare, colchicine, morphine, pilocarpine, reserpine, and cinchona alkaloids was accomplished. Isolation of these substances was an important scientific achievement because specific chemical constituents could be linked to the beneficial therapeutic effects of natural products. In addition, these chemicals provided leads to the development of even more specific therapeutic entities.

Nearly 150 years ago Louis Pasteur observed the specific microbial utility of (+)-tartaric acid from the medium containing the racemate. Many years later, Emil Fischer examined the steric preference of sugars in the biological system and introduced the 'lock and key' concept of interaction of micromolecules. By 1905, the concept of 'receptive substance' was independently postulated by Paul Ehrlich and John Newport Langley. Otto Loewi recognized the cholinergic, atropine sensitive actions of the transmitter released from the vagus nerve of the frog heart. Henry Dale classified the muscarinic and nicotinic actions of acetylcholine sensitive to block by atropine and a ganglionic blocker respectively. In 1946, Von Euler identified norepinephrine as the major transmitter of the sympathetic neuron. Within two years of this discovery, Raymond P. Ahlquist postulated alpha and beta-adrenoceptors in the nervous system. In the mid 1950s, Bernard Brodie and Arvid Carlsson discovered the serotonin, norepinephrine and dopamine depleting action of the reserpine alkaloid. Julius Axelrod investigated the fate of the transmitter. Functional role of biogenic amines in the central and peripheral nervous system was elucidated. New era in the discovery of therapeutic agents based on transmitters and their receptors was initiated. Thus, a rational pharmacological approach to the development of medicines replaced the earlier empirical approach.

2. LEGACY FROM THE PAST

Ancient plant products used by various civilizations and cultures provided a rich source of medicines and lead drugs [1-4]. The collaboration in the early 1800s between the pharmacist-chemist Pelletier and the physiologist Magendie, on the pharmacological evaluation of various alkaloids such as strychnine, emetine, and quinine may have set the stage for the science of medicinal chemistry [5]. Magendie understood the concept that certain substances could produce specific changes in organ function. Magendie's student, James Blake, extended this idea by demonstrating a relationship between the chemical properties of ionic salts and their biological effects [6]. Another famous student of Magendie, Claude Bernard, demonstrated that the neuromuscular junction was the site of action of curare. This approach to understanding the biological effects of natural products was continued with the work of Edgar Stedman, who elucidated the chemical structure of physostigmine, which was isolated from calabar beans. Stedman showed that this substance competitively inhibited the activity of the enzyme, cholinesterase. Subsequently, many synthetic cholinesterase inhibitors were synthesized. The elucidation of the mechanism of action of physostigmine is considered one of the first examples of the antimetabolite concept [7].

One of the earliest studies on optical isomerism was performed around ~ 1848 by Louis Pasteur, who separated racemic tartaric acid by manual, chemical and biological methods and demonstrated the absolute stereochemical preference of the metabolism of tartaric acid by yeast. In 1904, Arthur Cushny reported differences in the pharmacological activity of l-hyoscyamine and dl-hyoscyamine. To explain the stereochemical pharmacological effects of the alkaloids, he wrote that "there can hardly be any question that the action is on a single receptor, which embraces both isomers though with unequal warmth" [8]. These studies not only laid the foundations for investigations of the biological activity of stereoisomers but also suggested that stereoisomers in their interactions with receptors can produce unequal conformational changes [9-11]. Drugs marketed as racemates have been rationally questioned, since it has been shown that the pharmacologically inert isomer can be toxic to humans [12]. Therefore, only the therapeutically beneficial isomer should be used in medicine.

Research elucidating the active constituents of natural products with unique biological activities is conducted in many countries around the world (Table 1). Podophyllotoxin, vinca alkaloids, maytansine, and taxol have been shown to be clinically active in the treatment of cancer. More importantly, these unique substances may provide clues to additional drug development in the treatment of neoplastic diseases. In China, various plant products containing one or more of the following substances - rubescensine, oridonine, ponicidine, indirubin, curcumol, curdione, and hainanolide and harringtonine have been isolated. These compounds have been used to treat various forms of cancer [13]. The official use of these products in therapy in the United States and most countries will require verification of their effectiveness and critical evaluation of their side effects.

Table 1. Activities of newly discovered plant product

SUBSTANCE	SOURCE	BIOLOGIC ACTIVITY
Asclepin	Asclepias curassavica	Cardiotonic
Bacosides	Bacopa monniera	Improves learning in rats
Coleonol	Coleus forskohlii	Hypotensive
Conessine	Holarrhena antidysenterica	Anti-amoebic
Curcumin	Curcuma longa	Antiinflammatory
Flavonoid (Arjunalone)	Terminalia arjuna	Oxytocic
Huperzine-A	Huperzia serrata	Anticholinesterase
Kutkin, A & B	Picrorhiza kurroa	Antiallergy
O-Acetyl Solanoside	Vallaris solanacea	Cardiotonic
Peruvoside	Thevetia neriifolia	Cardiotonic
Qinghaosu	Artemisia annua	Antimalarial
Resin (Guggulipid)	Commiphora mukul	Hypolipidemic
Toddasin	Toddalia asiatica	Diuretic
Tutin	Loranthus parasiticus	Antischizophrenia

Information is condensed from Rastogi and Dhawan [14], Klayman [15], Hansch [16], Sun and Xiao [13], Sun and Xiao [17].

Table 2. Activities of some substances of non-botanical origin

SUBSTANCE	SOURCE	BIOLOGIC ACTIVITY
Acetylcholine	Buccinum undatum (marine gastropod)	Cholinomimetic
Avermectins	Streptomyces avermitilis	Insecticides
Bombykol	Bombyx mori	Pheromone
Civetone	Civet cat (Viverra civetta) also obtained synthetically	Perfume
Dendrolasin	L.(Dendrolasius) fuliginosus (Formicine ant)	Toxic to other species
Iridomyrmecin	Dolichoderine ants	Toxic to other species
Gonyleptidine (Mixture of volatile substances)	Some spiders	Antibiotic
Murexine	Marine gastropod	Cholinomimetic
Homobatrachotoxin	Pitohuis dichrous (Birds)	Chemical defense
Batroxobin	Cobra venom factor	Hemostatic agent

Information condensed from Crescitelli and Geissman [21], Atkins [22], Fisher and Mrozik [23], Stocker [24], Dumbacher et al . [25].

2.1. Insecticides and Pheromones

Biologically active chemicals found in plants have also been used to control the insect population, thereby protecting agriculture. Natural products used for this purpose should be less toxic to humans than synthetic pesticides, like the organophosphate inhibitors [18]. Insecticides from pyrethrum are still popular and the insecticide Azadirachtin isolated from Azadirachta indica (neem tree) has been used with success in India [19].

Pheromones are substances secreted by animals that can stimulate a specific response by another individual of the same species. For example, the sensitivity of the antennal chemoreceptors of the male silkmoth (Bombyx mori) to the chemical, bombykol, which is secreted by the female, is extraordinary [20]. These types of higly specific substances can be used to control the movement and reproduction of invertebrate species that can produce profound damage to agriculture. Table 2 lists the biological activity of some of the interesting chemicals isolated from invertebrates.

2.2. Natural Products of Non-Botanical Origin

Vertebrate tissue extracts have also been used as medicines for thousands of years. In fact, many prescriptions containing animal products are listed in Egyptian papyrus. In recent years, the pharmacology of various extracts, particularly of endocrine organs, has been investigated. Extracts from pituitary, thyroid and adrenal glands have been shown to exhibit dramatic effects when administered to vertebrates, and the active principles of these extracts have been isolated. The chemistry and biological activity of many important endogenous constituents of endocrine organs, such as epinephrine, thyroxine, corticosterone, insulin, prostaglandins, leukotrienes, endorphins, atrial natriuretic factor, nitric oxide, and endothelin have been elucidated, and information obtained from these studies has been used as a basis for the development of many clinically useful drugs.

Natural products from marine organisms have also been shown to have marked biological effects [26]. Didemnin B and bryostatin I are being tested for antitumor activity. Ciguatoxin is responsible for the human illness caused by the consumption of ciguateric fishes. Manoalide, which is isolated from the sponge, Luffariella variabilis is an irreversible inhibitor of phospholipase A_2 and has been investigated for the treatment of various skin diseases.

Several natural products produce chemicals that are effective as ionophores [27]. These molecules were first recognized in mitochondrial membrane-linked energy functions. Ionophores can form lipid-soluble complexes with polar cations like potassium, sodium, calcium, and magnesium, and these complexes can diffuse through cell membranes and carry these cations into the cytosol along their concentration gradient. The ionophore, monensin, selectively binds sodium ion; valinomycin and nigericin are selective for potassium ion; and calcimycin is selective for calcium ion. In contrast, lasalocid is a nonselective ionophore that transports all three ions, sodium, potassium and calcium, across membranes. Although many ionophores are too toxic to be used in humans, a new synthetic ionophore that is chemically related to calcimycin and selectively complexes with calcium is being tested as a cardiotonic drug. It is also interesting that topically applied valinomycin lowers the intraocular pressure in laboratory animals and may be useful for the treatment of glaucoma.

2.3. Synthetic Medicines from Some Lead Molecules

The classical molecule that has been used as a lead for many other molecules is salicylic acid. In 1899, Hoffmann, whose father suffered from rheumatoid arthritis but could not tolerate salicylic acid, modified salicylic acid to produce the acetyl analogue, acetylsalicylic acid. This analogue, which was subsequently called aspirin, was found to be more effective than salicylic acid and became one of the first popular semisynthetic pharmaceutical products of the 19th century [28].

Preparations containing atropine have been used for centuries. After atropine was isolated, it was used both as a medicine and as a lead alkaloid to develop many useful synthetic preparations. Permanently charged quaternary ammonium analogs of atropine did not cross the blood brain barrier but retained peripheral activity. Therefore, these products could be used without central side effects. Ipratropium, a quaternary ammonium antimuscarinic agent, has been used in the treatment of asthma. Liposoluble cyclopentolate is an effective antimuscarinic agent used as a mydriatic. The basic tropine structure has been exploited in the synthesis of the antimuscarinic compound, benztropine, which is used to treat the symptoms of Parkinson's Disease.

Several hundred histamine H_2 blockers were synthesized before the receptor specificity of SKF 71448 was observed in a pharmacological test. This observation led eventually to the development of the therapeutically useful drug, cimetidine, an H_2 receptor antagonist used to inhibit acid secretion in the stomach. Nearly 14 years elapsed between the development of the initial lead antagonist and cimetidine [29].

Studies on the chemical structure and biological effects of the β-adrenoceptor agonist, isoproterenol, resulted in the synthesis of analogues, such as metaproterenol and salbutamol, which are selective agonists for β_2-adrenoceptors, are more potent, and have a longer duration of action.

The pioneering work of Ehrlich on microbial strain trypan red, which cured trypanosoma infected mice, opened a new era of antibacterials. Domagk was awarded the Nobel Prize for developing the clinically effective synthetic antibacterial, prontosil. Discovery of sulfa drugs also set the stage for the introduction of antitubercular drugs, like isoniazid, and synthetic anti-leprosy drugs, like dapsone. The concept of antibiotics was described originally by Louis Pasteur who, on the basis of microbial fermentations, stated that life hinders life. Waksman and Fleming introduced the antibiotics, streptomycin and penicillin. Synthetic and semisynthetic production of many antibiotics followed [30].

2.4. Interaction of Drugs with Endogenous Cellular Constituents

Drugs can produce a variety of effects, some of which are desirable and therapeutic while others are undesirable and may be adverse. In order for a drug to produce either type of effect, the drug molecule must interact with some endogenous chemical constituent of a cell. This interaction may produce an alteration in cellular function that may ultimately cause a change in the pathological or physiological state of the organism. The endogenous target molecule for the drug has been termed the receptor, and the changes in cellular function that occur as a consequence of the interaction of the drug with this endogenous substance constitute the drug response or effect.

A variety of endogenous cellular constituents can act as receptors for drugs. For example, drugs may produce their effects by interacting with the following cellular proteins:

- Acetylcholinesterase inhibitors - physostigmine
- Monoamine oxidase inhibitors - tranylcypromine
- Ion channels - local anesthetics
- Transporter molecules - cocaine
- Structural proteins - colchicine
- Nucleic acids - anticancer drugs

In addition to drugs and circulating hormones, neurotransmitters, and autacoids are known to interact with specific receptors, which can be located on the surface of cells or within cells. These receptors are usually macromolecules, which serve to regulate cell function. Their activation can elicit one or several biochemical signals, and when these signals are integrated, the response to the drug is observed. It is this latter category of receptor that will be discussed in the remainder of the chapter.

2.5. Cell Membrane Receptors

Cell surface receptors are located within the cell membrane. These molecules are specific components of the cell membrane, which transmit information from the outside the cell through the cell membrane phospholipids to inside the cell. Chemicals, as well as sensory modalities, can cause the activation of specific membrane receptors. This process may involve the binding of a drug to chemical groups on the part of the receptor located on the outer surface of the cell membrane, forming a drug-receptor complex. If the chemical is an agonist, the formation of this complex will cause specific changes in the conformation of the receptor that can trigger biochemical changes inside the cell. For example, the conformational changes resulting from the drug-receptor interaction may cause an ion channel to open, leading to a change in ionic flux across the membrane. The activity of enzymes may be regulated by the influx of ions.

The conformational change in the receptor promotes the interaction of the receptor with a G protein found on the cytoplasmic surface of the cell membrane. G proteins, which are named because of their interaction with guanine nucleotides (GTP), provide a link between the drug-receptor interaction and the cellular response. Changes in G proteins, as a consequence of their interaction with a stimulated receptor, may activate certain enzymes, thereby enhancing specific chemical reactions within the cell. G proteins can regulate ion flux as well.

In summary, cell membrane receptors are macromolecules, usually composed of protein, which are located in the cell membrane and which contain various domains or regions that can interact with extracellular chemicals, such as hormones and neurotransmitters, and intracellular molecules, such as G proteins. In some cases the receptor itself may be an ion channel that spans the cell membrane and regulates the entry of ions into the cell. An example of the latter is the nicotinic receptor that regulates sodium entry into skeletal muscle. Many organs contain more than one type of receptor within their membranes with the number and proportions of the receptor types varying among different tissues (See Figure 1).

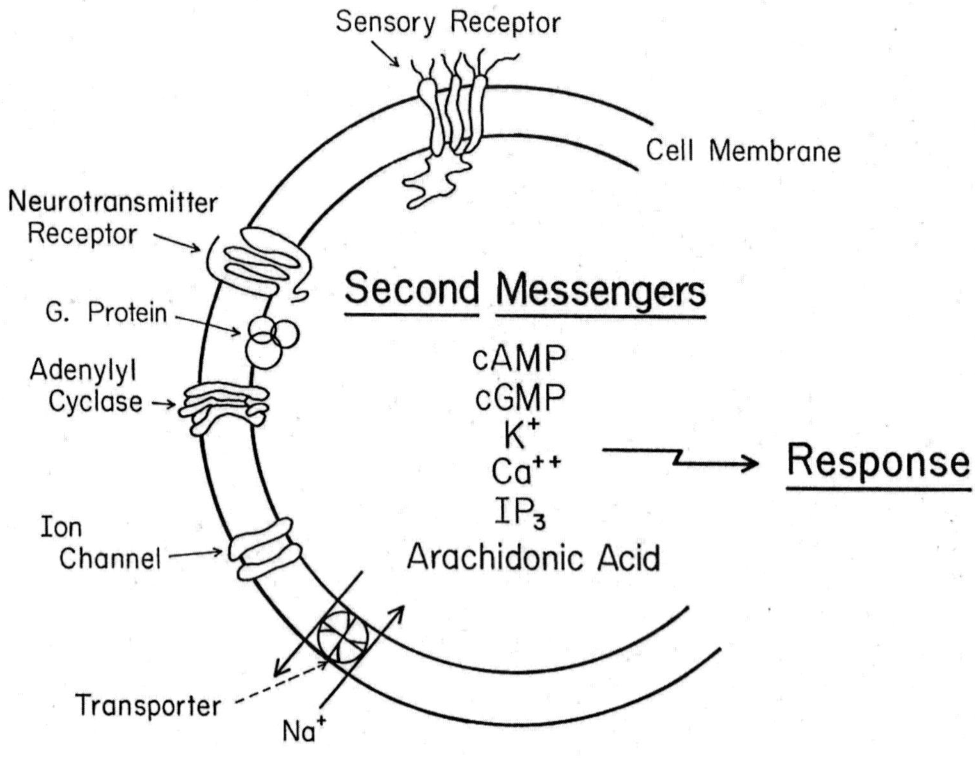

Figure 1. Diagram of a cell with its membrane containing various receptors. Some receptors for neurotransmitters are linked to G-proteins. The interaction between the activated receptor and G-protein results in the activation or inhibition of intracellular enzymes that alter the rate of synthesis of second messengers within the cell. An example is adenylyl cyclase, an enzyme which catalyzes the formation of the second messenger, cyclic AMP. These second messenger molecules are important in the regulation of specific chemical reactions within the cell. Some receptors regulate the opening of ion channels. When these receptors are activated, ions pass through the channels down an electrical or concentration gradient, crossing the cell membrane. Transporters can complex with ions or molecules and transport them across the cell membrane. Certain cells in the body contain specialized sensory receptors. These transmit information related to such sensations as touch, taste, smell, and light.

2.6. Isolation of Receptors from Membranes

Receptors have been removed from cell membranes in order to characterize their molecular properties [31]. Solubilization of receptors involves the use of a detergent to separate the receptor protein from membrane. The receptor is then purified and digested into peptide fragments by proteolytic enzymes, and the peptide fragments can be sequenced to determine the amino acid composition of the macromolecule. Photoaffinity agents are occasionally used in these studies. When exposed to light, an irreversible covalent bond is formed between the compound and the receptor, and this allows the receptor to be chemically characterized.

To determine whether a solubilized protein is the functional receptor, reconstitution experiments have been conducted in which the receptor is incubated with an artificial lipid bilayer. This type of study has been particularly valuable in providing information on receptors associated with ion channels. Thus, in the case of the nicotinic receptor, which is a protein consisting of multiple subunits, the binding of acetylcholine to the recognition units of the receptor has been found to open the ion channel to sodium ion, increasing ionic conductance through the lipid membrane barrier.

2.7. Molecular Nature of Receptors

As indicated above, the effects of many drugs can be explained by their interaction with specific receptors, which can trigger a biochemical response within the cell. Recently, there has been an intense interest in the molecular nature of receptors. This work has been, in part, stimulated by the belief that a better understanding of the molecular nature of different receptors may lead to the development of new drugs which can more specifically interact with receptors or receptor-related mechanisms and produce highly selective drug effects.

Data from molecular techniques such as receptor cloning and solubilization have provided evidence that receptors are proteins with distinctive peptide sequences and transduction mechanisms. For example, G protein linked receptors contain membrane spanning regions. These transmembrane spanning units contain amino acids with non polar side chains that interact with the phospholipid matrix of the membrane as well as specific amino acids with chemical groups interacting with drugs, hormones and neurotransmitters. Receptors have been classified according to their transduction mechanisms, and there are at least 4 different major classes of receptors. These include: a) ion channel receptor; b) G protein regulated receptors; c) tyrosine kinase linked receptors and d) steroid hormones receptors.

A brief overview of these different types of receptor is presented below.

2.8. Ion Channel Receptors

The ion channel receptors regulate the flow of ions through cell membranes [32]. Thus, these receptors respond to the binding of specific agonists by changing their conformations, resulting in the opening of an ion channel and allowing the passage of specific ions through cell membranes into the cytosol. Depending upon the nature of the charge on the ions entering the cell, the cell may become depolarized and excited or hyperpolarized and inhibited. These receptors are responsible for the rapid transfer of information across the cell membrane and operate within milliseconds. The receptors are very specific for particular ligands. These include:

a) Acetylcholine nicotinic receptors, which increase the passage of sodium ions; b) glutamate receptors, which increase the passage of sodium and/or calcium ions; c) GABA-A receptors and glycine receptors, which increase the passage of chloride ions; and d) ATP receptors, which increase the passage of sodium ion into smooth muscle cells.

The ion channel that has been most extensively studied is the nicotinic receptor [33]. This receptor is found in vertebrates at the neuromuscular junction, at autonomic ganglia, and in the CNS. While the receptors at these sites share some similarities, the pharmacological properties of the receptors at the neuromuscular junction are different from those at the other sites, suggesting that subtypes of receptor proteins are involved. Much of the studies of this receptor have been done on receptors solubilized from the electric fish called Torpedo californica or the electric eel called Electrophorus electricus, both of which closely resemble the receptor at the neuromuscular junction. This receptor has a molecular weight of approximately 290,000 daltons and consists of 5 subunits: there are 2 α subunits and one β γ, and a δ subunit. The extracellular region of each of the 5 subunits is long and has multiple glycosylation sites. Each subunit has 4 domains, consisting of 20 -30 amino acids that are hydrophobic and consequently are believed to be associated with the lipid membrane. These subunits are arranged in a symmetry around a central pore or opening, which forms the ion channel. In the absence of acetylcholine, the channel remains closed, and ions cannot pass into the cells. When two molecules of acetylcholine bind to each of the α subunits, a conformational change occurs, and the channel opens, propelling sodium ions into the cytoplasm [33].

2.9. G Protein Linked Receptors

As mentioned above, many receptors during transduction are linked with G proteins, which are regulated by the binding of guanine nucleotides. This class includes receptors for muscarine, serotonin, histamine, norepinephrine and epinephrine, adenosine, dopamine, GABA, opioid and neurokinins [34-35-36]. When an agonist binds to receptors of this family, the conformation of the receptor changes, and the intracellular domain of the receptor interacts with G protein. The bound G protein undergoes changes which regulate the activity of enzymes or ion channels. The enzymes, as a result of their interaction with G proteins, can increase or decrease the formation of intracellular second messengers which can initiate a chain of events including phosphorylation reactions, which ultimately result in a biological response. G proteins may be further linked with ion channels, which as a result of phosphorylation reactions, can alter the movement of ions into or out of the cell.

Much of the information about these receptors is derived from studies of β_2-adrenoceptors and rhodopsin, the photosensitive molecule of the rod-cells found in the retina [37,38]. The photoreceptor molecule, rhodopsin, consists of a protein, opsin, which contains seven transmembrane units, and 11-cis-retinal, a chromophore that is bound to the lysine residue of the receptor located within the membrane. Light is absorbed by the 11-cis-retinal group, which undergoes isomerization to all-trans-retinal, producing a change in the conformation of rhodopsin. This change activates a G protein called transducin. This receptor activated cascade further activates the enzyme, phosphodiesterase, which hydrolyzes 3',5' cyclic GMP. The resultant decrease in cyclic GMP levels causes the closing of sodium channels, which is the electrical signal that hyperpolarizes retinal cells. Thus, this light-transducing receptor is a G protein that resembles receptors that respond to certain neurotransmitters, hormones, and drugs.

The general molecular architecture of rhodopsin resembles that of the β_2-adrenoceptor, which is coupled to G proteins [35,39]. Several G protein-linked receptors consist of a single polypeptide chain, with seven hydrophobic membrane spanning domains of between 20 to 28 amino acids; each of the domains are arranged in an α-helix. The carboxyl terminal of the receptor protein is located on the cytoplasmic side of the membrane, while the amino terminal is located extracellularly. There are three connecting loops on the cytosolic side of the membrane; the third loop contains the largest number of amino acids and interacts with the G protein.

The identity of the ligand-interacting groups on the receptor protein is being actively investigated. It is thought that the transmembrane spanning segments of the receptor form a central pocket within the membrane, and ligands for these receptors bind to certain specific amino acids within this formed pocket. For biogenic amines, the side chain amine group is thought to bind to the receptor in the protonated form. Site-directed mutagenesis studies on the interactions of ligands with cloned receptors that do not contain specific amino acids suggest that the positively charged group may interact with the carboxyl group of an aspartate113 residue [40], in the third hydrophobic domain (Figure 2). This amino acid residue is conserved in all G protein receptors that bind to biogenic amines. It is also thought that the catechol groups of ligands interact with serine residues at positions 204 and 207 in the fifth membrane spanning domain of the receptor. Site-directed mutagenesis studies of the β-receptor have also provided evidence that the phenylalanine at position 290 of the sixth membrane spanning unit is important for ligand binding. Patil et al [41] used bovine rhodopsin crystal structure as a model to elucidate the agonist interacting amino acid residues for α_{1A} and β_1- adrenoreceptors. It was observed that both -$^+NH_3$ and chiral-OH group of (-)-epinephrine interacted with Asp ^{106} transmembrane domain III of α_{1A} adrenoreceptor. Similar interaction of the agonist was observed at β_1- adrenoreceptor. Some differences, particularly the interaction of catechol group with serine residues of domain V of two adrenoreceptors were observed. Refer to figure 2. The aromatic group of (+) - epinephrine fails to interact with the PHE ^{289} or PHE ^{319} of the ligand α_{1A} and β_1- adrenoreceptor respectively. The chiral –OH group of the unnatural enantiomer may not sustain the interaction with the aspartate residue of these adrenoreceptors. This explains the low potency of the enantiomer. Oxymetazoline exhibited strong complimentary interactions of its imidazoline, phenyl and the OH group with the respective sites of the α_{1A} adrenoreceptor. But the ligand had only aspartate interaction of the imidazoline with the β_1- adrenoreceptor. The drug does not activate β_1- adrenoreceptor. Recently crystal structure of human β_1 and β_2 adrenoreceptors has been determined [42,43].

The third intracellular loop and part of the carboxyl terminal cytoplasmic tail is thought to interact with the G protein. These regions also contain many threonine and serine residues, which are phosphorylation sites believed to be important for receptor desensitization. G proteins, which provide the link between the activated receptor and the effector systems within the cell, are composed of three subunits termed α, β and γ [44-46]. β and γ subunits are hydrophobic and are associated with cell membrane, while the α subunit interacts with the receptor and the effector systems. In addition, guanine nucleotides bind to the α subunit. G proteins exist in inactive and active forms. In the inactive form, the α subunit binds GDP. When the agonist binds to the receptor, change in receptor conformation allows it to interact with the α subunit of the G protein. This interaction decreases the affinity of the α subunit for GDP and increases its affinity for GTP; consequently, GTP displaces GDP from the α subunit. The binding of GTP causes α subunit to dissociate from the β, γ subunit of the G protein complex. The α subunit-GTP complex can then move within the membrane and bind to enzymes or ion channels, which in turn mediate the biological response.

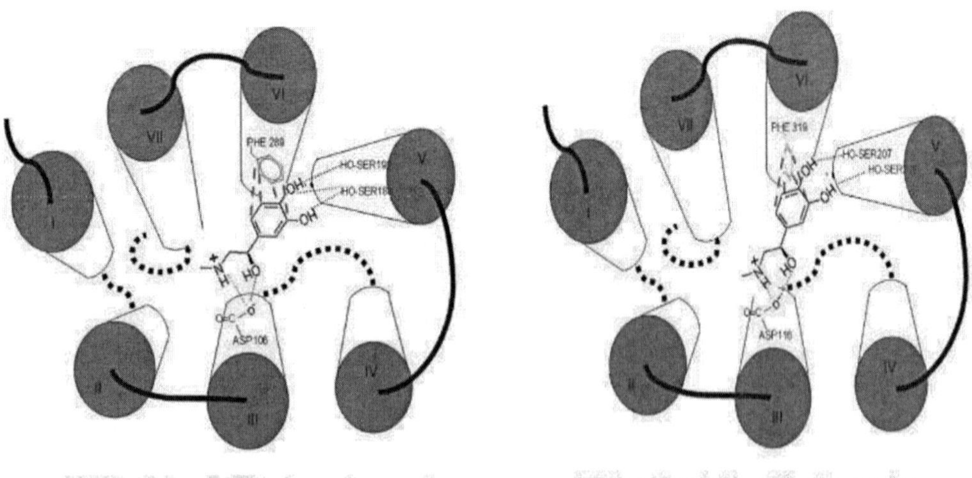

(-)Epinephrine docking at α_{1A}adrenoceptor (-)Epinephrine docking at β_1-adrenoceptor

Figure 2. Potential interacting sites at α_{1A} and β_1- adrenoreceptors for (-)- Epinephrine from *Patil et al.* [41]. Images reproduced with permission from Wiley InterScience.

In the case of the G_s protein that stimulates the adenylyl cyclase system, the α subunit with GTP attached activates adenylyl cyclase, which can then catalyze the formation of cAMP from ATP [34]. As long as GTP remains attached to the α subunit, this will be in the active form, and will continue to activate adenylyl cyclase. Adenylyl cyclase activation will be terminated when the attached GTP on the α_s molecule is hydrolyzed to GDP by a GTPase which is part of the α subunit. The α_s subunit then becomes inactive and will recombine with the β,γ subunit reforming the trimeric G complex. The slow rate of hydrolysis of GTP bound to the α_s subunit may cause a relatively prolonged activation of adenylate cyclase relative to the duration of receptor activation, and this may serve to amplify the effector response to receptor activation.

Receptors linked to G proteins also regulate other second messenger systems besides adenylate cyclase [47]. The activity of the enzyme phospholipase C, which mediates intracellular phosphorylation reactions and the release of Ca^{2+} from intracellular stores, can be increased by agonists of α_1-adrenergic receptors, H_1 histamine receptors, 5-HT_1 serotonergic receptors, and muscarinic receptors. This receptor mediated enzyme activation is regulated by G proteins.

Phospholipase C is an enzyme bound to the cell membrane [48,49]. It catalyzes the hydrolysis of phosphatidylinositol-4,5-bisphosphate in the cell membrane, resulting in the formation of inositol-1,4,5-triphosphate and diacylglycerol. Inositol-1, 4,5-triphosphate is a water soluble compound that can bind to a receptor located on the membrane of the endoplasmic reticulum, which stores Ca^{2+} ions. Activation of this receptor causes the intracellular release of Ca^{2+} ions into the cytosol, where they can diffuse and activate a variety of enzymes, such as Ca^{2+} calmodulin-dependent protein kinases and protein kinase C. Resultant phosphorylation reactions are believed to be essential to many types of cellular responses. Diacylglycerol, is very lipophilic and consequently will accumulate in the cell membrane, where it can bind to and increase protein kinase C activity. Protein kinase C can then catalyzes phosphorylation of several proteins, such as those associated with muscle contractions, ion channels, receptors, or transport processes, producing biological effects. Inositol-l,4,5-trisphosphate and diacylglycerol are cycled back to reform phosphatidyl inositol-4,5-bisphosphate.

As indicated above, ion channels can also be regulated by G proteins [47]. In this case, receptor activation by drugs and neurotransmitters causes the activation of G proteins which in turn activates ion channels that regulate ion conductance through the cell membrane. Second messenger systems such as cyclic AMP or phosphatidyl inositol-4,5-bisphosphate are not generally required for ion channel activation. Both calcium and potassium ion channels have been implicated in this type of reaction. For example, it has been shown in patch clamp experiments using cardiac cells that the activation of muscarinic receptors increases potassium conductance, which was inhibited by pertussis toxin. This indicates that a G_i protein is involved in producing this response.

2.10. Receptors Associated with Tyrosine Kinases

Specific receptor binding sites are found on the extracellular side of the membrane, which can interact with various growth factors, including insulin, platelet derived growth factor, colony stimulating factors, epidermal growth factor, among others [50]. Binding of the growth factor to these receptors can cause a conformational change, which is transmitted through a single hydrophobic transmembrane segment to the tyrosine kinase domain located on the intracellular side of the membrane. Activation of this receptor stimulates the activity of this enzyme, which catalyzes the phosphorylation of tyrosine residues of specific cytosolic proteins as well as the tyrosine residues in its own peptide structure (termed autophosphorylation). When tyrosine kinase, itself, is phosphorylated, it exerts greater catalytic activity. However, it is still unclear how increased activity of the enzyme, tyrosine kinase, leads to an alteration in cellular function. Initially, there seems to be an increase in the membrane transport of ions and nutrients into the cell, and these events are followed later by an increase in polyamine synthesis and the activation of DNA, RNA, and protein synthesis, which eventually leads to an increase in cell division.

2.11. Intracellular Receptors for Drugs and Hormones

Receptors for certain hormones are located intracellularly and can regulate the synthesis of specific proteins within the cell [51,52]. Ligands are thought to gain access to these receptors by diffusing through the cell membrane. These receptors contain a single polypeptide chain of between 400- and 1000 amino acid residues. In addition, they contain a binding site for hormones at the C-terminal region and a binding site for DNA located in the middle of the molecule. When cells are homogenized and centrifuged, these receptors are found in the soluble fraction, suggesting that they are cytosolic. Binding to the receptor requires that the hormone be sufficiently lipophilic to pass through the cell membrane. Hormones bind to receptors at the C-terminal portion of the peptide which contains hydrophobic amino acid residues. Hormone-receptor interaction produces a conformational change in the molecule, which exposes the hydrophilic DNA binding domain, which is positively charged. The hormone-receptor complex then passes into the nucleus, where the positively charged exposed region binds to specific chromatin regions of DNA. The newly formed complex serves to regulate DNA transcription, leading to an increased formation of specific proteins that mediate hormones different effects. Estrogens, glucocorticoids, mineralocorticoids, 1,25-vitamin D_3, and thyroid hormones activate specific intracellular receptors of this type.

2.12. Types of Interaction Between Drug and Receptor

The interaction of drugs or endogenous substances (hormones, neurotransmitters and autacoids) with receptors constitutes a chemical reaction. Drugs contain reactive groups which can chemically interact with complementary chemical groups on the receptor, and different types of chemical bonds may form between drugs and receptors. Most drug-receptor interactions are reversible. However, there are exceptions: for example, certain anticancer drugs are known to form reactive species that covalently bind to DNA. It is thought that most drug-receptor interactions involve ionic, hydrogen, hydrophobic, and van der Waals bonds and that these types of interaction can occur concurrently and reinforce binding of drugs to receptors. One consequence of this reinforced binding is that it produces a slower rate of dissociation of the drug-receptor complex than only a single type of bonding. Interaction of a ligand with a receptor depends not only on the presence of reactive groups on the ligand molecule but also on the correct spatial geometry between reactive groups, which enables multiple interactions between drug and receptor to occur. Both, reactive groups and their spatial arrangement on the drug molecule are important in determining the potency and specificity of the drug receptor interaction.

Structure-activity relationship studies, including the evaluation of the activity of chiral molecules, have been very valuable in determining which reactive groups on the drug molecule are important in their interaction with specific receptors. Information from structure activity relationship studies has also been used to infer the topography of the complementary reactive groups on receptors. Small changes in the chemical structure of a drug can produce marked changes in potency. For example, acetylcholine is 20 times more potent than propionilcholine in causing contraction of the guinea pig ileum. Biological responses to drugs also display stereoselectivity with one stereoisomer frequently being much more active than another. Binding studies have shown that d-isoproterenol has approximately 1/400 the affinity of the l-isomer for β-adrenoceptors. Studies have also indicated that the interaction of a drug with a receptor is a dynamic process and that many drugs particularly agonist molecules possess the property of flexibility.

Currently, computational modeling is probably one of the most commonly used approaches to study drug-receptor interactions at a molecular level. For example, using ligand-based computational modeling, McKay and colleagues [53] developed a pharmacophore for the binding site of noncompetitive antagonists of the α3β4* neuronal nicotinic acetylcholine receptors, thus describing the binding requirements for the interaction with this binding site. Furthermore, the same laboratory developed a computational dynamic homology model of the ligand-binding domain of α3β4 neuronal nicotinic acetylcholine receptors from the crystal structure of homopentameric acetylcholine binding proteins of three molluskan species, along with an α1 ligand-binding domain monomer, identifying the binding site of a potent negative allosteric modulator of these receptors by blind docking. Also, the amino acid residues involved in the allosteric binding pocket were identified, allowing the study of possible types of interactions occurring in this drug-receptor complex [54].

2.13. Agonists and Antagonists

The nature of the interaction between the reactive groups of the drug and the complementary sites on the receptors will determine whether a drug is an agonist or an antagonist. Both an agonist and antagonist drug can bind to a receptor, but only the agonist-receptor interaction produces the specific change in receptor conformation that elicits a biological response. In this case, conformational changes induced by the agonist-receptor complex acts as a stimulus, which initiates a sequence of events, culminating in the response. Hormones, neurotransmitters, or autacoids that normally interact with receptors are agonists because their binding to receptors initiates a response. In contrast, an antagonist drug may interact with some of the sites on the receptor, which also interact with the agonist, but the antagonist-receptor complex does not produce the specific conformational change required to produce a response. However, since the antagonist is bound to the receptor, it can prevent the binding of an agonist and thus can inhibit the agonist-induced response. Antagonists that interact with the receptor by an allosteric mechanism may bind to a site on the receptor that does not react with an agonist.

2.14. Quantitation of the Drug-Receptor Interaction

The biological effects of a drug can be quantified by evaluating the ability of the drug to produce a response. This depends upon the capacity of the drug to gain access to the biophase, the site where the drug acts on its target, the receptor, to produce its effects, and its ability to interact with the receptor. The interaction between drug and receptor depends on many factors, and these are discussed below.

2.15. Characteristics of the Dose- or Concentration-Response Curve

Pharmacologists generally express the ability of a drug to induce a response in terms of a dose-response (in vivo study) or concentration-response (in vitro study) curve (Figure 3). At low concentrations, the intensity of the response to the drug (D) is proportional to the concentration of drug, while at higher concentrations, the response does not change with a further increase in the drug concentration. The linear central part of the curve is expanded and extends throughout approximately a 10 fold range, making it easier to determine the concentration that produces a 50% effect, the half maximal response, referred to as ED_{50} (effective dose - 50%) or EC_{50} (effective concentration – 50%). These parameters are most frequently used to compare the potency of different drugs. The potency comparison is based on geometric mean EC_{50} values instead of arithmetic values [55].

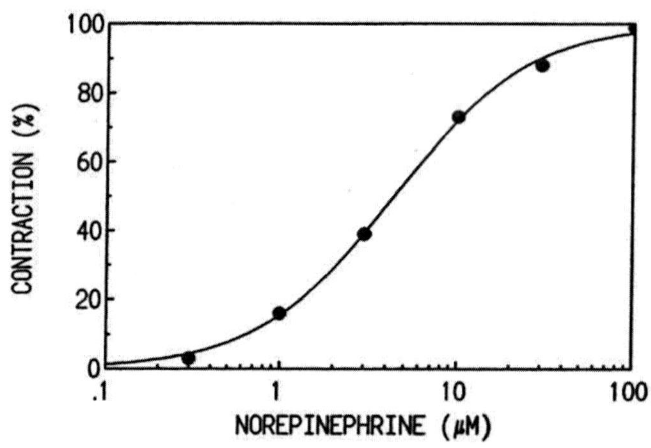

Figure 3. Concentration-response curve for the effects of norepinephrine in producing contraction of isolated rat vas deferens tissue preparation *in vitro*. Note that the X axis, which indicates the concentrations of norepinephrine added to the medium bathing the vas deferens, is in the log form. As a result, the central portion of the curve is linear. The ED_{50} **for norepinephrine is 4.3 μM. The maximum** effect occurs at a concentration of ~ 100 μM.

The potency of a drug is the concentration required to produce a given response. The relative potency of two agonists is usually expressed as the ratio of their EC_{50} values which are obtained from their concentration-effect curves. The threshold concentration is the lowest dose that will produce a measurable response. The relative order of potency of a series of drugs in producing effects in different tissues indicates whether the receptors mediating these effects are the same for each tissue. Thus, if the drug effects are mediated by a single receptor type in each tissue, then the relative order of potency of the drugs should be the same. For example, the relative order of potency of catecholamines differs in tracheal smooth muscle and heart of the guinea pig (Figure 4).

It should be noted that depending upon the drug tested and the tissue in which the effect is studied, the slope of the dose-response curve can vary. Drugs that stimulate the rat uterus in vitro are known to produce extremely steep concentration-response curves. The slope of the dose-response curve may indicate whether the same or different receptor types mediate a given response. Thus, if the response produced by two drugs in the same tissue is mediated by the same receptor, then the slopes of their normalized dose-response curves should be parallel. On the other hand, different slopes of the dose-effect curves may indicate that different receptors may be involved [58].

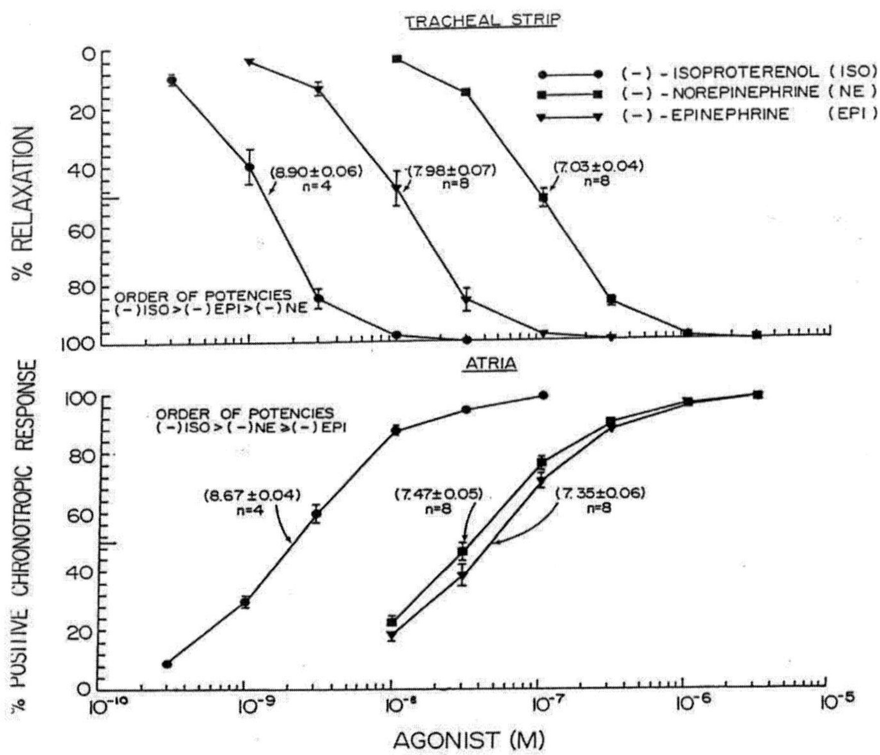

Figure 4. Concentration-response curves for three different β-adrenoceptor agonists on the tracheal strip and atria of the guinea pig. Numbers in the parentheses are the ED_{50} values for each agonist expressed in logarithmic form. The relative order of potencies of the agonists on the two receptor subtypes show obvious differences. Whereas epinephrine is approximately 10 times more potent than norepinephrine in causing relaxation of tracheal strips, it is equal in potency to norepinephrine in causing contraction of the atria. This is because two different receptor subtypes mediate the agonist effects, as postulated by *Lands et al.* [56]. Relaxation of tracheal strips is mediated by the activation of β_2 adrenoreceptors, while **the positive chronotropic response of the atria is mediated by the activation of** β_1 adrenoreceptors, data from *Buckner and Patil* [57] with permission from American Society for Pharmacology and Experimental Therapeutics.

Another parameter of the dose response curve is the maximum effect of the drug. This is the largest effect that can be produced by the drug in a given tissue and is also called the ceiling effect. When there are spare receptors (discussed further below) and the number of receptors do not limit the response, the maximum response of the tissue to the drug may be limited by some other tissue event occurring after receptor occupation. The relative maximal responses of two agonists that act through the same type of receptor would reflect their receptor mediated intrinsic activities. In a given tissue, if one of these drugs produces a lower maximum effect than the other, it is termed a partial agonist. That is, the drug has a lower ability to initiate a biological response after binding to the receptor. Experimentally, the maximum effects of several agonists can be expressed as a percentage of the maximum effect produced by a full agonist.

2.16. Relationship Between Drug Concentration, Receptor Occupancy and Response

The relationship between the interaction of an agonist with a receptor to produce a response was initially described by A. J. Clark (1937) using mass action principles [59]. According to Clark, an agonist drug, D, would interact with a receptor, R, forming a drug receptor complex [DR]. At equilibrium, the rate of drug-receptor complex formation would be equal to the rate of dissociation of the complex. The agonist, after binding to the receptor, would produce a conformational change in it, which would initiate the sequence of events that culminate in a biological response. The relationship between the concentration of agonist and the response reflected the mass action principles used to describe the interaction between two chemicals e.g. the drug and the receptor. The reversible interaction of a drug (D) and a receptor (R) to form a drug-receptor (DR) complex was represented as:

$$D+R \leftrightarrows DR$$

At equilibrium and according to mass action principles:

$$K_D = \frac{[D][R]}{[DR]}$$

K_D represents the dissociation constant of the drug-receptor complex.

While the relationship between receptor occupancy and the response was not known, Clark [59] assumed that the ratio of the intensity of the drug effect (E) as a fraction of the maximum effect was proportional to the fraction of receptors occupied. He further assumed that when all the receptors were occupied, the maximal effect of the drug would be produced and there were no spare receptors, refer to Ruffolo [60]. If E_{max} is defined as the maximum effect that can be produced by the drug, this relationship between response and effect which may apparently resemble Michaelis-Menten postulate for enzyme substrate plot.

However, there are important differences between these two types of reactions. While the interaction between substrate and enzyme leads to the formation of product, the drug-receptor interaction initiates a series of molecular events that eventually culminate in a physiological response. Thus, the drug-induced response reflects the amplification of a series of chemical reactions that are initially triggered by the formation of the drug-receptor complex. These differences in drug-receptor interaction become important when attempts are made to determine the affinity of the drug for the receptor based on measuring a distant parameter called the response to the drug.

As indicated above, the concentration-response curve produced by an agonist will indicate EC_{50} values as well as a concentration which will produce a maximal drug effect. These response parameters to the percentage of receptors occupied by a drug. Thus, the EC_{50} of a drug should be produced when 50% of the receptors are occupied, and the maximum response should occur when the drug occupies all the receptors. Furthermore, differences in the EC_{50} among several drugs should reflect their affinity differences for the receptor being studied.

However, it has been found that the relationship between the effect of a drug, as a percentage of the maximal effect, and the percentage of receptors occupied by the drug is frequently not linear as expected from the assumptions underlying the kinetic parameters [60,61]. The problem is that equations derived by Clark [59] do not take into account several factors. For example, it is clear that receptor occupation by a drug is necessary but not sufficient to induce a response, since competitive antagonists can occupy receptors without inducing physiological responses. Thus, in order for a drug to produce a response, the receptor must not only be occupied by the drug but also be activated. That is, after receptor occupation, the conformation of the receptor has to be altered in a specific way in order to initiate the subsequent events that lead to the response. Drugs can show marked differences in their ability to activate the receptor and produce a response. Thus, many drugs can bind to the receptor, but the maximum response is lower than that produced by a full agonist. In practical terms, any drug that produces a maximum response that is 95% or less than that produced by a full agonist can be considered a partial agonist. The maximum response produced by a partial agonist may vary depending upon the number of spare receptors in a given tissue. For example, pilocarpine produces about 80% of the maximum response produced by carbachol when tested in the human iris, whereas it produces only 11% of maximum when tested in rabbit iris (Figure 5). These data are consistent with a greater proportion of spare receptors or second messenger systems that contribute to the functional reserve in the human iris compared to the rabbit iris.

The partial agonist concept, which encompasses an agonist whose degree of activity is lower than that produced by a full agonist, can be added to the initial reaction described for drug-receptor interaction as follows:

$$D\ (\text{agonist}) + R \leftrightarrows DR \rightarrow DR^* \rightarrow \text{effect}$$

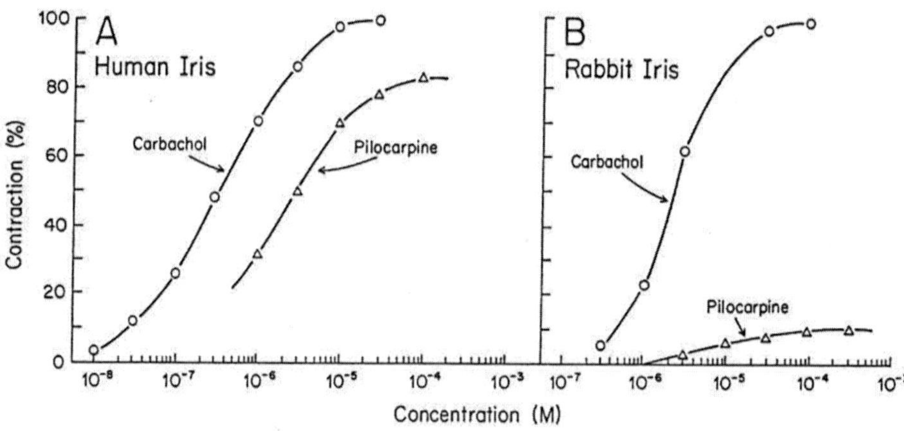

Figure 5. Effect of carbachol and pilocarpine in human (A) and rabbit (B) iris. Note that the effects of carbachol and pilocarpine are different in the two irides. Carbachol, a full agonist, appears to be more potent in the human iris than in the rabbit iris. Pilocarpine, a partial agonist, produces in the human iris a maximum response, which was 80% of that produced by carbachol. In contrast, the maximum response for pilocarpine in the rabbit iris was only 11% of that for carbachol. Figure is reproduced from *Akesson et al.* [62] with permission from Springer Verlag.

This reaction illustrates that after the drug (either agonist o partial agonist) binds to the receptor forming the drug-receptor complex (DR), the receptor becomes activated (forming DR*). The activated receptor then provides the stimulus that can trigger the subsequent intracellular steps leading to the response. According to this model, partial agonists differ from full agonists in that they are less able to convert the drug-receptor complex (DR) into the "activated drug-receptor complex" (DR*) and, therefore, produces a weaker effect.

The existence of drugs with partial agonist activity indicates that the effect produced by an agonist depends not only on the number of receptors occupied by the drug but also on the ability of the drug receptor complex to produce a significant conformational change in the receptor triggering a receptor mediated response. This latter effect has been termed intrinsic activity by [63] and has also been given the symbol α. If intrinsic activity (α) is expressed in the dose-response equation, the ratio of the effect (E) of the drug to the maximum effect (E_{max}) that can be produced is:

$$\frac{E}{E_{max}} = \frac{\alpha[DR]}{[R_t]}$$

This equation indicates that the effect of the drug is equal to the product of the proportion of receptors occupied and the intrinsic activity. R_t represents total receptors in the organ. Substituting for R_t:

$$\frac{E}{E_{max}} = \frac{\alpha[D]}{[D]+K_D}$$

As shown previously, the curve produced by plotting the response to an agonist as a fraction of the maximal response against a log concentration of the agonist is sigmoidal. The curve for a partial agonist would be similar to that for a full agonist except that the maximum response would be smaller. That is, when the partial agonist occupies 100% of receptors, the response that it produces would be less than the full tissue response. The intrinsic activity (α) of a partial agonist is evaluated as the ratio of the maximal response of a full agonist. For full agonists, α equals 1, while for partial agonists, $0 < \alpha < 1$. For antagonists, which can bind to the receptor but do not produce any response and, therefore, have no intrinsic activity, α equals 0.

The concept of intrinsic activity is based on the assumption that the intensity of the drug-induced response is proportional to the percentage of the receptors occupied and that the maximum response to a full agonist will be produced when all the receptors are occupied. Thus, this equation does not take into account the existence of spare receptors, which are present in many tissues [61]. This concept will be discussed below.

2.17. Spare Receptors

Many tissues have spare receptors or a functional receptor reserve. In these tissues there is a greater number of receptors present than are necessary to produce a maximum response. Under these conditions, the maximum response of a potent drug may be produced when only a small percentage of the total receptors are occupied by a drug. When spare receptors are present, the effect produced by a full agonist as a percentage of the maximum effect may not be linearly related to the percentage of the total number of receptors occupied by the drug. Furthermore, the EC_{50} would be lower than the concentration of drug that occupies 50% of receptors (K_D) [60,61]. Obviously then, the EC_{50} value for a drug may not be equivalent to the K_D of the drug. The total receptor saturation at the maximal effect of the drug does not occur.

2.18. Efficacy

To account for the presence of spare receptors in tissues and variations in the capacity of agonists to activate receptors, Stephenson [64] introduced into his equation the concept of a pharmacological stimulus (S). The stimulus is the product of the fraction of receptors occupied (DR/R_t) multiplied by the efficacy (e). Efficacy, which was not given dimensions by Stephenson [64] indicates the ability of a drug to cause a change in the receptor conformation and produce a response. The meaning of the term intrinsic activity introduced by E.J. Ariens is essentially the same as efficacy. The equation relating the pharmacological stimulus to the fraction of receptors occupied by the drug and to efficacy is:

$$S = \frac{e[DR]}{[R_t]} = \frac{e[D]}{[D] + K_D}$$

In Stephenson's description, the effect produced by the drug as a percentage of the maximal effect (e. g. E / E_{max}) is some function of the stimulus, but the actual relationship between the effect and the stimulus was considered unknown.

$$\frac{E}{E_{max}} = f(S) = \frac{e[D]}{[D] + K_D}$$

2.19. Estimation of Dissociation Constant of Agonists

Furchgott [65] and Furchgott and Bursztyn [66] introduced the term intrinsic efficacy (ε) which describes the ability of an agonist to produce a response after its binding to a *single* set of receptors. This is expressed in the following equation:

$$\varepsilon = \frac{e}{R_t}, \; e = \varepsilon\,[R_t]$$

Thus, intrinsic efficacy describes the ability of a drug to act as an agonist *per molecule of receptor.* Its value would not vary with receptor densities of different tissues. According to Furchgott, the stimulus produced when the receptor is occupied by an agonist is equal to the product of intrinsic efficacy multiplied by the concentration of agonist-receptor complex.

The effect of the drug as a fraction of the maximal effect is a function of efficacy, the fraction of total receptors occupied by the drug and the number of receptors in the tissue. A given drug may appear as a full agonist, partial agonist, or antagonist depending upon the tissue being studied. An agonist may exhibit different potencies in different tissues even though it is binding to the same type of receptor.

The dissociation constant (K_D) of the agonist-receptor complex, which is the concentration of agonist occupying 50% of receptors, is a characteristic of the agonist being studied. As mentioned above, this value is not equivalent to the EC_{50} because of spare receptors presence in the tissues [60]. However, a procedure has been developed for the calculation of the relative K_D of an agonist.

Furchgott [65] and Furchgott and Bursztyn [66] presented a method for estimating the dissociation constant of an agonist, which is based upon the addition of an irreversible receptor antagonist, which permanently inactivates a fraction of the total number of receptors. As indicated in equations, the relationship between the drug effect and the biological stimulus produced by a drug is:

$$\frac{E}{E_{max}} = f\,(S) = f\,(\varepsilon[DR])$$

And as shown above:

$$[DR] = \frac{R_t[D]}{[D] + K_D}$$

$$\frac{E}{E_{max}} = f\ (\varepsilon[DR]) = \frac{\varepsilon[D][R_t]}{[D] + K_D}$$

The maximum response produced by the agonist will be dependent upon the number of receptors inactivated and the amount of spare receptors present in the tissue. Lower concentrations of irreversible antagonist may decrease the number of receptors in the tissue, but the number of receptors may still be in excess and allow the agonist to produce the same maximal response as before the antagonist. Thus, although there are fewer receptors, the maximum response to the agonist remains the same as that without the antagonist. However, under these conditions, there may be a shift of the concentration-response curve to the right without changing the maximal response and the curve may resemble that produced by a competitive antagonist. At higher concentrations of irreversible antagonist, the number of receptors available for activation may be insufficient to produce a maximal response, agonist response is reduced [Figure 6].

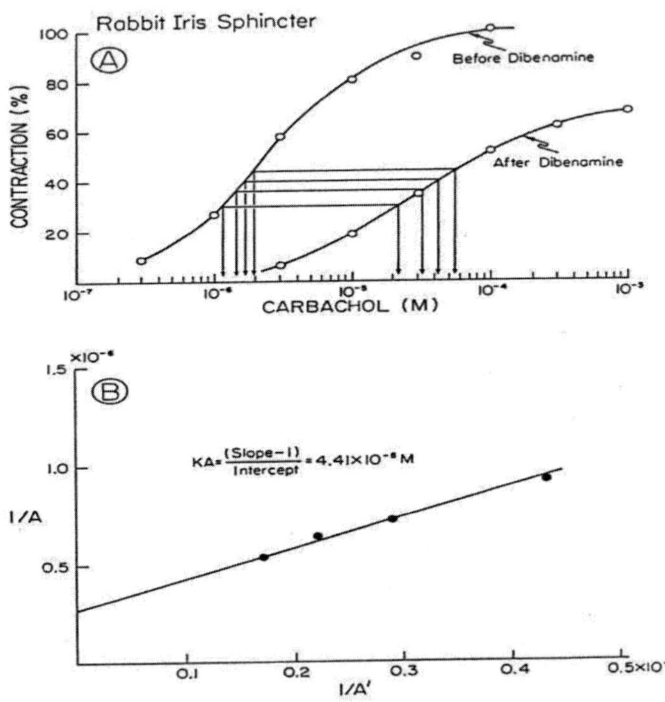

Figure 6. Effect of carbachol on the rabbit iris sphincter before and after dibenamine, an irreversible antagonist. (A) Dibenamine produced a shift to the right and reduced the maximum of the carbachol concentration-response curve. (B) A plot of the reciprocals of the concentrations of agonist that produced the same response before (A) and after (A') irreversible antagonism yields a straight line. Parameters of this line are used to calculate the K_D of carbachol for interaction with the iris sphincter muscle. Graph from Akesson et al. [62] with the permission from Springer Verlag.

Another approach that has been used to evaluate agonist activity is the operational model developed by Black and Leff [67]. This model is based on an analysis of the relationship between the concentration of agonist and the effect produced. The shape of the curve describing this relationship is usually a rectangular hyperbola, which converts to a sigmoidal curve when the log of the agonist concentration is plotted against the effect. The generation of the agonist-induced effect involves two major steps. The agonist must first interact with the receptor to form an agonist-receptor complex (AR), which follows the law of mass action. The K_D for this reaction is the concentration of agonist which occupies 50% of the receptors, and maximum concentration of the complex occurs when all the receptors are occupied by the agonist.

2.20. Competitive and Noncompetitive Antagonists

Antagonists interact with receptors, but unlike agonists, have no intrinsic activity. Thus, the drug-receptor complex does not produce a biological response. However, since antagonists occupy the receptor, they will inhibit the interaction between the agonist and the receptor according to the principles of mass action, resulting in an inhibition of the agonist-induced pharmacological response. Endogenous compounds such as hormones, neurotransmitters and autacoids are agonists and elicit a response by activating receptors. Therefore, specific antagonists of those receptors are able to inhibit a variety of physiological and pathological processes in which endogenous compounds are involved as agonists.

Pharmacologic antagonists can either interact with the receptor reversibly or irreversibly depending on their rate of dissociation from the receptor. The inhibitory effects of reversible competitive antagonists can be overcome by increasing agonist concentration. This type of blockade is also referred as surmountable. Thus, the dose-response curve for an agonist in the presence of a reversible antagonist is shifted to the right, but there is no change in the maximum response.

The K_D value, the equilibrium dissociation constant for an antagonist, can be determined by evaluating the response of the agonist, D, in the absence and in the presence of the antagonist [68,69]. If the response produced by the agonist in the presence of a competitive antagonist, I, is the same as that produced in the absence of the antagonist, then the fraction of receptors occupied by the agonist ($[DR]/R_t$) under both conditions should be the equal. Therefore, the equation for the proportion of receptors occupied by the agonist alone should equal that for the proportion of receptors occupied by the agonist in the presence of the antagonist.

The ratio of the agonist concentrations in the presence and absence of antagonist (inhibitor or blocker) that will produce equivalent responses is termed the dose-ratio. This equation relates the dose-ratio to the K_D and the concentration of inhibitor. The dose-ratio for the agonist depends on the concentration of antagonist and the affinity of the antagonist. It does not depend upon the size of the agonist-induced response. The dose-ratio is evaluated graphically to determine the K_D of the antagonist. The equation for calculation of K_D is presented below.

To determine the dose-ratio, concentration-effect curves for the agonist are plotted against different concentrations of the competitive antagonist. As the concentration of the competitive antagonist is increased, the agonist-response curves will exhibit a parallel shift to the right with no change in the maximum response. From these agonist-response curves, the dose ratios at EC_{50} of the agonist can be determined (Refer to the figure 7).

$$\text{Dose Ratio } (r) = 1 + \frac{[B]}{K_D}$$

rearrange
B= Concentration
K_D=Dissociation constant

$$K_D = \frac{[B]}{r-1}$$

$\log K_D = \log [B] - \log (r-1)$
$pA_2 = -\log [B] + \log (r-1)$

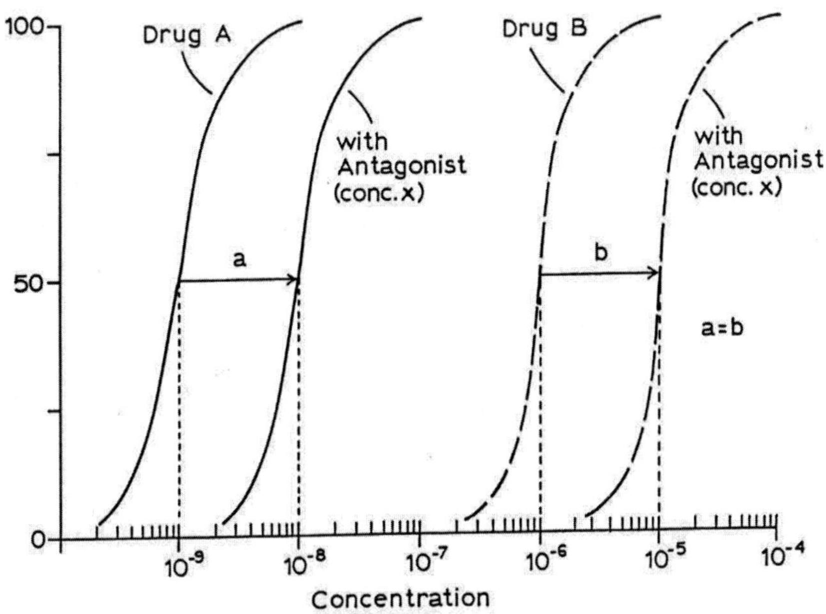

Figure. 7. The concentration-effect curves of two drugs A & B of different potencies are presented. Percent Effect is on vertical axis. In the presence of antagonist, the curve of each agonist is shifted to an equal extent i.e., the dose-ratio (r) = ED_{50} with blocker/ ED_{50} control, are equal or (a=b). Equal ratios generally indicate that both agonists act on a single type of receptor [69,70]. Figure reproduced from *Patil* [70] with permission credits to Indian Journal of Experimental Biology of the National Institute of Science, Communication and Information Resources (CSIR), India.

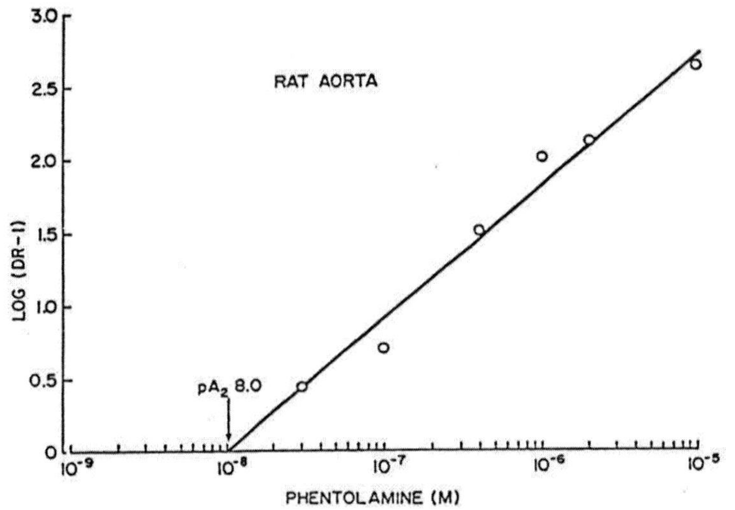

Figure 8. Schild plot for the antagonistic effects of phentolamine on α receptors of rat aorta. The agonist used was norepinephrine. The Y axis is log (DR - 1). X axis shows the concentration of the competitive antagonist, phentolamine. Note that the plot produces a straight line with a slope of 1. The X intercept is called the pA_2 and is equal to -log K_D , Figure from *Patil et al.* [71] with permission from Elsevier.

A Schild plot (Figure 8) refers to the regression line generated by graphing log (DR-1) on the Y axis against log concentration of antagonist on the X axis. This plot gives a straight line with a slope of 1. When the dose ratio is 2, then DR -1 = 1, and the log of 1 = 0. The X intercept is log K_D which is the equilibrium dissociation constant of the competitive antagonist.

The term pA_2, which has been used to describe the potency of a competitive antagonist, is the negative logarithm of the molar concentration of antagonist that produces a dose ratio $(D]_i/[D])$ of 2. Thus, pA_2 = -log K_B, K_B or K_D represents the affinity of the blocker for the receptor and is expressed in molarity units. It is independent of the agonist being used and the size of the response. If two agonists act on the same type of receptor, a fixed concentration of an antagonist will antagonize the response of both agonists to the same extent, producing the same K_B. Antagonists are useful for characterization of the receptor activated by the agonists. Similarities in the chemical structures of agonists do not mean that a single type of receptor is activated. The receptor type can be characterized by the K_B values of the competitive blocker against agonists.

In contrast to a competitive reversible antagonist, an irreversible antagonist tightly binds to the receptor usually by covalent bonding and dissociates very slowly, if at all, from the receptor. A nonequilibrium type of block is produced. Consequently, the receptor is occupied by the inhibitor, and the inhibition cannot be overcome by increasing the concentration of agonist. In other words, the blocking action is unsurmountable. As mentioned above, under conditions in which the response of an agonist is limited by the number of receptors, an irreversible antagonist will decrease the slope and the maximum response of the agonist-response curve. However, when spare receptors are present, the irreversible antagonist can inhibit the response of the agonist, which is reflected by a shift in the agonist-response curve to the right without decreasing the maximum response. In this case, although there is a decline in the number of receptors, there are still sufficient receptors available to produce a maximum response. The maximum response produced by a partial agonist, however, may be easily reduced by the irreversible inhibitor. This is because partial agonists have lower efficacies than full agonists and, therefore, must occupy a higher fraction of receptors to produce the same response as a full agonist. Consequently, for a partial agonist there is less receptor reserve.

Receptors can be protected from inactivation by irreversible antagonists by administering a competitive antagonist or an agonist prior to the exposure of the tissue to the irreversible antagonist. The competitive antagonist and the agonist will compete for the receptor sites and prevent the irreversible antagonist from interacting with these receptors.

It can be difficult in the laboratory to distinguish between a competitive reversible antagonist and an irreversible antagonist on the basis of their duration of action. A lipid soluble competitive antagonist may be retained by the tissue for a prolonged period, resulting in long term receptor antagonism. For example, the β-receptor antagonism produced by the competitive antagonist, propranolol, may have a long duration of action, giving propranolol the appearance of pseudo-irreversibly blockade of β - adrenoceptors.

Reversible noncompetitive antagonists inhibit the agonist induced response by acting at a site that is separate from the receptor. This site may be one of the steps in the sequence of events that occurs after receptor activation. Alternatively, the site affected may allosterically cause agonist binding inhibition. Noncompetitive antagonists would be expected to produce effects that are similar graphically to irreversible antagonists, e.g. a decrease in the slope and the maximum response of the agonist-response curve. Therefore, the time-course of recovery of the sensitivity of the tissue to the agonists after the antagonists have washed out must be determined.

2.21. Chemical Differences Between Agonists and Antagonists

What are the chemical properties of an agonist that make its interaction with the receptor different from that of an antagonist? While the complete answer to this question is not known, there are several possibilities to consider, and these may vary for different types of receptors: 1) the agonist, but not the antagonist, may contain specific reactive groups that can interact with the receptor; 2) the spatial arrangement of the reactive groups on the agonist molecule may be more effective for interaction with the receptor; 3) the agonist molecule may be more flexible than the antagonist molecule, permitting the specific conformation of the receptor to change during agonist-receptor interaction; 4) the agonist and antagonist may interact with different sites on the receptor, with the antagonists producing an allosteric inhibition of agonist-induced conformational change; 5) since antagonists are generally larger and more bulky molecules than agonists, chemical groups on the antagonist molecule may stabilize the receptor, hindering changes in receptor conformation; and 6) the sequence of interaction of critical groups of the drug with the receptor molecule may determine the intensity of the receptor activation [72].

2.22. Thermodynamic Analysis

One approach that has the potential of distinguishing between the effects of agonists and antagonist on the receptor is to evaluate the thermodynamic properties of their receptor interactions [73-75]. This approach is based on the premise that differences in the interaction of agonists and antagonists with receptors would be reflected in differences in enthalpy and/or entropy, the factors that determine free energy change (G). Enthalpy (H) is the heat content of a substance per unit mass and reflects the internal energy in the system. Entropy (S) is a measure of the randomness or disorder of the system. Changes in enthalpy and entropy associated with a chemical reaction are related to the change in free energy by the following equation, which holds for conditions of constant temperature, volume and pressure:

$$\Delta G = \Delta H - T\Delta S$$

ΔH is the change in enthalpy; T is the absolute temperature; and ΔS refers to the change entropy. ΔG is the difference in free energy of the reactants and the products. A reaction will be favored if the free energy associated with the products is lower than that associated with the reactants. There is a net loss of free energy when the product is formed. ΔG can be related to the equilibrium constant (K_{equil}) which is the reciprocal of the dissociation constant, K_D, by the following equation:

$$\Delta G = \Delta G^{\circ} + RT \ln K_{equil}$$

ΔG is the free energy change under standard conditions. When a chemical reaction reaches equilibrium, $\Delta G = 0$ and the equation becomes:

$$\Delta G^{\circ} = - RT \ln K_{equil}$$

Substituting the dissociation constant (K_D) for K_{equjl}:

$$\Delta G^{\circ} = RT \ln K_D$$

At standard conditions:

$$\Delta G^{\circ} = \Delta H^{\circ} - T\Delta S^{\circ}$$

Combining equations [74] results in the Van't Hoff Equation:

$$\ln (K_D) = \left(\frac{\Delta H^{\circ}}{R}\right)\left(\frac{1}{T}\right) - \frac{\Delta S^{\circ}}{R}$$

When ln K_D is plotted against l/T, a straight line is formed with a slope of $\Delta H°/R$ and a Y intercept of $\Delta S°/R$. Since R is the gas constant which is known, $\Delta H°$ and $\Delta S°$ can be calculated.

One problem with interpreting changes in enthalpy and entropy is that changes in thermodynamic properties associated with the binding of a drug to a receptor may reflect a summation of chemical changes associated with the interaction process. These include such interactions as the binding of the drug to receptor recognition site, changes in the specific conformation of the receptor as a result of the binding of an agonist, changes in the nonspecific conformation of the receptor as a result of binding of an antagonist, the interaction of an agonist-receptor complex with a membrane component, hydrophobic bonding resulting from the interaction of lipophilic groups on the drug with the membrane, displacement of water molecules surrounding the drug and receptor, among others. Therefore, changes in enthalpy and entropy may reflect over all drug effects other than those mainly concerned with the drug-receptor interaction.

While a variety of receptors have been studied, it has been generally difficult to distinguish between the effects of agonists and antagonists on the basis of differences in their thermodynamic properties. The most interesting results are from studies of β-adrenoceptors [73,76,77]. Based on calculations from studies of the binding of ligands to membranes at different temperatures, it has been found that antagonist binding to the β-adrenoceptor produces a negative free energy that is due to large changes in entropy. The latter was postulated to be due to hydrophobic binding. Agonist binding was also associated with a decrease in free energy, but it was largely attributed to an increase in enthalpy, which was able to overcome a concomitant small decrease in entropy. Studies of the thermodynamic changes produced by agonists and antagonists were also done on solubilized β-adrenoceptors, and the changes in entropy and enthalpy for agonists and antagonists were similar to those reported for membrane bound receptors. This suggests that the enthalpy and entropy changes associated with agonist binding to receptors in the low affinity state is not related to interactions with the lipid membrane environment of the receptor or the interaction of the agonist-receptor complex with G protein.

2.23. Inverse Agonists

An inverse agonist binds to the same receptor as a classical agonist and produces a response, but the response is in the opposite direction to that produced by the classical agonist [78]. Presumably, the receptor changes its conformation in such a way that a qualitatively different type of response is produced. Certain compounds, like ethyl β-carboline-3-carboxylate (β-CCE), have been proposed to be inverse agonists of the benzodiazepine receptor. These compounds inhibit the interaction of endogenous GABA with its receptor, thereby decreasing chloride entry into the neuron. The proanxiety effect of β-CCE on the functional level parallels these proposed receptor related interactions, while the classical agonists, the benzodiazepines, exert an antianxiety effect. Furthermore, β-CCE has proconvulsant, while the benzodiazepines have anticonvulsant properties.

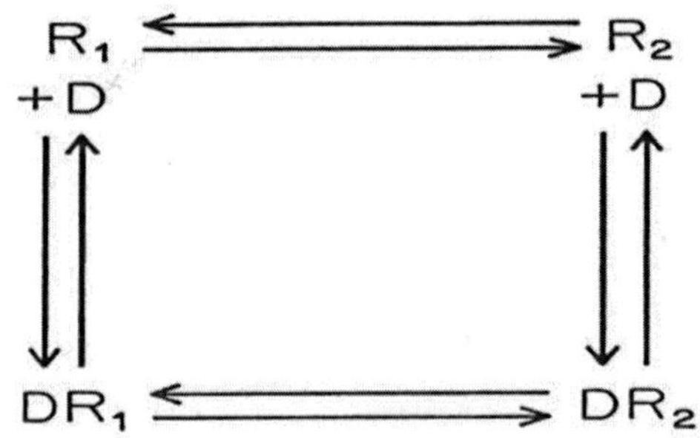

Figure 9. Two-state model of drug-receptor interaction. According to this model, the receptor exists in two interconvertible states, R_1, an inactive state and R_2, an active state. Drugs (D) can interact with either state. A full agonist will bind selectively to R_2, shifting the equilibrium to the R_2 state, the active conformation, producing a pharmacological response. A partial agonist will bind to both states but will have a higher affinity for R_2, producing only a small shift in equilibrium to the active form. Thus, the maximal effect of a high dose of partial agonist would be limited. An inverse agonist will have a higher affinity for the inactive R_1 state, causing a shift in equilibrium to the inactive conformation; therefore, the effect produced by the inverse agonist may be opposite to that of the agonist.

The interactions of full agonists, partial agonists and inverse agonists with receptors have also been explained using a dynamic model of the receptor (Figure 9) [47-79]. That is, the receptor is postulated as existing in two important interconvertible states, an active state (R_2) and a resting state (R_1). The conversion between the two states is regulated by an equilibrium constant (K). In the absence of agonist, most of the receptors are in the resting state. A full agonist binds preferentially to the active form of the receptor (R_2), which would shift the equilibrium to increase the proportion of receptors in the active form. Assuming no spare receptors, at saturating concentrations of agonist, 100% of the receptors would be in the active form, producing a maximum response. A partial agonist would bind to both receptor forms, R_2 and R_1, but has a higher affinity for the active form (R_2). When the receptors are saturated with the partial agonist, a smaller proportion of the receptors would be in the active form compared to that occurring with a full agonist. The binding of the partial agonist to the R_1 form would limit the amount of molecules formed and consequently, the maximal response that can be produced. In contrast to an agonist, an antagonist would bind equally well to both forms of the receptor, resulting in no change in the relative proportion of inactive and active forms of the receptors (the ratio of R_2/R_1 remains the same). The antagonist will, therefore, not produce an agonist action but will block the ability of an agonist to bind to either form of the receptor. An inverse agonist would have a higher affinity for the receptor in the resting state (R_1) and shift the equilibrium toward R_1. As a result, there would be an increase in the proportion of the receptors present in the state, which would produce effects that are opposite to those produced by an agonist.

2.24. Some Factors that Modify Drug-Response Relationships

The response to a drug is dependent on the free concentration of the drug in the vicinity of the receptor site. It may be influenced by a variety of factors. These factors include transport mechanisms, metabolic enzymes, and binding sites within the tissue that would lower the concentration of drug in the vicinity of the receptors. For example, acetylcholine would be hydrolyzed by acetycholinesterase and norepinephrine would be subject to transport mechanisms, which would decrease the concentration of these compounds that can interact with the receptor.

In the iris, lipid soluble drugs can bind extensively to melanin pigment, which would reduce the amount of drug available to interact with receptors. For example, the effectiveness of atropine in blocking muscarinic receptors of the sphincter muscle of the iris is lower than predicted because it binds to the melanin pigment. The free concentration of the blocker is lowered. As a result the slope of the pA_2 plots corresponds to a composed line in this tissue (Figure 10).

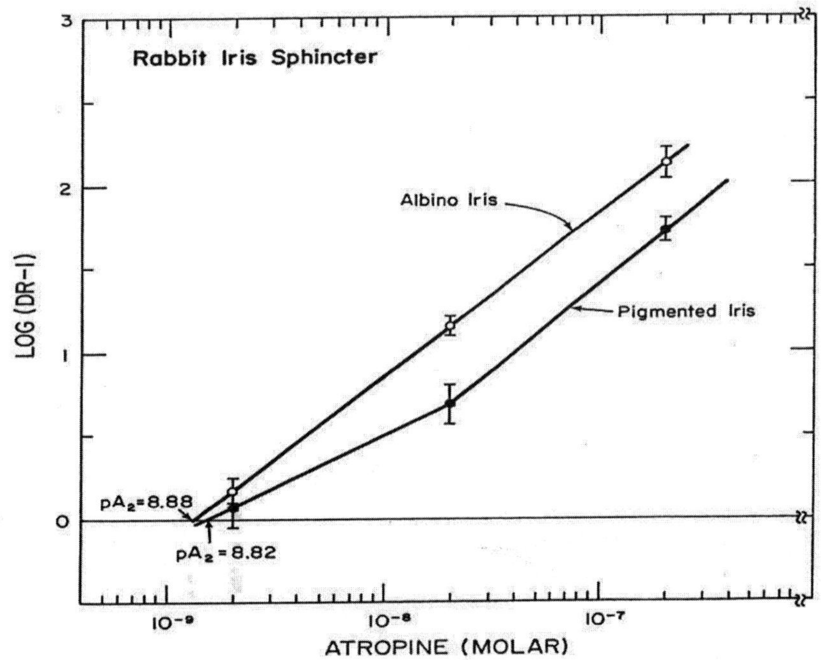

Figure. 10. Schild plot for the antimuscarinic effects of atropine on the sphincter muscle of albino and pigmented rabbit iris. The agonist used for this study was carbachol. The Schild plot for atropine in pigmented iris is not linear because at high concentrations the binding of atropine to the melanin pigment decreases the concentration available for interacting with the muscarinic receptor. Figure from Salazar et al. [80] reproduced with permission from the American Society for Pharmacology and Experimental Therapeutics.

The state of the tissue as a result of prior drug treatments or surgical manipulations may alter the responses to a drug. For example, if a sympathetically innervated tissue is pretreated with cocaine, the response of this tissue to norepinephrine will be enhanced. Cocaine, by inhibiting the high affinity transport of norepinephrine into the sympathetic nerve endings, would produce an increase in the concentration of norepinephrine available to activate noradrenergic receptors, causing the concentration-response curve to shift to the left. It is of interest that inhibition of the activity of the enzymes, monoamine oxidase and catechol-o-methyl transferase produces a much smaller enhancement of norepinephrine effects, because these enzymes are more concerned with long term changes in norepinephrine concentration at the synapse and less with acute changes. Similarly, other conditions that affect the noradrenergic synapse, such as postganglionic sympathetic denervation, chronic reserpine, and ganglionic blockade also can produce a shift to the left in the response to norepinephrine. Denervation, like cocaine, results in a loss of uptake sites, enhancing the synaptic concentration of norepinephrine. Reserpine, ganglionic blockade, and denervation attenuate the norepinephrine released from the sympathetic nerve endings, producing receptor supersensitivity. In these situations, there is an increase in the number of receptors along with other membrane changes and consequently the response to norepinephrine is enhanced.

A competitive antagonist can block the receptors for both endogenous neurotransmitter and exogenously administered agonist. However, it has been found at some synapses that the antagonist is less effective in inhibiting the response to the stimulated release of the endogenous neurotransmitter than that to exogenously administered drugs that activate the receptor. For example, while the competitive α_1 receptor antagonist, phentolamine, will block the effects of both exogenously administered and endogenously released norepinephrine, it produces a greater inhibition of the effects of exogenously administered norepinephrine. There are several possible explanations for this paradoxical effect of phentolamine. Substances may be released from neurons (in addition to norepinephrine) that produce a response that is similar to that of norepinephrine but mediated by a different receptor, which is not blocked by phentolamine. Alternatively, the endogenously released norepinephrine may act at different sites than exogenous norepinephrine, and phentolamine may not reach the required equilibrium concentration at the neuroeffector junction where high concentrations of norepinephrine are released. It is also possible that phentolamine by blocking autoreceptors on presynaptic nerve terminals markedly enhances norepinephrine release at the neuroeffector junction, which would tend to overcome the blockade by phentolamine at the postjunctional site. Since the response to exogenous norepinephrine does not require nerve activity, it would be less affected, if at all, by autoreceptor blockade. Thus, the effectiveness of a given dose of antagonist may depend on whether the response to an endogenous or exogenous substance is being antagonized.

When the effect of a drug is determined in vivo, the response may be modified by adaptive reflex mechanisms that oppose the action of the drug. For example, the response to vasoconstrictor drugs would be rapidly reduced by vasodilation and cardiac slowing, while the response to vasodilators would be offset by vasoconstriction and cardiac stimulation. Thus, while a drug may produce a response in one direction, compensatory mechanisms may be triggered that stimulate the opposite response.

2.25. Binding Studies

The interaction of a drug molecule with a receptor can be studied directly by measuring the ability of the drug to bind to the receptor. Kinetic analysis, involving the determination of the rates of association and dissociation, is used to determine dissociation constants [81-83].

For equilibrium analysis of the binding of a drug to a receptor, a compound with high radioactive specific activity and high affinity for the receptor is used as a ligand. Various concentrations of this radiolabeled drug are incubated with a membrane preparation suspended in buffer. Incubation is continued until binding equilibrium is attained. The membranes are then separated from the incubation medium, which contains free ligand, by either centrifugation or filtration, and the radioactivity in both fractions is determined by liquid scintillation counting. Some of the radiolabeled compound will bind to sites in the tissue other than the receptor that is being studied. This nonspecific binding is estimated by incubating the membrane preparation with the radioligand in the presence of a saturating concentration of an unlabelled drug that can specifically bind to the receptor being studied and will, therefore, compete with the ligand for binding. This unlabeled compound should inhibit the specific binding of the radioligand to the receptor, but will have low affinity for the nonspecific binding sites of the radioligand. The radioactivity that remains bound to the membrane preparation in the presence of this nonlabeled compound, which represents nonspecific binding, is subtracted from the total radioactivity, determined in the absence of the nonlabeled drug, to yield the specific binding of the radioligand to the receptor.

The binding of a selective ligand to a homogeneous class of receptors should display certain characteristic properties. The ligand should bind to the receptor with high affinity, and the binding should be saturable. It is essential that the binding properties of the ligand correlate with the pharmacological properties of the drug if the binding site is a receptor. Thus, receptor agonists and antagonists should be able to inhibit the binding of the ligand with relative affinities that parallel their order of potency in pharmacological studies.

The relationship between the concentration of ligand [L] to the amount of radioligand bound to the receptor [B] is expressed in the following equation:

$$B = \frac{B_{max}[L]}{[L] + K_D}$$

K_D is the equilibrium dissociation constant, and B_{max} is the maximum binding that can occur. This equation, which expresses a nonlinear rectangular hyperbola, can be rearranged to the following equation:

$$\frac{B}{[L]} = \frac{-B}{K_D} + \frac{B_{max}}{K_D}$$

This equation is in linear form. Plotting the bound/free radioligand on the Y axis against bound radioligand on the X axis is termed a Scatchard Plot and yields a straight line with slope equal to 1/ K_D and an X intercept equal to B_{max}. However, the Scatchard plot frequently is not linear, indicating that the interaction between ligand and receptor is complex. This may occur if the ligand binds to more than one class of binding sites or the interaction of the ligand with the receptor involves cooperativity.

Studies on the ability of drugs to inhibit radioligand binding to receptors are used frequently to characterize the interactions of drugs with receptors. Drugs whose pharmacological effects are mediated by their interactions with specific receptors, should specifically bind to these receptors, inhibiting the binding of the radioligand. An evaluation of the order of potency of drugs in inhibiting radioligand binding can provide information on the mechanism of drug action and can be used to test whether a given radioligand is binding to a specific receptor class.

In these types of experiments the binding of a specific radioligand is measured in the presence of various concentrations of unlabeled drug. If the unlabelled drug can bind to the receptor, then a curve can be produced with the following shape (Figure 11).

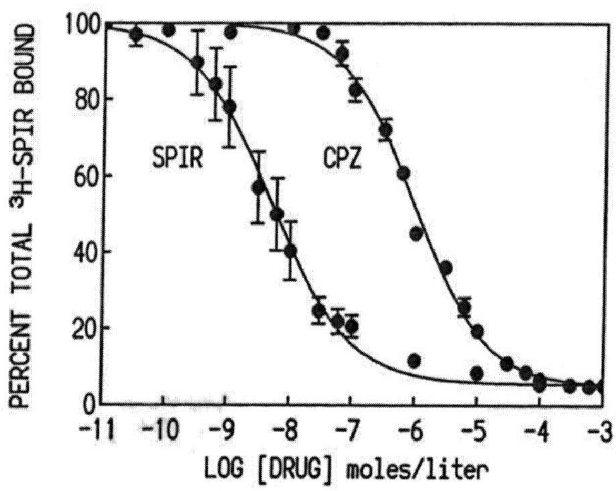

Figure. 11. Effect of spiperone (SPIR) and chlorpromazine (CPZ) on the binding of [^{3}H]-spiperone to a membrane preparation of rat striatum.

The IC_{50} (concentration that inhibits specific binding by 50%) can be determined from a nonlinear regression analysis using the equation:

$$[B] = \frac{[B_0]}{1 + \frac{[L]}{IC_{50}}}$$

In this equation, [B] = the concentration of bound ligand in the presence of an inhibitor; B_0 is ligand binding in the absence of the inhibitor; [L] is the concentration of the ligand; and IC_{50} is the concentration of inhibitor when binding is inhibited by 50%. The K_i of the inhibitor can then be calculated using the Cheng-Prusoff equation [84].

$$[K_i] = \frac{IC_{50}}{1 + \frac{[L]}{K_D}}$$

K_D is the dissociation constant of the radiolabeled ligand used for the study. The dissociation constant derived from binding studies and that derived from functional studies using the Schild plot describe affinity of the drug for its receptor.

The IC_{50} values for an inhibitor of radioligand binding can also be determined from a modification of the Hill plot [75]. This graph is drawn from the following equation, which was originally developed to describe the cooperative interaction of oxygen with hemoglobin [85]:

$$\text{Log}\left(\frac{[B]}{[B_0]-[B]}\right) = n_H \bullet \log[I] - n_H \bullet \log IC_{50}$$

A plot of log $[B]/[B_0]$ - [B] on the Y axis against log [I] on the X axis approximates a straight line in which the slope is n_H (called the Hill coefficient) and the X intercept IC_{50}. An n_H of one indicates that the ligand molecules bind independently of each other to one site. When the slope is different from one, it indicates that the binding of the inhibitor does not follow the laws of mass action and suggest multiple binding sites for the ligand and/or cooperativity in binding (the binding of ligand molecules at one site affects binding to another site).

The radioligand may bind to more than one class of receptors within the membrane. For example spiperone, which is used frequently to analyze D_2 dopaminergic receptors, also binds to the 5-HT_2 serotonergic receptor. In the case of two or more binding sites, the binding of the radioligand to the membrane preparation will depend upon its relative affinity for the different sites. One approach that is used to study binding when the radioligand binds to two sites is to add to the medium an unlabeled compound, which has a high affinity for the receptor(s) type that is not being studied. This compound, when added in sufficient concentration (100 fold higher than its for this receptor), should prevent the radioligand from binding to these receptors, but not affect the binding of the radioligand to the receptors being studied. Hence, the radioligand would only be able to bind to receptors not affected by the unlabeled compound. In the case of radiolabeled spiperone, cinanserin or ketanserin, which are 5-HT_2 receptor antagonists, are used to inhibit the binding of radiolabeled spiperone to the 5-HT_2 receptors. These compounds at the concentrations used have very low affinity for dopaminergic receptors but do not interfere with D_2 dopaminergic receptor binding, and hence allow the study of these dopaminergic receptors by radiolabeled spiperone.

Receptors that can bind to G proteins that are located within membranes can exist in high or low affinity states, which can be distinguished on the basis of agonist binding. In contrast antagonists appear to bind to the two states with equal affinity. These receptors first bind to agonists in their low affinity states forming the agonist-receptor complex. The binding of the agonist changes the conformation of the receptor, exposing a binding site for the G protein located within the membrane. The agonist receptor complex then binds to the G protein forming a ternary complex of agonist-receptor-G protein. The receptor in this form is in the high affinity state as the rate of dissociation of the agonist from the ternary complex is very low. The displacement of GDP by GTP (or guanine nucleotide analogs) from a site on the G-protein causes the G-protein to dissociate from the receptor, restoring the agonist-receptor complex binary complex, which is the low affinity state of the receptor. These changes can be observed by measuring the ability of an unlabeled agonist to inhibit the binding of a radiolabeled antagonist. The resultant inhibition curve for the agonist has a shallow slope with a Hill coefficient of less than 1, consistent with the presence of high and low affinity forms of the receptor. With the addition of GTP or other guanine nucleotides, the slope of the inhibition curve is increased, while the affinity of the agonist for the receptor is decreased. These changes are consistent with GTP causing the conversion of receptors from the high affinity state to the low affinity state.

The maximum binding (B_{max}) of the radioligand will occur at the concentration of radioligand that saturates the receptors being studied. This value is a function of the total number of receptors in the preparation and cannot be determined from functional studies in intact tissues where spare receptors can exist. The maximum binding of radioligand to receptors has been shown to increase under certain conditions, such as receptor denervation or hyperthyroidism and been used to explain the enhanced functional response to agonists. Similarly, a decrease in the number of receptors is usually associated with a decreased response to agonists. An increase in receptor number will produce an enhanced functional response if the receptors are linked to molecules, such as G proteins, that can mediate the response after receptor activation.

ACKNOWLEDGMENTS

Authors are delighted for the help from Dr. Tatiana González and Varathorn Bookaman M.S., Sarmistha Mazumder & Betsy Bulgrin who happily coordinated the preparation of this manuscript across the great distances.

REFERENCES

[1] Chopra, R.N., Chopra, I.C., Hanada, K.L. & Kapur, L.D. (1958) *Indigenous drugs of India.* Calcutta, India. U.N. Dhur & Sons Private Ltd.

[2] Aikman, L. (1977) *Nature's Healing Arts.* Washington D.C. National Geographic Society.

[3] Shen, M-L. (1983) The contribution of traditional Chinese medicine to modern pharmacology. *Trends Pharmacol Sci, 4*, 496-500.

[4] Schultes, R.E. & Raffauf, R.F. (1990) *The Healing Forest.* Portland, OR. Dioscorides Press.

[5] Olmsted, J.M.D. (1944) *Francois Magendie.* New York, NY. Schuman's.

[6] Leake, C.D. (1961) The scientific status of pharmacology. *Science, 134*, 2069-2079.

[7] Gaddum, J.H. (1962) The Pharmacologists of Edinburgh. *Annu Rev Pharmacol, 2*, 1-10.

[8] Cushny, A.R. (1926) *Biological relations of optical isomeric substances.* Baltimore, MD. Williams and Wilkins Co.

[9] Patil, P.N., Miller, D.D., & Trendelenburg, U. (1974) Molecular geometry and adrenergic drug activity. *Pharmacol Rev, 26*, 323-392.

[10] Ariens, E.J., Soudijn, W., & Timmermans, P.B.M.W.M. (1983) *Stereochemistry and Biological Activity of Drugs*. Oxford, UK. Blackwell Scientific Publication.

[11] Wainer, I.W. & Drayer, D.E. (1988) *Drug Stereochemistry.* New York, NY. Marcel Dekker.

[12] Hutt, A.J. (1991) Drug chirality: Impact on Pharmaceutical Regulation. *Chirality, 3*, 161-164.

[13] Sun, N.S. & Xiao, P. (1992) Personal Communication.

[14] Rastogi, R.P. & Dhawan, B.N. (1982) Research on medicinal plants at the Central Drug Research Institute, Lucknow (India). *Indian J Med Res, 76* (Suppl.), 27-45.

[15] Klayman, D.L. (1985) Qinghaosu (Artemisinin): An antimalarial drug from China. *Science, 228*, 1049-1055.

[16] Hansch, C. (1990) *Comprehensive Medicinal Chemistry.* Oxford, UK. Pergammon Press.

[17] Sun, N.S. & Xiao, P. (1989) Recent advances in studies of medicinal plants in China, presented at IX National Congress of Pharmacy, Bucharest, Romania, Sept. 1989.

[18] Ware, G.W. (1983) *Pesticides: theory and application.* San Francisco, CA. W.H. Freeman & Company.

[19] Balandrin, M.F., Klocke, J.A., Wurtele, E.S. & Bollinger, W.H. (1985) Natural plant chemicals: Sources of industrial and medicinal materials. *Science, 228*, 1154-1159.

[20] Eckert, R., Randall, D. & Augustine, G. (1988) *Animal Physiology.* New York, NY. W.H. Freeman and Company. p. 188.

[21] Crescitelli, F. & Geissman, T.A. (1962) Invertebrate pharmacology, selected topics. *Annu Rev Pharmacol, 2*, 143-192.

[22] Atkins, P.W. (1987) *Molecules.* New York, NY. Scientific American Library, W.H. Freeman Company.

[23] Fisher, M.H. & Mrozik, H. (1992) The chemistry and pharmacology of avermectins. *Annu Rev Pharmacol Toxicol, 32*, 537-553.

[24] Stocker, K.F. (1990) *Medical uses of snake venom proteins.* Boston, MA. CRC Press.

[25] Dumbacher, J.P., Beehier, B.M., Spande, T.F., Garraffo, H.M., & Daly, J.W., (1992) Homobatrachotoxin in the genus Pitohui: Chemical defense in birds? *Science, 258*, 799-801.

[26] Scheuer, P.J. (1990) Some marine ecological phenomenon: Chemical basis and biomedical potential. *Science, 248*, 173-177.

[27] Pressman, B.C. (1976) Biological applications of ionophores. *Annu Rev Biochem, 45*, 501-530.

[28] Wolff, H.P., Fleckenstein, A., & Philipp, E. (1989) Drug research and drug development in 21st century science and ethics, Bayer AG Centenary symposium (1988). *Arzneimittel Forschung, 39*, 935-1072.

[29] Spilker, B. (1989) *Multinational Drug Companies: Issues in drug discovery and development.* New York, NY. Raven Press.

[30] Sneader, W. (1985) *Drug Discovery: The Evolution of Modern Medicines.* New York, NY. John Wiley & Sons.

[31] Levitzki, A. (1984) *Receptors: A Quantitative Approach*. California. Benjamin/ Cummings, Menlo Park.

[32] Taylor, P. & Insel, P. (1990) *Principles of Drug Action: Molecular basis of drug action.* 3rd. Edition. New York, NY. Churchill Livingstone. pp 103-200.

[33] Changeux, J-P, Giraudat J., & Dennis, M. (1987) The acetylcholine receptor: molecular architecture of a ligand-regulated ion channel. *Trends Pharmacol Sci, 8*, 459-465.

[34] Gilman, A.G. (1989) G Proteins and regulation of adenylyl cyclase. *J Amer Med Assoc, 262*, 1819-1825.

[35] Lefkowitz, R.J., Kobilka, B.K., & Caron, M. (1989) The new biology of drug receptors. *Biochem Pharmacol, 38*, 2941-2948.

[36] Jackson, T. (1991) Structure and function of G protein coupled receptors. *Phamacol Therap, 50*, 425-442.

[37] Stryer, L. (1986) Cyclic GMP cascade of vision. *Annu Rev Neurosci, 9*, 87-119.

[38] Lefkowitz, R.J., & Caron, M.G. (1988) The adrenergic receptors: Models for the study of receptors coupled to guanine nucleotide regulatory proteins. *J Biol Chem, 263*, 4993-4996.

[39] O'Dowd, B.F., Lefkowitz, R.J. & Caron, M. (1989) Structure of the adrenergic and related receptors. *Annu Rev Neurosci, 12*, 67-83.

[40] Tota, M.R., Candelore, M.R., Dixon, R.A. & Strader, C. (1991) Biophysical and genetic analysis of the ligand-binding site of the β-adrenoceptor. *Trends Pharmacol Sci, 12*, 4-6.

[41] Patil P.N., Li, C., Kumari. V. & Hieble J.P. Analysis of efficacy of chiral adrenergic agonist. *Chirality*, 2008, 20, 529 -543.

[42] Rasmussen, S.G.F., Choi H.J., Rosenbaum D.M., Kobilka T.S. Tian F.S. Edwards P.C. Burghammer., M. Ratmala V.R.P. Sanishvilli R, Fischetti, R.E. Schertler G.F.K. Weis W.I. & Kobilka B.K. Crystal structure of human β_2 – adrenergic G-protein coupled receptor. *Nature*, 2007, doi 10.1038/nature 06325.

[43] Warne T. Serano-Vega M.J. Backer J.G., Moukhametzianow, R. Edwards P.C., Henderson, R. Lesley A.G.W., Tate, & C.G. Shertler G.F.X, Structure of a β_1 adrenergic G-protein coupled receptor, *Nature*, 2008, 454, 486-491.

[44] Stryer, L., & Bourne, H.R. (1986) G-proteins: a family of signal transducers. *Annu Rev Cell Biol, 2*,391-409.

[45] Gilman, A.G. (1987) G Proteins: transducers of receptor generated signals. *Annu Rev Biochem, 56*, 615-649.

[46] Neer, E.J. & Clapham, D.E. (1988) Roles of G-protein subunits in transmembrane signaling. *Nature, 333*, 129-134.

[47] Ross, E.M. (1990) *Goodman and Gilman's The Pharmacological Basis of Therapeutics: Pharmacodynamics: Mechanisms of drug action and the relationship between the drug concentration and effect.* 8th. Edition. New York, NY. Pergamon Press. pp 33-48.

[48] Michell, R.H. (1986) *Phosphoinositides and Receptor Mechanisms: Inositol lipids and their role in receptor function, history and general principles.* New York, NY. Alan Liss.

[49] Berridge, M. (1987) Inositol triphosphate and diacylglycerol: Two interacting second messengers. *Annu Rev Biochem, 56*, 159-193.

[50] Yarden, Y. & Ullrich, A. (1988) Growth factor receptor tyrosine kinases. *Annu Rev Biochem, 57*, 443-478.

[51] Evans, R.M. (1988) The steroid and thyroid hormone receptor superfamily. *Science, 240*, 889-895.

[52] Moore, D.D. (1989) Promiscuous behavior in the steroid hormone receptor superfamily. *Trends Neurosci, 12*, 165-168.

[53] McKay, D.B., Chang, C., González-Cestari, T.F., McKay, S.B., El-Hajj, R., Bryant, D.L., Zhu, M.X., Swaan, P.W., Arason, K.M., Pulipaka, A.B., Orac, C.B., & Bergmeier. S.C. (2007) Analogs of methyllycaconitine as novel noncompetitive inhibitors of nicotinic receptors: pharmacological characterization, computational modeling, and pharmacophore development. *Mol Pharmacol, 71*, 1288-1297.

[54] González-Cestari, T.F., Henderson, B.J., Pavlovicz, R., McKay S.B., El-Hajj, R.A., Pulipaka, A.B., Orac, C.M., Reed, D.D., Boyd, R.T., Zhu, M.X., Li, C, Bergmeier, S.C., & McKay, D.B. (2009) Effect of novel negative allosteric modulators of neuronal nicotinic receptors on cells expressing native and recombinant nicotinic receptors: implications for drug discovery. *J Pharmacol Exp Ther*, *328*, 504-515.

[55] Fleming, W.W., Westfall, D.P., De La Lands, I.S. & Jellett, L.B. (1972) Log-normal distribution of equieffective doses of norepinephrine and acetylcholine in several tissues. *J Pharmacol Exp Ther, 181*, 339-345.

[56] Lands, A.M., Arnold, A., McAuliff, J.P., Luduena, F.P. & Brown, T.G. (1967) Differentiation of receptor systems activated by sympathomimetic amines. *Nature, 214*, 597-598.

[57] Buckner, C.K. & Patil, P.N. (1971) Steric aspects of adrenergic drugs, XVI. Beta adrenergic receptors of guinea pig atria and trachea. *J Phamacol Exp Ther, 176*, 634-649.

[58] Venkataraman, B.V., Hamada, A., Shams, G., Miller, D.D., Feller, D.R. & Patil, P.N. (1989) Paradoxical effects of isothiocyanate analog of tolazoline on rat aorta and human platelets. *Blood Vessels, 26*, 335-346.

[59] Clark, A.J. (1937) *General Pharmacology*. Berlin, Germany. Springer.

[60] Ruffolo, R.R. (1982) Important concepts of receptor theory. *Autonomic Pharmacol, 2*, 277-295.

[61] Ruffolo, R.R. Jr., (1984) Use of isolated physiologically responding tissues to investigate neurotransmitter receptors. *Monographs Neural Sci, 10*, 53-84.

[62] Akesson, C., Swanson, C., & Patil, P.N. (1983) Muscarinic receptors of rabbit irides. *Arch Pharmacol, 322*, 104-110.

[63] Ariens E.J. Affinity of intrinsic activity in the theory of competitive inhibition. *Arch. Int. Pharmacodyn.*, 1954, 99, 32-49.

[64] Stephenson, R.P. (1956) A modification of receptor theory. *Br J Pharmacol Chemother, 11*, 379-393.

[65] Furchgott, R.F. (1966) *Advances in Drug Research: The use of β-haloalkylamines in the differentiation of receptors and in the determination of dissociation constants of receptor agonists complexes*. London, UK. Academic press. pp 21-55.

[66] Furchgott, R.F., & Bursztyn, P. (1967) Comparison of dissociation constants and of relative efficacies of selected agonists on parasympathetic receptors. *Ann N Y Acad Sci, 144*, 882-889.

[67] Black, J.W., Leff, P., & Shankley, N.P. An operational model of pharmacological agonsim: the effect of E/ [A] curve shape on agonist dissociation constant estimation. *Brit. J. Pharmacol.,* 1985, 84, 561-571.

[68] Schild, H.O. (1949) pAx and competitive drug antagonism. *Br J Pharmacol Chemother, 4*, 277-280.

[69] Arunlakshana, O. & Schild, H.O. (1959) Some quantitative uses of drug antagonists. *Br J Pharmacol Chemother, 14*, 48-58.

[70] Patil, P.N. Pharmacologic quantitation, *Indian J. Exptl. Biolo.,* 1996, 34, 615-633.

[71] Patil, P.N., Fudge, K. & Jacobowitz, D. (1972) Steric aspects of adrenergic drugs, XVIII. α-adrenergic receptors of mammalian aorta. *Euro J Pharmacol, 19*, 79-87.

[72] Patil, P.N., Feller, D. R., & Miller, D.D. (1991) Molecular aspects of the receptor activation by imidazolines: An overview. *Neural Trans Suppl, 34*, 187-194.

[73] Weiland, G., & Molinoff, P.B. (1981) Quantitative analysis of drug-receptor interactions: I. Determination of kinetic and equilibrium properties. *Life Sci, 29*, 313-330.

[74] Raffa, R. & Porreca, F. (1989) Thermodynamic analysis of the drug receptor interaction. *Life Sci, 44*, 345-258.

[75] Miklavc, A., Kocjan, D, Mavri, J., Koller, J., & Hadzi, D. (1990) On the fundamental difference in the thermodynamics of agonist and antagonist interactions with β-adrenergic receptors and the mechanism of entropy driven binding. *Biochem Pharmacol, 40*, 663-669.

[76] Contreras, M.L., Wolfe, B.B., & Molinoff, P.B. (1986a) Thermodynamic properties of agonist interactions with the beta adrenergic receptor-coupled adenylate cyclase system: I. High and low affinity states of agonist binding to membrane-bound beta adrenergic receptors. *J Pharmacol Exp Ther, 237*, 154-164.

[77] Contreras, M.L., Wolfe, B.B., & Molinoff, P.B. (1986b) Thermodynamic properties of agonist interactions with beta adrenergic receptor-coupled adenylate cyclase system. II. Agonist binding to soluble beta adrenergic receptors. *J Pharmacol Exp Ther, 237*, 165-172.

[78] Gardner, C.R. (1988) Functional in-vivo correlates of the benzodiazepine agonist-inverse agonist continuum. *Prog Neurobiol, 31*, 425-476.

[79] Rang, H.P. & Dale, M.M. (1991) *Pharmacology: How drugs act: General principles.* Edinburgh, UK. Churchill Livingstone. pp 3-23.

[80] Salazar, M., Shimada, K. & Patil, P.N. (1976) Iris pigmentation and atropine mydriasis. *J Pharmacol Exp Ther, 197*, 79-88.

[81] Titeler, M. (1983) *Multiple Dopamine Receptors Receptor Binding Studies in Dopamine Pharmacology.* New York, NY. Marcel Dekker.

[82] Limbird, L.E. (1986) *Cell Surface Receptors: A Short Course on Theory and Methods.* Boston, MS.

[83] McGonigle, P. & Molinoff, P.B. (1989) *Basic Neurochemistry: Quantitative aspects of drug-receptor interactions.* 4th. Edition. New York, NY. Raven Press. pp.183-201.

[84] Cheng, Y.C. & Prusoff, W.H. (1973) Relationship between the inhibition constant (K_i) and the concentration of inhibitor which causes 50% inhibition (I_{50}) of an enzymatic reaction. *Biochem Pharmacol, 22*, 3099-3108.

[85] Hill, A.V. (1910) The possible effects of the aggregation of the molecules of haemoglobin on its dissociation curves. *J Physiol, 40*, iv-vii.

In: Biogenic Amines
Editors: T. Farooqui, A. A. Farooqui
ISBN: 978-1-60876-625-3

Chapter VIII

What Has 6-Hydroxydopamine Taught Us About Parkinson's Disease?

Michael J. Zigmond[*]
University of Pittsburgh, Pittsburgh, PA 15260, USA.

Abstract

6-Hydroxydopamine was first identified in 1959. By the early 1970s, this selective toxin for norepinephrine and dopamine neurons had become the principal tool by which laboratory studies of Parkinson's disease were performed, due largely to the pioneering work of Urban Ungerstedt and Norman Uretsky. The toxin has taught us an great deal about the possible causes, pathophysiology, and treatments of the disease. Although improved models are clearly required, some of which are being developed, even now half a century later 6-hydroxydopamine continues to play an important role in studies of parkinsonism.

Abbreviations

6-OHDA – 6-hydroxydopamine
DOPA – 3,4-dihydroxyphenylalanine
DA – dopamine
NE – norepinephrine
PD – Parkinson's disease
SN – substantia nigra

[*] Correspondence concerning this article should be addressed to: Michael J. Zigmond, Pittsburgh Institute for Neurodegenerative Diseases, University of Pittsburgh, Pittsburgh, PA 15260. Phone: 412-624-4258; Fax: 412-624-7327; Email: zigmond@pitt.edu.

Introduction

Parkinson's disease (PD) is a neurodegenerative disease that affects some 6 million people throughout the world. This places it second only to Alzheimer's disease in its incidence. Differences in life span and in diagnostic criteria make it difficult to know the precise history or demographics of PD, though this might provide useful clues as to its cause. However, it is almost certain that the disease existed well before its first modern description by James Parkinson [1], and is likely to be closely related if not identical to the condition termed "Kampavata" in the Indian literature of Ayurvedic medicine some 3,000-5,000 years ago and to disorders mentioned in ancient Chinese and Greek texts and in the Old Testament. Current prevalence ranges widely from less than 10 to more than 300 per 100,000, and there is some evidence that it is greatest in areas in which there is a high incidence of environmental toxins. On the other hand, during the last 10 years examples of familial PD have been uncovered, as have a number of genetic mutations that serve as risk factors.

The first neuropathology to be associated with PD was the loss of pigmented cells of the substantia nigra (SN) of the ventral mesencephalon [2]. Then, almost 40 years later, Oleh Hornykiewicz and his colleagues found extensive loss of the neurotransmitter dopamine (DA) in the caudate and putamen [4,5], confirming a suggestion by Arvid Carlsson that the loss of DA would result in motor abnormalities [6]. Soon thereafter, the development by Bengt Falck and Nils-Åke Hillarp and their students of methods for visualizing DA and other biogenic amines [7] provided evidence to suggest that the loss of neurons in the SN and the loss of DA in the caudate and putamen (mixed to form the "striatum" in rat and mouse) were essentially two ends of the same pathology – the degeneration of the DA-containing neurons of the SN and their protections to the striatum [8], [see also 9,10].

This demonstration that PD was associated with the loss of DA led to a lengthy period that could be termed the "dopaminocentric" phase of PD research, one that remains active and productive today. However, it has long been known that PD is associated with many symptoms other than the classical motor deficits, which include tremor, bradykinesia, and rigidity. These can include autonomic deficits, depression, and cognitive deficits. And beginning in the mid-1980s it became increasingly clear that these non-motor symptoms were accompanied by additional neuropathology, including loss of selected peripheral neurons and degeneration within the brainstem, other mesencephalic nuclei, and telencephalon. Although a great many people have contributed to this literature, the most influential investigator has been Heiko Braak, whose landmark papers have greatly expanded our appreciation of the diversity of deficits associated with PD [11,12]. However, knowing this and modeling it in the laboratory are two quite different matters, as I will note at the end of this review.

6-Hydroxydopamine Meets PD

The first animal models of PD involved drugs such as reserpine that depleted the brain of biogenic amines and surgical lesions that caused the destruction of the medial forebrain bundle through which the nigrostriatal projection runs. With the realization that the motor symptoms caused by these manipulations were due to the loss of DA came the wish to produce such a loss in a more specific manner. The tool that would allow the field to accomplish this came quite by accident with the discovery of 6-hydroxydopamine (6-OHDA). 6-OHDA, an analogue of DA bearing an additional hydroxyl group in the ortho position may have first been observed as a metabolite of the amino acid 3,4-dihydroxyphenylalanine (DOPA) [13]. In 1963, Curt Porter and his colleagues reported that the systemic administration of 6-OHDA given to mice, caused a depletion of cardiac norepinephrine (NE) [14]. This in itself was probably not surprising since other structural analogues of biogenic amines, such as tyramine, also depleted NE from sympathetically innervated tissues, presumably by displacing the endogenous amines. What *was* surprising was the enormous amount of time required for NE levels to be restored: It took 80 days for a depletion level of 90% to be restored to 10% for 6-OHDA, as opposed to 0.7 days for tyramine. Richard Laverty reported a similar observation while working with Dennis Sharman and Marthe Vogt [15]. Clearly, something totally unexpected was going on. The answer came a few years later when Torbjörn Malmfors and Charlotte Sachs reported the NE terminals could no longer be seen by histochemical fluorescence [16], and Hans Thoenen and Jean-Pierre Tranzer used electron microscopy to demonstrate the loss of NE profiles that was not accompanied by the loss of cholinergic terminals, indicating the selective nature of the toxicity [17].

6-OHDA's neurotoxic actions result from two of its characteristics, as Richard Heikkila and Gerry Cohen emphasized early on [18]. First, 6-OHDA is highly electroactive and rapidly oxidizes to form reactive oxygen species and 6-hydroxyquinone. Second, 6-OHDA is a substrate for the high affinity NE and DA transporters located on NE and DA neurons, respectively. Thus, when administered in an appropriate dose, 6-OHDA is concentrated in catecholamine neurons, overwhelms intracellular defenses, and causes a wide variety of damage. (The nature of that damage, as well as the issue of "appropriate dose" will be discussed later.)

Because 6-OHDA, like the biogenic amines themselves, does not cross the blood brain barrier, central catecholaminergic neurons were presumably unaffected in these studies. However, almost simultaneously with the work of Thoenen and Tranzer, several laboratories began to introduce the toxin directly into the parenchyma or fluid compartments of the brain. Among the first to accomplish this were Urban Ungerstedt [19] and Norman Uretsky [20]. There were others who contributed at about this time, including Floyd Bloom [21], George Breese [22], Richard Heikkila [23], John Marshall [24], and myself working with Edward Stricker [25]. However, it was Ungerstedt and Uretsky who arguably made the most important initial contributions to the development of 6-OHDA as an important tool in neuroscience.

Urban Ungerstedt focused on the behavioral effects of the toxin, and developed the highly influential model of the "rotating rat," the rat which had a unilateral lesion of DA neurons caused by the intracerebral injection of 6-OHDA on one side of the brain. Such rats rotate away from the side of the brain that shows the greater amount of drug-induced DA-like activity – away from the lesioned side in the case of amphetamine (which releases more DA from the intact side) and toward the lesioned side in the case of apomorphine (which stimulates DA receptors directly, those receptors increasing in number after large DA depletions) [26,27]. However, it was Norman Uretsky, then a newly arrived postdoctoral fellow with Leslie Iversen at the University of Cambridge, who did the most in those early years to help us understand the neurochemical effects of 6-OHDA.

In his first paper [20], Norman injected the 6-OHDA into lateral ventricles of young adult rats. He made several critical observations: First, NE values fell within 24 hours (the shortest time examined) and showed no apparent recovery of NE levels by 32 days (the longest time interval examined). Second, except at extremely high doses of 6-OHDA, there were no behavioral deficits that he could observe even though NE levels were reduced by 60% (He did not initially measure DA.) Third, the uptake of radio-labeled NE into hypothalamic prepared from 6-OHDA-treated animals was greatly reduced both 2 and 16 hours. Over the next few years, Norman published 5 more papers on the neurochemical effects of 6-OHDA, in which he showed that DA terminals were also affected, albeit less than were NE terminals, and provided more precise information about the regional distribution and time course of the effects [28-32]. This flurry of creativity was followed by another, in which he and his colleague Ronald Schoenfeld used operant techniques to characterize some of the behavioral effects of 6-OHDA [33-35].

In short, Norman, together with Urban Ungerstedt and several of their contemporaries provided us with the first relatively specific tool for studying the loss of catecholamines. Soon George Breese showed that by blocking the uptake of 6-OHDA into NE neurons, its effects could be largely limited to DA neurons [36] and thereby providing an even more specific tool. This work on the development of 6-OHDA preceded by 15 years the emergence of the second such tool, MPTP [37]. Its discovery has led to an enormous outpouring of papers. (There are currently well over 9,000 citations on 6-OHDA listed in PubMed.)

During the 40 years since the first papers by Norman and his contemporaries, several new ways of exposing DA neurons to 6-OHDA have been developed. One approach, pioneered by Heinrich Sauer and Wolfgang Oertel has been to administer 6-OHDA directly into the striatum, after which striatal TH disappears over a period of 16 weeks or more [38], in contrast to the much more rapid disappearance that occurs after administration into the lateral ventricle, SN, or nigrostriatal projection [e.g., see 39,40]. A second approach to achieve a more progressive lesion has been used by Timothy Schallert and colleagues: administering multiple small intra-striatal injections of 6-OHDA over several weeks [41].

In the following section, we will focus on a single question: What have we learned about PD from studies of 6-OHDA. [For additional recent reviews on the use of 6-OHDA see [42,43-46,47 as ell as many others]. In most cases, I will be focusing on animal models (usually the young rat). However, 6-OHDA has also been used in a number of cell culture models, including PC12 [48], MN9D [49], and SH-SY5Y cells [50], and in some cases primary neurons from the mesencephalon [51]. Indeed, our own laboratory has employed 6-OHDA in many such in vitro models [52-54], and in some cases results from in vitro studies will be included in the following discussion.

One caveat before entering into a discussion of what we have learned from studies of 6-OHDA: the specificity of this toxin cannot be taken for granted. This is particularly true in the case of in vitro studies. Although 6-OHDA breaks down within a few minutes in culture medium [52], it is not unusual for investigators to report on the effects observed in cells incubated with 6-OHDA for 24 hours. Such effects may be of interest, but they cannot be assumed to result from the impact of 6-OHDA molecules acting from within the cells.

Precautions must also be taken in the use of 6-OHDA in animals. Here, too, non-specific effects can be produced [55-58]. In one particularly unfortunate instance, the amount of 6-OHDA normally injected into the lateral ventricle was administered directly into brain parenchyma, where a standard dose would normally be two orders of magnitude lower! Not surprisingly, the resulting tissue damage looked very much like what could be produced by H_2O_2, and the author concluded that "6-OHDA which may be used to induce a focal lesion cannot be regarded as a specific neurotoxic agent" [59]. The report, published in 1975, set the field back for a while – who would want to fund studies on the behavioral response to H_2O_2 – but we all recovered in due time.

What We Have Learned about PD from Studies of 6-OHDA

The first papers on the effects of intracerebral 6-OHDA made no mention of PD. Norman never used the term in his classic series of papers, and in fact six years appears to have elapsed between the first report of the effects of 6-OHDA on the brain and the explicit mention of PD in a discussion of those effects [60]. I suggest that there are at least four reasons for this. First, although the connection between DA and PD was made through post-mortem studies by Hornykiewicz and colleagues in 1962, it took a while (at least in the United States) before the general realization that DA was much more than a precursor of NE. Second, when given intraventricularly, as Norman (and I) initially did, 6-OHDA had a much greater effect on NE levels then on DA levels. Third, even when DA was affected, there were usually no gross behavioral changes, as greater than a 95% loss of striatal DA was required before such behavioral deficits would appear. Finally, and perhaps most interesting from a historical perspective, in the earlier days of the field there was a general hesitancy to suggest that one's research with cells or animals provided any connection to a clinical disorder. With notable exceptions, there was an invisible line between *basic research* and *clinical research*. Indeed, "translational research" does not appear as a term in PubMed in 1993. How things have changed!

1. Many of the Gross Motor Deficits can be Attributed to the Loss of Dopamine

This was the most immediate insight that came from studies of 6-OHDA: As long as the loss of the transmitter was extensive and bilateral, the most obvious consequence was one of severe motor deficits. The animals showed a hunched poster. Their motor behavior was slowed (bradykinesia), and in the extreme case, they were akinetic for hours at a time, requiring the intubation of water and nutrients to remain alive [61]. And, although some of these severely lesioned animals gradually recovered, many remained essentially akinetic for a year, the longest period examined.

2. Stress can Both Promote Temporary Motor Improvement and Lead a Worsening of Behavior

One of the paradoxes of PD is that when severely affected patients are exposed to a sudden stress they can respond with a temporary improvement in their motor behavior. This is exemplified by a number of anecdotes about patients responding to emergencies by catching a ball, running out of a burning building, or swimming out into the ocean to save a child [62]. Comparable findings were made in the 6-OHDA-treated rat. For example, Ed Stricker and I observed that if an akinetic rat were to be placed in a pail of water, it would swim with apparent ease to the edge to the pail and then climb out. However, as in the patient anecdotes, seconds later the rat would again freeze [unpublished observation; but see also [63,64]. The neurobiological basis of such phenomena is obscure. Although we found that stress increased DA release, even after 6-OHDA lesions [65], the DA receptor antagonist haloperidol failed to block stress-induced kinesia [63]. Equally mysterious is the common clinical observation of stress-induced exacerbation of symptoms [66], which can even lead to potentially harmful freezing. This, too, can be elicited in the 6-OHDA-leasioned rat [67].

3. The Partial Loss of DA Neurons Leads to a Variety of Compensatory Changes

The relationship between striatal DA levels and motor deficits in patients diagnosed with PD is not linear. As Oleh Hornykiewicz reported early on, DA levels in patients with PD are generally reduced by at least 80%. In the case of rats treated with intraventricular DA, we found the threshold to be even higher. This non-linearity is not unique to PD or animal models of DA deficiency, and has been described for other conditions, as well. Logically, there would seem to be several possible explanations for such non-linearity of neuronal loss versus symptoms. In the case of DA deficiency, we have identified three sets of explanations: First, it is possible, though unlikely in my opinion, that there is simply a considerable redundancy of DA neurons and, as a result, the loss of a substantial proportion of these neurons requires no adjustment. Second, compensations might occur that allow the remaining DA neurons to adjust to the loss and maintain normal neuron-neuron communication. These could be either passive or active. Third, non-DA regions of the brain may be able to compensate for the lesion. This might involve a transfer of function normally subserved by DA neurons, adjustments in other components of basal ganglia circuitry normally regulated by DA, or the development of wholly new ways to regulate motor behavior.

Our own work, which is reviewed in some detail elsewhere [see reviews 68,69,70], has focused on the second of these possibilities, which was suggested by the initial studies by Hornykiewicz of postmortem samples of caudate and putamen from patients with PD. In those studies he found elevated ratios of DA metabolites to DA, and from this he surmised that the brain had compensated for the loss of some DA neurons by increasing DA turnover (and presumably release) in the remaining DA neuron [71,72]. We have replicated the finding of increased metabolite/DA ratios in 6-OHDA-treated rats, and have gone on to identify several ways in which DA neurons and their targets adjust to the partial loss of nigrostriatal neurons:

Loss of high affinity DA uptake:

As already noted, once released into extracellular space, DA is typically inactivated by high affinity DA uptake via a specific membrane protein present within DA terminals. It follows that the loss of DA neurons automatically is accompanied by a loss of high affinity DA uptake sites. This, in turn, means that any DA diffusing into an area of denervation will not be readily removed, a phenomenon highlighted many years ago by Ullrich Trendelenburg in studies of sympathetic nervous system [73].

Increased DA Synthesis and Release:

Whereas the loss of uptake sites provides a passive means of compensating from the loss of some DA neurons, an active process also comes into play after the loss of DA neurons exceeds some threshold. This hypothesis is supported by two groups of observations, which in turn supplement the finding of increased DA turnover. First, there is an increase in the activity of the rate limiting enzyme in DA synthesis, tyrosine hydroxylase (TH), relative to the amount of DA that remains, i.e., an increase in the TH/DA ratio. This in turn may have two components – a slow increase in TH protein that would require increased protein synthesis followed by the transport of that protein to the nerve terminal and a rapid activation of preexisting TH molecules through the process of phosphorylation [74-76]. Second, studies of striatal slices prepared from 6-OHDA-treated animals indicated an increase in depolarization-induced DA release [77]. Moreover, the combination of increased release and decreased uptake is likely to explain two other sets of observations: that extracellular levels of DA are maintained relatively constant despite substantial loss of DA terminals [78-80], and that the loss of a considerable percentage of DA terminals fails to block the capacity of DA to influence the release of acetylcholine [81] and GABA [82] as assessed in studies of striatal slices.

Increased Target Cell Sensitivity to DA:

It is a common finding that following denervation, the normal target becomes supersensitive to the lost transmitter. This also occurs after 6-OHDA-induced lesions of DA projections to striatum. The phenomenon has been observed by monitoring the behavioral effects of direct DA agonists [84], the biochemical response to those agonists [83], and the number of DA binding sites on striatal targets [84]. It is instructive to note that these changes develop over several weeks and only after very large DA-depleting lesions [85,86]. The need for a very large lesion is consistent with microdialysis data indicating the concentration of DA in the extracellular space only begins to decrease when lesions become large [78-80].

Non-Synaptic Communication by DA Neurons:

Our hypothesis of "biochemical reinnervation" of denervated sites has an important corollary. If it is true that the combination of increased release and decreased uptake of DA serves to compensate for partial denervation, then precise point-to-point innervation of striatum by the SN must not be essential, at least for gross motor function. That is, DA must be capable of influencing its targets via non-synaptic communication somewhat as epinephrine acts in the periphery. The concept of non-synaptic or volume transmission has long been championed by E. Sylvester Vizi [87] and Kjell Fuxe [88]. It has played an important role in our thinking about compensations after 6-OHDA [89,90] and considerable support now exists for applying it to DA neurons in the striatum [e.g., 91-93], see also several reviews [e.g., 94-96].

4. The Pathophysiology of PD may Include Oxidative stress, Inflammation, and/or Excitotoxicity

Studies of 6-OHDA models have provided a number of hints regarding the etiology of PD. Perhaps the most immediate suggestion is that PD involves *oxidative stress*. Indeed, there is a good deal of evidence that an early abnormality in brains of patients with PD is an increase in oxidative stress (see reviews [97,98,99]). This includes reduced glutathione levels and glutathione-related enzymes [100,101], increased iron [102], decreased ferritin [103], and increased 5-S-cysteinyl-dopamine [104]. Oxidative stress can be accompanied by many types of other insults including lipid peroxidation, protein and DNA damage, elevated iron levels, and mitochondrial impairments, each of which has been observed in postmortem samples from patients with PD.

Although it is often assumed that oxidative stress is the primary cause of these changes, the actual causal sequence is not always clear. On the other hand, each of these effects has been observed in 6-OHDA-treated animals, strengthening the assumption that oxidative stress is usually the causal event.

These findings may be relevant to 6-OHDA in several other ways. First, 6-OHDA can be formed endogenously from DA [105-107] and from DOPA [108,109], the major drug used in the treatment of PD, and has been detected in human brain [110]. Second, like 6-OHDA, DA can rapidly oxidize [111], and in so doing can cause relatively specific lesions of DA neurons when injected into the striatum ([112]; see reviews [113,114,115]).

It has also been long been suspected that ***inflammation*** may play a role in the etiology of PD as well as other neurodegenerative diseases. Evidence includes an increase in microglial activation and an activation of the complement system [see reviews 116,117-119]. Thus, it seems of considerable interest that 6-OHDA causes an inflammatory response in the SN [120,121] and that the anti-inflammatory drug minocycline as well as non-steroidal anti-inflammatory drugs (NSAID) can reduce 6-OHDA neurotoxicity (see review [116,122,123]). Although some epidemiological evidence supports the concept of anti-inflammatory agents in the treatment of PD, a recent clinical pilot study has failed to find support for the use of minocycline in PD [124]. However, it is possible that this reflects a failure of the treatment to achieve a sufficient anti-inflammatory effect and/or the necessity to provide treatment at an earlier stage of disease progression.

A third factor that may contribute to PD and for which corroborating evidence has come from studies of 6-OHDA is ***excitotoxicity***. Glutamate plays a key role in the basal ganglia loops that regulate DA function, both in the striatum and in the SN [125]. Loss of DA neurons has been associated with an increase in glutamatergic activity in both of these regions [126]. Should this potentially excitotoxic transmitter be released in sufficient excess, a positive feedback leading to progressive neurodegeneration would be expected to result. This conjecture is support by the observation that lesioning of the glutamatergic subthalamic nucleus reduced the toxic effects of 6-OHDA [127] and by the neuroprotective effects of glutamate antagonists in such models [128].

5. Treatments that Reduce the Progression of PD may be Beginning to Emerge

Perhaps the most important contribution of 6-OHDA models is in the search for treatments. Most of this work has focused on drugs designed to reduce the motor *symptoms* of PD, commonly direct or indirect DA agonists. [44,129]. More recently, however, investigators have begun to use 6-OHDA models explore treatments that are designed to reduce *disease progression*, and it is these on which I will briefly comment.

Pharmacological Treatments:

Recently, the focus of much of the research on pharmacotherapy has switched its focus from symptoms to disease progression. The range of drugs that have, in most cases, been tested in 6-OHDA models is too wide to be reviewed here, but includes (in alphabetical order) antioxidants, anti-inflammatory agents, calcium channel antagonists, creatine, DA agonists, glutamate antagonists, free radical scavengers, iron chelators, melatonin, MAO-B inhibitors, nitric oxide synthase inhibitors, and polyphenols [see reviews 130,131].

To this list must be added a number of herbs, the extracts of which have not yet been characterized but that look promising, These include *Ganoderma lucidum* spore oil, *Ginkgo biloba*, ginseng, green tea, huperzine, *Nerium indicum* flowers, panax, polygonum, stepholidine, and *Tripterygium wilfordii hook* [132].

It is understood, of course, that the biological influence of a given drug that first led to its consideration as a treatment for PD is not necessarily the mechanism of it neuroprotective effects. Two examples are the MAO-B inhibitor rasagiline, which also seems to have anti-apoptotic effects [133], and the DA receptor agonist pramipexol which may exert its potential neuroprotective effects via its anti-oxidant properties [134].

Neurotrophic Factors; Proteins and Genes:

The discovery that the brain contained factors that could protect DA neurons was first made in tissue culture [135] and quickly focused on GDNF, which was shown to protect DA neurons from spontaneous cell death [136] and from 6-OHDA [137,138]. Protection from 6-OHDA has been shown to be a property of other neurotrophic factors, including BDNF [139] and IGF1 [140], and can also be shown to occur following the intracerebral administration of viruses that contain the genes for such factors, [e.g., 137,141]. The results of human trials of GDNF have been mixed [142] and a recent presentation of the administration of AAV containing the gene for a GDNF family member, neurturin, suggests only modest effects after 18 months [143]. Nonetheless, the delivery of molecules that can lead to the stimulation of receptors for neurotrophic factors remains a promising direction for the development of neuroprotective therapies.

Transplantation Approaches:

One of the earliest and most enduring approaches to moving beyond symptomatic relief has been the transplantation of cells into the brains of 6-OHDA-treated rats [see reviews 144,145,146]. Such approaches have included the transplantation of encapsulated cells capable of delivering DA or GDNF [147], fetal transplants of adrenal chromaffin cells [148] and SN [149], and most recently stem cells [150,151]. To date, virtually all the clinical work using transplantation has involved fetal SN tissue. Unfortunately, again the results are mixed. Although some very positive results have been reported [see review 152], a recent double blind study has not yielded any treatment effect. Moreover, many patients exhibited prominent dyskinesias [153]. Nonetheless, transplantation research is continuing and the 6-OHDA is likely to continue to be used to produce valuable preclinical models.

6. In the End, Maybe it is Largely about Life Style

The incidence of diagnosed PD appears to be on the rise [154,155]. There are likely to be many explanations for this, including better diagnoses, a rise in environmental toxins, and an increase in life span. However, one must also seriously consider the impact of the declining quality of our life style. At least within the so-called "developed" countries, we seem to be eating less well, getting less exercise, and spending less time in social activities. Consider the number of people who pick up fast food, return to an empty house, and then spend the evening watching television. It seems obvious that this is not the life style for which evolution has prepared us. Furthermore, consider studies of the effects of modifying the life style of the 6-OHDA rat:

Diet:

Feeding rats blueberry- and spirulina-enriched diets promoted recovery from 6-OHDA [156]. (It is interesting that improved diets also improved the motor performance of aged rats [157]). Somewhat analogous studies have been carried out in the MPTP-treated mouse model in which reduced toxicity was seen with resveratrol [158] and high dietary omega-3 polyunsaturated fatty acid [159], while a high fat diet increased the toxicity [160].

Environmental Enrichment and Exercise:

There have now been a number of studies to indicate that exercise can dramatically reduce the anatomical, neurochemical, and/or the behavioral effects of 6-OHDA in the rat ([161-165]; see review [166]. These studies have examined several different types of exercise including the forced motor activity that results from placing the "normal limb" in a cast after a unilateral lesion [161,162]; a running wheel [164], and a treadmill [163,167]. Environmental enrichment per se has not yet been studied in the 6-OHDA rat. However, the work of Richard Smeyne and colleagues using the MPTP mouse suggests that the neuroprotective effects of enrichment can be mimicked by exercise [168].

Future Directions

Where to go from here? First and foremost we need much better interventions. We need to develop treatments that reduce a more complete range of parkinsonian symptoms – physiological and psychiatric, as well as motor – and that do not show tolerance or dyskinesias. We need treatments that reduce disease progression. And ultimately we need treatments that actually prevent the appearance of the disease. If we are to accomplish these goals, then a second set of future requirements is a much more complete understanding of the causes of PD – genetic factors, environmental influences, and cellular mechanisms,

And this leads directly to a third need: better models. Giving 6-OHDA to young rats, no matter what the protocol, will only get us so far. The approach has been extremely fruitful, as I hope this brief review has shown, but much more is needed. We need older animals (24 month old rodents, 20 year old non-human primates), even more progressive models (many months for rodents; several years for primates), toxins that are more relevant to what might be causing PD (pesticides, heavy metals, industrial pollutants), and genetic models, including conditional models, that more completely emulate the pathophysiology of the disease [see reviews 44,169,170,171].

So, has the useful life span of 6-OHDA come to an end after 40 years? Certainly not. Motor deficits are only part of the symptomatology of PD, but they are an important part; oxidative stress is not the only factor that causes neuronal cell death in the disease, but it is likely to play a key role; and acute degeneration does not mimic the pattern of loss across the entire nigrostriatal projection, but it probably emulates the time course of the degeneration of each individual DA neuron.. This simple molecule, which has already instructed us about so much, is destined to teach us a great deal more.

Acknowledgments

This review is dedicated to the memory of my friend Norman Uretsky, with whom I was a graduate student at the University of Chicago from 1963 to 1968. The work from my own laboratory that is part of the review has been supported over many years primarily by the National Institutes of Health and the Department of Defense. My thanks for the editorial assistance of Rehana Leak and Emma Culligan.

References

[1] Parkinson J. *An Essay on the Shaking Palsy*. In: Sherwood N, and Jones, ed. London 1817.

[2] Foix C, Nicolesco J. Les noyaux gris centraux et la région Mésencéphalo-sous-optique. Suivi d'un appendice sur l'anatomie pathologique de la maladie de Parkinson. In: Cie Me, ed. *Anatomie cérébrale*. Paris 1925:508–39, as cited in [3].

[3] Saper CB. 'Like a thief in the night': the selectivity of degeneration in Parkinson's disease. *Brain*. 1999. 122 (Pt 8): 1401-2.
[4] Ehringer H, Hornykiewicz O. [Distribution of noradrenaline and dopamine (3-hydroxytyramine) in the human brain and their behavior in diseases of the extrapyramidal system.]. *Klin Wochenschr*. 1960. 38: 1236-9.
[5] Hornykiewicz O. [Dopamine (3-hydroxytyramine) in the central nervous system and its relation to the Parkinson syndrome in man.]. *Deutsche medizinische Wochenschrift* (1946). 1962. 87: 1807-10.
[6] Carlsson A, Lindqvist M, Magnusson T. 3,4-Dihydroxyphenylalanine and 5-hydroxytryptophan as reserpine antagonists. *Nature*. 1957. 180: 1200.
[7] Falck B, Hillarp NA. On the cellular localization of catechol amines in the brain. *Acta anatomica*. 1959. 38: 277-9.
[8] Anden NE, uxe K, Hamberger B, Hokfelt T. A quantitative study on the nigro-neostriatal dopamine neuron system in the rat. *Acta Physiol Scand*. 1966. 67: 306-12.
[9] Moore RY, Bhatnagar RK, Heller A. Anatomical and chemical studies of a nigro-neostriatal projection in the cat. *Brain Research*. 1971. 30: 119-35.
[10] Poirier LJ, Sourkes TL. [Influence of Locus Niger on the Concentration of Catecholamines in the Striatum.]. *Journal de physiologie*. 1964. 56: 426-7.
[11] Braak H, Braak E, Yilmazer D, Schultz C, de Vos RA, Jansen EN. Nigral and extranigral pathology in Parkinson's disease. *J Neural Transm Suppl*. 1995. 46: 15-31.
[12] Braak H, Ghebremedhin E, Rub U, Bratzke H, Del Tredici K. Stages in the development of Parkinson's disease-related pathology. *Cell Tissue Res*. 2004. 318: 121-34.
[13] Senoh S, Witkop, B, Creveling, CR, Udenfriend, S. 2,4,5-trihydroxyphenetylamine, a new metabolite of 3,4-dihydroxyphenetylamine. *J Am Chem Soc*. 1959. 91: 1768-9.
[14] Porter CC, Totaro, J.A. and Stone, C.A. Effect of 6-hydroxydoapmine and some other compounds on the concentration of norepinephrine in the hearts of mice. *Journal of Pharmacology & Experimental Therapeutics*. 1963. 140: 308.
[15] Laverty R, Sharman DF, Vogt M. Action of 2, 4, 5-Trihydroxyphenylethylamine on the Storage and Release of Noradrenaline. *British journal of pharmacology and chemotherapy*. 1965. 24: 549-60.
[16] Malmfors T, Sachs C. Degeneration of adrenergic nerves produced by 6-hydroxydopamine. *Eur J Pharmacol*. 1968. 3: 89-92.
[17] Thoenen H, Tranzer JP. Chemical sympathectomy by selective destruction of adrenergic nerve endings with 6-Hydroxydopamine. *Naunyn-Schmiedebergs Archives of Pharmacology*. 1968. 261: 271-88.
[18] Heikkila R, Cohen G. Further studies on the generation of hydrogen peroxide by 6-hydoxydopamine- potentiation by ascorbic acid. *Molecular Pharmacology*. 1971. 8: 241-8.
[19] Ungerstedt U. 6-Hydroxydopamine-induced degeneration of central monoamine neurons. *European Journal of Pharmacology*. 1968. 5: 107-10.
[20] Uretsky NJ, Iversen LL. Effects of 6-hydroxydopamine on noradrenaline-containing neurones in the rat brain. *Nature*. 1969. 221: 557-9.

[21] Bloom FE, Algeri S, Groppetti A, Revuelta A, Costa E. Lesions of central norepinephrine terminals with 6-OH-dopamine: biochemistry and fine structure. *Science*. 1969. 166: 1284-6.

[22] Breese GR, Traylor TD. Effect of 6-hydroxydopamine on brain norepinephrine and dopamine evidence for selective degeneration of catecholamine neurons. *J Pharmacol Exp Ther*. 1970. 174: 413-20.

[23] Heikkila RE, Cohen G. 6-Hydroxydopamine: evidence for superoxide radical as an oxidative intermediate. *Science*. 1973. 181: 456-7.

[24] Marshall JF, Teitelbaum P. A comparison of the eating in response to hypothermic and glucoprivic challenges after nigral 6-hydroxydopamine and lateral hypothalamic electrolytic lesions in rats. *Brain Research*. 1973. 55: 229-33.

[25] Zigmond MJ, Stricker EM. Deficits in feeding behavior after intraventricular injection of 6-hydroxydopamine in rats. *Science*. 1972. 177: 1211-4.

[26] Fuxe K, Ungerstedt U. Antiparkinsonian drugs and dopaminergic neostriatal mechanisms: studies in rats with unilateral 6-hydroxydopamine (=6-OH-DA)-induced degeneration of the nigro-neostriatal DA pathway and quantitative recording of rotational behaviour. *Pharmacol Ther B*. 1976. 2: 41-7.

[27] Ungerstedt U. 6-hydroxydopamine-induced degeneration of the nigrostriatal dopamine pathway: the turning syndrome. *Pharmacol Ther B*. 1976. 2: 37-40.

[28] Bell LJ, Iversen LL, Uretsky NJ. Time course of the effects of 6-hydroxydopamine on catecholamine-containing neurones in rat hypothalamus and striatum. *Br J Pharmacol.* 1970. 40: 790-9.

[29] Evetts KD, Uretsky NJ, Iversen LL, Iversen SD. Effects of 6-hydroxydopamine on CNS catecholamines, spontaneous motor activity and amphetamine induced hyperactivity in rats. *Nature*. 1970. 225: 961-2.

[30] Iversen LL, Uretsky NJ. Regional effects of 6-hydroxydopamine on catecholamine containing neurones in rat brain and spinal cord. *Brain Res*. 1970. 24: 364-7.

[31] Uretsky NJ, Iversen LL. Effects of 6-hydroxydopamine on catecholamine containing neurones in the rat brain. *J Neurochem.* 1970. 17: 269-78.

[32] Uretsky NJ, Simmonds MA, Iversen LL. Changes in the retention and metabolism of 3 H-l-norepinephrine in rat brain in vivo after 6-hydroxydopamine pretreatment. *J Pharmacol Exp Ther*. 1971. 176: 489-96.

[33] Schoenfeld RI, Uretsky NJ. Operant behavior and catecholamine-containing neurons: prolonged increase in lever-pressing after 6-hydroxydopamine. *Eur J Pharmacol.* 1972. 20: 357-62.

[34] Schoenfeld RI, Uretsky NJ. Enhancement by 6-hydroxydopamine of the effects of DOPA upon the motor activity of rats. *J Pharmacol Exp Ther*. 1973. 186: 616-24.

[35] Uretsky NJ, Schoenfeld RI. Effect of L-dopa on the locomotor activity of rats pretreated with 6-hydroxydopamine. *Nat New Biol.* 1971. 234: 157-9.

[36] Breese GR, Traylor TD. Depletion of brain noradrenaline and dopamine by 6-hydroxydopamine. *Br J Pharmacol*. 1971. 42: 88-99.

[37] Langston JW, Ballard P, Tetrud JW, Irwin I. Chronic Parkinsonism in humans due to a product of meperidine-analog synthesis. *Science*. 1983. 219: 979-80.

[38] Sauer H, Oertel WH. Progressive degeneration of nigrostriatal dopamine neurons following intrastriatal terminal lesions with 6-hydroxydopamine: a combined retrograde tracing and immunocytochemical study in the rat. *Neuroscience*. 1994. 59: 401-15.

[39] Jeon BS, Jackson-Lewis V, Burke RE. 6-Hydroxydopamine lesion of the rat substantia nigra: time course and morphology of cell death. *Neurodegeneration*. 1995. 4: 131-7.

[40] Zuch CL, Nordstroem VK, Briedrick LA, Hoernig GR, Granholm AC, Bickford PC. Time course of degenerative alterations in nigral dopaminergic neurons following a 6-hydroxydopamine lesion. *J Comp Neurol*. 2000. 427: 440-54.

[41] Fleming SM, Delville Y, Schallert T. An intermittent, controlled-rate, slow progressive degeneration model of Parkinson's disease: antiparkinson effects of Sinemet and protective effects of methylphenidate. *Behav Brain Res*. 2005. 156: 201-13.

[42] Deumens R, Blokland A, Prickaerts J. Modeling Parkinson's disease in rats: an evaluation of 6-OHDA lesions of the nigrostriatal pathway. *Exp Neurol*. 2002. 175: 303-17.

[43] Hirsch EC, Hoglinger G, Rousselet E, Breidert T, Parain K, Feger J, et al. Animal models of Parkinson's disease in rodents induced by toxins: an update. *J Neural Transm Suppl*. 2003. 89-100.

[44] Jenner P. Functional models of Parkinson's disease: a valuable tool in the development of novel therapies. *Ann Neurol*. 2008. 64 Suppl 2: S16-29.

[45] Willis GL, Kennedy GA. The implementation of acute versus chronic animal models for treatment discovery in Parkinson's disease. *Rev Neurosci*. 2004. 15: 75-87.

[46] Blandini F, Armentero MT, Martignoni E. The 6-hydroxydopamine model: news from the past. *Parkinsonism Relat Disord*. 2008. 14 Suppl 2: S124-9.

[47] Kostrzewa RM. *Evolution of neurotoxins: from research modalities to clinical realities*. Current protocols in neuroscience / editorial board, Jacqueline N Crawley [et al. 2009. Chapter 1: Unit 1 18.

[48] Walkinshaw G, Waters CM. Neurotoxin-induced cell death in neuronal PC12 cells is mediated by induction of apoptosis. *Neuroscience*. 1994. 63: 975-87.

[49] Oh YJ, Wong SC, Moffat M, O'Malley KL. Overexpression of Bcl-2 attenuates MPP+, but not 6-ODHA, induced cell death in a dopaminergic neuronal cell line. *Neurobiol Dis*. 1995. 2: 157-67.

[50] Seitz G, Stegmann HB, Jager HH, Schlude HM, Wolburg H, Roginsky VA, et al. Neuroblastoma cells expressing the noradrenaline transporter are destroyed more selectively by 6-fluorodopamine than by 6-hydroxydopamine. *J Neurochem*. 2000. 75: 511-20.

[51] Han BS, Hong HS, Choi WS, Markelonis GJ, Oh TH, Oh YJ. Caspase-dependent and -independent cell death pathways in primary cultures of mesencephalic dopaminergic neurons after neurotoxin treatment. *J Neurosci*. 2003. 23: 5069-78.

[52] Ding YM, Jaumotte JD, Signore AP, Zigmond MJ. Effects of 6-hydroxydopamine on primary cultures of substantia nigra: specific damage to dopamine neurons and the impact of glial cell line-derived neurotrophic factor. *J Neurochem*. 2004. 89: 776-87.

[53] Leak RK, Zigmond MJ, Liou AK. Adaptation to chronic MG132 reduces oxidative toxicity by a CuZnSOD-dependent mechanism. *J Neurochem*. 2008. 106: 860-74.

[54] Ugarte SD, Lin E, Klann E, Zigmond MJ, Perez RG. Effects of GDNF on 6-OHDA-induced death in a dopaminergic cell line: Modulation by inhibitors of PI3 kinase and MEK. *J Neurosci Res*. 2003. 73: 105-12.

[55] Agid Y, Javoy F, Glowinski J, Bouvet D, Sotelo C. Injection of 6-hydroxydopamine into the substantia nigra of the rat. II. Diffusion and specificity. *Brain Res*. 1973. 58: 291-301.

[56] Hedreen JC, Chalmers JP. Neuronal degeneration in rat brain induced by 6-hydroxydopamine; a histological and biochemical study. *Brain Res*. 1972. 47: 1-36.

[57] Hokfelt T, Ungerstedt U. Specificity of 6-hydroxydopamine induced degeneration of central monoamine neurones: an electron and fluorescence microscopic study with special reference to intracerebral injection on the nigro-striatal dopamine system. *Brain Res*. 1973. 60: 269-97.

[58] Javoy F, Sotelo C, Herbet A, Agid Y. Specificity of dopaminergic neuronal degeneration induced by intracerebral injection of 6-hydroxydopamine in the nigrostriatal dopamine system. *Brain Res*. 1976. 102: 201-15.

[59] Poirier LJ. Histopathologic changes associated with intracerebral injections of 6-hydroxydopamine (6-OHDA) and peroxide (H2O2) in the cat and the rat. *J Neural Transm*. 1975. 37: 209-18.

[60] Ungerstedt U, Ljungberg T, Steg G. Behavioral, physiological, and neurochemical changes after 6-hydroxydopamine-induced degeneration of the nigro-striatal dopamine neurons. *Adv Neurol*. 1974. 5: 421-6.

[61] Zigmond MJ, Stricker EM. Recovery of feeding and drinking by rats after intraventricular 6-hydroxydopamine or lateral hypothalamic lesions. *Science*. 1973. 182: 717-20.

[62] Rosenbaum RB. *Understanding Parkinson's Disease, a Personal and Professional View*: Praeger Publishers 2006.

[63] Keefe KA, Salamone JD, Zigmond MJ, Stricker EM. Paradoxical kinesia in parkinsonism is not caused by dopamine release. Studies in an animal model. *Arch Neurol*. 1989. 46: 1070-5.

[64] Marshall JF, Levitan D, Stricker EM. Activation-induced restoration of sensorimotor functions in rats with dopamine-depleting brain lesions. *J Comp Physiol Psychol*. 1976. 90: 536-46.

[65] Keefe KA, Stricker EM, Zigmond MJ, Abercrombie ED. Environmental stress increases extracellular dopamine in striatum of 6-hydroxydopamine-treated rats: in vivo microdialysis studies. *Brain Res*. 1990. 527: 350-3.

[66] Schwab RS, Zieper I. Effects of mood, motivation, stress and alertness on the performance in Parkinson's disease. *Psychiatry and Neurology*, Basel. 1965. 150: 345-57.

[67] Snyder AM, Stricker EM, Zigmond MJ. Stress-induced neurological impairments in an animal model of parkinsonism. *Ann Neurol*. 1985. 18: 544-51.

[68] Zigmond MJ. Do compensatory processes underlie the preclinical phase of neurodegenerative disease? Insights from an animal model of parkinsonism. *Neurobiol Dis*. 1997. 4: 247-53.

[69] Zigmond MJ, Abercrombie ED, Berger TW, Grace AA, Stricker EM. Compensations after lesions of central dopaminergic neurons: some clinical and basic implications. *Trends Neurosci.* 1990. 13: 290-6.

[70] Zigmond MJ, Berger TW, Grace AA, Stricker EM. Compensatory responses to nigrostriatal bundle injury. Studies with 6-hydroxydopamine in an animal model of parkinsonism. *Mol Chem Neuropathol.* 1989. 10: 185-200.

[71] Hornykiewicz O. Brain monoamines and parkinsonism. *National Institute on Drug Abuse research monograph series.* 1975. 13-21.

[72] Hornykiewicz O. Parkinson's disease and the adaptive capacity of the nigrostriatal dopamine system: possible neurochemical mechanisms. *Adv Neurol.* 1993. 60: 140-7.

[73] Trendelenburg U. Supersensitivity and subsensitivity to sympathomimetic amines. *Pharmacol Rev.* 1963. 15: 225-76.

[74] Acheson AL, Zigmond MJ. Short and long term changes in tyrosine hydroxylase activity in rat brain after subtotal destruction of central noradrenergic neurons. *J Neurosci.* 1981. 1: 493-504.

[75] Acheson AL, Zigmond MJ, Stricker EM. Compensatory increase in tyrosine hydroxylase activity in rat brain after intraventricular injections of 6-hydroxydopamine. *Science.* 1980. 207: 537-40.

[76] Wolf ME, Zigmond MJ, Kapatos G. Tyrosine hydroxylase content of residual striatal dopamine nerve terminals following 6-hydroxydopamine administration: a flow cytometric study. *J Neurochem.* 1989. 53: 879-85.

[77] Snyder GL, Keller RW, Jr., Zigmond MJ. Dopamine efflux from striatal slices after intracerebral 6-hydroxydopamine: evidence for compensatory hyperactivity of residual terminals. *J Pharmacol Exp Ther.* 1990. 253: 867-76.

[78] Abercrombie ED, Bonatz AE, Zigmond MJ. Effects of L-dopa on extracellular dopamine in striatum of normal and 6-hydroxydopamine-treated rats. *Brain Res.* 1990. 525: 36-44.

[79] Castaneda E, Whishaw IQ, Robinson TE. Changes in striatal dopamine neurotransmission assessed with microdialysis following recovery from a bilateral 6-OHDA lesion: variation as a function of lesion size. *J Neurosci.* 1990. 10: 1847-54.

[80] Robinson TE, Whishaw IQ. Normalization of extracellular dopamine in striatum following recovery from a partial unilateral 6-OHDA lesion of the substantia nigra: a microdialysis study in freely moving rats. *Brain Res.* 1988. 450: 209-24.

[81] MacKenzie RG, Stachowiak MK, Zigmond MJ. Dopaminergic inhibition of striatal acetylcholine release after 6-hydroxydopamine. *Eur J Pharmacol.* 1989. 168: 43-52.

[82] Harsing LG, Jr., Zigmond MJ. Dopaminergic inhibition of striatal GABA release after 6-hydroxydopamine. *Brain Res.* 1996. 738: 142-5.

[83] Mishra RK, Gardner EL, Katzman R, Makman MH. Enhancement of dopamine-stimulated adenylate cyclase activity in rat caudate after lesions in substantia nigra: evidence for denervation supersensitivity. *Proc Natl Acad Sci U S A.* 1974. 71: 3883-7.

[84] Creese I, Burt DR, Snyder SH. Dopamine receptor binding enhancement accompanies lesion-induced behavioral supersensitivity. *Science.* 1977. 197: 596-8.

[85] Marshall JF, Ungerstedt U. Supersensitivity to apomorphine following destruction of the ascending dopamine neurons: quantification using the rotational model. *European Journal of Pharmacology*. 1977. 41: 361-7.
[86] Zigmond MJ, Stricker EM. Supersensitivity after intraventricular 6-hydroxydopamine: relation to dopamine depletion. *Experientia*. 1980. 36: 436-8.
[87] Vizi ES. N*on-synaptic Interactions Between Neurons: Modulation of Neurochemical Transmission*. Chichester, UK: John Wiley & Sons, Ltd. 1984.
[88] Agnati LF, Fuxe K, Zoli M, Ozini I, Toffano G, Ferraguti F. A correlation analysis of the regional distribution of central enkephalin and beta-endorphin immunoreactive terminals and of opiate receptors in adult and old male rats. Evidence for the existence of two main types of communication in the central nervous system: the volume transmission and the wiring transmission. *Acta Physiol Scand*. 1986. 128: 201-7.
[89] Zigmond MJ. Chemical transmission in the brain: homeostatic regulation and its functional implications. *Prog Brain Res*. 1994. 100: 115-22.
[90] Zigmond MJ. When it comes to communications between neurons, synapses are overrated: insights from an animal model of parkinsonism. *Prog Brain Res*. 2000. 125: 317-26.
[91] Hersch SM, Yi H, Heilman CJ, Edwards RH, Levey AI. Subcellular localization and molecular topology of the dopamine transporter in the striatum and substantia nigra. *J Comp Neurol*. 1997. 388: 211-27.
[92] Maley BE, Engle MG, Humphreys S, Vascik DA, Howes KA, Newton BW, et al. Monoamine synaptic structure and localization in the central nervous system. *Journal of electron microscopy technique*. 1990. 15: 20-33.
[93] Szalay KS, Orso E, Juranyi Z, Vinson GP, Vizi ES. Local non-synaptic modulation of aldosterone production by catecholamines and ATP in rat: implications for a direct neuronal fine tuning. *Hormone and metabolic research = Hormon- und Stoffwechselforschung = Hormones et metabolisme*. 1998. 30: 323-8.
[94] Di Chiara G, Morelli M, Consolo S. Modulatory functions of neurotransmitters in the striatum: ACh/dopamine/NMDA interactions. *Trends Neurosci*. 1994. 17: 228-33.
[95] Rice ME, Cragg SJ. Dopamine spillover after quantal release: rethinking dopamine transmission in the nigrostriatal pathway. *Brain Res Rev*. 2008. 58: 303-13.
[96] Saul'skaia NB. [Volume transmission as a means of interneuronal interaction in the striatum]. *Zhurnal vysshei nervnoi deiatelnosti imeni I P Pavlova*. 1997. 47: 362-73.
[97] Fariello RG. Experimental support for the implication of oxidative stress in the genesis of parkinsonian syndromes. *Funct Neurol*. 1988. 3: 407-12.
[98] Gotz ME, Freyberger A, Riederer P. Oxidative stress: a role in the pathogenesis of Parkinson's disease. *J Neural Transm* Suppl. 1990. 29: 241-9.
[99] Hirsch EC. Does oxidative stress participate in nerve cell death in Parkinson's disease? *Eur Neurol*. 1993. 33 Suppl 1: 52-9.
[100] Sian J, Dexter DT, Lees AJ, Daniel S, Agid Y, Javoy-Agid F, et al. Alterations in glutathione levels in Parkinson's disease and other neurodegenerative disorders affecting basal ganglia. *Ann Neurol*. 1994. 36: 348-55.
[101] Sian J, Dexter DT, Lees AJ, Daniel S, Jenner P, Marsden CD. Glutathione-related enzymes in brain in Parkinson's disease. *Ann Neurol*. 1994. 36: 356-61.

[102] Jellinger K, Paulus W, Grundke-Iqbal I, Riederer P, Youdim MB. Brain iron and ferritin in Parkinson's and Alzheimer's diseases. *J Neural Transm Park Dis Dement Sect*. 1990. 2: 327-40.

[103] Dexter DT, Carayon A, Vidailhet M, Ruberg M, Agid F, Agid Y, et al. Decreased ferritin levels in brain in Parkinson's disease. *J Neurochem*. 1990. 55: 16-20.

[104] Fornstedt B, Rosengren E, Carlsson A. Occurrence and distribution of 5-S-cysteinyl derivatives of dopamine, dopa and dopac in the brains of eight mammalian species. *Neuropharmacology*. 1986. 25: 451-4.

[105] Slivka A, Cohen G. Hydroxyl radical attack on dopamine. *J Biol Chem*. 1985. 260: 15466-72.

[106] Jellinger K, Linert L, Kienzl E, Herlinger E, Youdim MB. Chemical evidence for 6-hydroxydopamine to be an endogenous toxic factor in the pathogenesis of Parkinson's disease. *J Neural Transm* Suppl. 1995. 46: 297-314.

[107] Linert W, Herlinger E, Jameson RF, Kienzl E, Jellinger K, Youdim MB. Dopamine, 6-hydroxydopamine, iron, and dioxygen--their mutual interactions and possible implication in the development of Parkinson's disease. *Biochimica et biophysica acta*. 1996. 1316: 160-8.

[108] Li CL, Werner P, Cohen G. Lipid peroxidation in brain: interactions of L-DOPA/dopamine with ascorbate and iron. *Neurodegeneration*. 1995. 4: 147-53.

[109] Maharaj H, Sukhdev Maharaj D, Scheepers M, Mokokong R, Daya S. l-DOPA administration enhances 6-hydroxydopamine generation. *Brain Res*. 2005. 1063: 180-6.

[110] Curtius HC, Wolfensberger M, Steinmann B, Redweik U, Siegfried J. Mass fragmentography of dopamine and 6-hydroxydopamine. Application to the determination of dopamine in human brain biopsies from the caudate nucleus. *J Chromatogr*. 1974. 99: 529-40.

[111] Graham DG, Tiffany SM, Bell WR, Jr., Gutknecht WF. Autoxidation versus covalent binding of quinones as the mechanism of toxicity of dopamine, 6-hydroxydopamine, and related compounds toward C1300 neuroblastoma cells in vitro. *Mol Pharmacol*. 1978. 14: 644-53.

[112] Hastings TG, Lewis DA, Zigmond MJ. Role of oxidation in the neurotoxic effects of intrastriatal dopamine injections. *Proc Natl Acad Sci U S A*. 1996. 93: 1956-61.

[113] Zigmond MJ, Hastings TG, Perez RG. Increased dopamine turnover after partial loss of dopaminergic neurons: compensation or toxicity? *Parkinsonism Relat Disord*. 2002. 8: 389-93.

[114] Hastings TG, Zigmond MJ. Loss of dopaminergic neurons in parkinsonism: possible role of reactive dopamine metabolites. *J Neural Transm Suppl*. 1997. 49: 103-10.

[115] Stokes AH, Hastings TG, Vrana KE. Cytotoxic and genotoxic potential of dopamine. *J Neurosci Res*. 1999. 55: 659-65.

[116] Esposito E, Di Matteo V, Benigno A, Pierucci M, Crescimanno G, Di Giovanni G. Non-steroidal anti-inflammatory drugs in Parkinson's disease. *Exp Neurol*. 2007. 205: 295-312.

[117] Hirsch EC, Hunot S. Neuroinflammation in Parkinson's disease: a target for neuroprotection? *Lancet neurology*. 2009. 8: 382-97.

[118] Liu B, Hong JS. Role of microglia in inflammation-mediated neurodegenerative diseases: mechanisms and strategies for therapeutic intervention. *J Pharmacol Exp Ther*. 2003. 304: 1-7.
[119] McGeer PL, Yasojima K, McGeer EG. Inflammation in Parkinson's disease. *Adv Neurol*. 2001. 86: 83-9.
[120] Armentero MT, Levandis G, Nappi G, Bazzini E, Blandini F. Peripheral inflammation and neuroprotection: systemic pretreatment with complete Freund's adjuvant reduces 6-hydroxydopamine toxicity in a rodent model of Parkinson's disease. *Neurobiol Dis*. 2006. 24: 492-505.
[121] de Meira Santos Lima M, Braga Reksidler A, Marques Zanata S, Bueno Machado H, Tufik S, Vital MA. Different parkinsonism models produce a time-dependent induction of COX-2 in the substantia nigra of rats. *Brain Res*. 2006. 1101: 117-25.
[122] He Y, Appel S, Le W. Minocycline inhibits microglial activation and protects nigral cells after 6-hydroxydopamine injection into mouse striatum. *Brain Res*. 2001. 909: 187-93.
[123] Quintero EM, Willis L, Singleton R, Harris N, Huang P, Bhat N, et al. Behavioral and morphological effects of minocycline in the 6-hydroxydopamine rat model of Parkinson's disease. *Brain Res*. 2006. 1093: 198-207.
[124] A pilot clinical trial of creatine and minocycline in early Parkinson disease: 18-month results. *Clin Neuropharmacol*. 2008. 31: 141-50.
[125] DeLong MR, Wichmann T. Circuits and circuit disorders of the basal ganglia. *Arch Neurol*. 2007. 64: 20-4.
[126] Schmidt WJ. Balance of transmitter activities in the basal ganglia loops. *J Neural Transm Suppl*. 1995. 46: 67-76.
[127] Benazzouz A, Piallat B, Ni ZG, Koudsie A, Pollak P, Benabid AL. Implication of the subthalamic nucleus in the pathophysiology and pathogenesis of Parkinson's disease. *Cell Transplant*. 2000. 9: 215-21.
[128] Vernon AC, Zbarsky V, Datla KP, Croucher MJ, Dexter DT. Subtype selective antagonism of substantia nigra pars compacta Group I metabotropic glutamate receptors protects the nigrostriatal system against 6-hydroxydopamine toxicity in vivo. *J Neurochem*. 2007. 103: 1075-91.
[129] Abdel-Salam OM. Drugs used to treat Parkinson's disease, present status and future directions. *CNS & neurological disorders drug targets*. 2008. 7: 321-42.
[130] Mandel S, Grunblatt E, Riederer P, Gerlach M, Levites Y, Youdim MB. Neuroprotective strategies in Parkinson's disease: an update on progress. *CNS drugs*. 2003. 17: 729-62.
[131] Schapira AH. Neuroprotection and dopamine agonists. *Neurology*. 2002. 58: S9-18.
[132] Chen LW, Wang YQ, Wei LC, Shi M, Chan YS. Chinese herbs and herbal extracts for neuroprotection of dopaminergic neurons and potential therapeutic treatment of Parkinson's disease. *CNS & neurological disorders drug targets*. 2007. 6: 273-81.
[133] Maruyama W, Akao Y, Youdim MB, Naoi M. Neurotoxins induce apoptosis in dopamine neurons: protection by N-propargylamine-1(R)- and (S)-aminoindan, rasagiline and TV1022. *J Neural Transm Suppl*. 2000. 171-86.

[134] Le WD, Jankovic J, Xie W, Appel SH. Antioxidant property of pramipexole independent of dopamine receptor activation in neuroprotection. *J Neural Transm.* 2000. 107: 1165-73.

[135] Engele J, Bohn MC. The neurotrophic effects of fibroblast growth factors on dopaminergic neurons in vitro are mediated by mesencephalic glia. *J Neurosci.* 1991. 11: 3070-8.

[136] Lin LF, Doherty DH, Lile JD, Bektesh S, Collins F. GDNF: a glial cell line-derived neurotrophic factor for midbrain dopaminergic neurons. *Science.* 1993. 260: 1130-2.

[137] Choi-Lundberg DL, Lin Q, Chang YN, Chiang YL, Hay CM, Mohajeri H, et al. Dopaminergic neurons protected from degeneration by GDNF gene therapy. *Science.* 1997. 275: 838-41.

[138] Winkler C, Sauer H, Lee CS, Bjorklund A. Short-term GDNF treatment provides long-term rescue of lesioned nigral dopaminergic neurons in a rat model of Parkinson's disease. *J Neurosci.* 1996. 16: 7206-15.

[139] Sauer H, Campbell K, Wiegand SJ, Lindsay RM, Bjorklund A. Brain-derived neurotrophic factor enhances striatal neuropeptide expression in both the intact and the dopamine-depleted rat striatum. *Neuroreport.* 1994. 5: 609-12.

[140] Guan J, Krishnamurthi R, Waldvogel HJ, Faull RL, Clark R, Gluckman P. N-terminal tripeptide of IGF-1 (GPE) prevents the loss of TH positive neurons after 6-OHDA induced nigral lesion in rats. *Brain Res.* 2000. 859: 286-92.

[141] Fjord-Larsen L, Johansen JL, Kusk P, Tornoe J, Gronborg M, Rosenblad C, et al. Efficient in vivo protection of nigral dopaminergic neurons by lentiviral gene transfer of a modified Neurturin construct. *Exp Neurol.* 2005. 195: 49-60.

[142] Sherer TB, Fiske BK, Svendsen CN, Lang AE, Langston JW. Crossroads in GDNF therapy for Parkinson's disease. *Mov Disord.* 2006. 21: 136-41.

[143] Bartus R. CERE-120 (AAV2-NTN) for Parkinson's Disease: Review of Progress and Future Plans (abstract). *American Soc Gene Therapy.* San Diego, CA 2009.

[144] Lindvall O. Transplants in Parkinson's disease. *Eur Neurol.* 1991. 31 Suppl 1: 17-27.

[145] Lindvall O, Kokaia Z. Prospects of stem cell therapy for replacing dopamine neurons in Parkinson's disease. *Trends Pharmacol Sci.* 2009. 30: 260-7.

[146] Yurek DM, Sladek JR, Jr. Dopamine cell replacement: Parkinson's disease. *Annu Rev Neurosci.* 1990. 13: 415-40.

[147] Lindner MD, Winn SR, Baetge EE, Hammang JP, Gentile FT, Doherty E, et al. Implantation of encapsulated catecholamine and GDNF-producing cells in rats with unilateral dopamine depletions and parkinsonian symptoms. *Exp Neurol.* 1995. 132: 62-76.

[148] Kamo H, Kim SU, McGeer PL, Tago H, Shin DH. Transplantation of cultured human adrenal chromaffin cells into 6-hydroxydopamine-lesioned rat brain. *Synapse.* 1987. 1: 324-8.

[149] Dunnett SB, Bjorklund A, Stenevi U, Iversen SD. Grafts of embryonic substantia nigra reinnervating the ventrolateral striatum ameliorate sensorimotor impairments and akinesia in rats with 6-OHDA lesions of the nigrostriatal pathway. *Brain Res.* 1981. 229: 209-17.

[150] Madhavan L, Daley BF, Paumier KL, Collier TJ. Transplantation of subventricular zone neural precursors induces an endogenous precursor cell response in a rat model of Parkinson's disease. *J Comp Neurol.* 2009. 515: 102-15.

[151] Sanchez-Pernaute R, Lee H, Patterson M, Reske-Nielsen C, Yoshizaki T, Sonntag KC, et al. Parthenogenetic dopamine neurons from primate embryonic stem cells restore function in experimental Parkinson's disease. *Brain.* 2008. 131: 2127-39.

[152] Lindvall O, Bjorklund A. Cell therapy in Parkinson's disease. *NeuroRx.* 2004. 1: 382-93.

[153] Olanow CW, Goetz CG, Kordower JH, Stoessl AJ, Sossi V, Brin MF, et al. A double-blind controlled trial of bilateral fetal nigral transplantation in Parkinson's disease. *Ann Neurol.* 2003. 54: 403-14.

[154] Imaizumi Y, Kaneko R. Rising mortality from Parkinson's disease in Japan, 1950-1992. *Acta Neurol Scand.* 1995. 91: 169-76.

[155] Lilienfeld DE, Chan E, Ehland J, Godbold J, Landrigan PJ, Marsh G, et al. Two decades of increasing mortality from Parkinson's disease among the US elderly. *Arch Neurol.* 1990. 47: 731-4.

[156] Stromberg I, Gemma C, Vila J, Bickford PC. Blueberry- and spirulina-enriched diets enhance striatal dopamine recovery and induce a rapid, transient microglia activation after injury of the rat nigrostriatal dopamine system. *Exp Neurol.* 2005. 196: 298-307.

[157] Joseph JA, Shukitt-Hale B, Denisova NA, Prior RL, Cao G, Martin A, et al. Long-term dietary strawberry, spinach, or vitamin E supplementation retards the onset of age-related neuronal signal-transduction and cognitive behavioral deficits. *J Neurosci.* 1998. 18: 8047-55.

[158] Blanchet J, Longpre F, Bureau G, Morissette M, DiPaolo T, Bronchti G, et al. Resveratrol, a red wine polyphenol, protects dopaminergic neurons in MPTP-treated mice. *Prog Neuropsychopharmacol Biol Psychiatry.* 2008. 32: 1243-50.

[159] Bousquet M, Saint-Pierre M, Julien C, Salem N, Jr., Cicchetti F, Calon F. Beneficial effects of dietary omega-3 polyunsaturated fatty acid on toxin-induced neuronal degeneration in an animal model of Parkinson's disease. *Faseb J.* 2008. 22: 1213-25.

[160] Choi JY, Jang EH, Park CS, Kang JH. Enhanced susceptibility to 1-methyl-4-phenyl-1,2,3,6-tetrahydropyridine neurotoxicity in high-fat diet-induced obesity. F*ree radical biology & medicine*. 2005. 38: 806-16.

[161] Cohen AD, Tillerson JL, Smith AD, Schallert T, Zigmond MJ. Neuroprotective effects of prior limb use in 6-hydroxydopamine-treated rats: possible role of GDNF. *J Neurochem.* 2003. 85: 299-305.

[162] Tillerson JL, Cohen AD, Philhower J, Miller GW, Zigmond MJ, Schallert T. Forced limb-use effects on the behavioral and neurochemical effects of 6-hydroxydopamine. *J Neurosci.* 2001. 21: 4427-35.

[163] Yoon MC, Shin MS, Kim TS, Kim BK, Ko IG, Sung YH, et al. Treadmill exercise suppresses nigrostriatal dopaminergic neuronal loss in 6-hydroxydopamine-induced Parkinson's rats. *Neurosci Lett.* 2007. 423: 12-7.

[164] Mabandla M, Kellaway L, St Clair Gibson A, Russell VA. Voluntary running provides neuroprotection in rats after 6-hydroxydopamine injection into the medial forebrain bundle. *Metab Brain Dis.* 2004. 19: 43-50.

[165] Tillerson JL, Caudle WM, Reveron ME, Miller GW. Exercise induces behavioral recovery and attenuates neurochemical deficits in rodent models of Parkinson's disease. *Neuroscience*. 2003. 119: 899-911.

[166] Smith AD, Zigmond MJ. Can the brain be protected through exercise? Lessons from an animal model of parkinsonism. *Exp Neurol.* 2003. 184: 31-9.

[167] Poulton NP, Muir GD. Treadmill training ameliorates dopamine loss but not behavioral deficits in hemi-parkinsonian rats. *Exp Neurol.* 2005. 193: 181-97.

[168] Faherty CJ, Raviie Shepherd K, Herasimtschuk A, Smeyne RJ. Environmental enrichment in adulthood eliminates neuronal death in experimental Parkinsonism. *Brain Res Mol Brain Res*. 2005. 134: 170-9.

[169] Fleming SM, Fernagut PO, Chesselet MF. Genetic mouse models of parkinsonism: strengths and limitations. *NeuroRx*. 2005. 2: 495-503.

[170] Harvey BK, Wang Y, Hoffer BJ. Transgenic rodent models of Parkinson's disease. *Acta Neurochir Suppl.* 2008. 101: 89-92.

[171] Moore DJ, Dawson TM. Value of genetic models in understanding the cause and mechanisms of Parkinson's disease. *Current neurology and neuroscience reports*. 2008. 8: 288-96.

In: Biogenic Amines
Editors: T. Farooqui, A. A. Farooqui

ISBN: 978-1-60876-625-3

Chapter IX

THE MOLECULAR AND GENETIC BASIS OF PARKINSON DISEASE*

Tahira Farooqui&

The Ohio State University, Columbus, OH, 43210, USA.

ABSTRACT

Dopamine is a major catecholamine neurotransmitter in the mammalian central nervous system. It plays a crucial role in regulating diverse functions, including motor control, emotional processing, learning, reinforcement, and cognitive attention. Dopamine acts through G-protein-coupled receptors (GPCRs), which cluster into D1-like and D2-like subfamilies. D1-like receptors comprising D_1 and D_5 are coupled to adenylyl cyclase (AC) through stimulatory G protein ($G\alpha_s$). Whereas D2-like comprising D_2, D_3 and D_4 are coupled to AC through inhibitory G protein ($G\alpha_i$/o). Such stimulation and inhibition of AC regulate intracellular cAMP levels, but less is known about the function of each dopamine receptor subtype. Dopamine-mediated signaling depends on its metabolism, activation of postsynaptic dopamine receptors, dopamine receptor interacting proteins (DRIPs), and phosphorylation and dephosphorylation status of dopamine- and cAMP-regulated phosphoprotein of 32 kDa (DARPP-32). Furthermore, the cross-talk between dopamine and other neurotransmitters (such as glutamate, serotonin, and adenosine) receptors critically modulates cellular signaling responses. Dysfunction in dopaminergic (DA-ergic) system in human brain is linked to various pathological conditions, including Parkinson disease (PD). Aging is considered as the most important risk factor for PD. The primary pathology of PD is degeneration of DA-ergic neurons in the substantia nigra pars compacta (SNpc). Therefore, loss of the

* This manuscript is written in the loving memory of my Ph.D adviser late Prof. Norman J. Uretsky, a beloved teacher, mentor and friend, who introduced me to the interesting field of 'Biogenic Amines'. Because of the impact Dr. Uretsky made on his graduate students, he will live undoubtedly in their thoughts and minds".

& Correspondence concerning this article should be addressed to: Tahira Farooqui, Department of Entomology/ Center for Molecular Neurobiology, The Ohio State University, Columbus, OH 43210-1220, Telephone: (614) 783-4369, Email: farooqui.2@osu.edu.

nigrostriatal pathway decreases striatal dopamine levels in the striatum, resulting in consequent extrapyramidal motor dysfunction. The etiology of PD is elusive, however a combination of genetic and environmental factors seem to play a critical role. Although, majority of cases of PD are sporadic but several genes (such as *α-synuclein*, *parkin*, *DJ-1*, *PINK-1* and *LRRK2*) have been shown to be linked with rare familial forms of the disease. Therefore, It is considered that PD is caused by both genetic and environmental factors, which cause a key molecular mechanism 'Oxidative Stress' that is responsible for the loss of DA-ergic neurons. Current therapeutic strategies for PD focus primarily on reducing the severity of its symptoms using DA-ergic medications, such as dopamine precursor, dopamine receptor agonists, dopamine uptake blockers, and inhibitors of dopamine catabolism. However, these medications produce side effects and help only in relieving some of symptoms but can not stop disease progression because of not addressing the underlying problem associated with the pathogenesis of PD. Thus, in addition to developing drugs with specific and selective affinities for each dopamine receptor subtype as a possible therapeutic agent, one should also examine the whole genome for the susceptibility of genes for PD.

Keywords: Dopamine, D1-like, D2-like, Neurodegeneration, Parkinson disease, Aging, ROS, Dopamine-mediated behavior.

1. Introduction

Dopamine (4-(2-aminoethyl)benzene-1,2-diol), a biogenic amine, is a member of the catecholamine family. In the central nervous system (CNS), dopamine plays a crucial role as a neurotransmitter and regulates a variety of vital brain functions, such as cognition, motor coordination, endocrine activity, reward circuitry, pain processing, mood, motivation, and learning and memory (Figure 1). Dopaminergic (DA-ergic) neurons of substantia nigra pars compacta (SNpc) are involved in the control of voluntary movements and postural reflexes. From SNpc, DA-ergic neurons project to the striatum (caudate and putamen), therefore, death of these neurons in SNpc causes dopamine deficiency in the striatum, leading to Parkinson disease (PD). This disease can be clinically characterized by movement disorders, including muscle rigidity, tremor at rest, slowness of physical movement (bradykinesia) and impaired balance and coordination (postural instability) in PD patients. DA-ergic neurons in the ventral tegmental area are also involved in modulation of rewarding and cognitive behaviors, and their dysfunction may lead to the pathogenesis of addiction and mood disorders.

Brain aging is associated with a decline in dopamine-mediated functions, implicating loss of DA-ergic neurons. A marked decline in dopamine system with aging includes a gradual loss of presynaptic markers such as dopamine transporters [1], and post-synaptic markers such as dopamine receptors [2-3]. Such decline in markers has been consistently observed in striatal, neocortical, and limbic areas [4]. Alterations between midbrain dopamine synthesis and prefrontal activity in cortical reward processing with human aging suggest that dopamine system involved with reward processing becomes vulnerable with aging [5]. The magnification of catechol-o-methyltransferase (COMT) gene with aging provides novel information on the inverted U-shaped relation linking DA-ergic neuromodulation in prefrontal cortex to cognitive performance [6]. Molecular imaging studies have demonstrated impairment in multiple cognitive tasks with age, implicating loss of dopamine in aged brain [4,7-8]. Thus, aging can be considered as one of the most important risk factor for PD. The etiology of PD involves both familial (genetic) and environmental (sporadic) factors. Predisposing factors in sporadic PD include trauma, diet and environmental factors, including exposures to metals (manganese, iron), drinking well water, herbicides and pesticides [9-10]. Association of several genes along with other risk factors, such as gender, race, and positive family history, support the occurrence of familial PD.

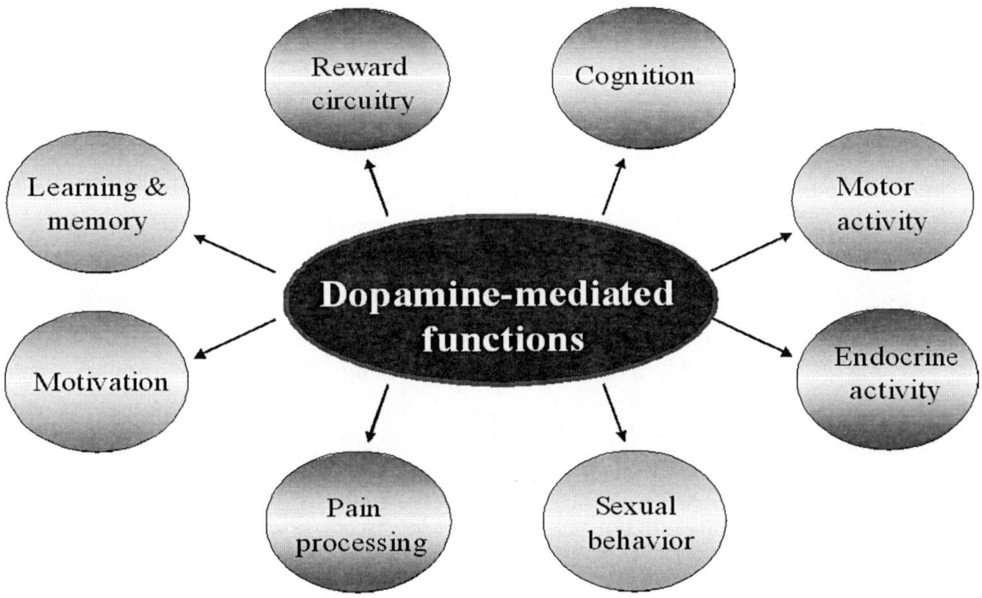

Figure 1. Dopamine-mediated functions.

PD is one of the most common neurodegenerative movement disorder. A prominent feature of this disease is the loss of striatal dopamine. Current treatment strategies for PD mainly involve pharmacotherapy using DA-ergic drugs like 3,4-dihydroxy-L-phenylalanine (L-DOPA, a precursor required for dopamine synthesis), and deprenyl (Selgiline), an inhibitor of monomanine oxidase B (MAOB; the enzyme that breaks down dopamine), as well as neuroablation or neurostimulation of the downstream brainstem nuclei affected by the DA-ergic deficit [11]. These strategies are associated with limited effectiveness and side effects but do not cure the disease. They do not address the underlying problem of the progressive loss of DA-ergic neurons. Therefore, more compelling strategies are required to expand treatment beyond the classical dopamine replacement towards the targeted reconstruction of nigrostrial pathway. This chapter summarizes dopamine metabolism, receptor classification, behavior, signaling, and possible molecular mechanisms involved with the nigrostriatal DA-ergic neurodegeneration, which may help in understanding the pathogenesis of PD.

2. Dopamine Metabolism

Dopamine metabolism in the brain is a function of presynaptic synthesis, release and reuptake of dopamine. Dopamine is a major neurotransmitter that plays a prominent role in a variety of vital brain functions. To perform these physiological functions, it is necessary to keep a balance between dopamine synthesis, degradation, release, and synaptic uptake. An alteration in dopamine metabolism disrupts dopamine homeostasis that results disturbances in the DA-ergic neurotransmission, leading to region-specific functional abnormalities.

2.1. Dopamine Synthesis and Release

Dopamine biosynthetic pathway starts with the dietary amino acid precursor, L-tyrosine (Figure 2). Tyrosine is first converted to 3,4-dihydroxy-L-phenylalanine (L-DOPA) by tyrosine hydroxylase (TH), a rate limiting enzyme in catecholamine biosynthesis. Tyrosine hydroxylase is a mixed function oxidase that uses molecular oxygen and tyrosine as its substrates and tetrahydrobiopterin as its cofactor in catalyzing the addition of a hydroxyl group to the meta position of tyrosine, thus forming L-DOPA. It is then rapidly decarboxylated to dopamine by DOPA decarboxylase (DDC), a pyridoxine dependent enzyme, that catalyzes the removal of carboxyl group from L-DOPA and forms dopamine. In DA-ergic neurons, the pathway stops here (Figure 2). TH is the rate limiting enzyme in dopamine biosynthesis, and is particularly susceptible to physiological regulation and pharmacological manipulation. TH exists in two kinetic forms that exhibit different affinities for its cofactor tetrahydrobiopterin (BH-4). The conversion from low to high affinity form for BH-4 involves the phosphorylation and activation of TH. Enzyme activity of TH is subjected to four major regulatory mechanisms (Figure 2), including (1) end-product inhibition by intraneuronal dopamine, (2) availability of its cofactor BH-4, (3) activation of presynaptic D_2 dopamine receptors, and (4) rate of impulse flow in the nigrosriatal pathway [12-13].

In brain, the striatum receives two major projections: (1) DA-ergic from the subtantia nigra and ventral tegmental area and (2) glutamatergic from most of the cerebral cortex. Under normal physiological conditions, dopamine is released in the striatum from dopamine nerve terminals via an exocytotic, calcium-dependent mechanism. Calcium-dependent release of dopamine occurs in response to invasion of the terminal by nerve impulse action potential (Figure 2). The extent of this release appears to be a function of the rate and pattern of firing. The burst-firing mode leads to an enhanced release of dopamine. Dopamine release is also modulated by presynaptic dopamine receptors (also referred as release-modulating autoreceptors) (Figure 2). In general, dopamine agonists inhibit, while dopamine antagonists enhance the evoked release of dopamine. Presynaptic dopamine receptors expressed along the somatodendritic extent of midbrain DA-ergic neurons modulate impulse activity, whereas dopamine receptors expressed at the nerve terminals regulate both dopamine synthesis and release [14].

Glutamate and dopamine interactions in the striatum also play a major role in the normal function of the corticostriatal system, involved in motor and cognitive functions. Abnormalities in these interactions are considered to play a role in the pathophysiology of PD

[15]. The repetitive transcranial magnetic stimulation (TMS) of the human dorsolateral prefrontal cortex leads to dopamine release in the ipsilateral caudate nucleus, which supports corticostriatal control of dopamine release in humans [16]. The repetitive TMS of the motor cortex induces dopamine release in the ipsilateral putamen [17], confirming that dopamine release is focally restricted in healthy humans. However, measurement of dopamine release following cortical stimulation in early PD patients with evidence of unilateral motor symptoms shows that the amount of TMS-induced striatal dopamine release is less in the symptomatic hemisphere than in the contralateral hemisphere [18], which may be due to advanced degeneration of DA-ergic nerve terminals and the failure of functional compensatory mechanisms. Thus, a possibility of *in vivo* expression of a loss of functional segregation of cortical information to the striatum gives an indirect evidence of abnormal corticostriatal transmission in early PD. Collectively, impairement in dopamine synthesis and release are directly attributable to PD.

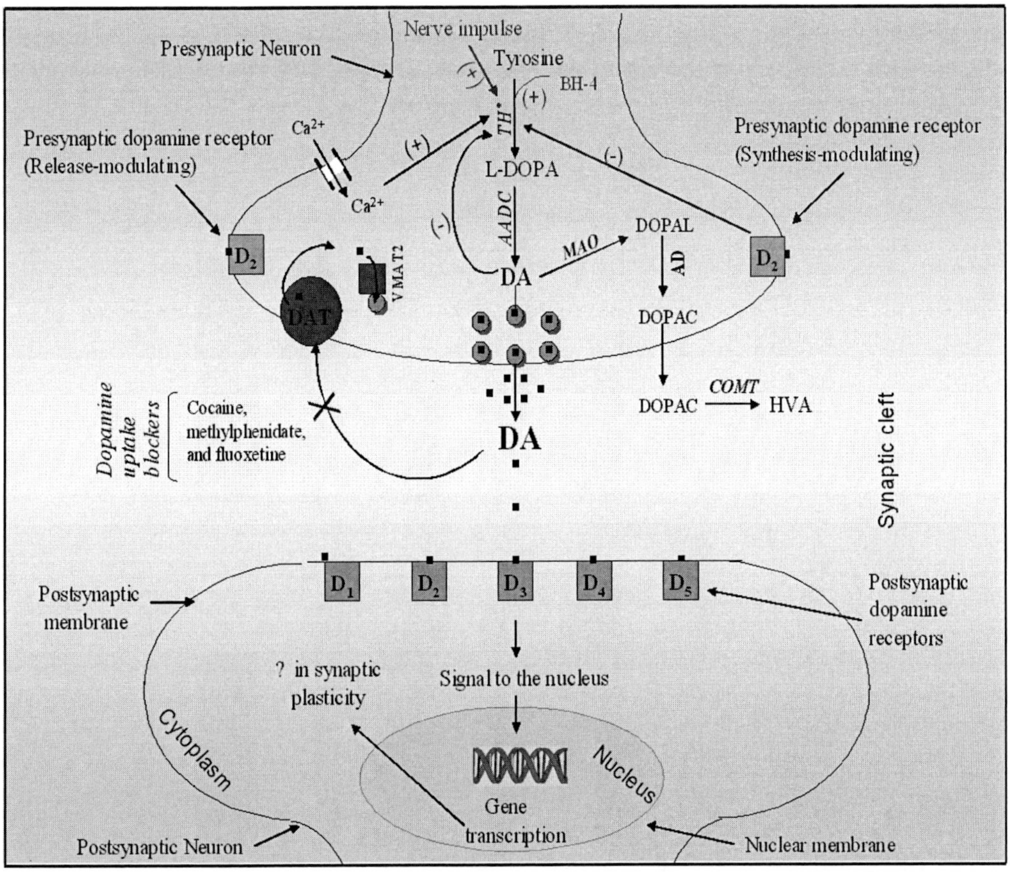

Figure 2. DA-ergic presynaptic and postsynaptic neurons: involvement in dopamine metabolism. Tyrosine hydroxylase (TH); Aromatic amino acid decarboxylase (AADC); Dopamine receptor subtypes (D_1-D_5); Dopamine uptake blockers (cocaine, methylphenidate and fluoxetine).

2.2. Dopamine Inactivation

Dopamine is inactivated by two major processes: (1) uptake by the nerve terminal and (2) enzymic degradation. Dopamine can be taken up from the synaptic cleft back into the presynaptic terminal via dopamine transporter (DAT) that terminates its synaptic action [19-20]. Dopamine transporter belongs to the SLC6 family of Na^{+}/Cl^{-} dependent symporters [21]. Dopamine uptake is an energy dependent process where sodium/potassium ATPase uses energy from adenosine 5'-triphosphate (ATP) hydrolysis to create a concentration gradient of ions across the presynaptic membrane, driving the opening of DAT, which exhibits a channel mode of conduction that directly modulates membrane potential and neuronal function. Dopamine uptake plays an important role in maintaining synaptic homeostasis. Following uptake into the cytoplasm, dopamine undergoes a second transport step that is required for its efficient cycling. This step is performed by a vesicular monoamine transporter type-2, VMAT2, driven by a H^{+} electrochemical gradient. Dopamine is re-packaged into vesicles for their release by exocytosis (Figure 2). Psychostimulants, such as cocaine, methylphenidate, and fluoxetine, block dopamine uptake by binding to DAT and incease dopamine levels in the synapse, leading to addiction or neurological symptoms.

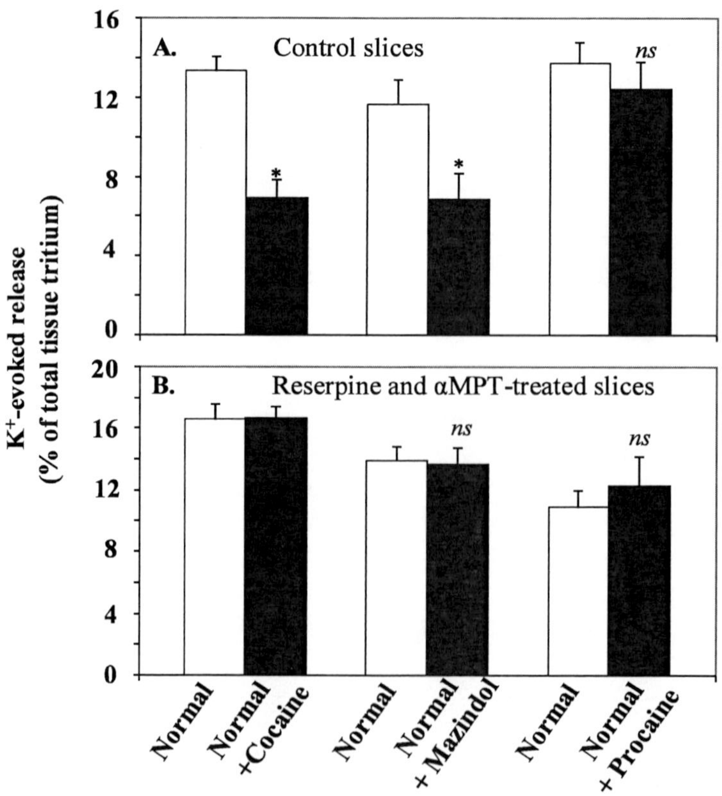

Figure 3. The effect of cocaine, mazindole and procaine on K^{+}-evoked release of [^{3}H]-acetylcholine in mouse striatal slices. (A) control slices, (2) reserpine and α-methyl-*p*-tyrosine (α-MPT)-treated slices. Modified from reference [27].

Alterations in the density, function and relative expression of DAT compared to VMAT2 may result in neurological and psychiatric disorders, including PD, schizophrenia, and addiction [22-24]. Both cocaine and mazindol are known as monoamine uptake inhibitors. Nigrostriatal DA-ergic fibers exert a direct inhibitory influence on striatal cholinergic interneurons. *In vitro* perfusion studies using striatal slice preparation indicate that the release of acetylcholine (Ach) from slices is modulated in an inhibitory fashion by the stimulation of dopamine receptors, particularly the D_2 subtype [25-26]. Cocaine and mazindol, but not procaine (local anesthetic), produce a marked reduction in K^+-evoked release of $[^3H]$Ach in mice striatal slices (Figure 3A). Inhibition is overcome in reserpine (depletes stores of amines) - and dopamine synthesis inhibitor, α-methylparatyrosine (α-MPT)-treated slices (Figure 3B), suggesting the dependence of endogenous dopamine [27]. Both cocaine and mazindol produce ~ 8-fold shift in concentration-response curve for dopamine, but not for dimethyl analog of dopamine, DMDA [27]. Furthermore, IC_{50}'s of DMDA and dopamine for the inhibition of $[^3H]$dopamine uptake are 26.31 μM and 0.83 μM, respectively, suggesting that dopamine has higher affinity for dopamine uptake transporter than DMDA.

Dopamine, DMDA, and permanently charged analogs of dopamine (quaternary ammonium (DA-N^+), selenium (DA-Se^+), dimethylsulfonium (DA-S^+), inhibit the K^+-evoked release of $[^3H]$Ach but with different EC50s (Figure 4A) [28]. The inhibitory effects of DMDA and permanently charged analogs of dopamine on K^+-evoked release of $[^3H]$Ach are not influenced by cocaine [29,27], implicating that dopamine transporter has a very high affinity for dopamine compared to its analogs. The striatonigral pathway also plays an important role in maintaining rotational behavior. Rats treated with permanently charged analogs (DA-N^+, DM-S^+) show significant increase in the circling behavior (Figure 4B), but permanently uncharged analog of dopamine (DM-S) or amphetamine (AMP) produce no response [29-30]. These findings suggest that permanently charged analogs of dopamine are not taken up by dopamine transporter, and atom on α-carbon of dopamine should have a positive charge in aqueous solution to interact with the D_2 DA-ergic receptor.

The two major enzymes of dopamine catabolism are monoamine oxidase and catechol-*O*-methyltransferase (COMT). MAO, a flavin containing enzyme, is located in the outer membrane of mitochondria. It converts dopamine to its corresponding aldehyde (dihydroxyphenyl acetaldehyde, DOPAL) by oxidative deamination. DOPAL is rapidly oxidized to dihydroxyphenylacetic acid (DOPAC) by aldehyde dehydrogenase (Figure 2). COMT is a cytosolic enzyme that converts DOPAC to 3-methoxy-4-hydroxypheny acetic acid (homovanillate, HVA). The outcome of dopamine inactivation is the conversion of dopamine into HVA. Dopamine deficiency in PD, can be compensated by increasing dopamine synthesis and dopamine release, or reducing dopamine catabolism.

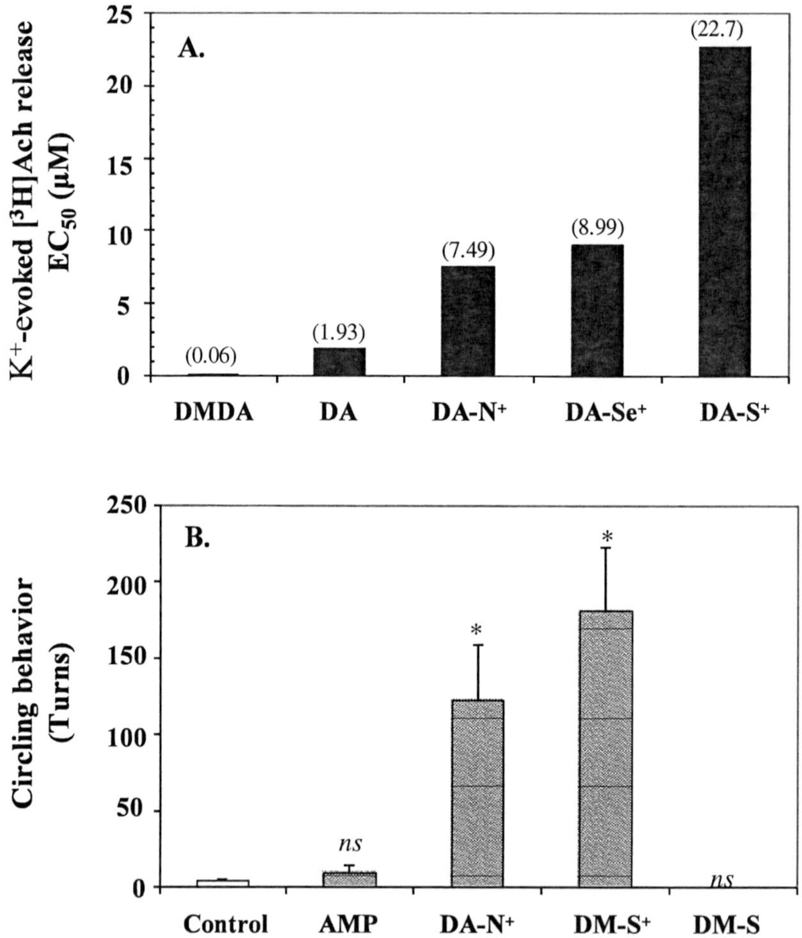

Figure 4. Effects of charged and uncharged analogs of dopamine on K^+-evoked release of [^{3}H]-acetylcholine in mouse striatal slices, and on the circling behavior in rats. (A) Effect of dopamine, dimethyldopamine, and permanently charged analogs of dopamine. Before decapitation, mice have been injected with reserpine (5mg/kg) and α-methyl-*p*-tyrosine (250 mg/kg) 20 and 2 hr, respectively. Number in parenthesis represents EC50 (μM) values. The order of potency for inhibition of the K^+-evoked stimulation of [^{3}H]acetylcholine release is: dimethyldopamine (DMDA) > dopamine (DA) > quaternary ammonium analog (DA-N^+) = dimethylselenonium analog (DA-Se^+) > dimethyl-sulfonium analog (DA-S^+). Modified from reference [28]. (B) Effect of direct intrastriatal injection of drugs on the circling behavior of rats previously lesioned unilaterally with 6-hydroxydopamine (6-OH). Dose for each drug injected was 100 nmol in 2μl volume. Turns in ipsilateral and contralateral directions were determined at 5 min intervals at 15, 30, 45, and 60 min, respectively. For getting the net number of turns for each animal, contralateral minus ipsilateral is added together. Permanently uncharged monomethylsulfide (DM-S) analog does not elicit significant contralateral circling, whereas, DA-N^+ and DA-S^+ at the same dose as amphetamine elicit significant contralateral circling. *p <0.05 when compared to controls. *ns*, not significantly different from control. Modified from reference [29].

3. Classification of DA-ergic Receptors

The original classification of DA-ergic receptors in the CNS is primary based on biochemical and pharmacological criteria. According to this classification, dopamine receptors are divided into two subclasses: the D1 receptors are positively coupled to AC and mediate cyclic adenosine monophosphate (cAMP) formation, whereas the D2 receptors are negatively coupled to AC and therefore reduce synthesis of cAMP [31-32]. The application of modern molecular biology techniques revealed the existence and identification of five distinct genes for DA-ergic receptors [33]. Based on their structural similarities, homology and conservation of signature amino acids, signal transduction pathways and pharmacological features, DA-ergic receptor family in mammals is again divided in two subfamilies: (1) D1-like family consisting of D_1 and D_5 subtypes, and (2) D2-like family consisting of D_2, D_3 and D_4 subtypes of dopamine receptors [34-36]. Protein sequence alignment of human dopamine receptor subtypes (Figure 5A), identified by Genbank accession number (D_1: AC#: NP_000785; D_{2Long}: AC#: AAB26274; D_{2Short}: NP-057658; D_3: NP_000787; D_4: NP_000788; D_5:NP_00789), reveals the existence of two such subfamilies (Figure 5B).

D1-like receptors are coupled to stimulatory G protein ($G\alpha_s$), thereby activate AC through stimulatory G protein ($G\alpha_s$), whereas D2-like (D_2, D_3 and D_4) receptors are coupled to AC via inhibitory **G protein** ($G\alpha_i$/o) leading to inhibition of AC. In D1-like subfamily, both receptor subtypes are located post-synaptically and share similar pharmacological patterns, whereas in D2-like subfamily, receptors are located pre- and- post-synaptically and show variable effects (Table 1). In general, all subtypes in D2-like subfamily have higher affinity for anti-psychotic drugs and are involved in PD. There are major molecular differences between D1-like and D2-like subfamilies. Receptors in D1-like subfamily are intronless, contain short 3rd intracytosolic loops and have long C terminal tail. Receptors in D2-like subfamily contains introns in their respective coding regions thereby have a possibility for alternative splicing. In addition, they have long 3rd intracytosolic loops and short C terminal tails (Table 1). In D2-like subfamily, alternatively spliced forms of D_2, D_3 receptors have been identified. These repeat sequences found in D_4 receptor mRNA sequence suggest many possible forms of splice variants.

The mammalian D_2 receptor gene encodes two molecularly distinct isoforms (D_{2L}, long isoform, and D_{2S}, short isoform), which differ from each other by the presence of additional 29 amino acids within the third intracellular loop [37-38]. Both isoforms bind to pertussis toxin-sensitive G proteins, therefore, inhibit AC. However, these isoforms bind to distinct Gi proteins, which may account for their structural differences [39-40]. D_{2L} mainly acts at the postsynaptic sites, whereas, D_{2S} acts at the presynaptic sites, suggesting differential coupling of D_{2L} and D_{2S} due to binding with different G proteins. The cataleptic effects of antipsychotic drug, haloperidol, are absent in D_{2L} deficient mice, confirming that D_{2L} postsynaptic receptors are targeted by antipsychotic drugs. In contrast to haloperidol, D_1 receptor antagonist, SCH23390 can produce cataleptic effects in D_{2L} deficient mice, suggesting that motor inhibition induced by dopamine receptor antagonists is governed by two independent pathways [41]. These isoforms have different (probably antagonistic) functions *in vivo,* and there is a possibility of cross-regulatory mechanism between D_1-and-D_2-mediated signal transduction pathways.

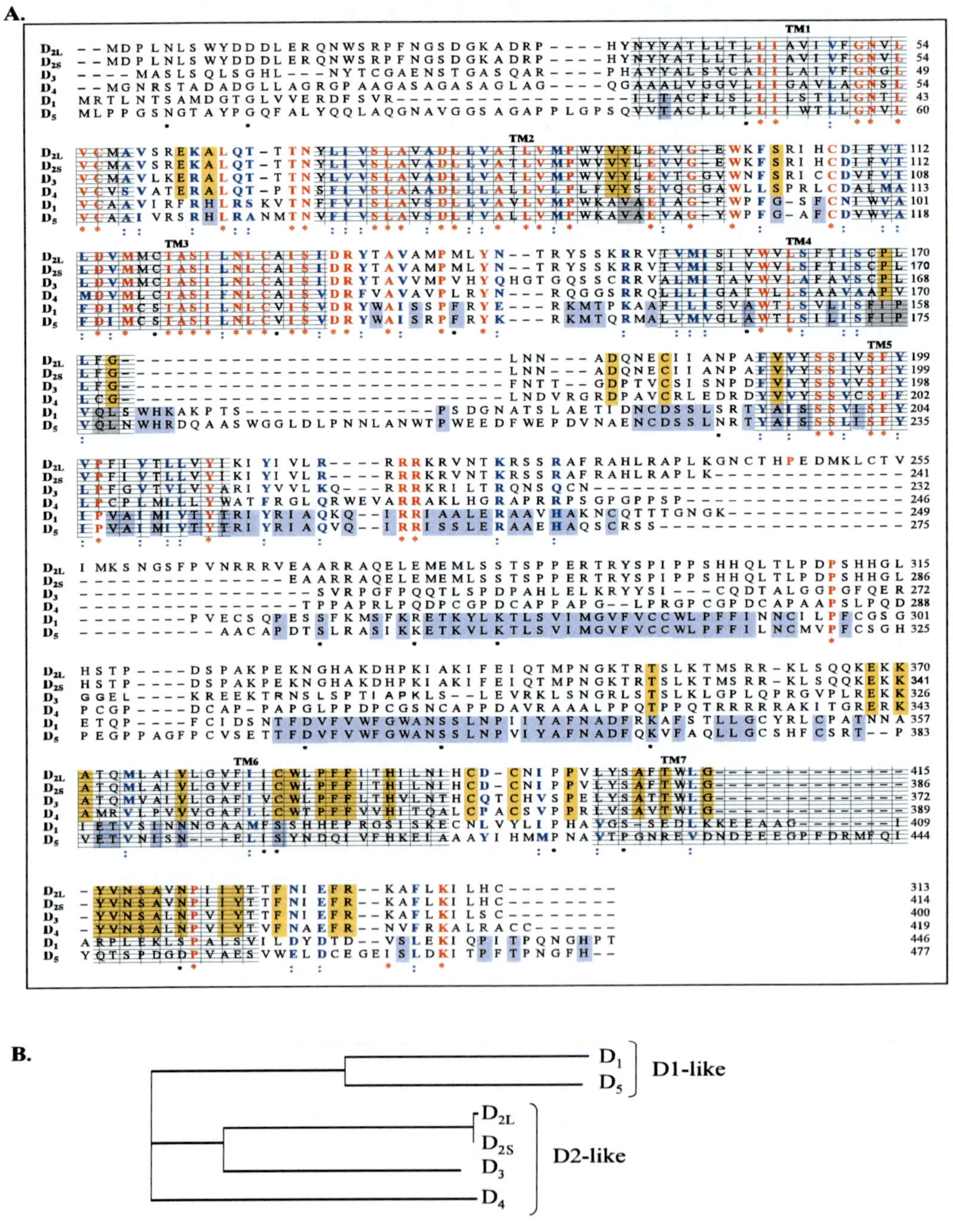

Figure 5. Protein sequence alignment of human dopamine receptor subtypes. (A). Produced by ClustalW, of five human dopamine receptor subtypes, identified by Genbank accession number (D_1: AC#: NP_000785; D_{2Long}: AC#: AAB26274; D_{2Short}: NP_057658; D_3 : NP_000787; D_4: NP_000788; D_5:NP_00789). Seven transmembrane regions (TM1-TM7) are indicated. Key: Single letters: amino acids. "*": identical in all sequences; ":": amino acids in similar group; ".": semi-conserved substitution (similar shapes). Grey background: amino acids conserved in D_1 and D_5. Orange background: amino acids conserved in D_{2Long}, $D_{2\ Short}$, D_3 and D_4 receptor sequences. Amino acid numbers are shown at the right. (B). A cladogram depicting the hypothesized branching order of dopamine receptor subtypes (D_1-D_5) sequences.

Table 1. Differences between D1-like and D2-like dopamine receptors in the CNS

	D1-Like Receptors (Intronless)		D2-Like Receptors (Introns)		
	D_1	D_5	D_2	D_3	D_4
G protein	$G\alpha_s$, G_{olf}	$G\alpha_s$, G_{olf}	$G\alpha_{i/o}$	$G\alpha_{i/o}$	$G\alpha_{i/o}$
Synaptic location	Post-synaptic	Post-synaptic	Post- and presynaptic	Post- and presynaptic	Post- and presynaptic
AC activity	Increase	Increase	Decrease	Decrease	Decrease
PLC	G_q, ↑Ca^{2+}	G_q,↑Ca^{2+}			
Selective agonists⁎	SKF83822 SKF38393 Dihydrexidine	KF83822 SKF38393 Dihydrexi-dine	(+)PHNO Ropinirol Pramipexole	(+)-PD128907	PD168077
Selective antagonists♠	SKF83566 SCH23390	SKF83566 SCH23390	Raclopride, L-41626	GR103691 NGB2904	L741742 U101958
Coding region	Intron -less	Intron -less	Introns	Introns	Introns
Intracytosolic loop	Short	Short	Long	Long	Long
C-terminal	Long	Long	Short	Short	Short
Isomers			D_{2L} & D_{2S}		
Homology with D_1	100	82	44	44	42
Homology with D_{2S}	44	49	100	76	52

Both D_{2L} & D_{2S} inhibit adenylyl cyclase (AC) **by distinct** $G\alpha_i$ proteins. Information has been adapted from references [35-36].

4. DOPAMINE-MEDIATED SIGNALING

Dopamine-mediated signaling activity depends on availability of pre-and-postsynaptic dopamine receptors, homomers formation between dopamine receptor subtypes, and heteromers formation between dopamine and other GPCRs, such as adenosine$_{A2A,\ A1}$, subtype 5 metabotropic glutamate (mGluR5), *N*-methyl-d-aspartate (NMDA), γ-aminobutryic acid (GABA)-A, and α-amino-3-hydroxy-5-methyl-4-isoxazole propionic acid (AMPA) [42]. D1-like receptor signaling is linked with mobilization of intracellular calcium $(Ca^{2+})_i$ via phospholipase C (PLC) enzyme coupled to G_q protein [43]. D1-like receptor is pharmacologically distinct from the classic D1 receptor that is coupled to $G_{\alpha s}$ protein. D2-like receptor signaling is also linked with mobilization of intracellular calcium $(Ca^{2+})_i$, via PLC enzyme coupled to G_q protein and cAMP/PKA cascade through heteromer (A_{2A}-D2) formation. The phosphorylation and dephosphorylation of a cAMP-regulated phosphoprotein, molecular mass 32 kDa (DARPP-32), a dual function protein kinase/phosphatase inhibitor, integrates information arriving at dopaminoceptive neurons (expressing dopamine receptor) in brain through a variety of neurotransmitters, neuromodulators, and neuropeptides [44-45]. Based on the sensitivity to a receptor antagonist (SCH23390) in striatonigral neurons, D1-like receptors are coupled to three signaling cascades: (1) SCH23390-sensitive $G_{s/olf}$/AC/PKA, (2) SCH23390-insensitive adenosine$_{A2A}$ receptor-dependent $G_{s/olf}$/AC/PKA, and (3) G_q/PLC signaling (Kuroiwa et al., 2008). In SCH23390-sensitive $G_{s/olf}$/AC/PKA cascade, stimulation of D1-like receptors activates AC that increases cAMP formation. The increased activity of PKA phosphorylates DARPP-32 on a single threonine residue, Thr^{34}, and converts it into a potent inhibitor of protein phosphatase-1 (PP-1). In SCH23390-insensitive adenosine$_{A2A}$ receptor-dependent $G_{s/olf}$/AC/PKA cascade, D1-like dopamine receptors require interaction with adenosine A_{2A} receptors to induce the phosphorylation of DARPP-32 at Thr^{34} [46]. In G_q/PLC signaling cascade, activation of G_q-coupled D1-like receptors stimulate PLC signaling, resulting in the suppression of DARPP-32 Thr^{34} phosphorylation. However, the interaction of G_q/PLC signaling with SCH23390-sensitive $G_{s/olf}$/AC/PKA through phosphorylation at $Thr\text{-}^{75}$ by Cdk5 inhibits $G_{s/olf}$/AC/PKA signaling [46].

In D_2-like receptor signaling, dopamine receptors are coupled to G proteins ($G_{\alpha i}/G_{\alpha o}$) and inhibit AC activity. These receptors can also be coupled with multiple effector systems, including calcium channels, potassium channels, phospholipases, protein kinases, and tyrosine kinases, through specific G-proteins. In striatopallidal neurons, the stimulation of D2-like receptor induces dephosphorylation of DARPP-32 [44-45]. Phosphorylation of DARPP-32 is also modulated by adenosine$_{A2A}$ receptors, which are predominantly expressed in enkephalin-expressing GABAergic striatopallidal neurons. Adenosine acts for A_{2A} receptors induces phosphorylation of DARPP-32 at Thr^{34}, opposing dopamine D2 receptor signaling [47]. This suggests that A_{2A} receptors interact structurally with D2 receptor and form receptor heteromers that exhibit reciprocal antagonistic effect by triggering the cAMP-PKA signaling cascade. This allows basal ganglia to regulate movement, motor learning, motivation and reward. A_{2A} receptors also act as pre-and/or postsynaptic modulators of glutamatergic synapses, including D2 receptors and group I metabotropic mGlu5 glutamate receptors. Adenosine A_{2A} and dopamine D_2 receptors associate in homo- and heteromeric complexes in the striatum, suggesting differential alteration in A_{2A}/D_2 receptor oligomerization after long-term drug exposure [48]. Dissociation of an actin binding protein, spinophilin, from actin filaments results in its phosphorylation. Both PP-2A and PP-1 play a role in dephosphorylation of spinophilin at serine (Ser^{94}) in response to activation of both D2 and NMDA receptors [49]. DARPP-32 is regulated by several other modulatory neurotransmitters, such as serotonin, adenosin, GABA, and glutamate.

The term signalplex, an assembly of G protein coupled receptors into signaling complex, represents the primary mechanism for establishing and maintaining receptor-mediated signaling [50]. In the the dopamine D_2 receptor signalplex within the brain, D_2 receptor is coupled with a variety of proteins and forms a complex. This signaling is regulated by dopamine receptor interacting proteins (DRIPs) such as calcyon and neuronal Ca^{2+} sensor 1 (NCS^{-1}) that participate in Ca^{2+} homeostasis [51], cytoskeleton proteins including neurofilament-M (NF-M) and postsynaptic density-95 (PSD-95) that interact with receptor and regulate its expression [52], adaptors or chaperones such as calnexin that is involved in protein folding or DRiP78 that regulates receptor transport [52], and sodium pump (Na^+-K^+-ATPase) that regulates cellular ion balance [53] (Table 2). Collectively, intramembrane dopamine receptor interactions are important in the information handling in the basal ganglia. Therefore, unraveling the interaction of DRIPs with each dopamine receptor subtype may help in understanding the molecular mechanisms underlying neurological and neuropsychiatric diseases.

Table 2. Dopamine receptor interacting proteins involved in dopamine signaling

Receptor	DRIPs	Role
D_1	NF-M	Modification of receptor expression and regulation
D_1	γ-COP	Receptor trafficking
D_1	PSD-95	Regulates G protein coupling to the receptor
D_1	DRiP78	Receptor transport from ER to the PM
D_1 and D_2	Calnexin	Glycoprotein folding and ER quality control
D_1	Calveolin-1	Receptor internalization
D_1	Arrestin	Receptor desensitization
D_1 and D_5	NSF	Post-endocytic recycling of receptor back to PM
D_5	SNX-1	Targets receptor to lysosomal degradation
D_1 and D_2	α1 subunit of Na^+, K^+-ATPase	Modulation of receptor function in the presence or in the absence of ligand
D_1	D_2	D_1- D_2 receptor heterooligomer activates PLC, ↑ Ca^{2+} signaling
D_1	Adenosine A_1	Heteromerization
D_1	NMDA	Direct protein-protein interaction facilitates the functional interaction
D_5	GABA	Direct protein-protein interaction facilitates the functional interaction
D_{2L}	Zinc finger protein 491, **Calumenin, α1 subunit of** Na^+, K^+ ATPase, ubiquitin b, mono carboxylic acid transporter, & kinesin family member 11	Several proteins have been identified by mass spectroscopy, which co-immunoprecipitated with D_{2L} dopamine receptor in HEK293 T cells transfected with rat FLAG-D_{2L} receptor

Dopamine receptor interacting proteins (DRIPs); Neurofilament protein (NF-M); **γ subunit of the coatomer protein complex** (γ-COP); Postsynaptic density-95 (PSD-95); N-ethylmaieimide-sensitive factor (NSF) and sorting nexin 1(SNX-1); Endoplasmic reticulum (ER); Plasma membrane (PM), N-methyl-D-aspartic acid (NMDA); γ-Aminobutyric acid A (GABA). Information is summarized from [52-53].

5. Localization and Functional Implications of Dopamine Receptor Subtypes

A wide range of behavioral functions mediated by dopamine receptor subtypes in the CNS suggest that along with motor programming dopamine modulates non-motor activities (Table 3). D1-like dopamine receptors (D_1, D_5) have been located in the cerebral cortex, limbic and niagrostriatal systems. In the prefrontal cortex, D_1 is enriched in spines and D_5 in dendrites with considerable overlap in their distribution within neuronal compartments [54]. Within the striatum, D_5 receptor is expressed predominantly in cholinergic interneurons, whereas D_1 receptor is expressed predominately in GABAergic neurons [55-56]. D_1 and D_5 receptors cannot be pharmacologically distinguished, but have important functional differences. Unlike the D_1 receptor, D_5 receptor does not form functional complexes with A_1 adenosine receptors [57]. Recent study in D_5 receptor knockout mice has demonstrated that dopamine and D1-like receptor agonists produce a complete loss of the ability in inducing inositol phosphate accumulation or diacylglycerol production in brain tissues, suggesting that D_5 subtype is required for inducing PLC-mediated phosphoinositide signaling [58]. D1-like (D_1, D_5) dopamine receptors are generally postsynaptic and regulate locomotion, and the expression of activity-dependent synaptic plasticity at corticostriatal synapses [56].

Table 3. Localization and functional implications of dopamine receptor subtypes in human brain

Brain regions	Subtypes	Neural pathways	Dopamine-mediated functions
Caudate putamen, nucleus accumbens, ofactory tubercle, hypothalamus, thalamus, frontal cortex	D_1, $^RD_4$, $^MD_5$	Nigrostriatal	Motor function
Caudate Putamen, nucleus accumbens, ofactory tubercle, Thalamus, cerebral cortex (low)	D_2	Nigrostriatal	Motor function
Nucleus accumbens, olfactory tubercle, cerebral cortex (low), islands of calleja	D_3	Mesocortical	Cognition, reward and reinforcement
Hippocampus, hypothalamus, frontal cortex, mid brain, amygdala, retina	D_4	Mesocortical	Cognition, reward and reinforcement
Hippocampus, thalamus, Cerebral cortex, lateral memory nucleus	D_5	Mesocortical	Cognition, reward, and reinforcement

Caudate putamen, dorsal striatum; $^RD_4$ = D_4 mRNA expression at extremely low levels has been reported in rat striatum, but its expression is a controversial issue in the human striatum; $^MD_5$ = D_5 receptor mRNA expression at low levels in the old-world monkey striatum. Nigrostriatal pathway (connects the substantia nigra with striatum) is particularly involved in the production of movement, therefore any disturbance in nigrostriatal dopamine function has been implicated in PD. The mesolimbic dopamine system originates in the ventral tegmental area and projects to several forebrain regions, including the nucleus accumbens. Mesocortical pathway connects the ventral tegmentum to the cerebral cortex, particularly the frontal lobes. Mesocortical dopamine system is essential for the normal cognitive function, motivational and emotional responses. Thus, disturbances in mesocortical dopamine functions have been implicated in both neuropsychiatric and neurodegenerative diseases.

In the nigrostriatal pathway, cooperativity between D_1 and D_2 receptor subtypes is involved for the autoregulation of dopamine [59]. In response to high dopamine efflux, stimulation of striatal D_1 receptors contributes to the negative feedback regulation of dopamine release from nigrostriatal neurons [59-62], which may be due to activation of an inhibitory long loop pathway. At basal dopamine concentrations, presynaptic D_2 receptors reduce the functioning of DA-ergic neurons by three mechanisms: (1) dopamine synthesis, (2) reduction in impulse generation, and (3) dopamine release [63-65]. Furthermore, somatodendritically released dopamine binds with D_2 receptors on presynaptic DA-ergic neurons in the substantia nigra that results in reduction of dopamine neuronal activity and subsequent dopamine release in the striatum [66].

Postsynaptic D_2 receptors are present in DA-ergic projection areas, such as the striatum, limbic areas (nucleus accumbens, olfactory tubercle), hypothalamus and pituitary (Table 3). D_2 receptors are also present presynaptically in the substantia nigra, ventral tegmental area and striatum. Bromocriptine has been proven to be efficacious in treating PD because of its D_2 agonist activity, whereas, D_2 antagonists (e.g. haloperidol and chlorpromazine) are commonly used in the treatment of psychiatric illness. The D_3 and D_4 dopamine receptors are differently distributed, and are much less abundant than the D_2 subtype. D_3 receptors are primarily expressed in limbic regions in mammals with low density in the striatum [67-68]. The selective D_3 receptor antagonists (SB-277011 and NGB-2904) block cocaine- or nicotine-enhanced brain stimulation reward, implicating their role in drug dependence and addiction [67,69]. A necessary role for D_3 receptor has been suggested for ethanol sensitization but not amphetamine sensitization [70]. Some evidence links D_3 function with PD and schizophrenia [71]. D_4 receptors are found in the frontal cortex, amygdala, mid-brain and medulla, but their functions are not known [36]. Collectively, the understanding of location and molecular pharmacology of dopamine receptor subtypes may offer clues regarding potential therapies of PD.

6. Dopamine-Mediated Oxidative Stress and Neuroinflammation in PD

PD is a progressive neurodegenerative disorder of nigrostriatal DA-ergic system. It is characterized by the loss of DA-ergic neurons in the substantia nigra pars compacta (SNpC) that control muscle movement [72], 50-70% loss of striatal dopamine, presence of Lewy bodies (intracytoplasmic protein aggregates mainly composed of α-synuclein in the midbrain [73], and mitochondrial complex I deficiency in the SNpc [74-76]. Although, molecular mechanism associated with the pathogenesis of PD is not fully understood. However, involvement of oxidative stress, inflammation, excitotoxicity, cytokine activation, mitochondrial dysfunction, and ubiquitin-proteasomal impairment has been suggested in PD [77-81]. In this review, I will emphasize the involvement of oxidative stress and neuroinflammation, because these processes have recently taken center stage for neurodegeneration in PD (Figure 6).

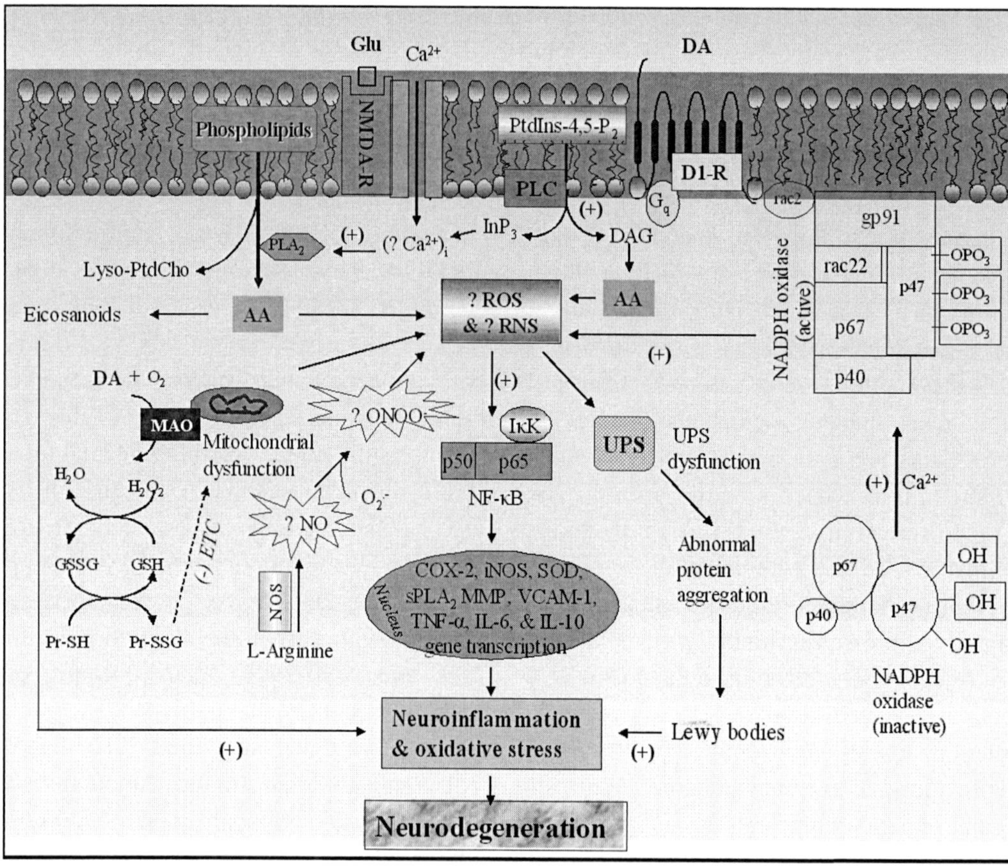

Figure 6. Possible ROS/RNS-mediated oxidative stress pathways implicating neurodegeneration of DA-ergic neurons in Parkinson disease: (1) mitochondrial dysfunction, (2) nicotinamide adenine dinucleotide phosphate-oxidase (NADPH oxidase), (3) increase in nitric oxide ($NO^{\cdot}$) formation, (4) Ubiquitin-proteasome system (UPS) dysfunction. NADPH oxidase is comprised of cytoplasmic subunits (p47phox, p67phox, and p40phox) and upon phosphorylation by specific protein kinases, these subunits can form a complex and translocate to the plasma membrane to dock with the plasma membrane subunits (p91phox, p22phox, and Rac2). Catalysis of NADPH oxidase occurs in the p91phox subunit, and is initiated by transferring of electrons from molecular oxygen through redox coupling with NADPH, FAD, and heme, producing superoxide. diacylglycerol (DAG); nitric oxide synthase (NOS); (ONOO); ($O2^{\cdot}$); (InP_3); (PtdIns-4,5-P_2).

Oxidative stress is caused by a disturbance in the cellular prooxidant/antioxidant ratio. In this phenomenon, antioxidant mechanisms are overwhelmed by reactive oxygen species (ROS), which include (superoxide, hydroxyl, peroxyl, and hydroperoxyl radicals) and nonradical oxidizing agents, such as hydrogen peroxide, that can produce hydroxyl radicals [82]. Dopamine receptors are coupled with enzymes of phospholipid catabolism via G-proteins. These enzymes include phospholipase A_2 (PLA_2) and phospholipase C (PLC) (Figure 6). Activation of dopamine receptors leads to the stimulation of PLA_2 and PLC, resulting in the production of arachidonic acid, AA (Figure 6). DA-ergic nigrostriatal neurons are also rich in glutamate receptors, which may contribute to excitotoxicity in PD [83]. Hyperstimulation of glutamate receptors results in extensive influx of Ca^{2+} that again activates PLA_2, resulting in production of AA from neural membrane glycerophospholipids. Glutamate-mediated release of AA also involves participation of PLC/diacylglycerol lipase pathway (Figure 6). Enzymic oxidation of AA through cyclooxygenase results in the production of eicosanoids that promote inflammation. Non-enzymic oxidation of AA produces ROS that activates NF-κB, a key regulator of neuronal death (Figure 6). ROS interacts with p50 and p65 subunits of NF-κB and promotes the translocation of NF-κB from the cytoplasm to the nucleus [84]. In the nucleus, NF-κB promotes transcription of genes that encode number of proteins, including many enzymes (cyclooxygenase-2 (COX-2), inducible nitric oxide synthase (iNOS), superoxide dismutase (SOD), soluble phospholipase A_2 ($sPLA_2$), matrix metalloprotease (MMP), vascular cell adhesion molecule-1 (VCAM-1), and cytokines, such as TNF-α, IL-6, & IL-10. These mediators not only contribute to oxidative stress, but also induce neuroinflammation that may harm DA-ergic neurons.

In addition, different nicotinamide adenine dinucleotide phosphate oxidase (NADPH oxidase) isoforms are present in neurons and glial cells. NADPH oxidase is a complex that consists of two membrane-bound and three cytosolic components, plus rac 1 or rac 2 (Figure 6). Its activation involves the phosphorylation of one of the cytosolic components. In the presence of O_2 and NADPH, this enzyme catalyzes the production of superoxide ($O_2^{\bullet -}$), implicating its role in altering synaptic plasticity [85]. In astrocytes and microglial cells, ROS are mainly produced by NADPH oxidase. Thus, mitochondrial dysfunction and NADPH oxidase produce ROS, which result in an imbalance in the cellular oxidative status (Figure 6). Similarly, activation of iNOS generates NO·, which reacts with $O_2^{\bullet -}$ and generates peroxynitrite ($ONOO^-$, a potent oxidant). $ONOO^-$-mediated protein nitration has been reported in PD [86]. $ONOO^-$ not only interacts with proteins and DNA but also reduces mitochondrial respiration, inhibits membrane pumps, and depletes cellular glutathione levels [87], contributing to death of DA-ergic neurons. The ubiquitin–proteasomal system (UPS) is responsible for homeostatic degradation of intact protein substrates as well as the elimination of damaged or misfolded proteins. ROS impair the UPS, which exacerbates protein aggregation, resulting in the formation of Lewy bodies (Figure 6). Collectively, ROS/RNS-mediated processes lead to oxidative and/or nitrosative damage to cellular proteins, lipids and DNA, implicating their major role in the neurodegeneration of dopaminergic neurons in PD.

7. Dopamine-mediated Learning, Aging and PD

Sequential learning, involved with the processing of a variety of cognitive functions, such as linguistic expression, semantic sequencing, working memory, and procedural memory, is an important aspect of cognitive processing [88]. Sequential learning is impaired in PD patients, suggesting that altered dopamine neurotransmission may be responsible for serial reaction time learning deficits associated with this disease [89-90]. In animal studies, following neurotoxin, 1-methyl-4-phenyl-1,2,3,6-tetrahydropyridine (MPTP)-mediated unilateral lesions in the striatum to deplete dopamine unilaterally in the striatum, show loss in learning sequential tasks [91-92]. Recent study using molecular imaging supports the notion that striatal dopamine is critical for motor sequential learning [88].

In aging, the deficiency of dopamine in the prefrontal cortex results in the deficit in working memory, whereas, in acute stress, excessive dopamine neurotransmission leads to impairment in working memory. These impairments can be attenuated by D1 receptor agonist and antagonist treatment, respectively [93-94]. Age-related loss of dopamine synthesis has been reported in the striatum as well as in the extrastriatal regions except midbrain in the postmortem brain [95]. Age-related loss of D_2 and D_3 receptors has been observed in the striatal and several extrastriatal regions in normal human brain [96]. Age-related decline in brain performance, including learning, is due to deterioration of synaptic contact and changes in neurotransmitters/neuromodulators concentrations [93-94,97].

In PD, there is a decline in some forms of memory while leaving others relatively intact [98-99]. These patients suffer from impaired working memory and have problems in organizing and using new materials as well as applying strategies. Evidence indicates that these problems are probably due to dysregulation in prefrontal cortex as a result of decreased subcortical input into the frontal cortex [100-101]. In the prefrontal cortex, D_1 and D_5 receptors play an essential role in mediating working memory functions, whereas less is known about the importance of other dopamine receptor subtypes (D_2, D_3, D_4) in the prefrontal cortex [94]. D_1-like antagonists that inhibit D_1 and D_5 receptors (but not D_2 receptors) suppress prefrontal cortex, disrupting working memory, whereas prefrontal infusion of low levels of D_1 agonists (but not D_2 agonists) improves working memory. Interestingly, D_4 receptors, are also abundantly expressed in the prefrontal cortex, and seem to be involved with working memory of rats [102]. Moreover, patients with schizophrenia show a variety of impairments in cognitive and executive functions subserved by the prefrontal cortex, which may be due to alterations in mesocortical dopamine activity in the prefrontal cortex, impairement in prefrontal calcium signaling, and altered D_1 receptor signaling within prefrontal cortex [103-104]. Collectively, dopamine dysregulation disrupts synaptic plasticity that leads to deficits in working memory.

8. Mitochondrial Electron Transport Chain and Oxidative Stress in PD

Mitochondria are primary organelles responsible for ATP generation for cellular activities. The mitochondrial generation of energy is regulated by five electron transport chain complexes located in the inner mitochondrial membrane. The redox energy from reduced electron carriers, nicotinamide adenine dinucleotide (NADH) and flavin adenine dinucleotide ($FADH_2$), is transferred to O_2 (terminal electron acceptor in the chain) to produce H_2O via four respiratory chain complexes. The establishment of a proton gradient across the inner mitochondrial membrane forms the basis of the inner mitochondrial transmembrane potential ($\Delta\Psi_m$). The energy released from the proton gradient is used to drive ATP synthesis via fifth complex in the electron transport chain.

Inner mitochondrial membrane contains five electron transport chain complexes and two mobile carriers, coenzyme Q and cytochrome c (Figure 7). Complex I is NADH dehydrogenase that transfers one electron from NADH to coenzyme Q and two protons to the mitochondrial intermembrane space. Complex II is succinate dehydrogenase that catalyzes oxidation of succinate to fumarate, reducing FAD to $FADH_2$, allowing electron flow from succinate to coenzyme Q. No proton transport occurs in this complex. Complex III is coenzyme Q-cytocrome c reductase that drives proton transport through Q cycle where reduced coenzyme QH_2 passes its electrons to cytochrome c. Complex IV is cytochrome c oxidase that accepts electrons from cyt c and directs them to two electron reduction of O_2 to form H_2O. Complex V is ATP synthase that uses the energy of proton gradient to synthesize ATP from ADP and P_i in the matrix (Figure 7).

Deficiencies in mitochondrial electron transport chain underlie defects in energy metabolism and have been implicated in the neurodegenerative process. A reduction in complex I activity in PD is thought to cause bioenergetic dysfunction with subsequent loss of DA-ergic neurons [105]. Experimental evidence also points the involvement of certain genes, such as *SNCA*, *Parkin, DJ-1, PINK1, GSTO1, LRRK2,* and *HTRA2* that encode corresponding proteins, including α-synuclein, E3 ubiquitin-protein ligase (parkin), transcriptional co-activator DJ-1, phosphatase and tensin homolog (PTEN)-induced kinase 1 (PINK-1), glutathione S-transferase omega 1 (GSTO1), leucine-rich repeat kinase 2 (LRRK2), and serine protease (HTRA2), in the pathogenesis of PD (Table 4).

8.1. α-Synuclein

α-Synuclein is a small 140 amino acid presynaptic protein especially abundant in the brain. This protein is characterized by repetitive imperfect repeats (KTKEGV) throughout most of the amino-terminal half of the polypeptide, a hydrophobic middle region, and acidic carboxy-terminal region [106]. It is found in both cytosolic and membrane bound forms, but mainly located in the cytoplasm (Figure 7). α-Synuclein is a major constituent of Lewy bodies [73]. An important property of α-Synuclein's is that in the presence of specific phospholipids conformational changes occur in this protein [107-108]. Aggregation of α-synuclein is toxic to vulnerable neurons (Table 4), as observed in the brains of patients with PD. When expressed in nM concentration, α-synuclein protects neurons against cellular stress through the phosphatidylinositol 3-kinase (PI_3K)/Akt (protein kinase B) signaling pathway, whereas its overexpression (μM concentration) exerts cytotoxic effect [109]. These data suggest dual roles of α-synuclein (neuroprotection and neurotoxicity) depending on its expression levels. In MN9D cells (an immortalized dopamine-containing neuronal hybrid cell line) transfected with wild-type or A53T mutant gene, overexpression does not alter TH protein levels but markedly reduces TH activity, suggesting that α-synuclein regulates dopamine biosynthesis [110].

Mutations in α-synuclein gene are associated with mitochondrial abnormalities and cell death due to increased susceptibility to oxidative stress [111-112]. Furthermore, the overexpression of α-synuclein in human embryonic kidney (HEK) cells results in increased mitochondrial susceptibility to rotenone-induced toxicity compared to control HEK cells [113]. A significant reduction in intracellular ATP levels in response to subtoxic concentrations of rotenone observed in human embryonic kidney (HEK) cells overexpressing α-synuclein suggests that at high concentration this protein is translocated from cytoplasm to mitochondria, causing enhanced toxicity. It is associated with an increased rate of neuronal cell death when gene is either in its mutant or highly expressed form. The exact cellular function of α-synuclein remains unclear, but it may regulate synaptic plasticity.

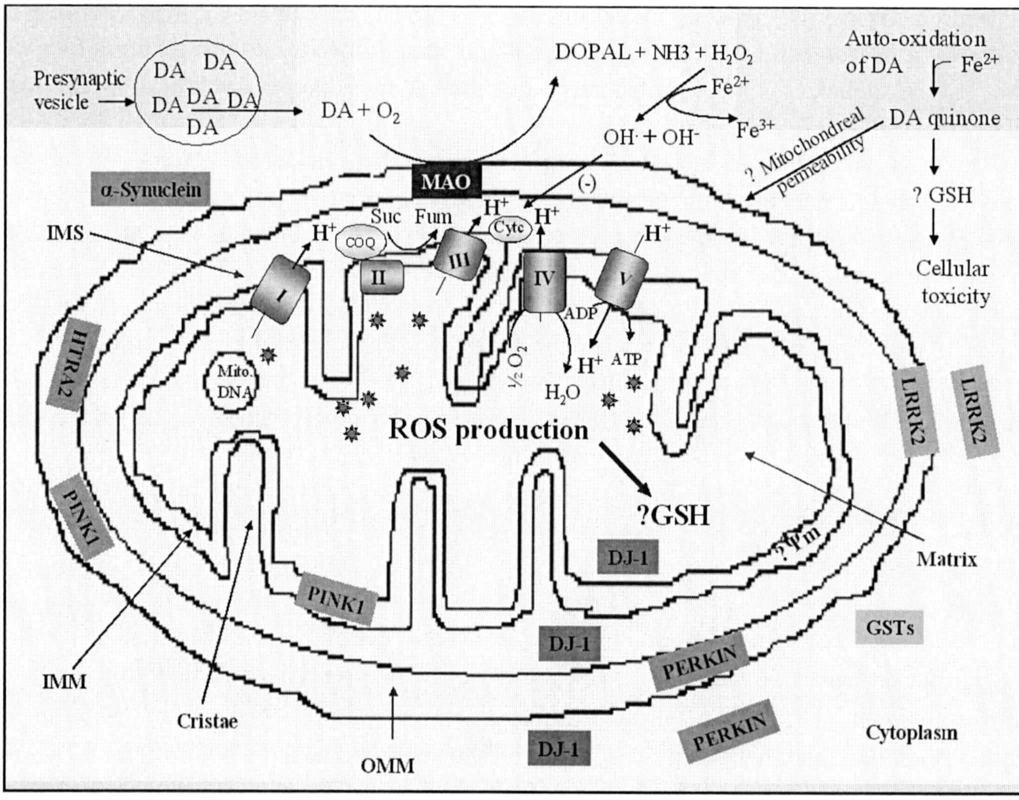

Figure 7. Schematic diagram of a mitochondrion showing the electron transport chain and location of selected proteins involved in oxidative stress. IMM, intermitochondrial membrane; OMM, outer mitochondrial membrane; IMS, inner matrix surface; ☼, ROS; Suc, succinate; Fum, Fumurate, DOPAL, 3,4-dihydroxyphenyl- acetaldehyde; mito. DNA, mitochondrial DNA; DA, dopamine; MAO, monoamine oxidase. PINK1, PTEN-induced kinase 1; LRKK2, Leucine-rich repeat kinase 2 ; DJ-1, mitochondrial antioxidant protein; HTRA2, high temperature requirement protein A2; GST, glutathione S-transferase; parkin, E3 ubiquitin-protein ligase; α-synuclin, a presynaptic protein. Auto-oxidation of dopamine produces dopamine quinones, and deficiency of glutathione results in cellular toxicity.

8.2. Parkin

Parkin is a 465-amino acid protein that has E3 ubiquitin-protein ligase activity [114]. Parkin is mainly located in the cytoplasm and in the outer mitochondrial membrane (Figure 7). It is involved in the UPS to clear misfolded proteins and other target molecules. In PD, death of nigrostriatal DA-ergic neurons is accompanied by the accumulation of a wide range of poorly degraded proteins and formation of Lewy bodies, implicating that mutation in *parkin* gene causes impairment in protein clearance (Table 4). A wide variety of mutations, such as exon deletions and duplications as well as point mutations in the *parkin* gene result in the autosomal recessive early-onset parkinsonism [115-118]. Mutations in genes of two enzymes in the ubiquitin-proteasome system (parkin and ubiquitin C-terminal hydrolase) are associated with neurodegeneration in some familial forms of PD, supporting the notion that failure of the UPS may contribute to neurodegeneration and formation of Lewy bodies in PD [119]. Parkin serves as a neuroprotective role by forming polyubiquitin chains in substrate proteins that target them for proteasomal degradation [120]. Two of its subsrates are glycoslyated form of α-synuclein and synphilin-1 (α-synuclein interacting protein), thus parkin may be involved in modulating the turnover and function of these presynaptic proteins.

Table 4. Genes involved in Parkinson Disease (PD)

Gene	Role	Gene mutation- linked to diseases	Reference
SNCA	Binding with lipids promotes formation of oligomers, ↑ sensitivity to oxidative stress	An inherited form of PD	[188, 107-108]
Parkin	E3-ubiquitin ligase controls protein degradation via UPS, & improves mitochondrial dysfunction	Autosomal recessive PD of early onset. Selective neural cell death without forming Lewy bodies	[137]
DJ-1	A neuroprotective transcriptional co-activator that blocks oxidative stress, and maintains mitochondrial function	Early onset autosomal recessive form of PD	[127, 134, 137]
PINK1	Mitochondrial protein kinase has a neuroprotective role in the mitochondria against oxidative stress	Autosomal recessive form of PD	[140-141]
HTRA2	Serine protease maintains mitochondrial integrity	Autosomal recessive form of PD	[168-169]
LRRK2	The increased kinase activity in mutants, but its function not clearly elucidated	Familial and sporadic PD	[151-152, 156]
GSTO1	Post-translational modification of IL-1β protects against inflammation	Linked to PD	[173-174]

Inactivation of the parkin gene in mice may cause motor and cognitive deficits, inhibition of amphetamine-induced dopamine release and inhibition of glutamate neurotransmission, suggesting that parkin facilitates dopaminergic neurotransmission [121]. This finding is supported by a study reporting that parkin-expressed substantia nigra in mice shows a significant increase in the survival of tyrosine hydroxylase-positive neurons [122]. Intracellular oxidation of dopamine plays a critical role in the degeneration of DA-ergic neurons [123]. The overexpression of parkin results in reduction of DOPAC levels, produced during dopamine oxidation by MAO, further supporting a stimulatory role of parkin for the presynaptic dopamine neurotransmission. Parkin suppresses the expression of monoamine oxidases [124]. Overexpression of parkin has been reported to protect human dopamine neuroblastoma (SH-SY5Y) cells against dopamine-induced apoptosis by decreasing ROS produced during dopamine metabolism [125]. Thus, parkin may limit the production of ROS generated by MAO during dopamine oxidation, suggesting its protective role for the survival of DA-ergic neurons.

8.3. DJ-1

DJ-1 is a putative mitochondrial antioxidant protein of 189 amino acids. DJ-1 acts as a H_2O_2 sensor [126]. It is found in a variety of tissues, including brain, and partially localized to the mitochondrial matrix and intermembrane space (Figure 7). Homodimerization of DJ-1 is critical for its catalytic activity, therefore, absence or inactivation of DJ-1 may result in PD [127-128]. A number of missense mutations in DJ-1, such as L166P, disrupt homodimerization resulting in a poorly folded protein, and other mutations A104T, E163K, and M26I subtly perturb DJ-1 structure and reduce thermal stability, implicating rare forms of familial PD [129].

The downregulation of DJ-1 increases ROS-mediated cell death in neuronal cell lines [130]. DJ-1-deficient mice have normal number of DA-ergic neurons in the substantia nigra, but evoked dopamine overflow in the striatum is significantly reduced primarily due to increased reuptake [131]. Nigral neurons lacking DJ-1 are also less sensitive to the inhibitory response of presynaptic D_2 dopamine receptor stimulation. Nigrostriatal DA-ergic dysfunction, motor deficits, and hypersensitivity to MPTP in DJ-1-deficient mice suggest an essential role for DJ-1 in dopaminergic neurotransmission [131-132]. Another independent study showing age-dependent progression of motor deficits and DA-ergic dysfunction in nigrostriatal pathway in DJ-1 null mice supports the importance of DJ-1 in the pathogenesis of PD [133]. In *Drosophila melanogaster,* inhibition of DJ-1a (a homolog of the human DJ-1) results in cellular accumulation of ROS, hypersensitivity to oxidative stress and degeneration of DAergic neurons [134]. Furthermore, DJ-1A RNAi flies show reduction in Akt phosphorylation causing impairment in PI3K/Akt signaling, implicating that DJ-1 promotes neuronal survival [134]. Double knockout flies for DJ-1α and DJ-1β display selective sensitivity to other toxins, paraquat and rotenone, further confirming its protective role against oxidative stress [135]. Collective findings suggest that DJ-1 plays an important player (Table 4) in cellular defense against oxidative stress and maintains mitochondrial function [127,135-137]. Therefore, loss of DJ-1 may lead to PD by conferring hypersensitivity to dopaminergic insults.

8.4. PINK1

PTEN-induced kinase 1 (PINK1) protein is a serine-threonine kinase, which is localized in mitochondria. It is not yet clear whether PINK1 is selectively localized (Figure 7) within the outer or the inner mitochondrial membrane [138]. PINK1 protects cells from stress-induced mitochondrial dysfunction and is required for long-term survival of DA-ergic neurons [139-141].

Mutations in *PINK1* gene (Table 4) cause autosomal recessive early-onset in PD [142]. Some missense mutations in *PINK1* are associated with down-regulation of the protein serine/threonine kinase activity [143-144]. *Drosophila* bearing null mutation for *dPINK1* show fragmented cristae, loss of outer membrane, and ATP depletion accompanied with DA-ergic neuronal degeneration, supporting that PINK1 plays functional and structural roles in mitochondria [145-146]. *PINK1* knockout mice have been reported to show impairment in evoked dopamine release without any alteration in number of DA-ergic neurons, DA-ergic receptors or dopamine synthesis [147]. Direct activation of appropriate postsynaptic DA-ergic receptors compensates for a presynaptic defect in dopamine release, suggesting that impaired presynaptic release of dopamine may be a common pathophysiological mechanism in PD [147]. In contrast to *PINK1* knockout fly, mitochndria in *PINK1* knockout mice appear to be structurally intact and preserved in total number, however, mitochondrial key function is impaired [138]. There is a possibility that mitochondrial functional defect in *PINK1* knockout mice may be a causal and early pathogenic event of PD. Loss-of-function mutations, including degeneration of DA-ergic neuronal, locomotor defects and mitochondrial defects, in *Drosophila PINK1* model system reproduce some aspects of PD. Collectively, PINK1 is a neuroprotective protein and therefore loss of PINK1 function causes PD.

8.5. LRRK2

Leucine-rich repeat kinase 2 (LRRK2) is a multidomain protein kinase with autophosphorylation activity. It is an extraordinarily large complex of proteins that contains multiple domains, including a leucine-rich repeat (LRR), serine-threonine kinase mitogen-activated protein kinase kinase kinase (MAPKKK), Ras of complex proteins (ROC) that may act as GTPase (ROC-GTPase), and WD40 domain [148-149]. Each domain of this complex is targeted by pathogenic mutations in familial PD. Although, most of LRRK2 is localized in the cytosol, but ~ 10% is associated with the mitochondrial outer membrane (Figure 7). ROC domain dimer may act as a GTPase to regulate its LRRK2 kinase activity [150]. GTPase can be activated independently, but LRRK2 kinase activity strictly requires activation of GTPase [151]. PD-associated LRRK2 is a dimer that undergoes intramolecular autophosphorylation, and its intact C terminus is required for autophosphorylation activity [152].

Dominant missense mutations in *LRRK2* gene are the most common genetic cause of PD, but the mechanisms by which these mutations disrupt neuronal function causing loss of DA-ergic neurons remain poorly understood. It is the most common cause of familial autosomal dominant and also sporadic forms of PD [153-155]. Mutations in LRKK2 (autophosphorylation and the phosphorylation of a generic substrate) augment kinase activity [156]. Thus, mutations increase in kinase activity (Table 4) that may be associated with apoptotic cell death in dopaminergic cell lines and primary neurons [151].

In primary neuronal culture, mutated *LRRK2*-mediated neurodegeneration is prevented by the functional inhibition of Fas-associated protein with death domain (FADD) or depletion of caspase-8, two key elements of the extrinsic cell death pathway. This finding suggests that mutated LRRK2 may induce death signaling by interacting with FADD and caspase-8, which establishes a direct link between a mutant *LRRK2* gene and programmed cell death [157]. Furthermore, heat shock protein 90 (Hsp90) has been reported to interact with LRRK2. This interaction is critical for maintaining the stability of LRRK2. Therefore, any disruption in the association between Hsp90 and LRK2 results in proteasomal degradation of LRRK2 [158]. Destabilization of mutated LRRK2 (by blocking Hsp90-mediated chaperone activity) may be a novel way to limit its detrimental effects. These new findings of LRRK2 research warrant further investigation.

8.6. HTRA2

High temperature requirement protein A2, (HTRA2), is a mitochondrial serine protease that is released from mitochondria to the cytoplasm and inhibits the function of X chromosome-linked inhibitor of apoptosis (XIAP) [159]. A monomeric structure of HTRA2 consists a trimerization motif, a C-terminal PDZ domain, and a central serine protease domain that contains the His-198, Asp 228, and Ser-306 in its active site HTRA2 protein contains a central serine protease domain and a C-terminal PDZ domain [160-162]. HTRA2 is localized in intermembrane space within the mitochondria (Figure 7). HTRA2 is released into the cytosol in response to apoptotic stimuli, where it can induce apoptosis in a caspase dependent or independent manner [163-164].

Mutational analysis reveals that the phenylalanine 149 in HTRA2 at the homotrimerization motif is crucial for the formation of homotrimeric assembly of HTRA2 [165]. In the absence of this assembly, HTRA2 monomeric form abolishes its autoproteolytic activity and proteolytic activity against XIAP, a mammalian inhibitor of apoptosis protein having caspase inhibitory activity, suggesting that the homotrimeric structure of protein is required for executing its serine protease activity. The missense mutation of the HTRA2 protease domain (substitution of cysteine for serine at residue 276) in mice exhibits progressive loss of striatal neurons, leading to motor dysfunction [166]. Another missense mutation (substitution of serine for glycine at residue 399) is rare in individuals with PD, but absent in healthy humans [167]. HTRA2 phosphorylation decreases in brain tissues of parkinsonian patients carrying *PINK1* mutations, suggesting that HTRA2 protease activity is regulated by PD-associated PINK1 [168]. Similarly, *HTRA2* mutants share some phenotypic similarities with *Parkin* and *PINK1* mutants in fruit fly [169]. Taken together, HTRA2 plays an important role in maintaining the mitochondrial homeostasis.

8.7. GSTs

The glutathione-S-transferases (GSTs) gene family encodes cytosolic proteins (Figure 7), which are involved in detoxification of toxic chemical compounds, drugs, oxidized metabolites of catecholamines, and environmental pollutants, through conjugation of reduced glutathione. Human cytosolic GST genes (*GSTO1*, *GSTM1*, *GSTM3*, *GSTP1*, *GSTT1*, and *GSTZ1*) are polymorphic. The encoded proteins in GST family may vary in their substrate specificity and tissue distribution, and to some extent may even show some overlap. GSTs are a family of inducible phase II enzymes and possess detoxification and antioxidative functions. Various mechanisms have been implicated in the pathogenesis of PD. Among them, oxidative toxicity (excessive formation ROS and/or absence of detoxification of ROS) is suggested to play an important role. In the DA-ergic system, oxidation of dopamine to *o*-quinone, its subsequent product, aminochrome (2,3dihydroindole-5,6-dione), may elicit redox cycling, toxicity and apoptosis [170-171].

Omega class of cytosolic GST possesses dehydroascorbate reductase and thioltransferase activities and also catalyzes the reduction of monomethylarsonate, an intermediate in the pathway of arsenic biotransformation. There are two functional Omega class glutathione transferase genes in humans: *GSTO1* and *GSTO2*. *GSTO1* encodes glutathione S-transferase omega 1 protein family (GSTO1). GSTO1 modulates ryanodine receptors (a class of intracellular calcium channels in the endoplasmic reticulum of various cells, including neurons), and interacts with cytokine release inhibitory drugs [172-173]. GSTO1 modifies the age-at-onset of PD [173]. It is involved in the post-translational modification of the inflammatory cytokine IL-1β, a proinflammatory cytokine that is overexpressed in the brains of parkinsonian patients (Table 4), suggesting its protective role against inflammation [173-174]. GSTO1 uses glutathione in the process of the biotransformation of drugs, xenobiotics and oxidative stress, therefore mutation in *GSTO1* is linked to PD. Biological function of GSTO2 is not known, but it is suggested to protect against oxidative stress by recycling ascorbate [175], implicating its role in influencing the age-at-onset of PD. Association of GSTM1 null polymorphism with PD is reported to be strongest in the earlier age range [176]. Under physiological conditions, this enzyme catalyzes the conjugation reaction between glutathione and quinones [177-178]. The deficiency/absence of GSTM1 influences an individual's risk to PD.

Collectively, GSTs protect DA-ergic neurons in PD through their direct antioxidant activity, and facilitate the elimination of endogenous toxins from the cell. In general, they have cytoprotective properties and are hypothesized to protect against neurodegeneration.

9. PHARMACOGENETICS/PHARMACOGENOMICS IN THE TREATMENT OF PD

Pharmacogenetics, deals with the influence of genetic factors on drug response based on interindividual differences considering one or few candidate genes while pharmacogenomics considers the entire genome using genomic technologies to develop new drugs as well as to characterize older drugs. For example, a large interindividual variability occurs in response to anti-PD drugs, such as L-DOPA, with respect to adverse effects. Some PD patients exhibit L-DOPA- induced dyskinesias, whereas, others remain free despite treatment with L-DOPA. This may be due to changes in neuroplasticity among patients depending on their age and genetics [179]. In fact, genetic variability among individuals may affect differences in the expression of drug-metabolizing enzymes, receptors, and signaling proteins involved in the signal transduction pathways, resulting variability in drug responses [180]. The role of pharmacogenetics in the treatment of PD is relatively unexplored. Well-recognized drug-related complications (dopamine dysregulation syndrome) include hallucinations and psychosis [181]. The etiology of these complications is not clearly dose related, while the management can be difficult and needs to be tailored to the individualized therapy. Cholinergic and DA-ergic drugs may both influence cognitive function. Therefore, use of pharmacogenetics may improve the therapeutic ratio of medical approaches to PD in future [181]. Pharmacogenetics aims at identifying genetic markers (polymorphic genes) associated with drug response. Thus, the primary goal of future research should be to gain knowledge on genetic variability of polymorphic genes (genes encoding drug-metabolizing enzymes, drug receptors and proteins involved in dopamine signaling pathway) to determine interindividual variability in drug response.

Pharmacogenomics aims to improve the therapeutic efficacy of a drug based on the genetic profile of a patient. Pharmacogenomic, genomic, proteomics and bioinformatic analyses can identify genes whose expression is associated with the pathology of a specific disease. In addition, microarrays have also emerged as key analytical tools in the field of pharmacogenomics. Extensive validation studies performed in parallel with drug development are crucial for the selection of appropriate target genes. Advances in neuroimaging and pharmacogenomics may help in early detection of sporadic PD in patients that may extend the quality of life of PD patients due to availability of in-time and selective gene therapy. Early intervention may likely spare more DA-ergic neurons. Furthermore, the successful integration of genomic technologies into the drug discovery will provide the promise of increased efficiency from pharmaceutical industries as well as improved efficacy and fewer side-effects of anti-PD drugs that will be beneficial for the patients. It is hoped that rapid advances in dopamine research at the pharmacogenetic/pharmacogenomic and behavioral pharmacology level may help in developing more selective drugs for treating PD.

10. Conclusion and Direction for Future Studies

Dopamine, a major modulatory transmitter in the CNS, plays a crucial role in controlling multiple brain functions, including voluntary movement and a broad array of behavioral processes. DA-ergic neurons in the substantia nigra pars compacta are the main source of dopamine in the brain, and the selective loss of these neurons in this region leads to PD. This condition results in impairment in multiple motor and nonmotor features that significantly affect the quality of life. Current pharmacological treatments for PD are limited because they only offer symptomatic relief, but do not provide a cure. Therefore, these drugs are only effective in early stage of the disease. However, as disease progresses then complications predominate because current drugs do not prevent the continuing degeneration of DA-ergic neurons in the substantia nigra. Therefore, understanding of the exact molecular mechanism underlying the neurodegeneration of DA-ergic neurons in PD will promote the development of therapies that may cure the disease without involving any severe side effects.

With progress in molecular biology and application of pharmacogenetics/ pharmacogenomics in medicine, scientists may provide new strategies to understand the underlying molecular mechanisms for PD that may shed the light in optimizing drug development with higher efficacy and fewer side effects. *In situ* studies, the microglia, as well as PD microroglia, are immunoreactive for D_1-D_4 but not D_5 dopamine receptors, suggesting that activated microglia play a role in the selective vulnerability of dopamine neurons in PD [182]. Due to increased physiological significance of dopamine in our mental and physical health, it is very important to develop subtype selective agonists and antagonists for its receptor subtypes to improve the efficacy and safety of a drug for therapeutic purpose of treating various dopamine-related disorders. Dopamine signaling can be regulated via DRIPs. Therefore, unraveling the involvement of DRIPs in dopamine-mediated signaling may also help in understanding the molecular mechanisms of DA-ergic system, which may help in identifying novel therapeutic targets.

Therapies for PD largely rely on long-term dopamine receptor stimulation with L-DOPA to compensate for the loss of striatal DA-ergic neurons and are often accompanied with dyskinesias. A_{2A} antagonists also have recently been applied with reduced L-DOPA doses in clinical studies and are shown to prevent and alleviate L-DOPA-induced dyskinesias [183-184]. The precised molecular mechanism underlying this adjunctive therapy is lacking, but long-term L-DOPA treatment may be involved in altering A_{2A} and D_2 homo- and heteromerization in striatal neurons [185]. The D_2 agonist-induced decrease in A_{2A}/D_2 heteromers relative to A_{2A} homomers may alleviate the constitutive D_2 antagonism of A_{2A} signaling. Such increase in A_{2A} signaling may sensitize A_{2A}-mediated cAMP accumulation after D_2 receptor activation [186].

Mitochondrial dysfunction is believed to stimulate the production of ROS with subsequent induction of neuronal cell death. Advances in genetics have demonstrated a prominent role for mitochondrial dysfunction in the pathogenesis of the PD. Several proteins associated with PD are localized in the mitochondria, implicating oxidative stress a major culprit for the degeneration of DA-ergic neurons. Mutations in several genes, including *PINK1*, *PARKIN*, and *DJ-1*, cause autosomal recessive forms of PD [138,187-188], suggesting that mitochondria can be considered as a promising target for the development of PD biomarkers. However, the protein complex of these three genes (*PINK1*, *PARKIN*, and *DJ-1*) in the cytosol may or may not be related to the regulation of oxidative stress *in vivo.* Point mutations, duplications and triplications in the *SNPCA* gene cause a rare dominant form of PD in families. Whereas, mutations in *LRRK2* and *HTRA2* genes cause a most common dominant form of PD in families, showing association with prominent intracellular protein inclusions [188]. Mitochondrial localization of the products of several PD-associated genes, mitochondrial dysfunction causing oxidative stress, its integral role in the apoptotic cell death, and oxidative stress leading to neuronal loss, suggest a direct molecular link between mitochondria and the pathogenesis of PD. Thus, the future gene therapy to mitigate oxidative stress and neuroinflammation holds significant promise for the treatment of PD.

Collective studies have also pointed out a large interindividual variability in response to anti-PD drugs with respect to both drug efficacy and toxicity [180]. Therefore, future research should be dealt with candidate genes that encode dopamine-metabolizing enzymes, dopamine receptors signaling proteins involved in DA-ergic pathway, and mitochondrial neuroprotective proteins. Their interindividual variability should be determined so that in future each individual suffering from PD can be treated for their motor and mental complications with an effective dose according to their profile.

References

[1] Erixon-Lindroth N, Farde L, Wahlin TB, Sovago J, Halldin C, Backman L. The role of the striatal dopamine transporter in cognitive aging. *Psychiatry Research,* 2005;138:1-12.

[2] Kaasinen V, Nagren K, Hietala J, Oikonen V, Vilkman H, Farde L, Rinne J. Extrastriatal dopamine D_2 and D_3 receptors in early and advanced Parkinson. *Neurology,* 2000;54:1482-1487.

[3] Suhara T, Fukuda H, Inoue O, Itoh T, Suzuki K, Yamasaki T, Tateno Y. Age-related changes in human D1 dopamine receptors measured by positron emission tomography. *Psychopharmacology (Berl),* 1991;103:41-45.

[4] Bäckman L, Nyberg L, Lindenberger, U, Li SC, Farde L. The correlative triad among aging, dopamine, and cognition: current status and future prospects. *Neuroscience and Biobehavioral Reviews,* 2006;30:791–807.

[5] Dreher JC, Meyer-Lindenberg A, Kohn P, Berman KF. Age-related changes in midbrain DA-ergic regulation of the human reward system. *Proceedings of the National Academy of Sciences of the United States of America,* 2008;105:15106-15111.

[6] Nagel IE, Chicherio C, Li C, von Oertzen T, Sander T, Villringer A, Heekeren HR, Bäckman L, Lindenberger U. Human aging magnifies genetic effects on executive functioning and working memory. *Frontiers in Human Neuroscience,* 2008;2:1.

[7] Volkow ND, Wang GJ, Fowler JS, Ding YS, Gur RC, Gatley J, Logan J, Moberg PJ, Hitzemann R, Smith G, Pappas N. Parallel loss of presynaptic and postsynaptic dopamine markers in normal aging. *Annals of Neurology,* 1998;44:143-147.

[8] Bäckman L, Ginovart N, Dixon RA, Wahlin TB, Wahlin A, Halldin C, Farde L. Age-related cognitive deficits mediated by changes in the striatal dopamine system. *American Journal of Psychiatry,* 2000;157:635-637.

[9] Tanner CM, Hubble JP, Chan P. Epidemiology and genetics of Parkinson. In: Watt RL. Koller W. *Movement Disorders: Neurologic Principles and Practice*. NY: McGraw Hill, NY; 1997; 135-152.

[10] Prabhakara JPR, Feist G, Thomasson S, Thompson A, Schommer E, Othman G. Differential effects of 24-hydroxycholesterol and 27-hydroxycholesterol on tyrosine hydroxylase and α-synuclein in human neuroblastoma SH-SY5Y cells. *Journal of Neurochemistry,* 2008;107:1722-1729.

[11] Gross RE, Lozano M. Advances in neurostimulation for movement disorders. *Neurological Research,* 2000;22:247-258.

[12] Farooqui, T. *Interaction of permanentlycharged DA-ergic agonists and antagonists with D_2 dopamine receptors*. Ph.D. thesis, The Ohio State University; 1992.

[13] Cooper JR, Bloom FE, Roth RH. Dopamine. *In The Biochemical Basis of Neuropharmacology,* Oxford, UK: Oxford University Press; 2003; 225-270.

[14] Koeltzow TE, Xu M, Cooper DC, Hu X-T, Tonegawa S, Wolf ME, White FJ. Alterations in dopamine release but not dopamine autoreceptor function in dopamine D_3 receptor mutant mice. *The Journal of Neuroscience,* 1998;18:2231-2238.

[15] Carlsson M, Carlsson A. Interactions between glutamatergic and monoaminergic systems within the basal ganglia-implications for schizophrenia and Parkinson's disease. *Trends in Neuroscience,* 1990;13:272-276.

[16] Strafella AP, Paus T, Barrett J, Dagher A. Repetitive transcranial magnetic stimulation of the human prefrontal cortex induces dopamine release in the caudate nucleus. *Journal of Neuroscience,* 2001;21:RC157.

[17] Strafella AP, Paus T, Fraraccio M, Dagher A. Striatal dopamine release induced by repetitive transcranial magnetic stimulation of the human motor cortex. *Brain,* 2003;16: 2609-2615.

[18] Strafella AP, Ko JH, Grant J, Fraraccio M, Monchi O. Corticostriatal functional interactions in Parkinson's disease: a rTMS/[11C]raclopride PET study. *European Journal of Neuroscience,* 2005;22:2946-2952.

[19] Iversen LL. Uptake processes for biogenic amines: In Iversen LL, Iversen SD, Snyder SH (Eds.) *Handbook of Psycopharmacology.* Vol. 3, New York: Plenum Press; 1975;381-442.

[20] Iversen LL. The discovery of monoamine transporters and their role in CNS drug discovery. *Brain Research Bulletin,* 1999;50:379.

[21] Chen N, Reith ME. Structure and function of the dopamine transporter. *European Journal of Pharmacology,* 2000;405:329-339.

[22] Reith ME, Xu C, Chen NH. Pharmacology and regulation of the neuronal dopamine transporter. *European Journal of Pharmacology,* 1997;324:1-10.

[23] Giros B, Caron MG. Molecular characterization of the dopamine transporter. *Trends in Pharmacological Sciences,* 1993;14:43-49.

[24] Miller GW, Gainetdinov RR, Levey AI, Caron MG. Dopamine transporters and neuronal injury. *Trends in Pharmacological Sciences,* 1999;20:424-429.

[25] Stoof JC, Thieme RE, Vrijmoed-de Vries MC, Mulder AH. In vitro acetylcholine release from rat caudate nucleus as a new model for testing drugs with dopamine-receptor activity. *Naunyn-Schmiedeberg's Archives of Pharmacology,* 1979;309: 119-124.

[26] Cubeddu LX, Hoffmann IS, James MK. Frequency-dependent effects of neuronal uptake inhibitors on the autoreceptor-mediated modulation of dopamine and acetylcholine release from the rabbit striatum. *Journal of Pharmacology and Experimental Therapeutics,* 1983;226:88-94.

[27] Farooqui T, Wallace RA, Miller DD, Uretsky NJ. Effect of the inhibition of dopamine uptake on the dopamine- and dimethyldopamine-induced-inhibition of the potassium-evoked release of [^{3}H]acetylcholine from striatal slices. *General Pharmacology,* 1989;20:731-736.

[28] Wallace RA, Farooqui T, Wallace L, Ares J, Chang YA, Miller D, Uretsky N. Interaction of permanently charged analogs of dopamine with the D-2 DA-ergic receptor. *Biochemical Pharmacolology,* 1987;36:3903-3910.

[29] Sabol B, Boldry R, Farooqui T, Chang YA, Miller D, Uretsky N. Effect of permanently charged and uncharged DA-ergic agonists on the potassium-induced release of [^{3}H]acetylcholine from striatal slices. *Biochemical Pharmacology,* 1987;36:1679 - 1685.

[30] Uretsky NJ, Schoenfeld RI. Effect of L-dopa on the locomotor activity of rats pretreated with 6-hydroxydopamine. *Nature- New Biology,* 1971;234:157–159.

[31] Kebabian JW, Calne DB. Multiple receptors for dopamine. *Nature,* 1979;277:93-96.

[32] Spano PF, Govoni S, Trabucchi M. Studies on the pharmacological properties of dopamine receptors in various areas of the central nervous system. *Advances in Biochemical Psychopharmacology,* 1978;19:155-165.

[33] Sibley DR, Monsma FJ Jr. Molecular biology of dopamine receptors. *Trends in Pharmacological Sciences,* 1992;13:61-69.

[34] Neve KA, Seamans JK, Trantham-Davidson H. Dopamine receptor signaling. *Journal of receptor and Signal transduction research,* 2004;24:165-205.

[35] Alexander SPH, Mathie A, Peters JA. Dopamine receptors, in Guide to Receptors and Channels. *British Journal of Pharmacology,* 2004; 141: S25.

[36] Marsden CA. Dopamine: the rewarding years. *British Journal of Pharmacology,* 2006;147:S136–S144.

[37] Dal Toso R, Sommer B, Ewert M, Herb A, Pritchett DB, Bach A, Shivers BD, Seeburg PH. The dopamine D2 receptor: two molecular forms generated by alternative splicing. *EMBO Journal,* 1989;8:4025-4034.

[38] Giros B, Sokoloff P, Martres MP, Riou JF, Emorine LJ, Schwartz JC. Alternative splicing directs the expression of two D2 dopamine receptor isoforms. *Nature,* 1989;342:923-926.

[39] Senogles SE. The D2 dopamine receptor isoforms signal through distinct $G_{i\alpha}$ alpha proteins to inhibit adenylyl cyclase. A study with site-directed mutant $G_{i\alpha}$ proteins. *Jourbal of Biological Chemistry,* 1994;269:23120-23127.

[40] Guiramand J, Montmayeur JP, Ceraline J, Bhatia M, Borrelli E. Alternative splicing of the dopamine D2 receptor directs specificity of coupling to G-proteins. *Journal of Biological Chemistry,* 1995;270:7354-7358.

[41] Usiello A, Baik J-H, Rougé-Pont F, Picetti R, Dierich A, LeMeur M, Piazza PV, Borrelli E. Distinct functions of the two isoforms of dopamine D2 receptors. *Nature,* 2000;408:199-203.

[42] Agnati LF, Ferre S, Burioni R, Woods A, Genedani S, Franco R, Fuxe K. Existence and theoretical aspects of homomeric and heteromeric dopamine receptor complexes and their relevance for neurological diseases. *Neuromolecular Medicine,* 2005;7:61-78.

[43] Undie AS, Weinstock J, Sarau HM, Friedman E. Evidence for a distinct D1-like dopamine receptor that couples to activation of phosphoinositide metabolism in brain. *Journal of Neurochemistry,* 1994;62:2045-2048.

[44] Greengard P, Allen PB, Nairn AC. Beyond the dopamine receptor: the DARPP-32/protein phosphatase-1 cascade. *Neuron,* 1999;23:435 447.

[45] Svenningsson P, Nishi A, Fisone G, Girault JA, Nairn AC, Greengard P. DARPP-32: an integrator of neurotransmission. *Annual Review of Pharmacologicaly and Toxicology,* 2004;44:269 -296.

[46] Kuroiwa M, Bateup HS, Shuto T, Higashi H, Tanaka M, Nishi A. Regulation of DARPP-32 phosphorylation by three distinct dopamine D1-like receptor phosphorylation by three distinct dopamine D1-like receptor phosphorylation by three distinct dopamine D1-like receptor signaling pathways in the neostriatum. *Journal of Neurochemistry,* 2008;107:1014-1026.

[47] Schiffmann SN, Fisone G, Moresco R, Cunha RA, Ferré S. Adenosine A_{2A} receptors and basal ganglia physiology. *Progress in Neurobiology,* 2007;83:277-292.

[48] Vidi P, Chemel BR, Hu C-D, Watts VJ. Ligand-dependent oligomerization of dopamine D_2 and adenosine A_{2A} receptors in living neuronal cells. *Molecular Pharmacology*, 2008;74: 544-551.

[49] Uematsu K, Futter M, Hsieh-Wilson LC, Higashi H, Maeda H, Nairn AC, Greengard P, Nishi A. Regulation of spinophilin Ser94 phosphorylation in neostriatal neurons involves both DARPP-32-dependent and independent pathways. *Journal of Neurochemistry,* 2005;95:1642-1652.

[50] Kabbani N, Levenson R. A proteomic approach to receptor signaling: molecular mechanisms and therapeutic implications derived from discovery of the dopamine D2 receptor signalplex. *European Journal of Pharmacology,* 2007;572:83-93.

[51] Bergson C, Levenson R, Goldman-Rakic PS, Lidow MS. Dopamine receptor-interacting proteins: the Ca^{2+} connection in dopamine signaling. T*rends in Pharmacological Sciences,* 2003;24, 486-492.

[52] Wang M, Lee FJ. Dopamine receptor interacting proteins (DRIPs) of dopamine D1-like receptors in the central nervous system. *Molecular Cell,* 2008;25:149-157.

[53] Hazelwood LA, Free RB, Cabrera DM, Skinbjerg M, Sibley DR. Reciprocal modulation of function between the D1 and D2 dopamine receptors and the Na^+, K^+-ATPase. *Journal of Biological Chemistry,* 2008;283:36441-36453.

[54] Bordelon-Glausier, JR, Khan ZU, Muly EC. Quantification of D_1 and D_5 dopamine receptor localization in layers I, III, and V of Macaca mulatta prefrontal cortical area 9: coexpression in dendritic spines and axon terminals. *Journal of Comparative Neurology,* 2008;508:893-905.

[55] Rivera A, Alberti A, Martín AB, Narva´ez JA, de la Calle A, Moratalla R. Molecular phenotype of rat striatal neurons expressing the dopamine D_5 receptor subtype. *European Journal of Neuroscience,* 2002;16:2049–2058.

[56] Centonze D, Grande C, Saulle E, Martín AB, Gubellini P, Pavón N,. Distinct roles of D_1 and D_5 dopamine receptors in motor activity and striatal synaptic plasticity. *The Journal of Neuroscience,* 2003; 23:8506-8512.

[57] Le Crom S, Prou D, Vernier P. Autocrine activation of adenosine A1 receptors blocks D1A but not D1B dopamine receptor desensitization. *Journal of Neurochemistry,* 2002;82:1549–1552.

[58] Sahu A, Tyeryar KR, Vongtau HO, Sibley DR, Undieh AS. D_5 dopamine receptors are required for DA-ergic activation of phospholipase C. *Molecular Pharmacology,* 2009;75:447-453.

[59] Shi W, Pun C, Smith PL, Bunney BS. Endogenous DA-mediated feedback inhibition of DA neurons: involvement of both D1- and D2-like receptors. *Synapse,* 2000;35:111–119.

[60] Shi W, Smith PL, Pun C, Millet B, Bunney BS. D1–D2 interaction in feedback control of midbrain dopamine neurons. *Journal of Neuroscience,* 1997;17:7988–7994.

[61] Rahman S, McBride WJ. D1–D2 dopamine receptor interaction within the nucleus accumbens mediates long-loop negative feedback to the ventral tegmental area (VTA). *Journal of Neurochemistry,* 2001;77:1248–1255

[62] Saklayen SS, Mabrouk OS, Pehek EA. Negative feedback regulation of nigrostriatal dopamine release: mediation by striatal D1 receptors. *Journal of Pharmacology and Experimental Therapeutics,* 2004;311:342-348.

[63] Imperato A, Di Chiara G. Effects of locally applied D-1 and D-2 receptor agonists and antagonists studied with brain dialysis. *European Journal of Pharmacology,* 1988; 156:385-393.

[64] Westerink BH, de Vries JB. On the mechanism of neuroleptic induced increase in striatal dopamine release: brain dialysis provides direct evidence for mediation by autoreceptors localized on nerve terminals. *Neuroscience Letters,* 1989;99:197–202.

[65] Santiago M, Westerink BH. Characterization and pharmacological responsiveness of dopamine release recorded by microdialysis in the substantia nigra of conscious rats. *Journal of Neurochemistry,* 1991;57:738-747.

[66] Santiago M, Westerink BH. The regulation of dopamine release from nigrostriatal neurons in conscious rats: the role of somatodendritic autoreceptors. *European Journal of Pharmacology,* 1991;204:179-185.

[67] Heidbreder CA, Gardner EL, Xi ZX, Thanos PK, Mugnaini M, Hagan JJ, Ashby CR Jr. The role of central dopamine D_3 receptors in drug addiction: a review of pharmacological evidence. *Brain Research Brain Research Reviews,* 2005;49:77-105.

[68] Hammad H, Wagner JJ. Dopamine-mediated disinhibition in the CA1 region of rat hippocampus via D3 receptor activation. *Journal of Pharmacology and Experimental Therapeutics,* 2006;316:113-120.

[69] Spiller K, Xi ZX., Peng XO, Newman AH, Ashby CR Jr., Heidbreder C, Gaál J, Gardner EL. The selective dopamine D_3 receptor antagonists SB-277011A and NGB 2904 and the putative partial D_3 receptor agonist BP-897 attenuate methamphetamine-enhanced brain stimulation reward in rats. *Psychopharmacology,* 2008;196:533-542.

[70] Harrison SJ, Nobrega JN. Differential susceptibility to ethanol and amphetamine sensitization in dopamine D_3 receptor-deficient mice. *Psychopharmacology,* 2009;204:49-59.

[71] Luedtke RR, Mach RH. Progress in developing D_3 dopamine receptor ligands as potential therapeutic agents for neurological and neuropsychiatrc disorders. *Current Pharmaceutical Design,* 2003;9:643–671.

[72] Braak H, Ghebremedhin E, Rub U, Bratzke H, Del Tredici K. Stages in the development of Parkinson's disease-related pathology. *Cell Tissue Research,* 2004;318:121–134.

[73] Spillantini MG, Crowther RA, Jakes R, Hasegawa M, Goedert M. α- synuclein in filamentous inclusions of Lewy bodies from Parkinson's disease and dementia with

lewy bodies. *Proceedings of the National Academy of Sciences USA,*1998; 95:6469–6473.

[74] Mizuno Y, Ohta S, Tanaka M, Takamiya S, Suzuki K, Sato T, Oya H, Ozawa T, Kagawa Y. Deficiencies in complex I subunits of the respiratory chain in Parkinson's disease. *Biochemical and Biophysical Research Communications,* 1989;163:1450–1455.

[75] Swerdlow RH, Parks JK, Miller SW, Tuttle JB, Trimmer PA, Sheehan JP, Bennett JP Jr, Davis RE, Parker WD Jr. Origin and functional consequences of the complex I defect in Parkinson's disease. *Annals of Neurology,* 1996;40: 663–671.

[76] Schapira AH, Cooper JM, Dexter D, Clark JB, Jenner P, Marsden, CD. Mitochondrial complex I deficiency in Parkinson's disease. *Journal Neurochemistry,* 1990;*54:*823–827.

[77] Klegeris A, Pelech S, Giasson BI, Maguire J, Zhang H, McGeer EG, McGeer PL. Alpha-synuclein activates stress signaling protein kinases in THP-1 cells and microglia. *Neurobiology of Aging,* 2008;29:739-752.

[78] Hald A, Lotharius J. Oxidative stress and inflammation in Parkinson's disease: is there a causal link? *Experimental Neurology*, 2005;193:279-290.

[79] Hald A, Van Beek J, Lotharius J. Inflammation in Parkinson's disease: causative or epiphenomenal. *Subcellular Biochemistry*, 2007;42:249-279.

[80] Schulz JB. Update on the pathogenesis of Parkinson's disease. *Journal of Neurology,* 2008;255 (Suppl 5):3-7.

[81] Zhao B. Natural antioxidants protect neurons in Alzheimer's dsease and Parkinson's disease. *Neurochemical Research.* 2009;36:630-638.

[82] Farooqui T, Farooqui AA. Aging: an important factor for the pathogenesis of neurodegenerative diseases. *Mechanisms of Ageing and Development*, 2009;130: 203-215.

[83] Olanow CW, Tatton WG. Etiology and pathogenesis of Parkinson's disease. *Annual Reviews in Neuroscience,* 1999;22:123–144.

[84] McInnis J, Wang C, Anastasio N, Hultman M, Ye Y, Salvemini D, Johnson KM. The role of superoxide and nuclear factor-kappaB signaling in N-methyl-D-aspartate-induced necrosis and apoptosis. *Journal of Pharmacology and Experimental Therapeutics,* 2002;301:478-87.

[85] Sun GY, Horrocks LA, Farooqui AA. The roles of NADPH oxidase and phospholipases A_2 in oxidative and inflammatory responses in neurodegenerative diseases. *Journal of Neurochemistry,* 2007;103:1-16.

[86] Good PF, Hsu A, Werner P, Perl DP, Olanow CW. Protein nitration in Parkinson's disease. *Journal of Neuropathology and Experimental Neurology,* 1998;57:338-342.

[87] Beckman JS, Ischiropoulos H, Zhu L, van der Woerd M, Smith C, Chen J, Harrison J, Martin JC, Tsai M. Kinetics of superoxide dismutase-and iron-catalyzed nitration of phenolics by peroxynitrite. *Archives of Biochemistry and Biophysics,* 1992;298:438-445.

[88] Badgaiyan RD, Fischman AJ, Alpert NM. Striatal Dopamine Release in Sequential Learning. *Neuroimage,* 2007;38:549-556.

[89] Jackson GM, Jackson SR, Harrison J, Henderson L, Kennard C. Serial reaction time learning and Parkinson's disease: evidence for a procedural learning deficit. *Neuropsychologia,* 1995;33:577–593.

[90] Carbon M, Ma Y, Barnes A, Dhawan V, Chaly T, Ghilardi MF, Eidelberg D. Caudate nucleus: influence of DA-ergic input on sequence learning and brain activation in Parkinsonism. *Neuroimage,* 2004;21:1497–1507.

[91] Aosaki T, Graybiel AM, Kimura M. Effect of the nigrostriatal dopamine system on acquired neural responses in the striatum of behaving monkeys. *Science.* 1994;265:412–45.

[92] Matsumoto N, Hanakawa T, Maki S, Graybiel AM, Kimura M. Nigrostriatal Dopamine System in Learning to Perform Sequential Motor Tasks in a Predictive Manner. *Journal of Neurophysiology,* 1999;82:978-998.

[93] Williams GV, Castner SA. Under the curve: Critical issues for elucidating D1 receptor function in working memory. *Neuroscience,* 2006;139:263–276.

[94] Dash PK, Moore AN, Kobori N, Runyan JD. Molecular activity underlying working memory. *Learning & memory,* 2007;14:554–563.

[95] Ota M, Yasuno F, Ito H, Seki C, Nozaki S, Asada T, Suhara T. Age-related decline of dopamine synthesis in the living human brain measured by positron emission. *Life Science,* 2006;79:730-736.

[96] Kaasinen, V., Vilkman, H., Hietala, J., Nagren, K., Helenius, H., Olsson, H, Farde L, Rinne J. Age-related dopamine D2/D3 receptor loss in extrastriatal regions of the human brain. *Neurobiology of Aging,* 2008;21:683-688.

[97] Nieto-Sampedro M, Nieto-Diaz M. Neural plasticity: changes with age. *Journal of Neural Transmission,* 2005;112:3-27.

[98] Ohsawa I, Nishimaki K, MurakamiY, Suzuki Y, Ishikawa M, Ohta S. Age-dependent neurodegeneration accompanying memory loss in transgenic mice defective in mitochondrial aldehyde dehydrogenase 2 activity. *Journal of Neuroscience,* 2008;28:6239-6249.

[99] Hudson JM. Automatic memory processes in normal ageing and Alzheimer's disease. *Cortex,* 2008;44:345-349.

[100] Mattay VS, Tessitore A, Callicott JH, Bertolino A, Goldberg TE, Chase TN, Hyde TM, Weinberger DR. Dopaminergic modulation of cortical function in patients with Parkinson's disease. *Annals of Neurology,* 2002;51:156–164.

[101] Brooks DJ. Dopaminergic action beyond its effects on motor function: Imaging studies. *Journal of Neurology,* 253 Suppl: IV8-15.

[102] Zhang K, Grady CJ, Tsapakis EM, Andersen SL, Tarazi FI, Baldessarini RJ. Regulation of working memory by dopamine D4 receptor in rats. *Neuropsychopharmacology,* 2004;29:1648–1655

[103] Pantelis C, Barber FZ., Barnes TR, Nelson HE, Owen AM, Robbins TW. Comparison of set-shifting ability in patients with chronic schizophrenia and frontal lobe damage. *Schizophrenia Research,* 1999;37:251–270.

[104] Goldman-Rakic PS, Castner SA, Svensson TH, Siever LJ, Williams GV. Targeting the dopamine D1 receptor in schizophrenia: Insights for cognitive dysfunction. *Psychopharmacology,* 2004;174:3–16.

[105] Pathak RU, Davey GP. Complex I and energy thresholds in the brain. *Biochimica et Biophysica Acta,* 2008;1777:777-782.

[106] Eliezer D, Kutluay E, Bussell Jr. R, Browne G. Conformational properties of alpha-synuclein in its free and lipid-associated states. *Journal of Molecular Biology,* 2001;307:1061–1073.

[107] Perrin RJ, Woods WS, Clayton DF, George JM. Interaction of human alpha-synuclein and Parkinson's disease variants with phospholipids. Structural analysis using site-directed mutagenesis. *Journal of Biological Chemistry,* 2000;275:34393–34398.

[108] Perrin RJ, Woods WS, Clayton DF, George JM. Exposure to long-chain polyunsaturated fatty acids triggers rapid multimerization of synucleins. *Journal of Biological Chemistry,* 2001;276:41958–41962.

[109] Seo JH, Rah JC, Choi SH, Shin JK, Min K, Kim HS, Park CH, Kim S, Kim EM, Lee SH, Lee S, Suh SW, Suh YH. α-Synuclein regulates neuronal survival via Bcl-2 family expression and PI3/Akt kinase pathway. *The FASEB Journal,* 2002;16:1826-1828.

[110] Perez RG, Waymire JC, Lin E, Liu JJ, Guo F, Zigmond MJ. A role for alpha-synuclein in the regulation of dopamine biosynthesis. *Journal of Neuroscience,* 2002;22:3090-3099.

[111] Ostrerova N, Petruceli L, Farrer M, Mehta N, Choi P, Hardy J, Wolozin B. α-Synuclein shares physical and functional homology with 14-3-3 proteins. *Journal of Neuroscience,* 1999;19:5782–5791

[112] Sherer TB, Betarbet R, Stout AK, Lund S, Baptista M, Panov AV, Cookson MR, Greenamyre JT. An in vitro model of Parkinson's disease: linking mitochondrial impairment to altered α-synuclein metabolism and oxidative damage. *Journal of Neuroscience,* 2002;22:7006-7015.

[113] Shavali S, Brown-Borg HM, Ebadi M, Porter J. Mitochondrial localization of α-synuclein protein in α-synuclein overexpressing cells. *Neuroscience Letters,* 2008;439:125-128.

[114] Rankin CA, Joazeiro CAP, Floor E, Hunter T. E3 ubiquitin-protein ligase activity of parkin is dependent on cooperative interaction of RING finger (TRIAD) elements. *Journal of Biomedical Science,* 2001;8:421-429.

[115] Periquet M, Lücking C, Vaughan J, Bonifati V, Dürr A, De Michele G, Horstink M, Farrer M, Illarioshkin SN, Pollak P, Borg M, Brefel-Courbon C, Denefle P, Meco G, Gasser T, Breteler MM, Wood N, Agid Y, Brice A, French Parkinson's Disease Genetics Study Group. The European Consortium on Genetic Susceptibility in Parkinson's Disease. Origin of the mutations in the parkin gene in Europe: exon rearrangements are independent recurrent events, whereas point mutations may result from Founder effects. *The American Journal of Human Genetics,*2001;68:617-626.

[116] Hedrich K, Eskelson C, Wilmot B, Marder K, Harris J, Garrels J, Meija-Santana H, Vieregge P, Jacobs H, Bressman SB, Lang AE, Kann M, Abbruzzese G, Martinelli P, Schwinger E, Ozelius LJ, Pramstaller PP, Klein C, Kramer P. Distribution, type, and origin of Parkin mutations: review and case studies. *Movement Disorders,* 2004;19:1146–1157.

[117] Kitada, T., Asakawa, S., Hattori, N., Matsumine, H., Yamamura, Y., Minoshima, S, Yokochi M, Mizuno Y, Shimizu N. Mutations in the parkin gene cause autosomal recessive juvenile parkinsonism. *Nature,* 1998;392:605–608.

[118] Mata IF, Lockhart PJ, Farrer MJ. Parkin genetics: one model for Parkinson's disease. *Human Molecular Genetics,* 2004;13:R127–R133.

[119] McNaught KS, Mytilineou C, Jnobaptiste R, Yabut J, Shashidharan P, Jennert P, Olanow CW. Impairment of the ubiquitin-proteasome system causes DA-ergic cell death and inclusion body formation in ventral mesencephalic cultures. *Journal of Neurochemistry,* 2002;81:301-306.

[120] Liani E, Eyal A, Avraham E, Shemer R, Szargel R, Berg D. Ubiquitylation of synphilin-1 and α-synuclein by SIAH and its presence in cellular inclusions and Lewy bodies imply a role in Parkinson's disease. *Proceedings of the National Academy of Sciences USA,* 2004;101:5500-5505.

[121] Itier JM, Ibanez P, Mena MA, Abbas N, Cohen-Salmon C, Bohme GA, Laville M, Pratt J, Corti O, Pradier L., Ret G, Joubert C, Periquet M, Araujo F, Negroni J., Casarejos MJ, Canals S, Solano R, Serrano A, Gallego E, Sanchez M, Denefle P, Benavides J, Tremp G, Rooney TA, Brice A, Garcia de Yebenes J. Parkin gene inactivation alters behaviour and dopamine neurotransmission in the mouse. *Human Molecular Genetics,* 2003;12:2277–2291.

[122] Paterna JC, Leng A, Weber E, Feldon J, Büeler H. DJ-1 and Parkin modulate dopamine-dependent behavior and inhibit MPTP-induced nigral dopamine neuron loss in mice. *Molecular Therapy,*2007;15:698-704.

[123] Jenner P, Olanow CW. Oxidative stress and the pathogenesis of Parkinson's disease. *Neurology*, 1996;**47:**S161–S170.

[124] Jiang H, Jiang Q, Liu W, Feng J. Parkin suppresses the expression of monoamine oxidases. *Journal of Biological Chemistry,* 2006;281:8591–8599.

[125] Jiang H, Ren Y, Zhao J, Feng J. Parkin protects human dopaminergic neuroblastoma cells against dopamine-induced apoptosis. *Human Molecular Genetics,* 2004;13:1745-1754.

[126] Mitsumoto A, Nakagawa Y. DJ-1 is an indicator for endogenous reactive oxygen species elicited by endotoxin. *Free Radical Research,* 2001;35:885–893.

[127] Bonifati V, Rizzu P, Squitieri F, Krieger E, Vanacore N, van Swieten JC, Brice A, van Duijn CM, Oostra B, Meco G, Heutink P. DJ-1(PARK7), a novel gene for autosomal recessive, early onset parkinsonism. *Neurological Sciences,* 2003;24:159-160.

[128] Wilson MA, Collins JL, Hod Y, Ringe D, Petsko GA. The 1.1-A resolution crystal structure of DJ-1, the protein mutated in autosomal recessive early onset Parkinson's disease. *Proceedings of the National Academy of Sciences USA,* 2003;100:9256-9261.

[129] Lakshminarasimhan M, Maldonado MT, Zhou W, Fink AL, Wilson MA. Structural Impact of Three Parkinsonism-Associated Missense Mutations on Human DJ-1. *Biochemistry,* 2008;47:1381-1392.

[130] Yokota T, Sugawara K, Ito K, Takahashi R, Ariga H, Mizusawa H. Down regulation of DJ-1 enhances cell death by oxidative stress, ER stress, and proteasome inhibition. *Biochemical and Biophysical Research Communications,* 2003; 312:1342–1348.

[131] Goldberg MS, Pisani A, Haburcak M, Vortherms TA, Kitada T, Costa C, Tong Y, Martella G, Tscherter A, Martins A, Bernardi G, Roth BL, Pothos EN, Calabresi P, Shen J. Nigrostriatal dopaminergic deficits and hypokinesia caused by inactivation of the familial Parkinsonism-linked gene DJ-1. *Neuron,* 2005;45:489-496.

[132] Kim RH, Smith PD, Aleyasin H, Hayley S, Mount MP, Pownall S, Wakeham A, You-Ten AJ, Kalia SK, Horne P, Westaway D, Lozano AM, Anisman H, Park DS, Mak TW. Hypersensitivity of DJ-1-deficient mice to 1-methyl-4-phenyl-1,2,3,6-tetrahydropyrindine (MPTP) and oxidative stress. *Proceedings of the National Academy* of Sciences *USA,* 2005;102:5215-5220.

[133] Chen LN, Cagniard B, Mathews T, Jones S, Koh HC, Ding YM, Carvey PM, Ling ZD, Kang UJ, Zhuang XX. Age-dependent motor deficits and dopaminergic dysfunction in DJ-1 null mice. *Journal of biological Chemistry,* 2005;280:21418–21426.

[134] Yang Y, Gehrke S, Haque ME, Imai Y, Kosek J, Yang L, Beal MF, Nishimura I, Wakamatsu K, Ito S, Takahashi R, Lu B. Inactivation of *Drosophila* DJ-1 leads to impairments of oxidative stress response and phosphatidylinositol 3-kinase/Akt signaling. *Proceedings of the National Academy of Sciences USA,* 2005;38: 13670-13675.

[135] Meulener MC, Graves CL, Sampathu DM, Armstrong-Gold CE, Bonini NM, Giasson BI. DJ-1 is present in a large molecular complex in human brain tissue and interacts with alpha-synuclein. *Journal of Neurochemistry,* 2005;93:1524-1532.

[136] Meulener M, Whitworth AJ, Armstrong-Gold CE, Rizzu P, Heutink P, Wes, PD, Pallanck LJ, Bonini NM. *Drosophila* DJ-1 mutants are selectively sensitive to environmental toxins associated with Parkinson's disease. *Current Biology,* 2005;15:1572-15777.

[137] Biskup S, Gerlach M, Kupsch A, Reichmann H, Riederer P, Vieregge P, Wüllner U, Gasser T. Genes associated with Parkinson syndrome. *Journal of Neurology,* 2008;255: *Suppl 5,* 8-17.

[138] Gautier CA, Kitada T, Shen J. Loss of PINK1 causes mitochondrial functional defects and increased sensitivity to oxidative stress. *Proceedings of the National Academy of Sciences USA,* 2008;105:13364-11369.

[139] Marongiu R, Spencer B, Crews L, Adame A, Patrick C, Trejo M, Dallapiccola B, Valente EM, Masliah E. Mutant Pink1 induces mitochondrial dysfunction in a neuronal cell model of Parkinson's disease by disturbing calcium flux. *Journal of Neurochemistry,* 2009;108:1561-1574.

[140] Exner N, Treske B, Paquet D, Holmström K, Schiesling C, Gispert S, Carballo-Carbajal I, Berg D, Hoepken HH, Gasser T, Krüger R, Winklhofer KF, Vogel F, Reichert AS, Auburger G, Kahle PJ, Schmid B, Haass C. Loss-of-function of human PINK1 results in mitochondrial pathology and can be rescued by parkin. *Journal of Neuroscience,* 2007;27:12413-12418.

[141] Wood-Kaczmar A, Gandhi S, Yao Z, Abramov AS, Miljan EA, Keen G, Stanyer L, Hargreaves I, Klupsch K, Deas E, Downward J, Mansfield L, Jat P, Taylor J, Heales S, Duchen MR, Latchman D, Tabrizi SJ, Wood NW. PINK1 Is Necessary for Long Term Survival and Mitochondrial Function in Human DA-ergic Neurons. *PLoS ONE,* 2008; 3:e2455.

[142] Li Y, Tomiyama H, Sato K, Hatano Y, Yoshino H, Atsumi M, Kitaguchi M, Sasaki S, Kawaguchi S, Miyajima H, Toda T, Mizuno Y, Hattori N. Clinicogenetic study of *PINK1* mutations in autosomal recessive early-onset parkinsonism. *Neurology,* 2005;64:1955-1957.

[143] Beilina A, Van Der Brug M, Ahmad R, Kesavapany S, Miller DW, Petsko GA, Cookson MR. Mutations in PTEN-induced putative kinase 1 associated with recessive parkinsonism have differential effects on protein stability. *Proceedings of the National Academy of Sciences USA,* 2005;102:5703–5708.

[144] Sim CH, Lio DS, Mok SS, Masters CL, Hill AF, Culvenor JG, Cheng HC. C-terminal truncation and Parkinson's disease-associated mutations down-regulate the protein serine/threonine kinase activity of PTEN-induced kinase-1. *Human Molecular Genetics,* 2006;15:3251-3262.

[145] Park J, Lee SB, Lee S, Kim Y, Song S, Kim S, Bae E, Kim J, Shong M, Kim JM, Chung J. Mitochondrial dysfunction in *Drosophila PINK1* mutants is complemented by parkin. *Nature,* 2006;441:1157–1161.

[146] Clark IE, Dodson MW, Jiang C, Cao JH, Huh JR, Seol JH, Yoo SJ, Hay BA, Guo M. *Drosophila* pink1 is required for mitochondrial function and interacts genetically with parkin. *Nature,* 2006;441:1162–1166.

[147] Kitada, T., Pisani, A., Porter, D.R., Yamaguchi, H., Tscherter, A., Martella, G, Bonsi P, Zhang C, Pothos EN, Shen J. Impaired dopamine release and synaptic plasticity in the striatum of PINK1-deficient mice. *Proceedings of the National Academy of Sciences USA,* 2007;104:11441–11446.

[148] Li C, Beal MF. Leucine-rich repeat kinase 2: a new player with a familiar theme for Parkinson's disease pathogenesis. *Proceedings of the National Academy of Sciences USA,* 2005;102:16535-16536.

[149] Mata IF, Wedemeyer WJ, Farrer MJ, Taylor JP, Gallo KA. LRRK2 in Parkinson's disease: protein domains and functional insights. *Trends in Neurosciences,* 2006;29:286-293

[150] Deng J, Lewis PA, Greggio E, Sluch E, Beilina A, Cookson MR. Structure of the ROC domain from the Parkinson's disease-associated leucine-rich repeat kinase 2 reveals a dimeric GTPase. *Proceedings of the National Academy of Sciences USA,* 2008;105:1499-1504.

[151] Lu YW, Tan EK. Molecular biology changes associated with LRRK2 mutations in Parkinson's disease. *Journal of Neuroscience Research,* 2008;86:1895-18901.

[152] Greggio, E., Zambrano, I., Kaganovich, A., Beilina, A., Taymans, J.M., Daniëls, V, Lewis P, Jain S, Ding J, Syed A, Thomas KJ, Baekelandt V, Cookson MR. The Parkinson disease-associated leucine-rich repeat kinase 2 (LRRK2) is a dimer that undergoes intramolecular autophosphorylation. *Journal of Biological Chemistry,* 2008;283:16906-16914.

[153] Paisán-Ruiz C, Jain S, Evans EW, Gilks WP, Simon J, van der Brug M, López de Munain A, Aparicio S, Gil AM, Khan N, Johnson J, Martinez JR, Nicholl D, Carrera IM, Pena AS, de Silva R, Lees A, Marti-Massó JF, Pérez-Tur J, Wood NW, Singleton AB. Cloning of the gene containing mutations that cause PARK8-linked Parkinson's disease. *Neuron,* 2004;44:595–600.

[154] Farrer, M., Stone, J., Mata, I.F., Lincoln, S., Kachergus, J., Hulihan, M, Strain KJ, Maraganore DM. LRRK2 mutations in Parkinson disease. *Neurology,* 2005;65:738–740.

[155] Zimprich A, Biskup S, Leitner P, Lichtner P, Farrer M, Lincoln S, Kachergus J, Hulihan M, Uitti RJ, Calne DB, Stoessl AJ, Pfeiffer RF, Patenge N, Carbajal IC, Vieregge P, Asmus F, Müller-Myhsok B, Dickson DW, Meitinger T, Strom TM, Wszolek ZK, Gasser T. Mutations in LRRK2 cause autosomal-dominant parkinsonism with pleomorphic pathology. *Neuron,* 2004;44:601–607.

[156] West AB, Moore DJ, Biskup S, Bugayenko A, Smith WW, Ross CA, Dawson VL, Dawson TM. Parkinson's disease-associated mutations in leucine-rich repeat kinase 2 augment kinase activity. *Proceedings of the National Academy of Sciences USA,* 2005;102: 16842–16847.

[157] Ho CC, Rideout HJ, Ribe E, Troy CM, Dauer WT. The Parkinson disease protein leucine-rich repeat kinase 2 transduces death signals via Fas-associated protein with death domain and caspase-8 in a cellular model of neurodegeneration. *Journal of Neuroscience,* 2009;29:1011-1016.

[158] Wang L, Xie C, Greggio E, Parisiadou L, Shim H, Sun L, Chandran J, Lin X, Lai C, Yang W-J, Moore DJ, Dawson TM, Dawson VL, Chiosis G, Cookson MR, Cai H. The chaperone activity of heat shock protein (Hsp)90 is critical for maintaining the stability of leucine-rich repeat kinase 2. *Journal of Neuroscience,* 2008;28:3384–3391.

[159] Suzuki Y, Imai Y, Nakayama H, Takahashi K, Takio K, Takahashi RA. Serine protease, HtrA2, is released from the mitochondria and interacts with XIAP, inducing cell death. *Molecular Cell,* 2001;8:613-621.

[160] Faccio L, Fusco C, Chen A, Martinotti S, Bonventre JV, Zervos AS. Characterization of a novel human serine protease that has extensive homology to bacterial heat shock endoprotease HtrA and is regulated by kidney ischemia. *Journal of Biological Chemistry,* 2000;275:2581-2588.

[161] Gray CW, Ward RV, Karran E, Turconi S, Rowles A, Viglienghi D, Southan C, Barton A, Fantom KG, West A, Savopoulos J, Hassan NJ, Clinkenbeard H, Hanning C, Amegadzie B, Davis JB, Dingwall C, Livi GP, Creasy CL. Characterization of human HtrA2, a novel serine protease involved in the mammalian cellular stress response. *European Journal of Biochemistry,* 2000;267:5699-710.

[162] Li W, Srinivasula SM, Chai J, Li P, Wu JW, Zhang Z, Alnemri ES, Shi Y. Structural insights into the pro-apoptotic function. of mitochondrial serine protease HtrA2/Omi. *Nature Structural Biology,* 2002;9:436-41.

[163] Hegde R, Srinivasula SM, Zhang Z, Wassell R, Mukattash R, Cilenti L, DuBois G, Lazebnik Y, Zervos AS, Fernandes-Alnemri T, Alnemri ES. Identification of Omi/HtrA2 as a mitochondrial apoptotic serine protease that disrupts inhibitor of apoptosis protein-caspase interaction. *Journal of Biological Chemistry*, 2002;277:432-438.

[164] Suzuki Y, Takahashi-Niki K, Akagi T, Hashikawa T, Takahashi R. Mitochondrial protease Omi/HtrA2 enhances caspase activation through multiple pathways. *Cell Death & Differentiation,* 2004;11:208-216.

[165] Nam MK, Seong YM, Park HJ, Choi JY, Kang S, Rhim H. The homotrimeric structure of HtrA2 is indispensable for executing its serine protease activity. *Experimental and Molecular Medicine,* 2006;38:36-43.

[166] Strauss KM, Martins LM, Plun-Favreau H, Marx FP, Kautzmann S, Berg D, Gasser T, Wszolek Z, Müller T, Bornemann A, Wolburg H, Downward J, Riess O, Schulz JB, Krüger R. Loss of function mutations in the gene encoding Omi/HtrA2 in Parkinson's disease. *Human Molecular Genetics.* 2005;14:2099-2111.

[167] Martins LM, Morrison A, Klupsch K, Fedele V, Moisoi N, Teismann P, Abuin A, Grau E, Geppert M, Livi GP, Creasy CL, Martin A, Hargreaves I, Heales SJ, Okada H, Brandner S, Schulz JB, Mak T, Downward J. Neuroprotective role of the Reaper-related serine protease HtrA2/Omi revealed by targeted deletion in mice. *Molecular and Cellular Biology,* 2004;24:9848-9862.

[168] Plun-Favreau H, Klupsch K, Moisoi N, Gandhi S, Kjaer S, Frith D, Harvey K, Deas E, Harvey RJ, McDonald N, Wood NW, Martins LM, Downward J. The mitochondrial protease HtrA2 is regulated by Parkinson's disease-associated kinase PINK1. *Nature Cell Biology,* 2007;9:1243-1252.

[169] Tain LS, Chowdhury RB, Tao RN, Plun-Favreau H, Moisoi N, Martins LM, Downward J, Whitworth AJ, Tapon N. *Drosophila* HtrA2 is dispensable for apoptosis but acts downstream of PINK1 independently from Parkin. *Cell Death and Differentiation* 2009 (in press).

[170] Baez S, Segura-Aguilar J, Widersten M, Johansson AS, Mannervik B. Glutathione transferases catalyse the detoxication of oxidized metabolites (o-quinones) of catecholamines and may serve as an antioxidant system preventing degenerative cellular processes. *Biochemistry Journal,* 1997;324:25-28.

[171] Berman SB, Hastings TG. Dopamine oxidation alters mitochondrial respiration and induces permeability transition in brain mitochondria: implications for Parkinson's disease. *Journal of Neurochemistry,* 1999;73:1127-1137.

[172] Whitbread AK, Masoumi A, Tetlow N, Schmuck E, Coggan M, Board PG. Characterization of the omega class of glutathione transferases. *Methods of Enzymology,* 2005;401:78-99.

[173] Li YJ,Oliveira SA, Xu P, Martin ER, Stenger JE, Scherzer CR. Glutathione S-transferase omega-1 modifies age-at-onset of Alzheimer disease and Parkinson disease. *Human Molecular Genetics,* 2003;12:3259-3267.

[174] Li YJ, Scott WK, Zhang L Lin P-I, Oliveira SA., Skelly T, Doraiswamy MP, Welsh-Bohmer KA, Martin ER, Haines JL, Pericak-Vance MA, Vance, JM. Revealing the role of glutathione S-transferase omega in age-at-onset of Alzheimer and Parkinson diseases. *Neurobiology of Aging,* 2006;27:1087-1093.

[175] Schmuck EM, Board PG, Whitbread AK, Tetlow N, Cavanaugh JA., Blackburn, AC, Masoumi A. Characterization of the monomethylarsonate reductase and dehydroascorbate reductase activities of Omega class glutathione transferase variants: implications for arsenic metabolism and the age-at-onset of Alzheimer's and Parkinson's diseases. *Pharmacogenetics and Genomics,* 2005;15:493-501.

[176] Perez-Pastene C, Graumann R, Díaz-Grez F, Miranda M, Venegas P, Godoy OT, Layson L, Villagra R, Matamala JM, Herrera L, Segura-Aguilar J. Association of GST

M1 null polymorphism with Parkinson's disease in a Chilean population with a strong Amerindian genetic component. *Neuroscience Letters,* 2007;418:181-185.

[177] Segura-Aguilar J, Baez S, Widersten M, Welch CJ, Mannervik B. Human class Mu glutathione transferases, in particular isoenzyme M2-2, catalyze detoxication of the dopamine metabolite aminochrome. *Journal of Biological Chemistry,* 1997;272:5727-5731.

[178] Dagnino-Subiabre A, Cassels BK, Baez S, Johansson AS, Mannervik B, Segura-Aguilar J. Glutathione transferase M2-2 catalyzes conjugation of dopamine and dopa o-quinones. *Biochemical and Biophysical Research Communications,* 2000;274:32-36.

[179] Linazasoro G. New ideas on the origin of L-dopa-induced dyskinesias: age, genes and neural plasticity. *Trends in Pharmacological Sciences,* 2008;26:391-397.

[180] Arbouw ME, van Vugt JP, Egberts TC, Guchelaar HJ. Pharmacogenetics of antiparkinsonian drug treatment: a systematic review. *Pharmacogenomics,* 2007;8:159-176.

[181] Burn DJ, Tröster AI. Neuropsychiatric complications of medical and surgical therapies for Parkinson's disease. *Journal of Geriatric Psychiatry and Neurology,* 2004;17:172-180.

[182] Mastroeni, D, Grover, A, Leonard B, Joyce JN, Coleman PD, Kozik B, Bellinger DL, Rogers J. Microglial responses to dopamine in a cell culture model of Parkinson's disease. *Neurobiology of aging,* 2008 (in press).

[183] Schwarzschild MA, Agnati L, Fuxe K, Chen JF, Morelli M. Targeting adenosine A_2A receptors in Parkinson. *Trends in Neuroscience,* 2006;29:647-654.

[184] Morelli M, Di Paolo T, Wardas J, Calon F, Xiao D, Schwarzschild MA. Role of adenosine A2A receptors in parkinsonian motor impairment and *l*-DOPA-induced motor complications. *Progress in Neurobiology,* 2007;83:293-309.

[185] Antonelli T, Fuxe K, Agnati L, Mazzoni E, Tanganelli S, Tomasini MC, Ferraro L. Experimental studies and theoretical aspects on A2A/D2 receptor interactions in a model of Parkinson. Relevance for L-dopa induced dyskinesias. *Journal of Neurological Sciences,* 2006;248:16-22.

[186] Vortherms TA, Watts V.J. Sensitization of neuronal A2A adenosine receptors after persistent D2 dopamine receptor activation. *Journal of Pharmacology and Experimental Therapeutics,* 2004;308:221-227.

[187] Henchcliffe C, Beal MF. Mitochondrial biology and oxidative stress in Parkinson disease pathogenesis. *Nature Clinical Practice Neurology,* 2008;4:600-609.

[188] Bogaerts V, Theuns J, van Broeckhoven C. Genetic findings in Parkinson's disease and translation into treatment: a leading role for mitochondria? *Genes, Brain and Behavior,* 2008;7:129–151.

In: Biogenic Amines
Editors: T. Farooqui, A. A. Farooqui
ISBN: 978-1-60876-625-3

Chapter X

MODULATION OF NEUROTRANSMISSION SIGNALING BY NEURAL MEMBRANE POLYUNSATURATED FATTY ACIDS

Akhlaq A. Farooqui*
The Ohio State University, Columbus, Ohio 43210, USA.

ABSTRACT

Neurotransmission is defined as an electrical movement within synapses mediated by a propagation of nerve impulses, which are essential for the propagation of signals. During neurotransmission, the released neurotransmitter (dopamine, glutamate, serotonin etc) from presynaptic neuron travels across the synaptic cleft and is taken up by post synaptic neuron where through their receptors, neurotransmitters initiate a cascade of neurochemical events involving signal transduction network that lead to the transfer of message from one neuron to another. In postsynaptic neural membranes, phospholipase A_2, (PLA_2), an enzyme that hydrolyzes fatty acid from sn-2 position of glycerol moiety of membrane glycerophospholipids, is linked to dopamine, glutamate, serotonin, P_2-purinergic, muscarinic, cytokine, and growth factor receptors through different coupling mechanisms. Dopamine and serotonin receptors are linked through G proteins, which mediate the release of arachidonic acid. In contrast, glutamate receptors are linked to the release of arachidonic acid and docosahexaenoic acid through G protein independent mechanism. Docosahexaenoic acid modulates dopaminergic and serotonergic neurotransmission in rat frontal cortex, and a chronic deficiency of docosahexaenoic acid changes dopamine metabolism by altering dopamine receptor and transporters in the nucleus accumbens. Dietary supplementation of docosahexaenoic acid not only changes the endogenous dopamine levels in frontal cortex, but also elevates binding of dopamine to the D_2 receptors. Docosahexaenoic acid also modulates glutamatergic

* Correspondence concerning this article should be addressed to: Akhlaq A. Farooqui, Molecular Cellular Biochemistry, The Ohio State University, Columbus, Ohio 43210, U.S.A., Phone: (614-488-0361), E.mail: farooqui.1@osu.edu.

neurotransmission. It is a component of phosphatidylserine, a neural membrane phospholipid needed for maintenance of NMDA receptor mediated processes. In addition, docohexaenoic acid differentially modulates glutamate transporters (GLT1, GLAST, and EAAC1) through different mechanisms. Thus, docosahexaenoic acid stimulates GLT1 and EAAC1 through a mechanism that requires extracellular Ca^{2+}, CaM kinase II, and protein kinase C but not protein kinase A. In contrast, the inhibitory effect of docosahexaenoic acid on GLAST does not require extracellular Ca^{2+} and does not involve CaM kinase II. In addition, arachidonic acid-derived prostaglandins and 2-arachidonylglycerol-derived prostaglandin glycerylesters modulate excitatory and GABAergic neurotransmission. Collective evidence suggests that docosahexaenoic acid and arachidonic acid status not only improves neurotransmitter-mediated cognitive development and behavior, but also increases neuroplasticity of neural membranes.

Keywords: Dopaminergic, glutamatergic, serotonergic, GABaergic neurotransmission, arachidonic and docosahexaenoic acids, phospholipase A_2, phospholipase C, cyclooxygenase, and prostaglandins.

Abbreviations

Phospholipase A_2 (PLA_2)
Phospholipase C (PLC)
Phospholipase D (PLD)
Docosahexaenoic acid (DHA)
Arachidonic acid (ARA)
a-Linolenic acid (ALA)
Linoleic acid (LA)
Blood brain barrier (BBB)
Fatty acid-binding protein (FABP)
Fatty acid-transport protein (FATP)
Acyl-CoA binding protein (ACBP)
Ethanolamine plasmalogen (PlsEtn)
Phosphatidylserine (PtdSer)
Phosphatidylcholine (PtdCho)
Diacylglycerol (DAG)
Phosphotidylinositol 4,5-bis phosphate (PtdIns(4,5)P_2)
Protein kinase C (PKC)
Vesicular monoamine transporter (VMAT-2)
Dopamine transporter (DAT)
5-Hydroxyindole acetic acid (5-HIAA)
Acetylcholine (ACh)
Peroxisome proliferator activated receptors (PPAR)
Sterol regulatory element binding protein (SREBP)
Carbohydrate response element binding protein (ChREBP)
2-Arachidonylglycerol (2-AG)

Mitogen-activated protein kinase (MAP kinase)
1-Methyl-4-phenyl-1,2,3,6-tetrahydropyridine (MPTP)
Alzheimer disease (AD)
Parkinson disease (PD)
5-Hydroxyindoleacetic acid (5-HIAA)
Tryptophan hydroxylase (TPH)
Monoamine oxidase A (MAOA)

1. Introduction

It is well known that the presynaptic neuron and the postsynaptic neuron are separated from each other by a small gap called synaptic cleft, which is a physical barrier for the electrical signal carried by one neuron to be transferred to another neuron. Neurotransmission is defined as communication between one neuron to another as accomplished by the movement of chemicals or electrical signals across a synapse. The function of neurotransmitter is to link the action potential of one neuron to another. Neurotransmission involves the release neurotransmitter, binding of neurotransmitter to its receptor and passage of message from one neuron to another. At the nerve terminal, neurotransmitter containing synaptic vesicle are docked at specific area of presynaptic plasma membrane called active zone. After maturation and priming, in response to action potential, neurotransmitter containing vesicles undergo Ca^{2+}-triggered exocytotic fusion to release neurotransmitters [1]. The neurotransmitter release mediates communication among neurons and this communication underlies virtually all brain functions, from sensory perception to learning and memory. In order to avoid depletion of synaptic vesicles and interruption of neurotransmitter, vesicles are reinternalized through endocytosis, recycled, and refilled with neurotransmitters in a rapid and precise manner. At the molecular level, interactions of neurotransmitter with neurotransmitter receptors result in enhancement of glycerophospholipid metabolism, which not only regulates the activities of receptors, membrane bound enzymes, ion-channels, and transporters, but also modulates many physicochemical properties of neural membranes such as fluidity, lateral pressure profile, bilayer thickness, and permeability [2-5].

Most of these processes are mediated by polyunsaturated fatty acids, which are exclusively esterified at the sn-2 position of glycerol moiety in glycerophospholipids [4,6-7]. Two major families of polyunsaturated fatty acids, namely n-3 (docosahexaenoic acid, DHA) and n-6 (arachidonic acid, ARA) (Figure 1) are known to occur in neural membrane glycerophospholipids. Because humans do not possess desaturases that insert either the n-3 or the n-6 double bonds, these fatty acids are derived from diet. DHA can be obtained from fish or from the precursor α-linolenic acid (ALA), which is found in nuts. ARA is found in vegetable oils (corn, sunflower seeds, safflower seeds, cottonseed, and soybeans), and its common precursor is dietary linoleic acid (LA), which is found in plant sources. Changes in the fatty acid composition of neural membranes affect endocytosis, exocytosis, membrane fusion, neurotransmitter uptake and release, and the activities of membrane-associated enzymes [8]. The purpose of this article is to discuss the modulation of various neurotransmitter receptors including, biogenic amine neurotransmitter receptors, by polyunsaturated fatty acids with the hope that this discussion would initiate more studies on modulation of neurotransmission by fatty acids, but also promote better understanding of therapeutic effects of n-3 fatty acids in neurodegenerative and neuropsychiatric disorders characterized by alterations in neurotransmitters and changes in neural membrane fatty acid composition.

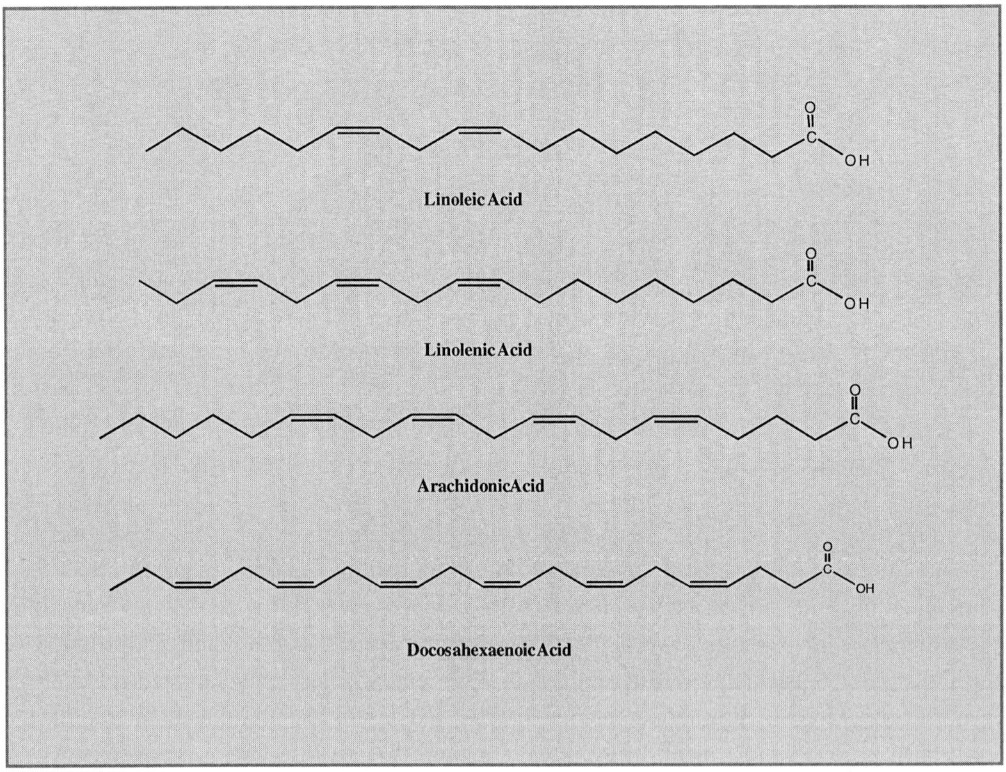

Figure 1. Structures of arachidonic and docosahexaenoic acids and their precursors.

2. Transport of Dietary Fatty Acids to the Brain

Human diet is enriched in fats that are enriched in polyunsaturated fatty acids. They are present in diet as triacylglycerols, which are hydrolyzed by lipases in gastric and intestinal lumen. DHA is released more slowly than ARA. Its intestinal absorption is delayed but not decreased [9]. These fatty acids are incorporated in chylomicron glycerophospholipids and transported to liver, where they associate themselves either to albumin or lipoproteins. From liver, these fatty acids are transported in the blood either bound to albumin or in the form of triacylglycerol associated with lipoproteins. Little is known about the mechanism associated with fatty acid entry into the brain tissue. Understanding the mechanisms by which fatty acids cross the blood brain barrier (BBB) and their utilization by neurons and glial cells is critical for knowing their role in brain development and function [10]. The transport of ARA and

DHA across BBB and other non-neural cellular membranes most likely occurs through passive diffusion. In addition, ARA and DHA transport is also facilitated by a number of membrane-bound and cytoplasmic proteins. These include membrane proteins fatty acid translocase (FAT/CD36), plasma membrane fatty acid-binding protein (FABPpm) and fatty acid-transport protein (FATP) [11]. For net ARA or DHA influx, these fatty acids must be desorbed from the inner leaflet of the neural membrane and should bind with FATP/$FABP_{pm}$ /acyl-CoA binding protein (ACBP) to prevent their repartitioning back into the membrane. When these events do not occur, fatty acids are repartitioned back into the outer leaflet and are desorbed back to the plasma to bind once again with serum albumin. ARA and DHA are incorporated and released from neural membrane glycerophospholipids through deacylation/reacylation cycle (Land's cycle), which involves the participation of acyl-CoA synthase, acyl-CoA lysophospholipid acyltransferase, and phospholipase A_2 [6].

3. Metabolism of Polyunsaturated Fatty Acids and its Effect on Neural Membranes Receptors

In neural membranes DHA mainly incorporates in ethanolamine plasmalogen (PlsEtn) and phosphatidylserine (PtdSer), whereas majority of ARA incorporates into phosphatidylcholine (PtdCho). From PlsEtn and PtdSer, DHA is released by the action of plasmalogen-selective PLA_2 and PtdSer hydrolyzing PLA_2, respectively [12-15]. The released DHA is metabolized into resolvins and neuroprotectins by 15-lipoxygenase [16-18]. These mediators are collectively known as docosanoids. They have anti-inflammatory and antiapoptotic properties and are involved in neuroprotection. In addition, docosanoids have antithrombotic, antiarrhythmic, hypolipidemic, and vasodilatory effects [4,16-18]. In contrast, ARA is released from neural membrane PtdCho by $cPLA_2$ and metabolized to prostaglandins, leukotrienes and thromboxanes. These metabolites are collectively known as eicosanoids. These metabolites have prothrombotic, proaggregatory, and proinflammatory properties. The levels of eicosanoids and docosanoids in neural and non-neural tissues are partly regulated by diet (see below).

In brain tissue, neurotransmitters not only carry information from one brain cell to another, but also transfer information from plasma membrane to nucleus via cytoplasm. This process is facilitated by other non-nuclear subcellular organelles, such as mitochondria and endoplasmic reticulum. Neuronal and glial cell membrane contains a series of receptors, which interact with neurotransmitters. Interactions of neurotransmitters (biogenic amine, glutamate, and cannabinoid) with their receptors result in enrichment of neural membrane glycerophospholipid metabolism through the stimulation of PLA_2 and PLC. Through cross talk between glutamate and biogenic amine receptors, the message is transferred from neural cell surface to the nucleus. This transfer of message is essential for maintaining neuronal and glial cell plasticity and growth. It should be noted that the products of one phospholipase modulate the activity of other phospholipase. Thus, the stimulation of PLC through biogenic amine receptor hydrolyzes PtdIns(4,5)P_2 and generates diacylglycerol (DAG). DAG activates protein kinase C (PKC), which in turn activates PLA_2 and PLD (Figure 2). PtdIns(4,5)P_2

stimulates PLA_2 and inhibits PLD [19-20]. Similarly, generation of ARA and eicosanoids activate PLD [21].

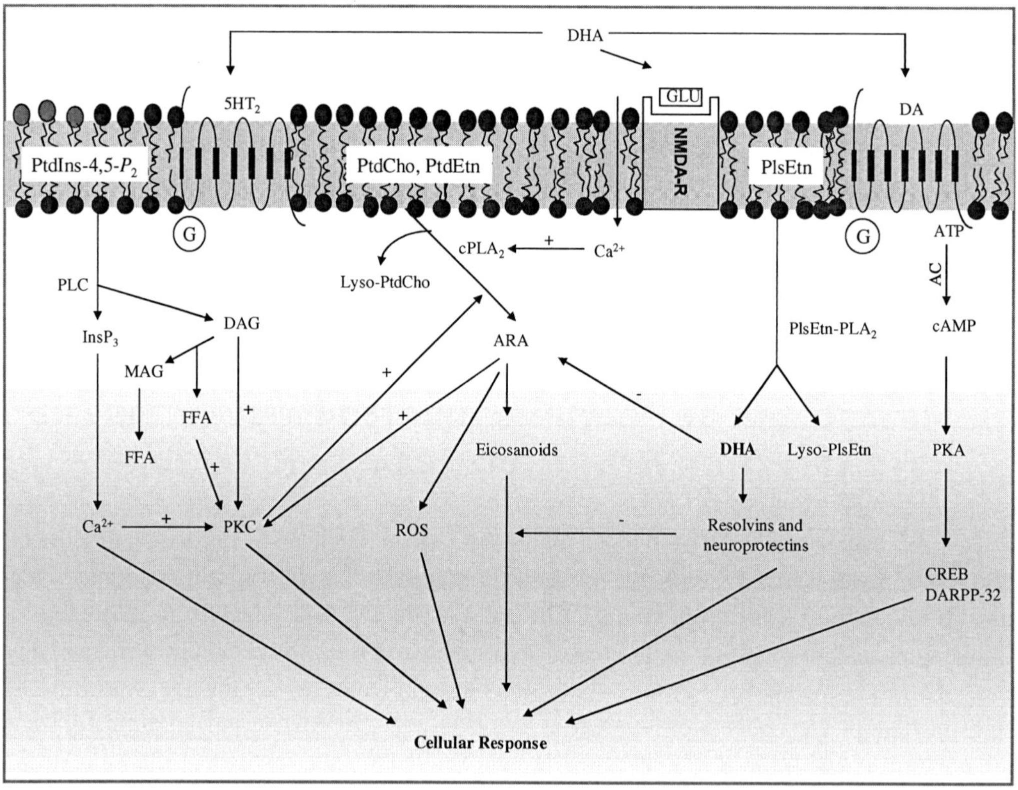

Figure 2. Degradation of neural membrane glycerophospholipid by receptor mediated phospholipases A_2 and C. Serotonin receptors ($5HT_2$) are linked to the degradation of PtdIns-4,5-P_2, dopamine (DA) receptors are linked to PLA_2 through G protein, dopamine receptors are also linked to Adenylyl cyclase and glutamate (GLU) receptors are linked to PLA_2 through G protein independent mechanisms. Phosphatidylinositol-4,5-bisphosphate (PtdIns-4,5-P_2); inositol 1,4,5-trisphosphate (Ins*P*3); phosphatidylcholine (PtdCho); phosphatidylethanolamine (PtdEtn); ethanolamine plasmalogen (PlsEtn); lyso-phosphatidylcholine (lyso-PtdCho); lyso-plasmalogen (lyso-PlsEtn); phospholipase C (PLC); cytosolic phospholipase A_2 ($cPLA_2$); plasmalogen-selective phospholipase A_2 (PlsEtn-PLA_2); Adenylyl cyclase (AC); protein kinase A (PKA); protein kinase C (PKC); arachidonic acid (ARA); diacylglycerol (DAG); monoacylglycerol (MAG); docosahexaenoic acid (DHA); reactive oxygen species (ROS); free fatty acid (FFA); cAMP response element binding protein (CREB); and cAMP-regulated phosphoprotein of 32,000 kDa (DARPP-32).

4. Modulation of Neurotransmission by Docosahexaenoic Acid

DHA modulates and facilitates neurotransmission through its effect on a variety of receptors including dopaminergic, **noradrenergic,** β-adrenergic glutamatergic, and serotonergic neurotransmitter receptors. In addition, DHA also modulates insulin, retinoid and TGF-β receptors in nuclear and non-nuclear compartments (Table 1) [22-25]. DHA modulates dopaminergic and serotonergic neurotransmission in rat frontal cortex, and a chronic depletion of DHA changes dopamine metabolism by altering dopamine receptor and transporters in the nucleus accumbens [22-23,26]. Dietary supplementation of DHA modulates the endogenous dopamine levels in the frontal cortex, and elevates binding of dopamine to D_2 receptors [23,26]. Free DHA is known to modulate PKC-mediated signal transduction pathways associated with synaptic plasticity [27-28]. During dopamine neurotransmission, increase in PKC activity not only alters D_2 receptor G-protein coupling and trafficking [29-30], but also induces changes in dopamine transporter (DAT) activity and trafficking [31-33], and filamentous actin-dependent dopamine vesicular trafficking and release efficacy [34]. These observations suggest that alterations in synaptic membrane DHA may have a significant impact on dopamine synaptic neurotransmission and plasticity. In addition, effects of DHA deficiency on amphetamine-induced locomotor activity and sensitization in DBA/2J mice indicate that DHA deficiency and elevation in the AA:DHA ratio augment amphetamine-induced sensitization in DBA/2J mice. This augmented response is associated with selective alterations in the mesolimbic dopamine pathway [35]. Quantitative real time-PCR studies show an overexpression of dopamine receptor genes and their products in the postnatal rat brain following maternal n-3 fatty acid dietary deficiency [36]. Dietary deficiency of maternal linolenic acid (LNA) results not only in a decrease in tyrosine hydroxylase (4.5-fold), but also in a down-regulation (7.5-fold) of the vesicular monoamine transporter (VMAT-2) and a depletion of VMAT-associated vesicles in the hippocampus of deficient offspring compared with adequately fed controls [37]. LNA deficiency has no effect on dopamine transporter (DAT) suggesting not only an increase in DAT/VMAT-2 ratio, but enhancement in the risk of damage of the dopaminergic terminal. These studies also show a robust increase in dopamine receptors (D_1 and D_2) levels in the cortex and_striatum possibly to compensate for the low levels of Dopamine in synaptic clefts. LNA deficiency also induces the activation of microglia and nuclear internalization of p65 NFkappaB [37]. Similarly, studies on n-3 fatty acid deficiency on serotonin neurotransmission in the adult female rat brain indicate that perinatal DHA-deficient rats show significantly lower prefrontal cortex serotonin (PFC 5-HT) content, significant greater 5-hydroxyindole acetic acid (5-HIAA) content and a significant greater 5-HIAA/5-HT ratio. Conversely, post-weaning DHA-deficient rats have significantly greater PFC 5-HT content, no change in 5-HIAA content, and a significantly smaller 5-HIAA/5-HT ratio (Mcnamara et al., 2008b). No alterations in PFC 5-HT or 5-HIAA content have been observed in post-pubertal DHA-deficient and ovariectomized rats [35]. Perinatal DHA-deficient rats also exhibit a significant reduction in midbrain tryptophan hydroxylase-2 mRNA expression. These observations support the view that there is a link between perinatal n-3 fatty acid deficiency and reduced central serotonin synthesis in adult female rats, which is independent of ovarian hormones including estrogen [35].

Studies on the effect of DHA deficiency on the central cholinergic neurotransmission system indicate that the cholinergic pathway is only modified in the hippocampus and not in the frontal cortex. The basal acetylcholine (ACh) release in the hippocampus of deficient rats is significantly higher than in controls compared to the KCl-mediated Ach release. The DHA deficiency also produces a 10% decrease in muscarinic receptor binding [38]. In contrast, acetylcholinesterase activity and the vesicular ACh transporter in both brain regions are not affected. Thus, DHA-deficient diet modulates cholinergic neurotransmission, probably via changes in the glycerophospholipid polyunsaturated fatty acid composition [38].

Table 1. Effect of docosahexaenoic acid deficiency on receptor activities in neural and non-neural tissues

Receptor/transporters	Effect	Reference
Dopamine receptor	Downregulation	[26]
Serotonin receptor	Downregulation	[130]
NMDA receptor	Downregulation	[98]
TGF-□receptor	Downregulation	[131]
TP receptor	Downregulation	[131]
GABA receptor	Downregulation	[132]
Insulin receptor	Downregulation	[133]
Nicotinic acetylcholine receptor	Upregulation	[134]
Cholinergic receptor	Upregulation	[38]
Retinoid receptor	Upregulation	[42]
PPAR receptor	Downregulated	[47]
Ca^{2+}-transport proteins	Stimulated	[40]
Carbohydrate response element	Stimulated	[46]
Glucose transporter	Decreased	[45]
GLT1 and EAAC1	Stimulated	[40]
GLAST	Inhibited	[40]

DHA also regulates glutamatergic neurotransmission. It is esterified at the sn-2 position of PtdSer, a glycerophospholipid essential for the NMDA receptor mediated processes [39]. DHA differentially modulates glutamate transporters (GLT1, GLAST, and EAAC1) through different mechanisms [40]. Thus, DHA stimulates GLT1 and EAAC1 through a mechanism that requires extracellular Ca^{2+}, CaM kinase II, and protein kinase C but not protein kinase A. In contrast, the inhibitory effect of DHA on GLAST does not require extracellular Ca^{2+}, and does not involve CaM kinase II [40]. Collective evidence suggests that DHA improves neuroplasticity, and synaptic transmission. These processes are closely associated with cognitive and behavioral changes in the brain tissue.

DHA is a ligand for receptors that are found in the nucleus. Thus, it modulates the retinoid X receptor (RXR) in brain [41-43]. RXR activation is closely associated with nuclear signaling and gene expression. RXR also serves as a fatty acid sensor *in vivo*. A number of proteins act as co-activators [44]. DHA supplementation not only modulates genes, but also mediates changes in the stability of the mRNA for several lipogenic enzymes. DHA deficit produces downregulation of brain glucose transporter expression, resulting in a decrease in glucose utilization in the cerebral cortex of DHA- deficient rats [45]. DHA also interacts with the peroxisome proliferator activated receptors (PPAR), sterol regulatory element binding protein (SREBP), and carbohydrate response element binding protein (ChREBP) [44, 46-47].

In addition, DHA modulates expression of genes for cytokines and their receptors, cell adhesion molecules, cytoskeleton proteins, and hormone receptors. The effect of DHA is sustained as long as DHA remains in the diet. Thus, DHA acts like a hormone to control the activity of key transcription factors, and modulates a variety of neural cell functions [44, 48-50]. Collective evidence suggests that DHA modulates the expression of a number of genes with such broad functions as DNA binding, transcriptional regulation, transport, cell adhesion, cell proliferation, raft formation, membrane protrusions and membrane localization. These effects, in turn, may significantly modify cell function, development and/or maturation. Collective evidence suggests that n-3 fatty acid-deficient diet through interactions with various neurotransmitter receptors produces symptoms of chronic locomotor hyperactivity, disturbance in melatonin rhythm, and striatal hyperdopaminergia. It is suggested that an n-3 fatty acid deficient diet lessens the melatonin rhythm, weakens endogenous functioning of the circadian clock, and plays a role in nocturnal sleep disturbances as occurs in attention deficit/hyperactivity disorder.

5. Interactions between Dopaminergic and Glutamatergic Neurotransmission

The prefrontal cortex mediates a range of neurotransmitter-mediated processes associated with planned behavior, which are controlled and directed according to shifting environmental demands. In prefrontal cortex, dopaminergic systems are modulated by glutamatergic neurotransmission, and this modulation is closely associated with the pathogenesis of schizophrenia [51]. Studies on the interaction between dopaminergic and glutamatergic neurotransmitter systems in prefrontal cortex indicate that application of the D_4 receptor agonist, PD168077 produces a reversible decrease in the NMDA receptor-mediated current, and this effect can be blocked by selective D_4 receptor antagonists [52]. Modulation of NMDA receptors in prefrontal cortex by dopamine D_4 receptors is associated with the inhibition of protein kinase A, activation of protein phosphatase 1, and the ensuing inhibition of active Ca^{2+}-calmodulin-dependent kinase II (CaMKII). Moreover, PD168077 not only inhibits the expression of NMDA receptor, but also triggers the internalization of NMDA receptor in a CaMKII-dependent manner. These studies identify a mechanistic link between D_4 dopamine receptors and NMDA glutamate receptors in prefrontal cortex pyramidal neurons [52, 53].

D_4 receptor activation also produces a persistent suppression of AMPA receptor-mediated synaptic transmission in prefrontal cortex interneurons. This effect of D_4 receptors on AMPA receptor-excitatory postsynaptic currents (EPSC) is mediated by a mechanism that is dependent on actin/myosin V motor-based transport of AMPA receptors. This transport is regulated by cofilin, a major actin depolymerizing factor. It is also demonstrated that the major cofilin-specific phosphatase Slingshot, which is activated by calcineurin downstream of D_4 signaling is necessary for the D_4 receptor regulation of glutamatergic transmission [54]. Collective evidence suggests that D_4 receptors, by using the unique calcineurin/Slingshot/cofilin signaling mechanism, regulate actin dynamics and AMPAR trafficking in PFC GABAergic interneurons. These observations provide a potential mechanism for D_4 receptors to control the excitatory synaptic strength in local-circuit neurons and GABAergic inhibition in the PFC network, which may be closely associated with the role of D_4 receptors in normal cognitive processes and mental disorders [54].

Modulation of the dopaminergic system by glutamatergic neurotransmission in schizophrenia is also supported through the involvement of neuregulin and dysbindin genes. These genes have an important impact in the glutamatergic system. Hypofunction of the glutamatergic system is mediated by NMDA receptor antagonism [55]. Kynurenic acid, an endogenous antagonist, not only blocks NMDA receptors, but also inhibits the nicotinergic acetylcholine receptor [51]. Thus, an increase in kynurenic acid induces cognitive deterioration due to the inhibition of NMDA receptor and also promotes psychotic symptoms because of the inhibition of nicotinergic acetylcholine receptor [51, 55].

The glutamatergic system is also linked to the cholinergic neurotransmitter system resulting in formation of the glutamatergic/aspartatergic-cholinergic circuit. This circuit is not only modulates learning and memory but facilitates cognition [56].

The cognitive abnormalities in Alzheimer disease may be caused by the loss of neurons in the cholinergic and glutamatergic neural systems and by irregular functioning of the surviving neurons in these two systems [57]. Glutamate and acetylcholine are ligands for the glutamatergic/aspartatergic-cholinergic neurotransmitter systems. Interactions among glutamatergic, dopaminergic, and cholinergic neurotransmissions are critical to normal learning and memory. All the above neurotransmitter systems are modulated by neural membrane DHA levels. Deficiency of this fatty acid suppresses glutamatergic, serotonergic, and dopaminergic neurotransmission, and supplementation of DHA restores normal neurotransmission [26]. In addition, DHA induced modifications of physical properties of the neuronal membrane modulate activities of proteins (receptors, transporters) that are enclosed in the membranes. In addition, DHA-mediated gene expression and/or transcription also contribute not only to the modulation of neurotransmission, but to behavioral changes. These changes may be related to clinical finding in patients suffering from neurological and psychiatric disorders.

6. Modulation of Neurotransmission by Arachidonic Acid

ARA and its enzymically oxidized products are associated with many physiological processes in the brain including synaptic signaling, neuronal firing, neurotransmitter release neuronal gene expression, cerebral blood flow, and sleep/wake cycle. As stated above, $cPLA_2$ releases ARA from glycerophospholipids. ARA is metabolized to prostaglandins, leukotrienes, and thromboxanes by cyclooxygenases, lipoxygenases, and epoxygenases [58]. ARA regulates glutamate uptake by astroglia. ARA also inhibits glutamate transport in several model systems. In the anterior uvea of the eye, another non-enzymically oxidized product of ARA, 8-isoprostane produces both excitatory and inhibitory effects on sympathetic neurotransmission in isolated mammalian iris ciliary bodies. Furthermore, stimulation of thromboxane receptors by isoprostane mediates the norepinephrine release from sympathetic nerves (4). In addition, other ARA containing endogenous molecules, such as virodhamine (O-arachidonylethanolamine), noladin, arachidonyldopamine, 2-arachidonylglycerol (2-AG) and arachidonylethanolamide (anandamide) have been reported to occur in brain. 2-AG and anandamide (Figure 3) are derived from the non-oxidative metabolism of ARA. 2-AG and anandamide are synthesized through two distinct pathways. Transfer of ARA from sn-1 position of 1,2-arachidonyl-PtdCho to the N-position of PtdEtn results in generation of 1-lyso-2-arachidonyl-PtdCho and N-arachidonyl-PtdEtn. This reaction is catalyzed by a Ca^{2+}-dependent, membrane-associated N-acyltransferase. 1-Lyso-arachidonyl-PtdCho is converted to 2-AG by PLC and N-arachidonyl-PtdEtn is transformed into anandamide by N-acylphosphatidylethanolamine specific PLD (NAPE-PLD), a member of the metallo-β-lactamase family, which specifically hydrolyzes N-acylphosphatidylethanolamine among glycerophospholipids, and appears to be constitutively active [59-60] (Figure 4). The recombinant enzyme hydrolyzed various N-acylphosphatidylethanolamines, including the anandamide precursor N-arachidonoylphosphatidylethanolamine at similar rates, but is inactive with phosphatidylcholine and phosphatidylethanolamine. NAPE-PLD is expressed in hippocampus, cortex, thalamus, hypothalamus, but the intensity of immunostaining in these regions is weaker than in mossy fibers. It is suggested that NAPE-PLD is expressed by specific populations of neurons in the brain and targeted to axonal processes. NAEs generated by NAPE-PLD in axons may act as anterograde synaptic signaling molecules that regulate the activity of postsynaptic neurons [59-60].

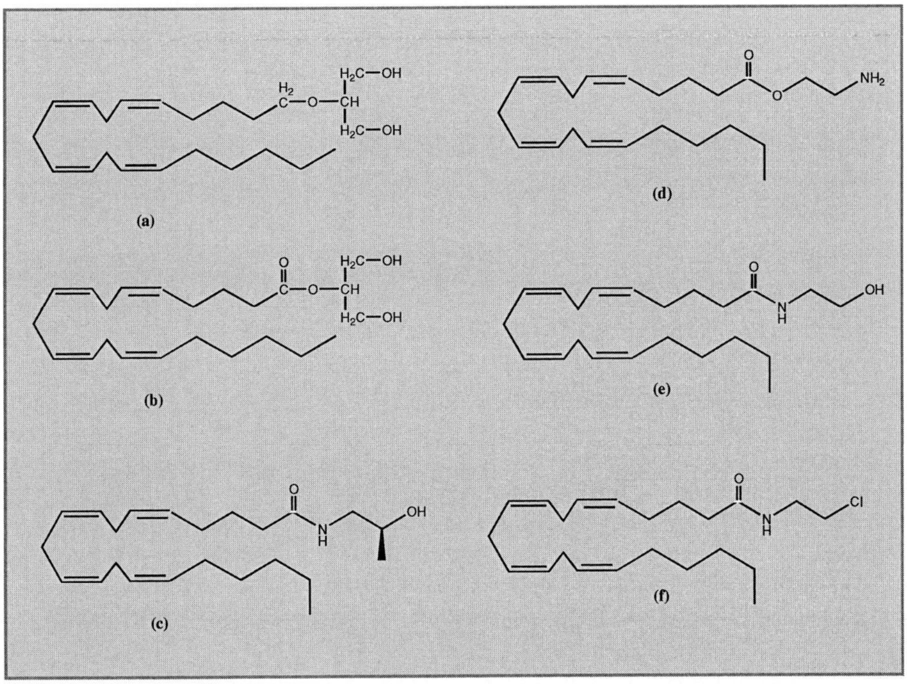

Figure 3. Structures of structures of some cannabinoid receptor agonists. Noladin ether (a); 2-arachidonylglycerol (b); Anandamide (c); virodhamine (d); and docosatetraenoyl ethanolamide (e); Arachidonyl-2-'chloroethylamide (f).

An alternative pathway for the generation of 2-AG involves the hydrolysis of 1,2-arachidonyl-PtdCho by PLC, followed by the action of DAG-lipase on 1-acyl-2-arachidonylglycerol. Unlike traditional neurotransmitters, ARA-derived endocannabinoids are not stored in vesicles. They are generated from ARA-containing glycerophospholipids within the neural membranes. Two types of cannabinoid receptor (CB_1 and CB_2) have been reported to occur in mammalian tissues. The CB_1 receptors are abundantly expressed in the brain, whereas CB_2 receptors are limited to lymphoid organs. 2-AG and anadamide nonselectively bind to both CB_1 and CB_2 receptors and act as neurotransmitter or neuromodulators in brain, immune and cardiovascular systems. Both receptors inhibit cAMP formation *via* Gi/o proteins, and activate mitogen-activated-protein kinase [61].

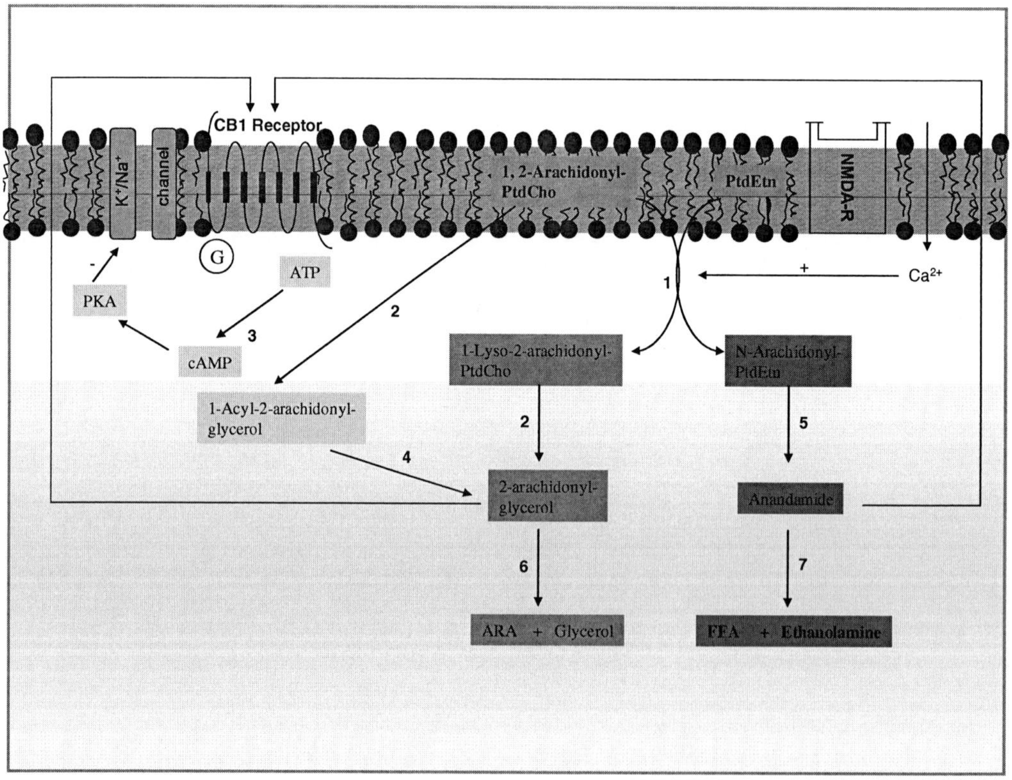

Figure 4. Pathway showing the generation of anandamide and 2-arachidonylglycerol in brain. Phosphatidylcholine (PtdCho); phosphatidylethanolamine (PtdEtn); protein kinase A (PKA); Cannabinoid receptor1 (CB_1) N-methyl-D-aspartate receptor (NMDA-R); N-acyltransferase (1); phospholipase C (2); Adenylyl cyclase (3); diacylglycerol lipase (4); N-acylphosphatidylethanolamine-specific phospholipase D (5); monoacylglycerol lipase (6); and amidase (7). Activation of receptors coupled to the phosphatidylinositol-specific phospholipase C and diacylglycerol lipase pathway leads to increases in 2-AG production.

Cannabinoid receptors may activate the extracellular signal-regulated kinase cascade through ceramide signaling. In addition, endocannabinoids also mediate their effects that are independent of cannabinoid receptors. Thus, in pharmacologically relevant concentrations, endocannabinoids modulate the functional properties of voltage-gated ion channels including

P/Q-type Ca^{2+} channels, Na^+ channels and inwardly rectifying K^+ channels, and ligand-gated ion channels such as 5-HT_3, and nicotinic ACh receptors [62]. Furthermore, functional modulations of ion-transporting membrane proteins such as transient potential receptor-class channels, gap junctions, and neurotransmitter transporters by endocannabinoids have also been demonstrated. Although the molecular mechanisms associated with these effects are unclear, but it is likely that these direct actions of endocannabinoids may be due to their lipophilic structures. It is proposed that additional molecular targets for endocannabinoids may also exist and that these targets represent important sites for cannabinoids to alter either the excitability of the neurons or the response of the neuronal systems [62]. Anandamide mediated signal is deactivated through a two step process, whereby the lipid mediator is transported into cells by a presently uncharacterized entity, and then degraded by the membrane-bound enzyme fatty acid amide hydrolase to produce ethanolamine and ARA [63-64].

A large body of preclinical data supports the view that either CB_2-selective agonists or CB_1 agonists acting at peripheral sites, or with limited CNS exposure, retard pain and neuroinflammation without side effects within the CNS [65]. CB_1 receptors involve Gi/o protein participation and are coupled the signal transduction pathways in presynaptic nerve terminal. The G proteins are linked to the cannabinomimetic stimulation of MAP kinase and adenylate cyclase, thereby modulating the generation of cAMP. In the absence of cannabinoids, PKA phosphorylates potassium channel protein and decreases outward potassium current, whereas in the presence of cannabinoids, the phosphorylation of the potassium channel protein increases outward current. Furthermore, cannabinoids also mediate closing of sodium channels. These processes are closely associated with the regulation of neurotransmetter release [66]. In addition, anandamide also acts as a ligand for vanilloid or capsaicin receptor [67]. Direct modulation of NMDA receptor activity by arachidonylethanolamide has also been reported [68]. Cannabinoids also mediate emotional responses by interacting with μ- and δ-opioid receptors. These interactions may produce either anti- or pro-anxiety effects which can explain bi-directional action of cannabinoids on anxiety [69].

Endocannabinoids not only mediate retrograde signaling, but also modulate synaptic transmission in various regions of the brain. Depolarization-mediated elevation of intracellular Ca^{2+} concentration causes endocannabinoid-mediated suppression of excitatory/inhibitory synaptic transmission. Activation of G(q/11)-coupled receptors including group I metabotropic glutamate receptors (mGluRs) also produces endocannabinoid-mediated suppression of synaptic transmission [70]. In the hippocampus, CB_1 receptors are expressed on axon terminals of GABAergic inhibitory interneurons [71]. A well-known effect of cannabinoids is the impairment of cognitive processes, including short-term memory formation, by altering hippocampal and neocortical functions reflected in network activity. Acting on presynaptically located G protein-coupled receptors in the hippocampus, cannabinoids modulate the release of neurotransmitter molecules. Activation of CB_1 receptors reduces GABA release from presynaptic terminals, thereby increasing the excitability of principal cells [71]. The molecular mechanism associated with the inhibition of GABA release is not fully understood. However, in brain endogenous cannabinoid, 2-AG is metabolized by COX-2, which oxygenates 2-AG to generate prostaglandin glyceryl esters

(PGE_2-G) and prostaglandin ethanolamides [72]. PGE_2-G is the main COX-2 oxidative metabolite of 2-AG. It increases the frequency of miniature inhibitory postsynaptic currents (mIPSCs). Similarly, PGD_2-G, $PGF_{2\alpha}$-G and PGD_2-EA also increase the frequency of mIPSCs, but PGE_2-EA and $PGF_{2\alpha}$-EA have no effect. PGE_2-G also enhances hippocampal glutamatergic synaptic transmission and mediates neuronal injury/death through caspase-3 activation. The actions of PGE_2-G are not mediated via a cannabinoid receptor 1. This increase is not blocked by SR141716, a CB_1 receptor antagonist suggesting other mechanisms are involved in this process. Detailed investigations indicate that PGE_2-G mediates its effect through ERK, p38 mitogen-activated protein kinase, InsP_3, and NF-κB signal transduction pathways. In addition, the PGE_2-G-mediated neurotoxicity is attenuated by blockade of the NMDA receptors [72-73]. Thus, endocannabinoid-derived prostaglandin-mediate responses are different from the corresponding arachidonic acid-derived prostaglandin. The effects of endocannabinoid-derived prostaglandins are not mediated through known prostaglandin receptors. It is also shown that the inhibition of COX-2 activity decreases inhibitory synaptic activity and augments depolarization-mediated suppression of inhibition (DSI), whereas the increase in COX-2 augments the synaptic transmission and abolishes DSI. Collectively, these studies suggest that COX-2-derived oxidative metabolites of cannabinoid exert opposite effects to ARA-derived prostaglandin on inhibitory synaptic transmission, and alterations in COX-2 activity may have significant impact on endocannabinoid signaling in hippocampal synaptic activity [73].

Activation of 5-HT_{2A} receptor facilitates the formation and release of 2-AG. This release of 2-AG partially depends on PtdIns-specific phospholipase C activation (Parrish, & Nichols, 2006). The production of DAG downstream of 5-HT_{2A} receptor-mediated by phospholipase D or phosphatidylcholine-specific phospholipase C activation does not contribute to 2-AG formation in NIH3T3-5HT_{2A} cells. DAG is hydrolyzed by diacylglycerol lipase to 2-AG, which may be further degraded by monoacylglycerol lipase to free ARA, a representative substrate of cyclooxygenase.

It remains to be seen whether or not DAG derived from the PLD-mediated hydrolysis of PtdCho can be converted to 2-AG or only inositol phosphoglyceride derived DAG is used for the synthesis of 2-AG. These studies support the view that there is a functional relationship between serotonin and endocannabinoid receptor. Thus, serotonin may act as regulators of endocannabinoid tone at excitatory synapses through the activation of phospholipase C-coupled G-protein coupled receptors [74].

Recently another arachidonic acid containing metabolite, arachidonoyl-L-serine (ARA-S) has been isolated from brain [75]. In contrast to anandamide, ARA-S very weakly interacts with cannabinoid CB_1 and CB_2 or vanilloid TRPV1 (transient receptor potential vanilloid 1) receptors. It produces endothelium-dependent vasodilation of rat isolated mesenteric arteries and abdominal aorta and stimulates phosphorylation of p44/42 mitogen-activated protein (MAP) kinase and protein kinase B/Akt in cultured endothelial cells [75]. ARA-S also suppresses LPS-mediated secretion of TNF-α in a murine macrophage cell line and in wild-type mice, as well as in mice deficient in CB_1 or CB_2 receptors. Many of these effects parallel to cannabidiol (Abn-CBD), a synthetic agonist of a putative novel cannabinoid-type receptor. Thus, ARA-S may be an endogenous agonist for this receptor [75]. In contrast, in non-neural cells, ARA-S directly activates large-conductance Ca^{2+}- and voltage-activated K^+ (BK(Ca) channels). These interactions do not involve cannabinoid receptors or cytosolic factors, but depends on the presence of membrane cholesterol [76]. It is proposed that direct BK(Ca) channel activation probably contributes to the endothelium-independent component of ARA-S-mediated mesenteric vasorelaxation [76].

7. Polyunsaturated Fatty Acids, Neurotransmission, and Neurodegenerative Diseases

Alterations in the expression of neurotransmitter receptors and levels of neurotransmitters along with the deficiency of DHA have been reported to occur in several neurodegenerative diseases including Alzheimer disease (AD) and Parkinson disease (PD). Thus, decrease in NMDA and AMPA receptors, along with glutamate transporter activity occurs in several brain regions from AD patients compared to brain tissue from age-matched control subjects [77-80]. In addition, a marked reduction in the expression of NR2A and NR2B subunit mRNA occurs in the hippocampus and entorhinal cortex in brains from AD patients [81]. These processes may induce changes in glutamate homeostasis in AD that may cause a major disturbance in Ca^{2+} homeostasis [82-83], and the activity of Ca^{2+}-dependent enzymes. Anomalous glutamatergic activity and changes in Ca^{2+} homeostasis leads to dementia and neuronal injury [56]. Similarly, marked alterations in dopamine levels have been reported to occur in PD. The abnormality in dopaminergic pathway in PD mainly occurs in the basal ganglia (caudate nucleus, putamen and globus pallidus). Degeneration of dopaminergic neurons takes place in substantia niagra, resulting in a marked depletion of nigrostriatal dopamine, changing striatal synaptic plasticity [84]. These neurons regulate body movement. Degeneration of dopaminergic neurons results in resting tremor, rigidity, bradykinesia, postural instability, and gait disturbances in PD patients.

Determination of DHA levels in plasma and hippocampus of AD patients show that there is a significant decrease in DHA level in AD compared to age-matched control [85-88]. This decrease correlates with upregulation of PlsEtn-PLA_2 [89], and a significant reduction in plasmalogen levels in AD patient [90-94]. Together these studies indicate that deficiency of DHA, decrease in plasmalogens, and alterations in glutamate homeostasis may not only result in the loss of synapses, but also impairment in memory and cognitive function in AD [95].

Dietary DHA incorporates into neuronal membranes and restores neuronal functions [96]. Chronic pre-administration of DHA retards β-amyloid-mediated impairment of an avoidance ability-related memory function in a rat model of AD [97] and prevents synaptic loss and dendritic pathology in the mouse model of AD [98]. DHA and its metabolite neuroprotectin D_1 inhibit apoptosis and β-amyloid secretion from aging brain cells [17, 99-100]. Collectively, these studies indicate that DHA is beneficial in preventing learning deficiencies in animal AD models [100-102] Similarly, DHA protects against other types of dementia [103].

In PD, Levodopa is commonly used for the treatment of PD. The effectiveness of levodopa therapy declines after years of treatment. So the long-term treatment of PD should be carefully managed. Long-term levodapa therapy results in dyskinesias (LIDs) [104]. DHA administration decreases the severity or delay the development of LIDs in a nonhuman primate model of PD [105]. Determination of brain fatty acid profile in cerebral cortex of 1-methyl-4-phenyl-1,2,3,6-tetrahydropyridine (MPTP) model of monkeys indicates that levodopa increases ARA content, but reduces DHA concentration and total n-3:n-6 fatty acids ratio compared to drug-naive MPTP treated animals [106]. Similarly, PD patients who experienced motor complications to levodopa have higher ARA concentrations in the cortex compared to controls and to levodopa-treated PD patients devoid of motor complications. Collective evidence suggests that levodopa treatment produces changes in brain fatty acid concentrations not only in a non-human primate model of Parkinsonism, but also in PD patients [106]. Polyunsaturated fatty acids are precursors for endogenous cannabinoids that are associated with the control of movement through the modulation of dopaminergic activity in basal ganglia [107], supporting the view that the intake of polyunsaturated fatty acids influences the risk of PD.

8. Alterations in Polyunsaturated Fatty Acids, Neurotransmission in Depression

Many neuropsychiatric disorders, including depression are characterized by dysregulation of dopaminergic, glutamatergic, serotonergic, cholinergic, and GABAergic neurotransmission at many different levels including the synthesis, storage, release, reuptake, and inactivation of neurotransmitters [108]. Major depression is accompanied by major alterations in neurotransmitters and a significant decrease in n-3 fatty acids and/or an increase of the n-6/n-3 ratio in brain and plasma and/or in the membranes of the red cells. The severity of depression is proportional to alterations in neurotransmetters (serotonin, norepinephrine and dopamine) levels in brain and low levels of n-3 fatty acids or the ratio n-3/n-6 fatty acids in plasma and red cell membranes.

Chronic psychoemotional stress of social defeats produces development of experimental anxious depression in male mice similar to this disorder in humans. Studies on the determination of serotonin and 5-hydroxyindole acetic acid (5-HIAA) levels, tryptophan hydroxylase (TPH) and monoamine oxidase (MAO) A activities, 5-HT1A-receptors in different brain areas at different stages of development of experimental model of stress disorder indicate that in the initial stage (3 days of social stress) is accompanied by increase of serotonin level in some brain areas [109]. At the onset of depression (10 days of social stress), a decrease in 5-HIAA levels is observed in the hippocampus, amygdala and nucleus accumbens. At the pronounced depression stage (20 days of stress), no differences in serotonin and 5-HIAA levels are seen in various regions of brain except hypothalamus [109]. However, increase number of 5-HT1A-receptors, decrease affinity in amygdale, and decrease in TPH and MAOA activities in hippocampus have been reported in depressive mice. These finding suggest that depression is accompanied by hypofunction of serotonergic system. Similar changes in dopamine and its metabolites are also observed in dopaminergic system in mice brain. It is suggested that dynamic changes of brain monoaminergic activities are closely associated with the development of anxious depression in animals [109]. Supplementation of diet with polyunsaturated fatty acids modulates serotonin, and serotonin transporter in the brain [110]. Thus, feeding diet enriched either in high saturated fat, n-6 fatty acid enriched, n-3 fatty acid enriched or low fat (control) for eight weeks results in modulation of serotonin metabolism and serotonin receptor binding, however the most pronounced effects are seen when diets are compared with the low fat control group [110]. Serotonin (5-HT_{2A}) receptor binding is increased in the caudate putamen, but reduced in the mammillary nucleus in high saturated fat and high n-6 fatty acid diet groups. Another serotonin receptor (5-HT_{2C}) binding is reduced in the mamillary nucleus of saturated fat group and reduced in prefrontal cortex of the n-6 fatty acid and n-3 fatty acid fed groups and serotonin transport protein (5-HTT) binding is reduced in the hippocampus in the n-6 fatty acid fed group. Overall, the n-6 fatty acid diet exerts the most influence on serotonin receptor and transporter binding [110]. These results may explain the importance of n-6 fatty acids in depression and beneficial effects of n-3 fatty acid in depression [110].

In addition, there is evidence for the involvement of cannabinoids in depression. Cannabinoids block long-term potentiation in the hippocampus, a process that underlies memory formation at the cellular level. Studies on comparison of hippocampal synapses in CB_1 knockout mice and wild-type controls indicate that CB_1 knockout mice exhibit a half-larger long-term potentiation than the controls indicating capacity for stronger synaptic connections related to memory formation. During two-trial recognition tests, CB_1 knockout mice retained memory for at least 48 h, whereas spatial learning in rats can be impaired by agonizing cannabinoid receptors [111]. A significant decrease in 2-AG levels and CB_1 receptor protein is observed in hippocampus of rats subjected to 3 weeks of chronic unpredictable stress [112]. These rats also display an impairment of reversal learning in Morris water maze test, which can be corrected by treating animals with cannabinoid agonist HU210 supporting the view that this effect is mediated by endocannabinoid signaling [112].

CB_1 receptors are also involved in emotional behavior [113]. CB_1 knock-out mice display more aggression, anxiety, and depressive like responses than wild-type mice when subjected to chronic and unpredictable stressors. Thus, compared to the wild-type controls, the CB_1 Knockout mice respond more aggressively during the resident-intruder test, present an increase in anxiogenic behavior in the light/dark box, and show a higher sensitivity to exhibit depressive like responses in the chronic unpredictable mild stress procedure [114]. There is a relationship between dietary polyunsaturated fatty acids consumption and elevated cellular levels of corresponding N-acylethanolamine, which may act as endogenous cannabinoids. Enrichment of polyunsaturated fatty acids in diets results in the generation of several biologically active acylethanolamines. Thus, levels of arachidonylethanolamine are increased by 4-fold, whereas docosahexaenoylethanolamine levels are increased 9- to 10-fold [115]. This increase in acylethanolamine may be important neurochemically because acylethanolamine with at least 20 carbons and three double bonds are known to bind to CB_1 receptor and produce their effect on brain tissue. I speculate that diet enriched in ARA may increase the chances of depression and downing of mood because of the generation of arachidonylethanolamine and 2-AG, which bind to CB_1 receptor with higher affinity than the docosahexaenoylethanolamine. DHA enriched diet, which facilitate the generation of docosahexaenoylethanolamine, may result in mood lifting thereby decreasing the risk of depression because it binds to CB_1 receptor less tenaciously than arachidonylethanolamine.

Beside depression beneficial effects of n-3 fatty acids, alterations in neurotransmitter levels are also observed in a variety of neuropsychiatric disorders, including schizophrenia, bipolar disorders, and attention-deficit/hyperactivity disorder and they are described elsewhere in detail [116].

9. Conclusion

Levels of neurotransmitters are modulated by diet. Thus, serotonin and dopamine are derived from nutrient precursors. Serotonin is derived from the tryptophan while dopamine is derived from the tyrosine [117]. Similarly, fatty acid composition of neural membrane glycerophospholipids is regulated by diet. As stated earlier, ARA and DHA are enriched in human brain. However, despite their abundance, these fatty acids cannot be synthesized de novo by human brain. They, or their precursors, must be ingested from dietary sources and transported to the brain [118]. These days, our diet has a ratio of n-6 to n-3 fatty acids of about 20:1. The Paleolithic diet on which human beings evolved, and lived for most of their existence, had a ratio of 1:1with high antioxidants [119-123]. Changes in eating habits, natural versus processed food, and agriculture development within the past 100 years have resulted in changes in the n-6 to n-3 ratio. n-6 fatty acid enriched diet promotes not only the pathogenesis of cardiovascular disease, but also neurodegenerative and neuropsychiatric diseases. In contrast, diet enriched in n-3 fatty acids exerts cardioprotective, immunosuppressive, neuroprotective, and psychoprotective effects [120,123]. Although, changes in fatty acid composition occur throughout the body, but neural membranes are the most sensitive targets for fatty acid alterations. Thus, in neural membranes incorporation of DHA and EPA not only promotes and maintains synaptic plasicity and induces neurogenesis in the brain circuits, but also reduces oxidative stress and neuroinflammation through the generation of lipoxins, neuroprotectins, and resolvins [8,16-18,116,124]. In addition, DHA and EPA also facilitate optimal membrane fluidity, promote appropriate neurotrophic support, and inhibit production of cytokines and expression of adhesion molecules [8,125]. In contrast, ARA initiates the generation of cytokines, especially IL-1β, the most potent proinflammatory cytokine that induces stress and anxiety-like behavior in rodents [126-127]. IL1β also stimulates neurotransmitter systems, thereby increasing the turnover of noradrenaline, serotonin, and dopamine [127-129]. ARA containing cannabinoids agonists through presynaptic mechanisms, modulate the release of several neurotransmitters implicated in the control of anxiety and depression. They not only suppress the outflow of glutamate in the hippocampus, but also in periaqueductal grey and amygdala. Cannabinoids mediate inhibitory effect on corticolimbic release of norepinephrine, dopamine, serotonin and anxiogenic neuropeptides corticotropin-releasing factor and cholecystokinin. They also interfere with GABAergic transmission in the amygdala, hippocampus and prefrontal cortex. Accumulating evidence suggests that n-3 and n-6 fatty acids play different role in the neuroendocrine-immune network, and the ratio of n-6 to n-3 fatty acid is an important dietary factor in reducing oxidative stress and inflammation and promoting neuroprotective and psychoprotective effects on human brain. These observations support the view that diet may also contribute to the etiology of neuropsychiatric and neurodegenerative diseases.

REFERENCES

[1] Li L, Chin L-S. The molecular machinery of synaptic vesicle exocytosis. *Cellular and Molecular Life Sciences,* 2003;60:942-960.

[2] Ikeda M, Kihara A, Igarashi Y. Lipid asymmetry of the eukaryotic plasma membrane: Functions and related enzymes. *Biological & Pharmaceutical Bulletin,* 2006;29:1542-1546.

[3] Yamaji-Hasegawa A, Tsujimoto M. Asymmetric distribution of phospholipids in biomembranes. *Biological & Pharmaceutical Bulletin,* 2006;29:1547-1553.

[4] Farooqui AA, Horrocks LA (2007) Glycerophospholipids in Brain: Phospholipases A_2 in Neurological Disorders. Springer New York

[5] Tillman TS, Cascio M. Effects of membrane lipids on ion channel structure and function. *Cell Biochemistry and Biophysics,* 2003;38:161-190.

[6] Farooqui AA, Horrocks LA, Farooqui T. Deacylation and reacylation of neural membrane glycerophospholipids. *Journal of Molecular Neuroscience,* 2000;14:123-135.

[7] Palsdottir H, Hunte C. Lipids in membrane protein structures. *Biochimica Biophysica Acta- Biomembranes*, 2004;1666:2-18.

[8] Farooqui T, Farooqui AA. Aging: an important factor for the pathogenesis of neurodegenerative diseases. *Molecular Mechanisms of Ageing and Development,* 2009;130:203-215.

[9] Bezard J, Blond JP, Bernard A, Clouet P. The metabolism and availability of essential fatty acids in animal and human tissues. *Reproduction Nutrition Development,*1994;34:539-568.

[10] Hamilton JA, Brunaldi K. A model for fatty acid transport into the brain. *Journal of Molecular Neuroscience,* 2007;33:12-17.

[11] Utsunomiya A, Owada Y, Yoshimoto T, Kondo H. Localization of mRNA for fatty acid transport protein in developing and mature brain of rats. *Brain Research. Molecular Brain Research*, 1997;46:217-222.

[12] Farooqui AA, Horrocks LA. Plasmalogens: workhorse lipids of membranes in normal and injured neurons and glia. *Neuroscientist,* 2001;7:232-245.

[13] Garcia MC, Kim HY. Mobilization of arachidonate and docosahexaenoate by stimulation of the 5-HT2A receptor in rat C6 glioma cells. *Brain Research,* 1997;768:43-48.

[14] Garcia MC, Ward G, Ma YC, Salem N Jr, Kim HY. Effect of docosahexaenoic acid on the synthesis of phosphatidylserine in rat brain in microsomes and C6 glioma cells. *Journal of Neurochemistry*, 1998;70:24-30.

[15] Strokin M, Sergeeva M, Reiser G. Docosahexaenoic acid and arachidonic acid release in rat brain astrocytes is mediated by two separate isoforms of phospholipase A_2 and is differently regulated by cyclic AMP and Ca^{2+}. *British Journal of Pharmacology,* 2003;139:1014-1022.

[16] Bazan NG. Lipid signaling in neural plasticity, brain repair, and neuroprotection. *Molecular Neurobiology*, 2005a;32:89-103.

[17] Bazan NG. Neuroprotectin D_1 (NPD_1): a DHA-derived mediator that protects brain and retina against cell injury-induced oxidative stress. *Brain Pathology*, 2005b;15:159-166.
[18] Serhan CN. Novel ω-3-derived local mediators in anti-inflammation and resolution. *Pharmacology & Therapeutics*, 2005;105:7-21.
[19] Hirabayashi T, Murayama T, Shimizu T. Regulatory mechanism and physiological role of cytosolic phospholipase A_2. *Biological & Pharmaceutical Bulletin,* 2004;27:1168-1173.
[20] Pasquaré SJ, Salvador GA, Giusto NM. Phospholipase D and phosphatidate phosphohydrolase activities in rat cerebellum during aging. *Lipids*, 2004;39:553-560.
[21] Klein J, Chalifa V, Lisocovitch M, Loffelholz K. Role of phospholipase D activation in nervous system physiology and pathophysiology. *Journal of Neurochemistry*, 1995;*65*:1445-1455.
[22] Zimmer L, Delion-Vancassel S, Durand G, Guilloteau D. Bodard S, Besnard JC, Chalon S. Modification of dopamine neurotransmission in the nucleus accumbens of rats deficient in n-3 polyunsaturated fatty acids. *Journal of Lipid Research*, 2000;41:32-40.
[23] Chalon S, Delion-Vancassel S, Belzung C, Guilloteau D, Leguisquet AM, Besnard JC, Durand G. Dietary fish oil affects monoaminergic neurotransmission and behavior in rats. *Journal of Nutrition*, 1998;128:2512-2519.
[24] Högyes E, Nyakas C, Kiliaan A, Farkas T, Penke B, Luiten PG. Neuroprotective effect of developmental docosahexaenoic acid supplement against excitotoxic brain damage in infant rats. *Neuroscience*, 2003;119:999-1012.
[25] Joardar A, Sen AK, Das SJ. Docosahexaenoic acid facilitates cell maturation and beta-adrenergic transmission in astrocytes. *Journal of Lipid Research*, 2006;47: 571-581.
[26] Chalon S. Omega-3 fatty acids and monoamine neurotransmission. *Prostaglandins, Leukotrienes, and Essent. Fatty Acids*, 2006;75:259-269.
[27] McNamara RK., Ostrander M, Abplanalp W, Richtand NM, Benoit S, Clegg D. Modulation of phosphoinositide-protein kinase C signal transduction by omega-3 fatty acids: Implications for the pathophysiology and treatment of recurrent neuropsychiatric illness. *Prostaglandins, Leukotrienes, and Essential Fatty Acids,* 2006;74: 237–257.
[28] McNamara RK, Able J, Liu Y, Jandacek R, Rider T, Tso P, Lipton JW. Omega-3 fatty acid deficiency during perinatal development increases serotonin turnover in the prefrontal cortex and decreases midbrain tryptophan hydroxylase-2 expression in adult female rats: Dissociation from estrogenic effects. *Journal of Psychiatric Research*, 2009;43:656-63.
[29] Di Marzo V, Vial D, Sokoloff P, Schwartz JC, Piomelli D. Selection of alternative G-mediated signaling pathways at the dopamine D_2 receptor by protein kinase C. *Journal of Neuroscience*, 1993;13:4846-4853.
[30] Namkung Y, Sibley DR. Protein kinase C mediates phosphorylation, desensitization, and trafficking of the D_2 dopamine receptor. *Journal of Biological Chemistry*, 2004;279:49533-49541.
[31] Johnson LA, Guptaroy B, Lund D, Shamban S, Gnegy ME. Regulation of amphetamine-stimulated dopamine efflux by protein kinase Cβ. *Journal of Biological Chemistry*, 2005;280:10914-10919.

[32] Loder MK, Melikian HE. The dopamine transporter constitutively internalizes and recycles in a protein kinase C-regulated manner in stably transfected PC12 cell lines. *Journal of Biological Chemistry*, 2003;278:22168-22174.

[33] Miranda M, Dionne KR, Sorkina T, Sorkin A. Three ubiquitin conjugation sites in the amino terminus of the dopamine transporter mediate protein kinase C-dependent endocytosis of the transporter. *Molecular Biology of the Cell*, 2007;8: 313-323.

[34] Shoji-Kasai Y, Itakura M, Kataoka M, Yamamori S, Takahashi M. 02) Protein kinase C-mediated translocation of secretory vesicles to plasma membrane and enhancement of neurotransmitter release from PC12 cells. *European Journal of Neuroscience*, 2002;15:1390-1394.

[35] McNamara RK, Sullivan J, Richtand NM, Jandacek R, Rider T., Tso, P. et al. (2008b) Omega-3 fatty acid deficiency augments amphetamine-induced behavioral sensitization in adult DBA/2J mice: relationship with ventral striatum dopamine concentrations. *Synapse,* 2008;62:725-735.

[36] Kuperstein, F., Yakubov, E., Dinerman, P., Gil, S., Eylam, R., Salem, N. Jr, Yavin E. Overexpression of dopamine receptor genes and their products in the postnatal rat brain following maternal n-3 fatty acid dietary deficiency. *Journal of Neurochemistry*, 2005;95:1550-1562.

[37] Kuperstein F, Eilam R., Yavin, E. Altered expression of key dopaminergic regulatory proteins in the postnatal brain following perinatal n-3 fatty acid dietary deficiency. *Journal of Neurochemistry*, 2008;106:662-671.

[38] Aid S, Vancassel S, Poumes-Ballihaut C, Chalon S, Guesnet P, Lavialle M. Effect of a diet-induced n-3 PUFA depletion on cholinergic parameters in the rat hippocampus. *Journal of Lipid Research*, 2003;44:1545-1551.

[39] Calon F, Lim GP, Yang F, Morihara T, Teter B., Ubeda O, Rostaing P, Triller A, Salem N Jr, Ashe KH, Frautschy SA, Cole GM. Docosahexaenoic acid protects from dendritic pathology in an Alzheimer's disease mouse model. *Neuron*, 2004;43:633-645.

[40] Berry CB, Hayes D, Murphy A, Wiessner M, Rauen T, McBean GJ. Differential modulation of the glutamate transporters GLT1, GLAST and EAAC1 by docosahexaenoic acid. *Brain Research*, 2005;1037:123-133.

[41] Lengqvist J, Mata de Urquiza A, Bergman AC, Willson TM, Sjövall J, Perlmann T, Griffiths WJ. Polyunsaturated fatty acids including docosahexaenoic and arachidonic acid bind to the retinoid X receptor α ligand-binding domain. *Molecular and Cellular Proteomics*, 2004;3:692-703.

[42] Farooqui AA, Antony P, Ong WY, Horrocks LA, Freysz L. Retinoic acid-mediated phospholipase A_2 signaling in the nucleus. *Brain Research Review*, 2004;45:179-195.

[43] de Urquiza AM, Liu S, Sjöberg M, Zetterström RH, Griffiths W, Sjövall J, Perlmann T. Docosahexaenoic acid, a ligand for the retinoid X receptor in mouse brain. *Science*, 2000;290:2140-2144.

[44] Jump DB. The biochemistry of *n*-3 polyunsaturated fatty acids. *Journal of Biological Chemistry*, 2002a;277:8755-8758.

[45] Pifferi F, Roux F, Langelier B, Alessandri JM, Vancassel S, Jouin M, Lavialle M, Guesnet P. (n-3) polyunsaturated fatty acid deficiency reduces the expression of both

isoforms of the brain glucose transporter GLUT1 in rats. *Journal of Nutrition*, 2005;135:2241-2246.

[46] Clarke SD. Polyunsaturated fatty acid regulation of gene transcription: a mechanism to improve energy balance and insulin resistance. *British Journal of Nutrition*, 2000;83:*Suppl 1*, S59-S66.

[47] Nakamura MT, Cheon Y, Li Y, Nara TY. Mechanisms of regulation of gene expression by fatty acids. *Lipids*, 2004;39:1077-1083.

[48] Jump DB. Dietary polyunsaturated fatty acids and regulation of gene transcription. *Current Opinion in Lipidology*, 2002b;13:155-164.

[49] Jump DB. Fatty acid regulation of gene transcription. *Critical Reviews in Clinical Laboratory Sciences*, 2004;41:41-78.

[50] Valentine RC, Valentine DL. Omega-3 fatty acids in cellular membranes: a unified concept. *Progress in Lipid Research*, 2004;43:383-402.

[51] Muller N, Schwarz M. Schizophrenia as an inflammation-mediated dysbalance of glutamatergic neurotransmission. *Neurotoxicity Research*, 2006;10:131-148.

[52] Wang X, Zhong P, Gu Z, Yan Z. Regulation of NMDA receptors by dopamine D4 signaling in prefrontal cortex. *Journal of Neuroscience*, 2003;23:9852-9861.

[53] Tseng KY, O'donnell P. Dopamine-glutamate interactions controlling prefrontal cortical pyramidal cell excitability involve multiple signaling mechanisms. *Journal of Neuroscience*, 2004;24:5131-5139.

[54] Yuen EY, Yan Z. Dopamine D_4 receptors regulate AMPA receptor trafficking and glutamatergic transmission in GABAergic interneurons of prefrontal cortex. *Journal of Neuroscience*, 2009;29:550-562.

[55] MacDonald AWI, Chafee MV. Translational and developmental perspective on N-methyl-D-aspartate synaptic deficits in schizophrenia. Development *and Psychopathology*, 2006;518: 853-876.

[56] Wenk GL. Neuropathologic changes in Alzheimer's disease: potential targets for treatment. *Journal of Clinical Psychiatry*, 2006;67*: Suppl 3*. 3-7.

[57] Riederer P, Hoyer S. From benefit to damage. Glutamate and advanced glycation end products in Alzheimer brain. *Journal of Neural Transmission*, 2006;113:1671-1677.

[58] Phillis JW, Horrocks LA, Farooqui AA. Cyclooxygenases, lipoxygenases, and epoxygenases in CNS: their role and involvement in neurological disorders. *Brain Research Reviews*, 2006;52:201-243.

[59] Di Marzo V, De Petrocellis L, Sugiura T, Waku K. Potential biosynthetic connections between the two cannabimimetic eicosanoids, anandamide and 2-arachidonoyl-glycerol, in mouse neuroblastoma cells. *Biochemical and Biophysical Research Communications,* 1996;227:281-288.

[60] Ueda N, Okamoto Y, Tsuboi K. Endocannabinoid-related enzymes as drug targets with special reference to N-acylphosphatidylethanolamine-hydrolyzing phospholipase D. *Current Medicinal Chemistry*, 2005;12:1413-1422.

[61] Childers, SR, Breivogel, CS. Cannabis and endogenous cannabinoid systems. *Drug Alcohol Dependence*, 1998;51:173–187.

[62] Oz M. Receptor-independent effects of endocannabinoids on ion channels. *Current Pharmaceutical Design*, 2006;12:227-239.

[63] Desarnaud F, Cadas H, Piomelli D. Anandamide amidohydrolase activity in rat brain microsomes. Identification and partial characterization. *Journal of Biological Chemistry*, 1995;270:6030-6035.

[64] Mangieri RA, Piomelli D. Enhancement of endocannabinoid signaling and the pharmacotherapy of depression. *Pharmacological Research*,2007;56:360-366.

[65] Cheng Y, Hitchcock SA. Targeting cannabinoid agonists for inflammatory and neuropathic pain. *Expert Opinion on Investigational Drugs*,2007;16:951-965.

[66] Yamamoto T, Takada K. Role of cannabinoid receptor in the brain as it relates to drug reward. *Japanese Journal of Pharmacology*,2000;84:229-236.

[67] Olah Z, Karai L, Iadarola MJ. Anandamide activates vanilloid receptor 1 (VR1) at acidic pH in dorsal root ganglia neurons and cells ectopically expressing VR1. *Journal of Biological Chemistry*, 2001;276:31163-31170.

[68] Hampson AJ, Bornheim LM, Scanziani M, Yost CS, Gray AT, Hansen BM, Leonoudakis DJ, Bickler PE. Dual effects of anandamide on NMDA receptor-mediated responses and neurotransmission. *Journal of Neurochemistry*,1998;70:671-676.

[69] Berrendero F, Maldonado R. Involvement of the opioid system in the anxiolytic like effects induced by delta (9)-tetrahydrocannabinol. *Psychopharmacology*, 2002;163:111–117.

[70] Maejima T, Oka S, Hashimotodani Y, Ohno-Shosaku T, Aiba A, Wu D, Waku K, Sugiura T, Kano M. Synaptically driven endocannabinoid release requires Ca^{2+}-assisted metabotropic glutamate receptor subtype 1 to phospholipase Cβ-4 signaling cascade in the cerebellum. *Journal of Neuroscience*,2005;25:6826-6835.

[71] Hájos N, Freund TF. Distinct cannabinoid sensitive receptors regulate hippocampal excitation and inhibition. *Chemistry and Physics of Lipids*, 2002;121:73-82.

[72] Sang N, Zhang J, Chen C. COX-2 oxidative metabolite of endocannabinoid 2-AG enhances excitatory glutamatergic synaptic transmission and induces neurotoxicity. *Journal of Neurochemistry*, 2007;102:1966-1977.

[73] Sang N, Zhang J, Chen C. PGE_2 glycerol ester, a COX-2 oxidative metabolite of 2-arachidonoyl glycerol, modulates inhibitory synaptic transmission in mouse hippocampal neurons. *Journal of Physiology*, 2006;572(Pt 3); 735-745.

[74] Parrish JC, Nichols DE. Serotonin 5-HT_{2A} receptor activation induces 2-arachidonoylglycerol release through a phospholipase C-dependent mechanism. *Journal of Neurochemistry*,2006;99:1164-1175.

[75] Milman G, Maor Y, Abu-Lafi S, Horowitz M, Gallily R, Batkai S, Mo FM, Offertaler L, Pacher P, Kunos G, Mechoulam R. N-arachidonoyl L-serine, an endocannabinoid-like brain constituent with vasodilatory properties. *Proceedings of the National Academy of Sciences USA*,2006;103:2428-2433.

[76] Godlewski G, Offertáler L, Osei-Hyiaman D, Mo FM., Harvey-White J, Liu J, Davis MI, Zhang L, Razdan RK, Milman G, Pacher P, Mukhopadhyay P, Lovinger DM, Kunos G. The endogenous brain constituent N-arachidonoyl L-serine is an activator of large conductance $Ca2^{+}$-activated K^{+} channels. *Journal of Pharmacology and Experimental Therapeutics*, 2009;*328*:351-361.

[77] Greenamyre JT, Penney JB, D'Amato CJ, Young AB. Dementia of the Alzheimer's type: Changes in hippocampal L-[3H]glutamate binding. *Journal of Neurochemistry*, 1987;48:543-551.

[78] Penney JB, Maragos WF, Greenamyre JT, Debowey DL, Hollingsworth Z, Young AB. Excitatory amino acid binding sites in the hippocampal region of Alzheimer's disease and other dementias. *Journal of Neurolology, Neurosurgery, and Psychiatry*, 1990;53:314-320.

[79] Dewar D, Chalmers DT, Shand A, Graham DI, McCulloch J. Selective reduction of quisqualate (AMPA) receptors in Alzheimer cerebellum. *Annals of Neurology*, 1990;28:805-810.

[80] Masliah E, Alford M., DeTeresa R, Mallory M, Hansen L. Deficient glutamate transport is associated with neurodegeneration in Alzheimer's disease. *Annals of Neurology*, 1996;40:759-766.

[81] Bi H, Sze CI. *N*-methyl-D-aspartate receptor subunit NR2A and NR2B messenger RNA levels are altered in the hippocampus and entorhinal cortex in Alzheimer's disease. *Journal of Neurological Sciences*, 2002;200:11-18.

[82] Peterson C, Goldman JE. Alterations in calcium content and biochemical processes in cultured skin fibroblasts from aged and Alzheimer donors. *Proceedings of the National Academy of Sciences USA,* 1986;*83*:2758-2762.

[83] Peterson C, Gibson GE, Blass JP. Altered calcium uptake in cultured skin fibroblasts from patients with Alzheimer's disease. *New England Journal of Medicine,* 1985;312:*1063-1065.*

[84] Beal MF. Mitochondrial dysfunction in neurodegenerative diseases. *Biochimica et Biophysica Acta*, 1998;1366, 211-223.

[85] Conquer JA, Tierney MC, Zecevic J, Bettger WJ, Fisher RH. Fatty acid analysis of blood plasma of patients with Alzheimer's disease, other types of dementia, and cognitive impairment. *Lipids*, 2000;35:1305-1312.

[86] Tully AM, Roche HM, Doyle R, Fallon C, Bruce I, Lawlor B, Coakley D, Gibney MJ. Low serum cholesteryl ester-docosahexaenoic acid levels in Alzheimer's disease: a case-control study. *British Journal of Nutrition*,2003;89:483-489.

[87] Söderberg M, Edlund C, Kristensson K, Dallner G. Fatty acid composition of brain phospholipids in aging and in Alzheimer's disease. *Lipids*.1991;26:421-425.

[88] Schaefer EJ, Bongard V, Beiser AS, Lamon-Fava S, Robins SJ, Au R, Tucker KL, Kyle DJ, Wilson PW, Wolf PA. Plasma phosphatidylcholine docosahexaenoic acid content and risk of dementia and Alzheimer disease: the Framingham Heart Study. *Archives of Neurology*,2006;63:1545-1550.

[89] Farooqui AA, Ong WY, Horrocks LA. Inhibitors of brain phospholipase A_2 activity: Their neuropharmacological effects and therapeutic importance for the treatment of neurologic disorders. *Pharmacological Reviews*,2006;58:591-620.

[90] Söderberg M, Edlund C, Kristensson K, Dallner G. Lipid compositions of different regions of the human brain during aging. *Journal of Neurochemistry,*1990;54:415-423.

[91] Söderberg, M, Edlund C, Kristensson K, Dallner G. Fatty acid composition of brain phospholipids in aging and in Alzheimer's disease. *Lipids*, 1991;26:421-425.

[92] Wells K, Farooqui AA, Liss L, Horrocks LA. Neural membrane phospholipids in Alzheimer disease. *Neurochemical Research*,1995; 20:1329-1333.

[93] Guan ZZ, Wang YA, Cairns NJ, Lantos PL, Dallner G, Sindelar PJ. Decrease and structural modifications of phosphatidylethanolamine plasmalogen in the brain with Alzheimer disease. *Journal of Neuropathology and Experimental Neurology*,1999;58:740-747.

[94] Han XL, Holtzman DM, McKeel DW Jr. Plasmalogen deficiency in early Alzheimer's disease subjects and in animal models: molecular characterization using electrospray ionization mass spectrometry. *Journal of Neurochemistry*, 2001;77:1168-1180.

[95] Wishart TM, Parson SH, Gillingwater TH. Synaptic vulnerability in neurodegenerative disease. *Journal of Neuropathology and Experimental Neurology*, 2006;65:733-739.

[96] Kalmijn S, van Boxtel MPJ, Ocke M, Verschuren WMM, Kromhout D, Launer LJ. Dietary intake of fatty acids and fish in relation to cognitive performance at middle age. *Neurology*,2004;62:275-280.

[97] Hashimoto M, Hossain S, Agdul H, Shido, O. Docosahexaenoic acid-induced amelioration on impairment of memory learning in amyloid beta-infused rats relates to the decreases of amyloid β and cholesterol levels in detergent-insoluble membrane fractions. *Biochimica Biophysica Acta*, 2005;1738:91-98.

[98] Calon F, Lim GP, Morihara T, Yang FS, Ubeda O, Salem NJ Jr, Frautschy SA, Cole GM. Dietary n-3 polyunsaturated fatty acid depletion activates caspases and decreases NMDA receptors in the brain of a transgenic mouse model of Alzheimer's disease. *European Journal of Neuroscience*,2005;22:617-626.

[99] Lukiw WJ, Cui JG, Marcheselli VL, Bodker M, Botkjaer A, Gotlinger K, Serhan CN, Bazan NG. A role for docosahexaenoic acid-derived neuroprotectin D1 in neural cell survival and Alzheimer disease. *Journal of Clinical Investigation*, 2005;115:2774-2783.

[100] Lukiw WJ, Bazan NG. Survival signalling in Alzheimer's disease. *Biochemical Society Transactions*, 2006;34:1277-1282.

[101] Cole GM, Frautschy SA. Docosahexaenoic acid protects from amyloid and dendritic pathology in an Alzheimer's disease mouse model. *Nutrition Health*,2006;18:249-259.

[102] Freund-Levi Y, Eriksdotter-Jönhagen M, Cederholm T, Basun H, Faxén-Irving, G, Garlind A, Vedin I, Vessby B, Wahlund LO, Palmblad J. Omega-3 fatty acid treatment in 174 patients with mild to moderate Alzheimer disease: OmegAD study: a randomized double-blind trial. *Archives of Neurology*,2006;63:1402-1408.

[103] Lim WS, Gammack JK, Van Niekerk J, Dangour AD. Omega 3 fatty acid for the prevention of dementia. *Cochrane Database of Systematic Reviews*, 2006;1: CD005379.

[104] Schneider JC, Gonczi H, Decamp E. Development of levodopa-induced dyskinesias in parkinsonian monkeys may depend upon rate of symptom onset and/or duration of symptoms. *Brain Research*, 2003;990:38-44.

[105] Samadi P, Gregoire L, Rouillard C, Bedard PJ, Di Paolo T, Levesque D. Docosahexaenoic acid reduces levodopa-induced dyskinesias in 1-methyl-4-phenyl-1,2,3,6-tetrahydropyridine monkeys. *Annals of Neurology*, 2006;59:282-288.

[106] Julien C, Berthiaume L, Hadi-Tahar A, Rajput AH, Bedard PJ, Di Paolot T, Julien P, Calon F. Postmortem brain fatty acid profile of levodopa-treated Parkinson disease patients and parkinsonian monkeys. *Neurochemistry International*, 2006;48:404-414.

[107] Giuffrida A, Piomelli D. The endocannabinoid system: a physiological perspective on its role in psychomotor control. *Chemistry and Physics of Lipids*, 2000;108:151-158.

[108] Harrison PJ. Neurochemical alterations in schizophrenia affecting the putative receptor targets of atypical antipsychotics. Focus on dopamine (D_1, D_3, D_4) and 5-HT_2a receptors. *British Journal of Psychiatry,* 1999;174 Supplement:12-22.

[109] Avgustinovich DF, Alekseenko OV, Bakshtanovskaia IV, Koriakina LA, Lipina TV, Tenditnik MV, Bondar' NP, Kovalenko IL, Kudriavtseva NN. Dynamic changes of brain serotonergic and dopaminergic activities during development of anxious depression: experimental study. *Usp Fiziol Nauk,* 2004;35:19-40.

[110] Du Bois TM, Deng C, Huang XF. Membrane phospholipid composition, alterations in neurotransmitter systems and schizophrenia. Progress in *Neuro-Psychopharmacology and Biological Psychiatry,* 2005;29:878-888.

[111] Reibaud M, Obinu MC, Ledent C, Parmentier M, Bohme G, Imperato A. Enhancement of memory in cannabinoid CB1 receptor knock-out mice. *European Journal of Pharmacology*,1999;379:R1-R2.

[112] Hill MN, Ho WS, Meier SE, Gorzalka BB, Hillard C.J. Chronic corticosterone treatment increases the endocannabinoid 2-arachidonylglycerol in the rat amygdala. *European Journal of Pharmacology*, 2005;528:99-102.

[113] Martin BR. Identification of the endogenous cannabinoid system through integrative pharmacological approaches. *Journal of Pharmacology and Experimental Therapeutics*,2002;301:790-796.

[114] Martin M, Ledent C, Parmentier M, Maldonado R, Valverde O. Involvement of CB_1 cannabinoid receptors in emotional behaviour. *Psychopharmacology*, 2002;159:379-387.

[115] Berger A, Crozier G, Bisogno T, Cavaliere P, Innis S, Di Marzo V. Anandamide and diet: inclusion of dietary arachidonate and docosahexaenoate leads to increased brain levels of the corresponding N-acylethanolamines in piglets. *Proceedings of the National Academy of Sciences USA,* 2001;98:6402-6406.

[116] Farooqui A.A. (2009). Hot Topics in Neural Membrane Lipidology. Springer New York.

[117] Fernstrom JD. Dietary amino acids and brain function. *American Dietetic Association*, 1994;94:71–77.

[118] Horrocks LA, Farooqui AA. Docosahexaenoic acid in the diet: its importance in maintenance and restoration of neural membrane function. *Prostaglandins, Leukotrienes, and Essential Fatty Acids,* 2004;70;361-372.

[119] Simopoulos AP. Omega-3 fatty acids and antioxidants in edible wild plants. *Biological Research*, 2004;37:263-277.

[120] Simopoulos AP. Evolutionary aspects of diet, the omega-6/omega-3 ratio and genetic variation: nutritional implications for chronic diseases. *Biomedicine and Pharmacotherapy*, 2006;60:502-507.

[121] Cordain Eaton SB, Sebastian A, Mann N, Lindeberg S, Watkins BA, O'Keefe JH, Brand-Miller J. Origins and evolution of the Western diet: health implications for the 21st century. *American Journal of Clinical Nutrition*, 2005;81:341-354.

[122] Weylandt KH, King JX. Rethinking lipid mediators. *Lancet,* 2005;336:618-620.

[123] Farooqui AA, Ong WY, Horrocks LA, Chen P, Farooqui T. Comparison of biochemical effects of statins and fish oil in brain: The battle of the titans. *Brain Research Reviews*, 2007;956:443-471.

[124] Farooqui A.A. (2009) Beneficial Effects of Fish Oil on Human Brain. Springer New York

[125] Farooqui AA, Horrocks LA, Farooqui T. (2009) Choline and ethanolamine glycerophospholipids. In: Tettamanti G, Goracci G, editors. pp. 71-89, Handbook of Neurochemistry. Springer New York.

[126] Song C. The effect of thymectomy and IL-1 on memory: implications for the relationship between immunity and depression. *Brain, Behavior and Immunity*, 2002;16:557-568.

[127] Song C, Manku MS, Horrobin DF. Long-chain polyunsaturated fatty acids modulate interleukin-1β-induced changes in behavior, monoaminergic neurotransmitters, and brain inflammation in rats. *Journal of Nutrition*, 2008;138, 954-963.

[128] Song C, Li X, Leonard BE, Horrobin DF. Effects of dietary n-3 or n-6 fatty acids on interleukin-1beta-induced anxiety, stress, and inflammatory responses in rats. *Journal of Lipid Researh*, 2003;44:1984-1991.

[129] Kodas E, Galineau L, Bodard S, Vancassel S, Guilloteau D, Besnard JC, Chalon S. Serotoninergic neurotransmission is affected by n-3 polyunsaturated fatty acids in the rat. *Journal of Neurochemistry*, 2004;89:695-702.

[130] Hirafuji M, Machida T, Hamaue N, Minami M. Cardiovascular protective effects of n-3 polyunsaturated fatty acids with special emphasis on docosahexaenoic acid. *Journal of Pharmacological Sciences*, 2003;92:308-316.

[131] Hamano H, Nabekura J, Nishikawa M, Ogawa T. Docosahexaenoic acid reduces GABA response in substantia nigra neuron of rat. *Journal of Neurophysiology*, 1996;75:1264-1270.

[132] Delarue J, LeFoll C, Corporeau C, Lucas D. N-3 long chain polyunsaturated fatty acids: a nutritional tool to prevent insulin resistance associated to type 2 diabetes and obesity? *Reproduction Nutrition Development,*2004;44:289-299.

[133] Nah SY, Kim DH, Rhim H. Ginsenosides: are any of them candidates for drugs acting on the central nervous system? *CNS Drug Reviews*, 2007;13:381-404.

In: Biogenic Amines
Editors: T. Farooqui, A. A. Farooqui
ISBN: 978-1-60876-625-3

Chapter XI

Estrogen-Dopamine Interactions in Forebrain Modulation of the Sensorimotor Gating

John J. Byrnes, Jacqueline T. Draper and Elizabeth M. Byrnes*
Tufts University, Cummings School of Veterinary Medicine, North Grafton, MA, USA

Abstract

Background: Prepulse inhibition (PPI) of the acoustic startle response is modulated by endocrine factors in both humans and rats. PPI is generally attenuated during periods of elevated estrogens, suggesting that these hormones might have a disruptive influence on sensorimotor gating. However, we recently demonstrated significant attenuation of PPI in the early postpartum period during which extremely low levels of circulating estrogens are present. One explanation for these apparently discrepant findings is that the precipitous postpartum decline in estrogen is preceded by substantially elevated levels present at the end of pregnancy. Thus, the observation of reduced PPI during the early postpartum period may be a consequence of high levels of circulating estrogen followed by its rapid withdrawal.

Material and Methods: In the present study, we examined the effects of continuous 7-day treatment with 17β estradiol, and its 4-day withdrawal, on PPI in ovariectomized female rats. As several studies have proposed an interaction between estrogen and forebrain dopamine systems, we also examined the effects of the dopamine agonist apomorphine on PPI, motor activity, and associated changes in forebrain dopamine function.

Results: While estradiol exposure did not significantly alter baseline PPI, significant differences in the manner by which apomorphine disrupted PPI were observed. Specifically, in both estradiol and estradiol-withdrawn groups, apomorphine decreased PPI via an overall reduction in acoustic startle rather than a selective disruption of the inhibitory effect of the prepulse stimulus. In addition, while apomorphine significantly

* Correspondence concerning this article should be addressed to: jbyrnes@massmail.state.ma.us

decreased dopamine turnover in the nucleus accumbens of control animals, this effect was absent in estradiol and estradiol-withdrawn animals. Thus, estradiol exposure significantly altered the effects of apomorphine on PPI, a change which persists for up to four days of withdrawal.

Conclusion: These findings suggest that estradiol may modulate sensorimotor gating mechanisms via a shift in the response of the nucleus accumbens dopamine system.

INTRODUCTION

Sensorimotor gating serves as a mechanism for filtering out non-essential stimuli [1-3]. It is hypothesized that impaired sensorimotor gating leads to information overload, cognitive fragmentation, and disordered thought [4,5]. Such disruptions are often observed in patients with schizophrenia [6], Huntington's disease [7], Tourette's syndrome [8], and obsessive-compulsive disorder [9]. Accordingly, significant efforts have been made over the past two decades to measure sensorimotor gating, delineate its neural mechanisms, and characterize the underlying neuropathology of "gating disorders."

Prepulse inhibition (PPI) of the acoustic startle reflex (ASR) can be conveniently studied in the laboratory as an operational measure of sensorimotor gating [10,11]. The ASR is normally triggered by a strong auditory stimulus and is characterized by a rapid contraction of the skeletal muscles, eyelid closure, and initiation of the autonomic sympathetic flight response [2]. PPI of the ASR occurs when a non-startling, low intensity stimulus (often auditory) precedes the strong auditory stimulus [3]. Under these conditions, the magnitude of the ASR is dramatically suppressed. Due to its conservation and similarity across mammalian species, PPI of the ASR in rodents has been widely studied as a model for understanding the brain mechanisms of normal and deficient sensorimotor gating in humans [11,12].

From a clinical perspective, studies have shown that PPI deficits can be ameliorated by administration of antipsychotic drugs in both humans [13-16] and rats [4]. These findings have led to intense efforts to elucidate the role of dopamine (DA) systems in PPI. To date, numerous studies have demonstrated regulation of PPI by DA [17-21]. For example, both direct and indirect DA agonists have been consistently shown to disrupt PPI [22-25]. In the rat, it is generally agreed that some of the disruptive effects of DA on PPI are mediated by DA D2-like receptors in the nucleus accumbens [26,27]. Additional studies have implicated the prefrontal cortex in the regulation of sensorimotor gating as infusion of DA agonists into this region has also been shown to disrupt PPI [28]. Thus, the mesolimbic and mesocortical dopamine systems both appear to regulate PPI. In aggregate, these studies provide evidence for the involvement of forebrain D2 receptor function in normal and deficient sensorimotor gating.

In addition to the role of DA in sensorimotor gating, there is some evidence implicating estrogen as a possible mediator in the phenomenon. For example, differences in PPI across the menstrual cycle have been documented in women, with reduced PPI observed during the luteal phase, a period associated with elevated levels of both estrogen and progesterone [29,30]. Similar effects have been observed in rodents, with reduced PPI occurring during proestrus [31], the phase of the cycle associated with increased levels of estrogen. These studies suggest that high levels of circulating estrogen are associated with decreased PPI.

However, direct administration of estradiol to ovariectomized rats has been shown to either increase PPI [32] or have no effect on this measure [33]. Thus, the extent to which estrogen plays a role in modulating human sensorimotor gating remains unclear.

Studies in animals have provided substantial evidence for an interaction between estradiol and DA in the forebrain regions involved in sensorimotor gating. For example, studies in female rats have observed increased density of DA receptors following chronic estradiol treatment [34-36]. In addition, acute estradiol administration has been shown to potentiate DA release [37,38] and DA-mediated behaviors, including rotation and stereotypy [39,40]. Therefore, it is possible that estrogen-mediated alterations in PPI may occur as a result of altered DA function in forebrain structures.

In recent years, our laboratory has attempted to further characterize the interaction between ovarian hormones and DA in the modulation of PPI in female rats. Our initial studies observed a significant increase in apomorphine-induced disruption of PPI following ovariectomy with no corresponding change in baseline PPI [41]. These results suggest that the removal of ovarian hormones enhances DA receptor sensitivity resulting in increased disruption of PPI. We have also observed alterations in DA receptor sensitivity and reduced PPI during the early postpartum period [42]. As postpartum rats also have significantly reduced levels of estrogen, both of our findings would appear to be inconsistent with the suggestion that elevated estrogen levels are associated with PPI disruption. However, our paradigms essentially examined the effects of estrogen withdrawal, either following ovariectomy or during the natural decline in estrogens following parturition. The present study was designed to expand these initial findings and further define the effects of estradiol and estradiol withdrawal on PPI and forebrain DA function. For this study, ovariectomized virgin female rats were exposed to either 1) high circulating levels of estradiol, or 2) high circulating estradiol followed by complete withdrawal of the hormone. The effect of estradiol withdrawal on PPI and forebrain DA turnover was determined under basal conditions and following administration of apomorphine, the mixed D1/D2 receptor agonist. Two additional behavioral measures, locomotor activity and stereotypy, were also examined. Overall, the results indicate significant estradiol-DA interaction in the modulation of sensorimotor gating, an effect which appears to be related to decreased DA receptor sensitivity in the nucleus accumbens.

METHODS

Animals

Fifty-seven female Sprague-Dawley rats (200-250 g) were purchased from Taconic (Hudson, NY). Animals were housed in the Animal Behavior Research Core facility at Tufts University School of Medicine. All animals were allowed to acclimate to facility for 7 days prior to baseline testing. Food and water were available *ad libitum,* and animals were maintained on a 12:12 hr reverse light cycle with all testing conducted under dark conditions. Animals were housed in polypropylene cages (35 x 25 x 20 cm) at a constant room temperature (21-25°C). All procedures were conducted according to guidelines set forth by the Committee of Care and Use of Laboratory Animal Resources of the National Research Council. All procedures were approved by the Tufts-New England Medical Center Institutional Animal Care and Use Committee.

Experimental Design and Procedures

To ensure comparable levels of PPI across groups, baseline PPI testing was conducted on all animals prior to assignment to one of the three estradiol treatment groups; continuous estradiol (E2), estradiol withdrawal (E2/WD), or non-estradiol-exposed group (Control). The experimental design is presented in Table 1. On Experimental Day 1, animals were ovariectomized and implanted with either estradiol or blank (Control) silastic implants (2 mm in length). After 7 days, estradiol implants were removed from animals in the E2/WD group and replaced with a blank implant. At this time, E2 and Control groups underwent a sham surgery. Four days later, all animals underwent behavioral testing. For behavioral testing, animals were transferred to the testing room and weighed. Each animal was then administered 0.25 mg/kg of the D1/D2 receptor agonist apomorphine (Sigma, St. Louis, MO) or its vehicle (0.9% NaCl, 0.1% ascorbic acid) in a volume of 1 ml/kg, s.c. Ten minutes later, animals were placed in a startle chamber for acoustic startle and PPI testing. Total testing time was approximately 18 minutes. Following PPI testing (40 minutes post-drug or vehicle treatment), animals were placed in an open field and locomotor activity and stereotypy were monitored for 15 minutes. Following behavioral testing (55 minutes post-drug or vehicle treatment), animals were euthanized via CO_2 inhalation. Trunk blood was collected into heparinized tubes, and brains were rapidly removed and dissected on ice. Three brain regions were dissected bilaterally: the prefrontal cortex, nucleus accumbens, and dorsal striatum based on the atlas of Paxinos and Watson [43]. All tissue was stored at -80°C until processing for neurochemical analysis. Blood was centrifuged, and plasma was stored at -20°C until processing for estradiol content.

Table 1. Experimental Design

	Treatment Day and Procedure			
Group	**Day 1**	**Day 7**	**Day 11**	**n**
Control	OVX + Blank	Sham Surgery	Vehicle Apomorphine	11 8
E2	OVX + E2	Sham Surgery	Vehicle Apomorphine	10 8
E2/WD	OVX + E2	E2 Removal	Vehicle Apomorphine	11 8

Ovariectomy (OVX) and Estradiol (E2) Implantation

All animals were bilaterally ovariectomized under isoflurane anesthesia using standard aseptic procedures. Immediately following ovariectomy (OVX), a 2 cm silastic implant was inserted under the skin between the scapulas. Implants for animals in the E2 and E2/WD group contained 17β estradiol (Sigma, St. Louis, MO), while those for the Control group were empty (blank). Seven days after implantation, estradiol implants were removed from animals in the E2/WD group and were replaced with a blank implant. To control for any confounds associated with this procedure, all other animals in the study underwent a sham procedure in which the incision used for the implant was re-opened and closed.

Behavioral Measures

Prepulse Inhibition: PPI was measured using the Startle Monitor System (SRM100; Hamilton-Kinder, Poway, CA). The startle chamber contained a speaker, a piezoelectric sensing platform, and a plexiglass "holder" for the animal which was designed to restrain an animal on the sensing platform while allowing limited movement. Animals were exposed to 70 dBA broad band ("white") background noise during a 5 min acclimation period and throughout each test session. Animals were then presented with 10 consecutive Pulse stimuli (below) for habituation prior to PPI testing. PPI was determined immediately following habituation and utilized two trial types: The Pulse stimulus trial (startle-eliciting 40 ms, 118 dBA broad band noise stimulus); The Prepulse stimulus trial (40 ms, 76 dBA broad band noise stimulus preceding the Pulse stimulus by 100 ms). Whole-body startle responses were recorded in units of force (Newtons). For each testing session, animals received 51 stimuli presented in pseudo-random order and with a 15 s interstimulus interval. The percent of PPI expressed within each test session was calculated as follows: 100 – [(mean Prepulse response/mean Pulse response) x 100].

Locomotor Activity and Stereotypic Movements

Following PPI testing, overall locomotor activity was assessed in individual animals for 15 min in a novel open field environment. Locomotor activity was measured using the SmartFrame Activity System and MotorMoniter® software (Hamilton/Kinder, Poway, CA). The system consisted of individual PC-interfaced horizontal photobeam frames. Each frame contained 12 photobeams (8LX4W) which tracked movement of animals within standard plastic laboratory housing cages (40 cm X 20 cm X 18 cm). Data were collected in the form of photobeam breaks as an indication of activity. Repetitive breaking of individual photobeams was recorded as stereotypic movement.

Measurement of Dihydroxyphenylacetic Acid (DOPAC) and DA in Tissue Homogenates

Frozen tissue obtained from one hemisphere was homogenized in HPLC mobile phase (MDTM-70-1332; ESA, Chelmsford, MA) consisting of the following: NaH_2PO_4 75 mM; EDTA 25 μM; octanesulfonic acid sodium salt 1.7 mM; triethylamine 100 μl/L; 10% acetonitrile; pH 3.0 with phosphoric acid. Homogenates were centrifuged and supernatants were analyzed for DA and DOPAC content using a DECADE Digital Electrochemical Amperometric Detector (Antec Leyden, Netherlands). Mobile phase was delivered to the detector at a rate of 0.2 ml/min using an LC1150 Solvent Delivery System (GBC Scientific Equipment, Victoria, Australia). Catecholamines were separated on a ThermoHypersil-Keystone column (150 x 2.1 mm, pore size 5.0 μm). WinChrom Chromatography data management software (GBC Separations, Hubbardston, MA) was used for data analysis. Tissue protein content was determined using the Bradford method (Bio-Rad, Hercules, CA). DA and DOPAC (ng/mg protein) were determined based on comparison to an external standard curve. Turnover of DA for each brain region was calculated using the DOPAC/DA ratio.

Plasma Estradiol Radioimmunoassay

Plasma estradiol was assayed using a Coat-A-Count Kit for estradiol (Diagnostic Products Corporation, Los Angeles, CA). The procedure was carried out in accordance with the package insert. All samples were assayed in duplicate. Assay sensitivity was equivalent to 1 pg/ml. All samples were measured in a single assay. Plasma E2 levels for ovariectomized rats are routinely < 2 pg/ml.

Statistical Analysis

For all behavioral, physiological, and neurochemical measures, two-way analyses of variance (ANOVA) were conducted with Treatment Group (Control, E2, E2/WD) and Drug (vehicle, apomorphine) as factors. All post hoc analyses were conducted using the Student-Newman-Keuls method. Statistical significance was defined as $p \leq 0.05$.

Results

Plasma Estradiol and Body Weight

Plasma estradiol and body weight data are illustrated in Figure 1. Separate two-way ANOVA indicated significant main effects of Treatment Group on both plasma estradiol ($F[2,52] = 8.62$, $p < 0.01$; Panel A) and body weight ($F[2,56] = 13.22$, $p < 0.01$; Panel B). As expected, animals in the E2 group had significantly elevated estradiol levels compared to both E2/WD and Control groups (both p's <0.01). In addition to high circulating levels of estradiol, body weight was also reduced in the E2 group compared to both E2/WD and Control groups (both p's < 0.05). This finding is consistent with the previously reported anorexic effect of estradiol [44]. As expected no effect of apomorphine on estradiol levels or body weight was observed.

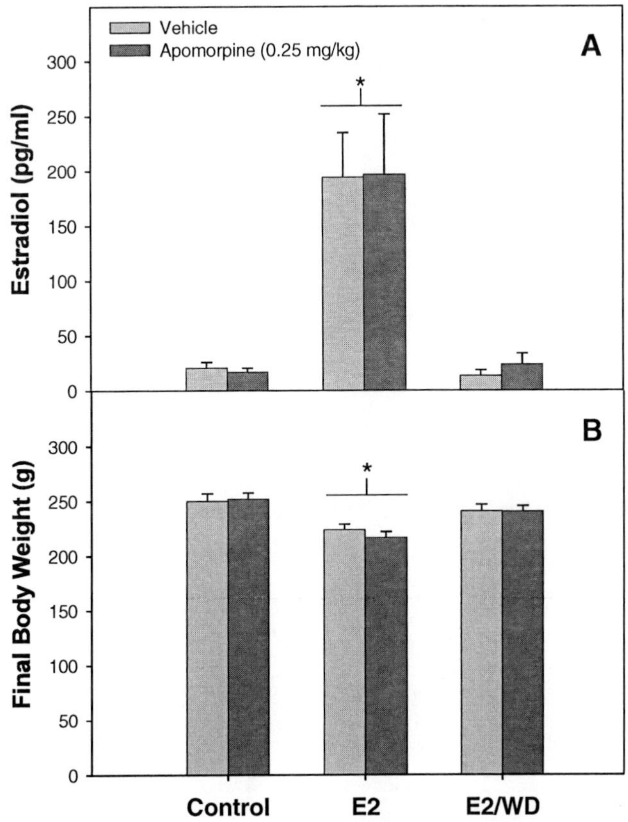

Figure 1. Effect of estradiol exposure on plasma concentration and body weight. Panel A: Data are mean (+SEM) plasma estradiol (pg/ml) for all experimental groups (n=8-11). Panel B: Data are mean (+SEM) body weight (g) for all experimental groups (n=8-11). * p < 0.05 vs. Control and E2/WD groups (collapsed across Drug condition).

Prepulse Inhibition

Two-way ANOVA revealed no effects of Treatment Group or Treatment Group X Drug interaction on percent PPI. However, there was a significant main effect of Drug ($F[1,55] = 21.53$, $p < 0.001$) due to reduced PPI in apomorphine-treated animals when compared to vehicle-treated animals (Figure 2). This apomorphine-induced disruption of PPI was significant for Control and E2 treatment groups (both p's < 0.05), but did not attain significance in the E2/WD group ($p = 0.11$).

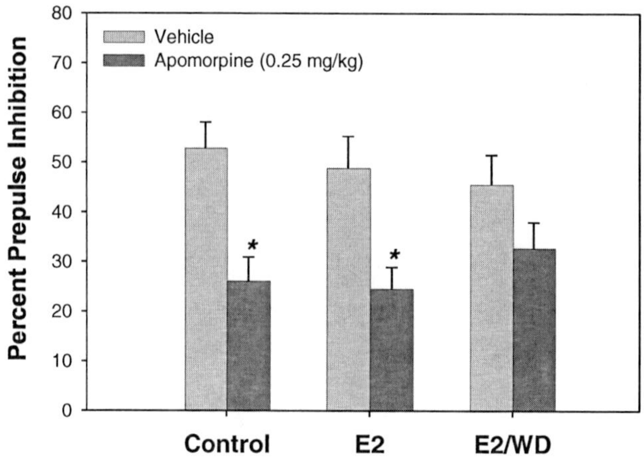

Figure 2. Effect of apomorphine on PPI in estradiol-exposed rats. Data are mean (+SEM) percent PPI for groups of 8-11 animals. * p < 0.05 vs. vehicle within Treatment Group.

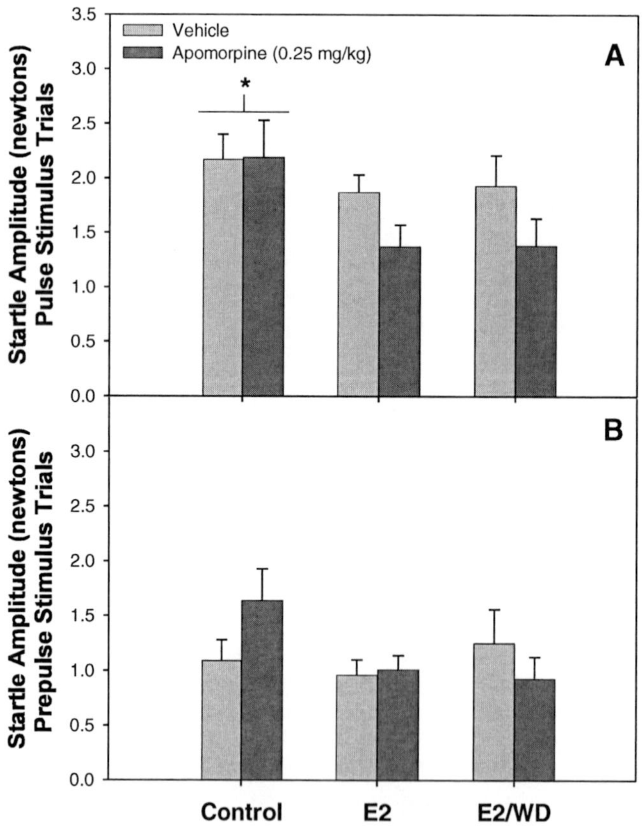

Figure 3. Effect of apomorphine on startle amplitude in estradiol-exposed rats. Data are mean (+SEM) startle amplitude (Newtons) for Pulse (Panel A) and Prepulse (Panel B) stimulus trial types for groups of 8-11 animals. * p < 0.01 vs. E2 and E2/WD groups (collapsed across Drug condition).

Acoustic Startle Response

When interpreting PPI data, it is important to consider the potential contribution of changes in startle amplitude (i.e. ASR to Pulse stimulus). In the current study, two-way ANOVA revealed a significant main effect of Treatment Group ($F[2,55] = 4.33$, $p < 0.02$) on startle amplitude in response to the Pulse stimulus (Figure 3, Panel A). Specifically, both the E2 and E2/WD groups demonstrated reduced startle amplitude when compared to the Control group (both p's < 0.03). Examination of Prepulse startle amplitude also suggests differential responses to apomorphine in Control and estradiol-treated groups (Figure 3, Panel B). While no significant effects were revealed, there was a trend toward a main effect of Treatment Group ($F[2,55] = 2.84$, $p = 0.07$) which is largely due to apomorphine-induced disruption in the ASR in the Control group compared to either the E2 or E2/WD groups. These findings suggest that while the magnitude of apomorphine-induced disruption of PPI is similar between experimental groups (Figure 2), the manner in which this occurs is altered by estradiol exposure.

Locomotor Activity

Two-Way ANOVA revealed no effects of Treatment Group or Treatment Group X Drug interaction on locomotor activity. However, there was a significant main effect of Drug ($F[1,56] = 7.94$, $p < 0.01$) due to reduced locomotion in apomorphine-treated animals when compared to vehicle-treated animals (Figure 4). This apomorphine-induced reduction of locomotion was significant for Control and E2/WD treatment groups (both p's < 0.07), but did not attain significance in the E2 group.

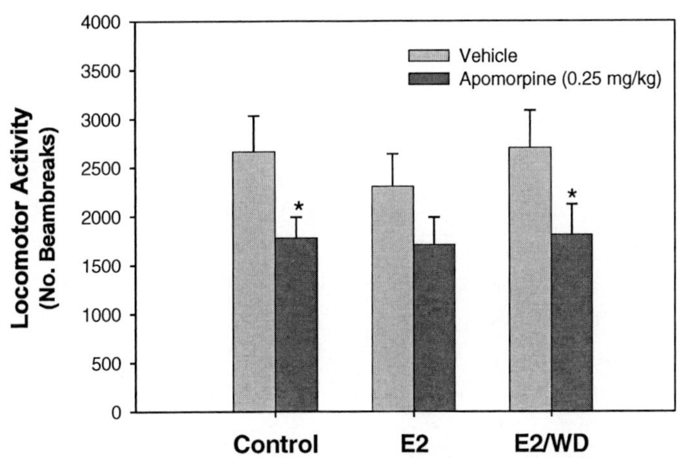

Figure 4. Effect of apomorphine on locomotor activity in estradiol-exposed rats. Data are mean (+SEM) photobeam breaks during 15 min of testing for groups of 8-11 animals. * $p < 0.05$ vs. vehicle within Treatment Group.

Stereotypy

Two-way ANOVA revealed significant effects of both Treatment Group (F[2,56] = 4.55, $p < 0.02$) and Drug (F[1,56] = 8.28, $p < 0.01$) on stereotypic activity (Figure 5).

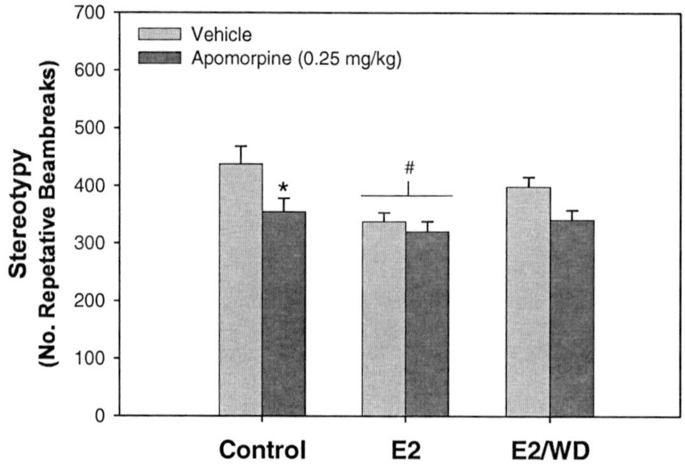

Figure 5. Effect of apomorphine on stereotypic movement in estradiol-exposed rats. Data are mean (+SEM) number stereotypic movements during 15 min of testing for groups of 8-11 animals. * $p < 0.05$ vs. Control group (collapsed across Drug condition). ** $p < 0.05$ vs. vehicle within Treatment Group.

Post hoc analyses indicate that stereotypic movements in the E2 group were reduced when compared to the Control group ($p = 0.015$), and to a lesser extent when compared to E2/WD group ($p = 0.078$). The effect of Drug was due to a significant apomorphine-induced reduction in stereotypic movement in the Control group ($p = 0.01$), with a trend toward an apomorphine effect in the E2/WD group ($p = 0.075$).

DOPAC, DA, and DA Turnover

Mean values and associated statistics for DOPAC and DA tissue concentrations (pg/mg protein) are presented in Table 2. Two-way ANOVA indicated no significant effects of Treatment Group, Drug, or their interaction on analyte content in the prefrontal cortex. However, a significant Treatment Group X Drug interaction was found for nucleus accumbens DA content. This effect was due to significantly decreased DA content in the E2 group treated with apomorphine when compared to both the E2 group treated with vehicle and the Control group treated with apomorphine. Significant apomorphine effects were also observed in the striatum. Across all three treatment groups, apomorphine administration was associated with decreased striatal DOPAC and increased DA content. Post hoc analyses indicate that the effects of apomorphine on striatal DA content were significant in both the Control and E2/WD groups, but not in the E2 group.

DA turnover (DOPAC/DA ratio) for all brain regions examined is illustrated in Figure 6.

Table 2. Forebrain DOPAC and DA

	DOPAC		DA	
Prefrontal Cortex	**Saline**	**Apomorphine**	**Saline**	**Apomorphine**
Control	43.6±10.8	41.9±7.7	45.1±19.7	33.4±13.4
E2	47.3±16.6	39.1±8.5	38.8±16.3	20.2±5.8
E2/WD	45.9±14.2	28.6±5.7	31.1±4.4	26.9±4.6
2-WAY ANOVA				
Treatment	F[2,44] = 0.29, p=0.74		F[2,44] = 0.73, p=0.49	
Drug	F[1,44] = 0.65, p=0.42		F[2,44] = 0.26, p=0.61	
Interaction	F[2,44] = 0.13, p=0.88		F[2,44] = 0.1, p=0.9	
Nucleus Accumbens				
Control	316.9±43.2	279.9±45.1	935.6±131.8	1425.7±237.9
E2	361.9±75.6	290.6±57.1	1355.1±332.6	785.9±145 *
E2/WD	287.8±90.2	284.3±39.7	910.4±156.31	1046.6±63.2
2-WAY ANOVA				
Treatment	F[2,44] = 0.22, p=0.79		F[2,44] = 0.56, p=0.57	
Drug	F[1,44] = 0.58, p=0.45		F[2,44] = 0.01, p=0.91	
Interaction	F[2,44] = 0.15, p=0.86		**F[2,44] = 3.9, p=0.02**	
Dorsal Striatum				
Control	289.2±27.6	223.4±22.2	279.0±48.2	488.3±92.2 **
E2	307.9±34.6	260.6±50.9	330.7±73.1	302.6±50.8
E2/WD	325.7±42.3	266.7±21.3	267.7±18.5	435.5±44.9 **
2-WAY ANOVA				
Treatment	F[2,44] = 0.73, p=0.48		F[2,44] = 0.66, p=0.52	
Drug	**F[1,44] = 4.15, p=0.049**		**F[2,44] = 5.9, p=0.02**	
Interaction	F[2,44] = 0.03, p=0.96		F[2,44] = 2.3, p=0.11	

* $p < 0.05$ as compared to the apomorphine-treated control group and saline-treated E2 group.

** $p \leq 0.05$ saline-treated as compared to apomorphine-treated within treatment group.

As above, two-way ANOVA indicated no significant effects of Treatment Group, Drug, or their interaction within the prefrontal cortex (Panel A). While no significant main effects or interactions were observed in the nucleus accumbens, it was clear that the groups were responding differently to apomorphine treatment (Panel B). Indeed, planned comparisons confirmed the expected reduction in DA turnover following apomorphine treatment in the Control group ($t[14] = -2.12$, $p < 0.05$), while no effect of apomorphine on DA turnover was observed in either the E2 or E2/WD groups (both p's > 0.7). However, apomorphine produced a robust and consistent decrease in striatal DA turnover (Panel C). This effect ($F[1,43] = 12.63$, $p < 0.001$) reached significance in both the Control and E2/WD groups (both p's < 0.05).

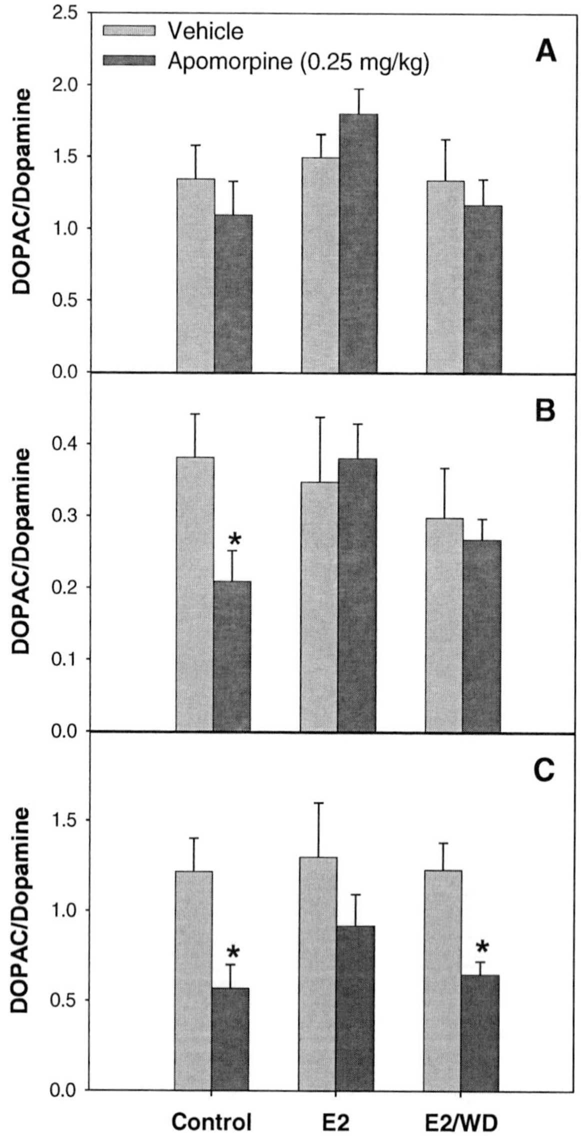

Figure 6. Effect of estradiol exposure and apomorphine treatment on forebrain DA turnover in rats. Data are mean (+SEM) DOPAC/dopamine ratio for prefrontal cortex (Panel A), nucleus accumbens (Panel B), and striatum (Panel C) for groups of 8-11 animals. * $p < 0.05$ vs. vehicle within Treatment Group.

Discussion

The present study was undertaken to better understand the effects of estradiol and estradiol withdrawal on dopaminergic modulation of sensorimotor gating. Compared to controls, estradiol-exposed animals (E2 and E2/WD groups) showed no significant group differences in percent PPI following treatment with the DA agonist apomorphine or its vehicle. However, separate analysis of raw acoustic startle data indicated a differential effect of estradiol treatment as a function of trial type (Pulse stimulus vs. Prepulse stimulus). As expected, apomorphine disrupted PPI in the Control group via dampening the effectiveness of the Prepulse stimulus. In contrast, the E2 group responded to apomorphine with an overall decrease in startle response to the Pulse stimulus, with no corresponding change in response to the Prepulse stimulus. This effect of estradiol exposure was similarly observed in the E2/WD group, suggesting persistence into the withdrawal period. Thus, it appears that estradiol exposure alters the underlying mechanism of DA receptor-mediated PPI disruption.

In addition to estradiol effects on startle amplitude following apomorphine, differential effects on DA turnover within the nucleus accumbens were also observed. As expected, apomorphine significantly decreased DA turnover in the Control group, an effect conventionally attributed to autoinhibitory effects of forebrain presynaptic DA D2 receptor stimulation [45]. In contrast, no effect of apomorphine on nucleus accumbens DA turnover was observed in either estradiol treatment group. This finding supports the notion that apomorphine disruption of PPI in E2 and E2/WD groups is mediated by an altered dopaminergic mechanism, perhaps involving DA D2-like receptor subsensitivity in the nucleus accumbens.

Several studies have demonstrated a critical role for DA D2-like receptors in DA-mediated disruption of PPI [24,27,46]. Additional findings suggest that release-regulating autoreceptors within the nucleus accumbens underlie these effects [47]. Autoreceptor regulation of DA output is mediated via both decreased release and increased uptake [45]. Of relevance to the current findings, several studies have demonstrated an effect of estradiol on DA release and reuptake within the nucleus accumbens. Specifically, estradiol priming (i.e. prolonged exposure) has been shown to decrease K^+-stimulated DA release [48], increase DA reuptake [49], and attenuate quinpirole-mediated potentiation of DA uptake [50]. These effects are in contrast to the acute effects of estradiol infusion which include potentiation of stimulated DA release [48,51] and decreased clearance [52]. Thus, the current findings are consistent with previous data demonstrating alterations in DA autoreceptor regulation following prolonged estradiol exposure. Furthermore, the present data suggest that these effects may be brain region specific and may persist even following estradiol withdrawal.

In addition to the failure of apomorphine to disrupt PPI in E2 and E2/WD groups, the other major finding in the current chapter is the decreased ASR observed in these two groups. While there was no significant interaction between treatment group and drug, apomorphine tended to reduce startle amplitude in both the E2 and E2/WD groups when compared to controls. The mechanism underlying this effect is unclear. However, a previous study examining effects of chronic estradiol also reported a reduction in the ASR following apomorphine when compared to ovariectomized controls [33]. One possible explanation for this effect may be a preferential activation of DA receptors within the amygdala in estradiol exposed females. A previous study has shown that DA infusion directly into the amygdala disrupts PPI via an overall decrease in startle amplitude [53], findings which are consistent with those of the current report. Such a shift could be related to the effects of estradiol on arousal. Indeed, a recent fMRI study demonstrated decreased arousal-induced activation within the amygdala of women during the luteal phase of the menstrual cycle [54]. This finding suggests that elevated levels of estrogen may be related to attenuation of arousal circuitry within this brain region. While an intriguing possibility, additional study is needed to determine whether the amygdala is a substrate for estradiol-DA interactions and their effects on the ASR.

The effects of estradiol exposure on locomotor activity and stereotypy did not parallel the effects on the ASR and PPI. For these behaviors, the effects of apomorphine were similar in Control and E2/WD groups, with both groups demonstrating decreased locomotor and stereotypic responses. In contrast, animals in the E2 group appeared to be subsensitive to these effects of apomorphine. Moreover, animals in the E2 group also exhibited decreased stereotypy compared to Control and E2/WD groups, and this effect was observed regardless of drug treatment. Previous studies have observed apomorphine-induced hypolocomotion using doses similar to those used in current study [55]. Other work has demonstrated attenuated apomorphine-induced locomotor activity following acute estradiol administration [56]. Since stereotypy is defined here as repetitive beam breaks, it is difficult to interpret this decrease. However, it is possible that this reflects decreased self-grooming in the E2 group, an effect which would be consistent with prior observations of increased grooming following ovariectomy and its reversal by estradiol replacement [57]. Collectively these data indicate that, unlike regulation of the ASR and PPI, the effects of estradiol exposure on locomotor activity and stereotypy return to control levels following estradiol withdrawal.

The effects of estradiol exposure on DA content and turnover were brain region specific and were similar to those reported previously in female rats [42]. There was no significant effect of apomorphine on DA turnover in the nucleus accumbens of E2 and E2/WD animals. However, examination of the treatment effects on DA content suggests that the decreased sensitivity to apomorphine in these two groups may be due to different mechanisms. In the E2 group, apomorphine administration resulted in a significant decrease in DA content within the nucleus accumbens, an effect not observed in either Control or E2/WD groups. The mechanism by which estradiol exposure leads to lower DA content following apomorphine is unclear. Since no significant changes in DOPAC content were observed, the apomorphine effect on nucleus accumbens DA is likely be related to altered DA synthesis or vesicular monoamine transport function.

In the striatum, apomorphine administration significantly increased DA content in both the E2/WD and Control groups, while having no effect in the E2 group. All groups demonstrated the expected apomorphine-induced decrease in DOPAC within the striatum. No effects of either treatment group or drug were observed in the prefrontal cortex. Taken together these findings indicate that apomorphine-induced alterations in DA turnover are one consequence of chronic estradiol and that the persistence of this effect following estradiol withdrawal appears to be brain region specific.

Conclusion

The present findings indicate that high circulating estradiol can produce both a shift in arousal as well as an alteration in the neural mechanisms underlying sensorimotor gating. This notion is supported by the differential effects of apomorphine on the ASR and nucleus accumbens DA turnover during and following estradiol treatment. Further, the findings suggest that these observed changes in sensorimotor gating and associated neurochemical alterations result from an underlying subsensitivity of D2 autoreceptors in the nucleus accumbens which does not recover during early estradiol withdrawal. The finding that estradiol treatment alters striatal DA receptor sensitivity and striatally-mediated behaviors is also intriguing given that these effects are normalized upon estradiol withdrawal. Collectively, the results of the present work have provided further elucidation of the role of estrogen-DA interactions in the regulation of sensorimotor gating and related dopaminergic behaviors.

REFERENCES

[1] Koch, M. and Schnitzler, H.U., The acoustic startle response in rats--circuits mediating evocation, inhibition and potentiation, *Behavioral Brain Research.* 1997;89:35-49.

[2] Koch, M., The neurobiology of startle, *Progress in Neurobiology.* 1999;59:107-28.

[3] Graham, F.K., Presidential Address, 1974. The more or less startling effects of weak prestimulation, *Psychophysiology.* 1975;12:238-48.

[4] Geyer, M.A., Krebs-Thomson, K., Braff, D.L. and Swerdlow, N.R., Pharmacological studies of prepulse inhibition models of sensorimotor gating deficits in schizophrenia: a decade in review, *Psychopharmacology (Berl).* 2001;156:117-54.

[5] Perry, W. and Braff, D.L., Information-processing deficits and thought disorder in schizophrenia, *American Journal of Psychiatry.* 1994;151:363-7.

[6] Braff, D.L., Grillon, C. and Geyer, M.A., Gating and habituation of the startle reflex in schizophrenic patients, *Archives of General Psychiatry.* 1992;49:206-15.

[7] Swerdlow, N.R., Paulsen, J., Braff, D.L., Butters, N., Geyer, M.A. and Swenson, M.R., Impaired prepulse inhibition of acoustic and tactile startle response in patients with Huntington's disease, *Journal of Neurology, Neurosurgery, & Psychiatry.* 1995;58:192-200.

[8] Castellanos, F.X., Fine, E.J., Kaysen, D., Marsh, W.L., Rapoport, J.L. and Hallett, M., Sensorimotor gating in boys with Tourette's syndrome and ADHD: preliminary results, *Biological Psychiatry.* 1996;39:33-41.

[9] Hoenig, K., Hochrein, A., Quednow, B.B., Maier, W. and Wagner, M., Impaired prepulse inhibition of acoustic startle in obsessive-compulsive disorder, *Biological Psychiatry.* 2005;57:1153-8.

[10] Geyer, M.A. and Swerdlow, N.R., Measurement of startle response, prepulse inhibition, and habituation, *Current Protocols in Neuroscience.* 2001;Chapter 8:Unit 8 7.

[11] Swerdlow, N.R., Braff, D.L., Taaid, N. and Geyer, M.A., Assessing the validity of an animal model of deficient sensorimotor gating in schizophrenic patients, *Archives of General Psychiatry.* 1994;51:139-54.

[12] Swerdlow, N.R., Braff, D.L. and Geyer, M.A., Animal models of deficient sensorimotor gating: what we know, what we think we know, and what we hope to know soon, *Behavioral Pharmacology.* 2000;11:185-204.

[13] Kumari, V. and Sharma, T., Effects of typical and atypical antipsychotics on prepulse inhibition in schizophrenia: a critical evaluation of current evidence and directions for future research, *Psychopharmacology (Berl).* 2002;162:97-101.

[14] Kumari, V., Soni, W. and Sharma, T., Prepulse inhibition of the startle response in risperidone-treated patients: comparison with typical antipsychotics, *Schizophrenia Research.* 2002;55:139-46.

[15] Quednow, B.B., Wagner, M., Westheide, J., Beckmann, K., Bliesener, N., Maier, W. and Kuhn, K.U., Sensorimotor gating and habituation of the startle response in schizophrenic patients randomly treated with amisulpride or olanzapine, *Biological Psychiatry.* 2006;59:536-45.

[16] Braff, D.L., Geyer, M.A. and Swerdlow, N.R., Human studies of prepulse inhibition of startle: normal subjects, patient groups, and pharmacological studies, *Psychopharmacology (Berl).* 2001;156:234-58.

[17] Swerdlow, N.R., Kuczenski, R., Goins, J.C., Crain, S.K., Ma, L.T., Bongiovanni, M.J. and Shoemaker, J.M., Neurochemical analysis of rat strain differences in the startle gating-disruptive effects of dopamine agonists, *Pharmacology, Biochemistry, & Behavior.* 2005;80:203-11.

[18] Swerdlow, N.R., Martinez, Z.A., Hanlon, F.M., Platten, A., Farid, M., Auerbach, P., Braff, D.L. and Geyer, M.A., Toward understanding the biology of a complex phenotype: rat strain and substrain differences in the sensorimotor gating-disruptive effects of dopamine agonists, *Journal of Neuroscience.* 2000;20:4325-36.

[19] Swerdlow, N.R., Platten, A., Kim, Y.K., Gaudet, I., Shoemaker, J., Pitcher, L. and , Auerbach, P., Sensitivity to the dopaminergic regulation of prepulse inhibition in rats: evidence for genetic, but not environmental determinants, *Pharmacology, Biochemistry, & Behavior.* 2001;70:219-26.

[20] Schwarzkopf, S.B., Bruno, J.P. and Mitra, T., Effects of haloperidol and SCH 23390 on acoustic startle and prepulse inhibition under basal and stimulated conditions, *Progress in Neuropsychopharmacology & Biological Psychiatry.* 1993;17:1023-36.

[21] Peng, R.Y., Mansbach, R.S., Braff, D.L. and Geyer, M.A., A D2 dopamine receptor agonist disrupts sensorimotor gating in rats. Implications for dopaminergic abnormalities in schizophrenia, *Neuropsychopharmacology.* 1990;3:211-8.

[22] Swerdlow, N.R., Mansbach, R.S., Geyer, M.A., Pulvirenti, L., Koob, G.F. and Braff, D.L., Amphetamine disruption of prepulse inhibition of acoustic startle is reversed by depletion of mesolimbic dopamine, *Psychopharmacology (Berl).* 1990;100:413-6.

[23] Varty, G.B. and Higgins, G.A., Dopamine agonist-induced hypothermia and disruption of prepulse inhibition: evidence for a role of D3 receptors?, *Behavioral Pharmacology.* 1998;9:445-55.

[24] Caine, S.B., Geyer, M.A. and Swerdlow, N.R., Effects of D3/D2 dopamine receptor agonists and antagonists on prepulse inhibition of acoustic startle in the rat, *Neuropsychopharmacology.* 1995;12:139-45.

[25] Sills, T.L., Amphetamine dose dependently disrupts prepulse inhibition of the acoustic startle response in rats within a narrow time window, *Brain Research Bulletin.* 1999;48:445-8.

[26] Wan, F.J., Geyer, M.A. and Swerdlow, N.R., Accumbens D2 modulation of sensorimotor gating in rats: assessing anatomical localization, *Pharmacology, Biochemistry & Behavior.* 1994;49:155-63.

[27] Wan, F.J. and Swerdlow, N.R., Intra-accumbens infusion of quinpirole impairs sensorimotor gating of acoustic startle in rats, *Psychopharmacology (Berl).* 1993;113:103-9.

[28] Lacroix, L., Broersen, L.M., Feldon, J. and Weiner, I., Effects of local infusions of dopaminergic drugs into the medial prefrontal cortex of rats on latent inhibition, prepulse inhibition and amphetamine induced activity, *Behavioral Brain Research.* 2000;107:111-21.

[29] Jovanovic, T., Szilagyi, S., Chakravorty, S., Fiallos, A.M., Lewison, B.J., Parwani, A., Schwartz, M.P., Gonzenbach, S., Rotrosen, J.P. and Duncan, E.J., Menstrual cycle phase effects on prepulse inhibition of acoustic startle, *Psychophysiology.* 2004;41:401-6.

[30] Swerdlow, N.R., Hartman, P.L. and Auerbach, P.P., Changes in sensorimotor inhibition across the menstrual cycle: implications for neuropsychiatric disorders, *Biological Psychiatry.* 1997;41:452-60.

[31] Koch, M., Sensorimotor gating changes across the estrous cycle in female rats, *Physiology & Behavior.* 1998;64:625-8.

[32] Van den Buuse, M. and Eikelis, N., Estrogen increases prepulse inhibition of acoustic startle in rats, *European Journal of Pharmacology.* 2001;425:33-41.

[33] Vaillancourt, C., Cyr, M., Rochford, J., Boksa, P. and Di Paolo, T., Effects of ovariectomy and estradiol on acoustic startle responses in rats, *Pharmacology, Biochemistry & Behavior.* 2002;74:103-9.

[34] Levesque, D. and Di Paolo, T., Chronic estradiol treatment increases ovariectomized rat striatal D-1 dopamine receptors, *Life Sciences.* 1989;45:1813-20.

[35] Hruska, R.E., Elevation of striatal dopamine receptors by estrogen: dose and time studies, *Journal of Neurochemistry.* 1986;47:1908-15.

[36] Levesque, D. and Di Paolo, T., Dopamine receptor reappearance after irreversible receptor blockade: effect of chronic estradiol treatment of ovariectomized rats, *Molecular Pharmacology.* 1991;39:659-65.

[37] Xiao, L. and Becker, J.B., Effects of estrogen agonists on amphetamine-stimulated striatal dopamine release, *Synapse.* 1998;29:379-91.

[38] Pasqualini, C., Olivier, V., Guibert, B., Frain, O. and Leviel, V., Acute stimulatory effect of estradiol on striatal dopamine synthesis, *Journal of Neurochemistry.* 1995;65:1651-7.

[39] Roy, E.J., Buyer, D.R. and Licari, V.A., Estradiol in the striatum: effects on behavior and dopamine receptors but no evidence for membrane steroid receptors, *Brain Research Bulletin.* 1990;25:221-7.

[40] Gordon, J.H., Modulation of apomorphine-induced stereotypy by estrogen: time course and dose response, *Brain Research Bulletin.* 1980;5:679-82.

[41] Byrnes, E.M., Byrnes, J.J. and Bridges, R.S., Increased sensitivity of dopamine systems following reproductive experience in rats, *Pharmacology, Biochemistry & Behavior.* 2001;68:481-9.

[42] Byrnes, E.M., Bridges, R.S., Scanlan, V.F., Babb, J.A. and Byrnes, J.J., Sensorimotor gating and dopamine function in postpartum rats, *Neuropsychopharmacology.* 2007;32:1021-31.
[43] Paxinos, G. and Watson, C., *The rat brain in stereotaxic coordinates* 6th ed edn., Elsevier, 2005.
[44] Butera, P.C., Estradiol and the control of food intake, *Physiology & Behavior.* 2009.
[45] Wu, Q., Reith, M.E., Walker, Q.D., Kuhn, C.M., Carroll, F.I. and Garris, P.A., Concurrent autoreceptor-mediated control of dopamine release and uptake during neurotransmission: an in vivo voltammetric study, *Journal of Neuroscience.* 2002;22:6272-81.
[46] Wan, F.J., Taaid, N. and Swerdlow, N.R., Do D1/D2 interactions regulate prepulse inhibition in rats?, *Neuropsychopharmacology.* 1996;14:265-74.
[47] Yamada, S., Harano, M. and Tanaka, M., Dopamine autoreceptors in rat nucleus accumbens modulate prepulse inhibition of acoustic startle, *Pharmacology, Biochemistry & Behavior.* 1998;60:803-8.
[48] Thompson, T.L. and Moss, R.L., Estrogen regulation of dopamine release in the nucleus accumbens: genomic- and nongenomic-mediated effects, *Journal of Neurochemistry.* 1994;62:1750-6.
[49] Thompson, T.L., Attenuation of dopamine uptake in vivo following priming with estradiol benzoate, *Brain Research.* 1999;834:164-7.
[50] Thompson, T.L., Moore, C.C. and Smith, B., Estrogen priming modulates autoreceptor-mediated potentiation of dopamine uptake, *European Journal of Pharmacology.* 2000;401:357-63.
[51] Becker, J.B. and Rudick, C.N., Rapid effects of estrogen or progesterone on the amphetamine-induced increase in striatal dopamine are enhanced by estrogen priming: a microdialysis study, *Pharmacology, Biochemistry & Behavior.* 1999;64:53-7.
[52] Karakaya, S., Kipp, M. and Beyer, C., Oestrogen regulates the expression and function of dopamine transporters in astrocytes of the nigrostriatal system, *Journal of Neuroendocrinology.* 2007;19:682-90.
[53] Swerdlow, N.R., Caine, S.B. and Geyer, M.A., Regionally selective effects of intracerebral dopamine infusion on sensorimotor gating of the startle reflex in rats, *Psychopharmacology (Berl).* 1992;108:189-95.
[54] Goldstein, J.M., Jerram, M., Poldrack, R., Ahern, T., Kennedy, D.N., Seidman, L.J. and Makris, N., Hormonal cycle modulates arousal circuitry in women using functional magnetic resonance imaging, *Journal of Neuroscience.* 2005;25:9309-16.
[55] Swerdlow, N.R., Krupin, A.S., Bongiovanni, M.J., Shoemaker, J.M., Goins, J.C. and Hammer, R.P., Jr., Heritable differences in the dopaminergic regulation of behavior in rats: relationship to D2-like receptor G-protein function, *Neuropsychopharmacology.* 2006;31:721-9.
[56] Kazandjian, A., Spyraki, C., Sfikakis, A. and Varonos, D.D., Apomorphine-induced behaviour during the oestrous cycle of the rat, *Neuropharmacology.* 1987;26:1037-45.
[57] Diaz-Veliz, G., Soto, V., Dussaubat, N. and Mora, S., Influence of the estrous cycle, ovariectomy and estradiol replacement upon the acquisition of conditioned avoidance responses in rats, *Physiology & Behavior.* 1989;46:397-401.

In: Biogenic Amines
Editors: T. Farooqui, A. A. Farooqui
ISBN: 978-1-60876-625-3

Chapter XII

STEREOSELECTIVITY OF NOREPINEPHRINE ENANTIOMERS AT PRE AND POST - JUNCTIONAL ADRENOCEPTORS[§]

Vandana Kumari[1], Miguel B. Farah[2], Salomon Z. Langer[3], Chenglong Li[1,*] and Popat N. Patil[1,*]
[1]The Ohio State University, U.S.A;
[2]Universidad de Buenos Aires, Argentina;
[3]L.E.R.S. Laboratoires d'Etudes et de Recherches Scientifiques, Paris, France.

ABSTRACT

The study was initiated to examine the degree of stereoselectivity of enantiomers of norepinephrine in regulating the release of the transmitter mediated by prejunctional adrenoceptor (α_2) in hypothalamus and vasa deferentia of rat. In vitro, Potassium (K^+) was used as a tool to investigate the release of the tritium labelled transmitter from the tissues. The α_2-adrenoceptors regulating norepinephrine (NE) release show stereoselectivity in favour of the (-)-enantiomer of NE in both tissues studied. The degree of stereoselectivity in both tissues is approximately 10-fold which perhaps reflects similar types of prejunctional α_2-adrenoceptors in the central and the peripheral nervous system. However, as compared to the high stereoselectivity, greater than 250 fold at the post junctional α_1-adrenoceptor mediated effects of the enantiomers on many organs including vasa deferentia, the selectivity at prejunctional α_2-adrenoceptor is low. Molecular modelling of prejunctional α_{2A} and the postjunctional α_{1A} adrenoceptors with the enantiomers shows additional hydrogen bond formation between hydroxyl group of catechol ring of (-)-NE and CYS201 of α_{2A}-AR as compared to α_{1A}-AR. Hydrophobic

§ This manuscript is submitted as a tribute to the long time colleague, an outstanding teacher, a scholar and a wonderful human being, late Professor Norman J. Uretsky.

* Correspondence concerning this article should be addressed to: Popat N. Patil (Phone: 614-292-7238 Email: patil.1@osu.edu); Chenglong Li (Phone: 614-247-8786 Fax: 614-292-2435 Email: cli@pharmacy.ohio-state.edu)

interactions between catechol ring of NE and TM VI of α_{1A}-AR and α_{2A}-AR differ by the presence of additional hydrophobic residue (Met292) in TM VI of α_{1A}-AR. This additional hydrophobic interaction leads to favorable packing of (-)-NE in the binding pocket and may also explains higher binding affinity of (-)-NE for α_{1A}-AR compared to α_{2A}-AR. (-)-NE interactions are more favorable leading to higher potency, in both α_{1A}-AR and α_{2A}-AR as compared to (+)-NE because of hydrogen bonding interactions β-OH of (-)-NE with the aspartate residue of TM III. The energetically unfavorable β-OH of (+)-NE lies in space towards hydrophobic pocket. Its low pharmacological potency is compared with dopamine which lacks the β-OH group.

Abbreviations

α_{1A}-AR,	Alpha1A adrenergic receptor;
α_{2A}-AR,	Alpha2A adrenergic receptor;
AR,	Adrenergic receptor/ Adrenoceptor;
β_2-AR,	Beta 2 adrenergic receptor;
β-OH,	Beta hydroxy;
NE,	Norepinephrine;
TM,	Transmembrane (helix).

Introduction

Extracellular potassium ions (K^+) have been widely utilized as a tool to study neurotransmitter release from either peripheral or central neurons. Although neurotransmitter released by pulses of the potassium ions appears to have a more limited specificity than that achieved with electrical stimulation, the method is useful in circumstances in which the electrical stimulus may not be adequate. A simple method for the study in vitro of ^{3}H-norepinephrine released from several areas of the central nervous system of the rat has been known for a long time [1]. Under these experimental conditions, the metabolic pathway of the ^{3}H-neurotransmitter resembles that described for ^{3}H-norepinephrine released under conditions of electrical stimulation in the peripheral nervous system. The involvement of neuronal α-adrenoreceptors in a negative feed-back mediated control of norepinephrine release elicited by electrical stimulation from the peripheral nerve endings has been well documented [2,3,4], however the molecular similarity and differences, if any, between the prejunctional receptors (α_2) of the central and the peripheral organs remain to be elucidated. Enantiomers of norepinephrine were used to gain some insight into the prejunctional α-Adrenoceptor of rat hypothalamus and the vas deferens. Molecular modeling of the prejunctional α_{2A}-adrenoceptor as well as that of the well known postjunctional α_{1A}-adrenoceptor was used to examine the degree of the selectivity for the receptors. Although no x-ray structure of α_{1A} and α_{2A}-AR have been published yet, but recently published crystal structure of human beta-2 adrenoceptor (hβ_2-AR) [5] combined with functional, structural and experimental data can produce reliable model of α_{1A} and α_{2A}-AR at atomic resolution. Atomic details of the receptors are needed to gain insight into conformational changes upon receptor activation. In previous studies, it has been shown that agonist binding site is located between TM III, TM V and TM VI (Figure 1A and B) and movement of these TM helices with respect to each other leads to activation of the receptor [6-10]. Previously, it has been also argued that a rotation of TM V is likely to be involved in receptor activation [11,12]. Figure 1B shows SiteMap's predicted ligand binding site at AR. SiteMap (v2.2, Schrödinger, LLC, New York, NY, 2008) uses a novel algorithm for rapid binding site identification and evaluation.

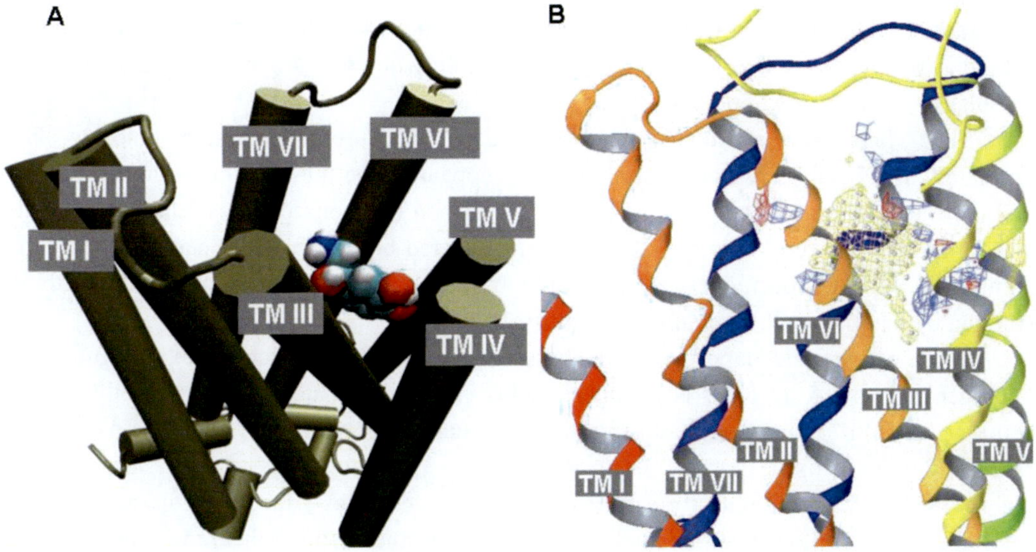

Figure 1. A. Representation of transmembrane helices of adrenoceptors (cartoon drawing of helices) and agonist binding pocket. Agoinst is represented as VWD form in the binding pocked formed by TM III, TM V and TM VI. B. SiteMap's predicted agonist binding pocket between TM III, TM V and TM VI. Yellow wireframe corresponds to hydrophobic pocket and red and blue wireframe corresponds to hydrogen bond acceptor and donor respectively.

Material and Methods

Experimental Methods

The hypothalamus and vas deferens of the Sprague-Dawley male rat (190-300g) were used for central and peripheral noradrenergic neurons, respectively. For the isolation of the hypothalamus and the vasa deferentia, the rats were sacrificed. Tissue slices of approximately 1 mm each were prepared and immediately transferred to the incubation medium in a cylindrical chamber, with a piece of nylon mesh fitted in the bottom as a small basket, as described by Farah et al [1]. The whole system was placed in a 10 ml beaker containing 2.5ml of slightly modified Kreb's solution previously equilibrated with 95% and 5% CO_2. The composition of the Kreb's solution was the following (millimolar concentrations): NaCl, 118; KCl, 4.7; $CaCl_2$, 2.5; $MgCl_2$, 0.54; NaH_2PO_4, 1.0, $NaHCO_3$, 25.0; glucose, 11.1; sodium ethylenediamine tetraacetic, 0.004 and ascorbic acid, 0.11. The temperature was maintained at 37 ^{0}C in a Dubnoff incubator with moderate shaking and the solution was oxygenated through a tube fitted to the bottom of the basket. The nylon mesh of the incubation system allowed the free passage of the bathing solution contained in the beaker, without loss of the tissue slices placed on it. A 40-minute period of stabilization was allowed to elapse before starting the experiment. During this period the Kreb's solution was replaced every 10 min.

Endogenous neuronal norepinephrine stores were labelled by incubating the tissues for 30 min with 0.1 μM of (-)-[7, 8-^{3}H (N)]-norepinephrine (New England Nuclear Corporation, Boston, Mass., specific activity 32.7 Ci/mmole. At the end of the incubation period, the tissue was washed eight times for 1 min each with fresh Kreb's solution. Subsequently, the Kreb's solution was replaced every 5 min during the following 60 min before the periods of stimulation by exposure to potassium began.

Tissue slices were stimulated for 1 minute by exposure to potassium chloride. The optimum 25mM and 75mM concentration of K^{+} appears to be the best for the hypothalamus and the vas deferens, respectively. The interval between two consecutive stimulation periods was 26 minutes. Three 1-minute samples were collected before and after each period of the stimulation. The radioactivity was monitored by counting 2.0 ml aliquots of the bathing solution by a liquid scintillation (Kew scientific).

At the end of the experiment the tissue was blotted on filter paper, weighed and homogenized with 5 ml of 0.4N perchloric acid. The homogenate was centrifuged at 7,000 rpm for 15 min and an aliquot of the supernatant was taken for estimation of total radioactivity in the tissue.

Overflow of radioactivity for each sample was expressed as "fractional release," i.e., the total nCi released divided by the total nCi remaining in the tissue at this time. The overflow of radioactivity elicited by potassium was calculated by subtracting the spontaneous outflow assumed to have occurred in each sample during and after the period of stimulation. The value of spontaneous outflow subtracted from these samples was the basal resting value obtained in the 1-min period immediately before exposure to potassium. The "fractional release induced by exposure to potassium" was the sum of all increases above spontaneous levels induced by the period of K^{+}-stimulation, and it represents the fraction of the radioactivity present in the tissue at the onset of stimulation released by exposure to potassium.

Only total radioactivity was measured, and no attempt was made to estimate the amount of radioactivity collected as unmetabolized ^{3}H-norepinephrine, since during ^{3}H-transmitter release elicited by exposure to K^{+} from central neurons [1] or peripheral tissues [13] uncharged ^{3}H-norepinephrine accounted for most of the increase in radioactivity. Additionally, when the stimulation is performed in the presence of an uptake inhibitor, practically all the radioactivity released is collected at unmetabolized ^{3}H-transmitter [1].

Statistical evaluations of the results were made according to conventional procedures [14]. The following drugs were used: (-)-norepinephrine (+)-bitartatehydrate and (+)-norepinephrine (+)-bitartate hydrate (Sterling-Winthrop Research Institute, Rensselaer, N.Y.), desmethylimipramine HCl (USV, Pharmaceutical Corporation, Tuckahoe, N.Y.), phentolamine HCl (Ciba Pharmaceutical Company, Summit, N.J.), phenoxybenzamine HCL (Smith Kine & French Labs., Philadelphia, PA), and clonidine (Boehringer Ingelheim, Germany).

Computational Methods

Homology modeling of α_{1A} (SWISS-PROT accession number P35348) and α_{2A} (SWISS-PROT accession number P08913) adrenergic receptors were done based on predicted active conformation model of human Beta-2 (hβ2) AR (template). Active hβ2 AR model was predicted based on 50ns NTP molecular dynamic simulation of crystal structure of hβ2 AR (PDBID: 2RH1) and endogenous ligand (-)-epinephrine (unpublished data). Set of initial protein models of the α_{1A} and α_{2A} were generated via MODELLER v9.2 [15]. Information on transmembrane helix residues were obtained from SWISS-PROT and alignment of transmebrane residues with template were done by Clustal W and manually (Figure 2). The backbone trace of the α_{1A} and α_{2A} models were kept with similar conformation to the hβ2 AR template, whereas the arrangement of the side chains of the binding-site residues were predominantly determined by available details about the interaction between protein and ligands. (-)-norepinephrine and (+)-norepinephrine were docked into the agonist binding sites by using AutoDock v4β [16]. The Lamarckian genetic algorithm (LGA) for ligand conformational searching was selected because it has enhanced performance relative to simulated annealing or the generic genetic algorithm [17]. Besides ligand being fully flexible, we allowed the side chains of ligand contacting residues to be flexible also, in order to simulate the binding process and the induced-fit effect. Energy minimization of the docked ligand and receptors were done using sander module of AMBER 9 [18].

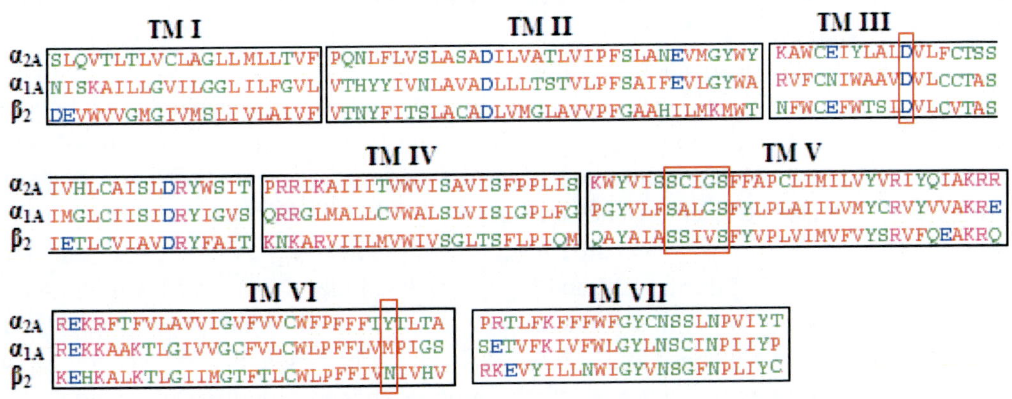

Figure 2. Alignment of amino acid residues: Transmembrane helix regions of human α_{2A}, α_{1A} and β_2 adrenergic receptors are shown. Red box in TM III indicates highly conserved aspartate residue. Red box in TM V indicates conserved serine residues. A cystine residue in α_{2A} corresponds to alanine residue in α_{1A} and Serine residue in β_2 adrenergic receptor. Red box in TM VI indicates difference in key residues in α_{2A} (Tyrosine), α_{1A} (Methonine) and β_2 (Asparagine) adrenergic receptors.

General Observations

In both tissues K^+ induced release of the labelled transmitter was enhanced by the competitive α-adrenoceptor blocker phentolamine (5μM, n=5) or the irreversible α-adrenoceptor blocker phenoxybenzamine (5μM - 8.7μM) indicating that inhibitory prejunctional receptor mediated activity of the transmitter norepinephrine was blocked.

Furthermore, in hypothalamus, the K^+ evoked release was reduced nearly 50% by the 0.15μM of the potent prejunctional adrenoceptor agonist clonidine (n=4). These observations indicate that in both tissues, selection of K^+ provided an essential tool to investigate the prejunctional receptor in the sympathetic nerve terminal.

Effects of (-)-and (+)-norepinephrine on the K^+-evoked release of ^{3}H-transmitter in rat vasa deferentia and hypothalamus

(a) Vasa Deferentia

Sixty-three minutes after the end of incubation of slices of the vasa deferentia with ^{3}H-norepinephrine, the fraction released spontaneously (0.36 ± 0.02%, n=15) represented a minor proportion of the total radioactivity retained in the tissue. There was a small, steady decline in the basal efflux of tritium as a function of time.

In all experiments, desmethylimipramine (DMI) was used to inhibit the neuronal uptake of the released norepinephrine. In the absence of neuronal uptake inhibitor, the exposure of tissue slices to very low concentration of norepinephrine induces a marked increase in the spontaneous outflow of radioactive products [19,20].

Even in the presence of DMI, the two higher concentrations of either (-) or (+)-norepinephrine increased the fraction of released spontaneously from vasa deferentia slices. The increase observed during exposure to norepinephrine (9μM) was slight and was approximately the same for both enantiomers. Perhaps norepinephrine displaces the labelled transmitter by neuronal membrane transport (uptake$_1$) mechanism. It was evident that a stereoselective increase of spontaneous outflow occurred when a 3-fold higher concentration of both enantiomers was employed. The higher concentrations of norepinephrine enantiomers were also able to override the blockade of neuronal uptake when the concentration of DMI was increased up to 5 μM. Concentrations of DMI above 5 μM increased the basal efflux of radioactivity. The α-Adrenoceptor blockade occurred when a higher concentrations of DMI was used [21]. These minor technical difficulties were unavoidable.

Concentrations of (-)-NE of 1 μM and higher inhibited the tritium overflow induced by exposure of tissue slices to K^+. In separate experimental groups the inhibitory potency of (-)-NE, 1 μM, on the release induced by K^+ at two different times of stimulation were compared. There were no differences in the degree of inhibition when the tissue was exposed to (-)-NE during the second period of stimulation (S_2: 38.5±8.0% of inhibition, n=4) consecutive to the first one performed in the absence of the drug, and its inhibitory effect during the fourth stimulus (S_4: 35.1±3.2%, n=3), after the tissue was previously exposed to other concentrations of (-)-NE during the stimuli.

In contrast to these results, there was no inhibition in ^{3}H-transmitter overflow when the tissue slices were exposed to 1 μM of the corresponding (+)-enantiomer. It was necessary to increase the concentration of (+)-NE to 27 μM in order to reduce to approximately one half the fractional release elicited K^+.

(b) Hypothalamus:

In the presence of DMI, the spontaneous release of radioactivity was a constant fraction of the tissue radioactivity from slices of the hypothalamus. Exposure of tissue slices to the highest concentration of either (-) or (+)-NE induced a moderate increase in the spontaneous outflow.

There was a small decline in the fractional release in four successive one-minute exposures to 25 mM K^+. Increasing concentrations of (-)-NE dose-dependently reduced the overflow induced by K^+. A concentration of 3 μM of the (-)-enantiomer was necessary to inhibit the fractional release by nearly 50%. (+)-NE was considerably less effective in inhibiting the K^+-evoked secretion of tritiated transmitter. At the highest concentration of (+)-NE, there was a significant degree of inhibition achieved.

As shown in Figure 3, a similar picture emerged when the inhibitory effects of both norepinephrine enantiomers were compared in the two systems, with the values being corrected for the variation in the corresponding control stimulus. In the both hypothalamus and vasa deferentia, the same concentration of (-)-NE was necessary to inhibit the release by nearly 50% (3 μM: 46.9 ± 5.2% in hypothalamus and 46.9 ± 8.5% in vasa deferentia). It required a 9 fold higher concentration of (+)-isomer to achieve equivalent inhibition in the vasa deferentia, with comparable results for the hypothalamus.

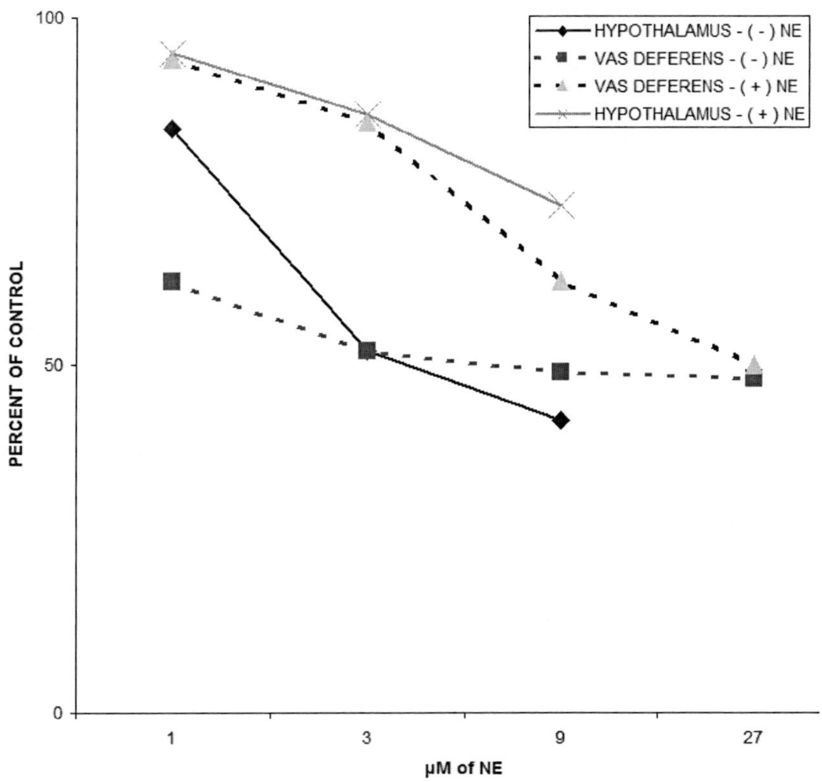

Figure 3. Comparison of the effect of (-) -and (+)-norepinephrine on ^3H-transmitter release induced by K^+ in central and peripheral neurons. Ordinates: fractional release induced by K^+ expressed as a percentage of the first stimulus (S_1). The values were corrected by the variation in the corresponding stimulus in the presence of DMI alone. The tissue slices were stimulated by 1- minute exposure to K^+, 75 mM (vasa defernetia) or 25 mM (hypothalamus).Abscissa: (-)-or (+)-norepinephrine (NE) at the concentration which is indicated was added 8 min before stimulation with K^+. Each point represents the mean value of four to seven experiments.

Norepinephrine interactions with α_{1A}- and α_{2A}- AR:

Asp106 (corresponding Asp113 of α_{2A} AR) of TM III of α_{1A}-AR is highly conserved residue among Adrenergic receptors (figure 2). It is involved in electrostatic as well as hydrogen bonding interactions with (-)-NE. The most important interaction is formed between positively charged amino group of NE and negatively charged Asp106 of TM III of both enantiomers. The β-OH group of (-)-NE forms hydrogen bond with the Asp106 and Asp113 of TM III of α_{1A}- and α_{2A}-AR respectively (Figure 4A, C), whereas this interaction of β-OH group of (+)-NE with aspartate residue of TM III can not take place (Figure 4B, D). Catechol ring of NE orients towards TM V of AR and make hydrogen bond network with Ser188 and Ser192 (corresponding Ser200 and Ser204 of α_{2A} AR) of TM V. Hydrophobic pocket is mainly formed by residues of TM VI and TM III. Phe288, Phe289 and Met292 of TM VI make one side of hydrophobic pocket while Val104 forms other side of hydrophobic pocket of α_{1A}-AR. These residues are involved in interaction with catechol ring of agonist. Phe390 and Phe391 of VI make one side of hydrophobic pocket while Val114 forms other side of hydrophobic pocket of α_{2A}-AR.

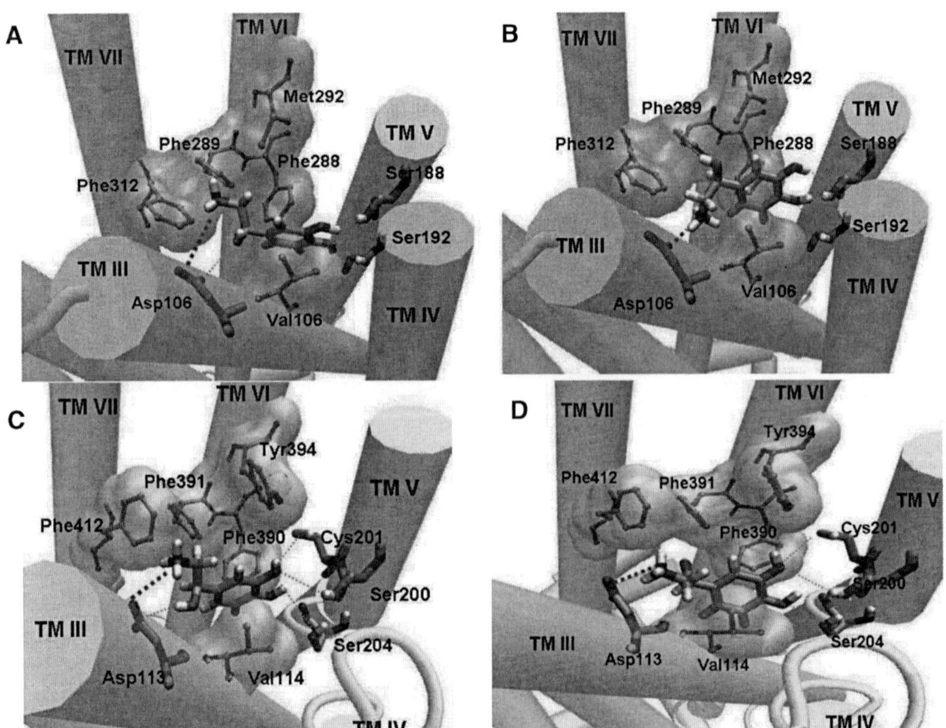

Figure 4. A. Interactions of (-)-NE with α_{1A}. B. Interactions of (+)-NE with α_{1A}. C. Interactions of (-)-NE with α_{2A}. D. Interactions of (+)-NE with α_{2A}. Transmembrane helices are shown in cartoon drawing. Red dash line denotes electrostatic interactions and green dash line denotes hydrogen bonding interaction. Surface and CPK drawing of residues involved in hydrophobic interactions are shown.

Differences in agonist binding pockets of between α_{1A}- and α_{2A}-AR

Transmembrane Helix V:

The α_{2A}-AR has a Cys201 residue in TM V whereas α_{1A} has alanine residue at corresponding position in TM V (figure 2). In β_2-AR receptors there are three conserved serine residues, which have been shown to be involved in hydrogen bond formation with hydroxyl of catechol ring. Our modeling and docking studies shows involvement of Cys201 in hydrogen bond formation with meta-hydroxyl of catechol ring. It has been shown that Cys201 is exposed in the binding cavity [22], and is important for agonist binding. SH groups of Cys201 residues can form interactions over long distance than OH groups. Thus, Cys201 of α_{2A}-AR appears to play important role in binding with agonist NE. This type of interaction is missing in α_{1A}-AR.

Transmembrane helix VI:

There are some differences in active site residues of TM VI involved in hydrophobic interactions with catechol ring of agonist in the α_{1A}- and α_{2A}-AR. In α_{1A}, Met292 is highly involved in hydrophobic interaction with the catechol ring of NE. In contrast, Tyr394 is present in α_{2A} at position equivalent to Met292 (α_{1A}) (figure 2). Tyr361 does not show any favourable interaction with NE.

Discussion

The inhibitory effect of extracellular norepinephrine in these experiments, in which pulses of elevated K^+ have been used as stimulus, closely resemble those obtained in numerous peripheral tissues under experimental conditions in which electrical stimulation was used [19,20,23,24,25,26], In contrast to the inhibitory effects of exogenous (-)-norepinephrine, an enhancement in transmitter release induced by electrical stimulation is observed when the presynaptic α-adrenoceptor are blocked by drugs [27,28,29].In the present experiments, both phentolamine and phenoxybenzamine enhanced the overflow of radioactivity elicited by K^+ in the presence of DMI from slices of the rat vasa deferentia.

The result obtained in the hypothalamus showed that in the same organ, under similar experimental conditions, a different ID_{50} can be obtained depending on the degree of depolarization and/or experimental conditions (i.e., presence of uptake inhibitor). The concentration of (-)-or (+)-norepinephrine required to inhibit the ^{3}H-transmitter release induced by K^+ by 50% in the vasa deferentia, however, are greater than the corresponding one obtained when the release was elicited by electrical stimulation in the same organ of the rat (unpublished).

Our results show that the α-adrenoceptor involved in modulating evoked norepinephrine release exhibit stereoselectivity characteristics both in central and peripheral neurons. In the present study, the isomeric activity ratio for the inhibitory effects of the isomers is approximately 10. This is small as compared to the ratio for the binding of enantiomers to the postjunctional α-adrenoceptor related membrane fragments from rat brain [30]. On the other hand, a ratio of 10 was also obtained for the reduction in the body temperature of the cat when the enantiomers were injected into the hypothalamus [31]. Although the relative importance of the pre- and postjunctional alpha receptor for the physiological regulation of the particular function in the central nervous system is yet to be clarified, the enantiomers and molecular modeling may aid in such an undertaking.

The α_{1A}- AR Agonist Binding and Differences from α_{2A}- AR

The agonist binding pocket of α_{1A} and α_{2A} are constituted by the electrostatic interaction of negatively charged Aspartate residue of TM III and positively charged amino group of norepinephrine. This Aspartate residue of TM III has been shown to form salt bridge with Lysine residue of TM VII. Initiation of activation process takes place by disruption of this salt bridge by the agonist [32], but this one site interaction does not account for the full activation of receptor. For full activation, ligand has to orient in a way that makes favourable interactions between catechol ring of agonist and Serine residues of TM V.

Modeling and docking studies of α_{1A} and α_{2A} with enantiomers of NE showed hydrogen bond formation between hydroxyl group of catechol ring and Ser188 (corresponding Ser200 of $\alpha_{2A)}$ and Ser192 (corresponding Ser204 of $\alpha_{2A)}$ of TM V of α_{1A}-AR. In addition, Cys201 residue (TM V) of α_{2A}-AR also shows hydrogen bond formation with meta-hydroxyl of catechol ring. This third residue (Cys201), involved in hydrogen bond formation is missing in $\alpha_{1A.}$ Ser204 in TM V of β_2 is present at the corresponding Cys201 of α_{2A}-AR. Ser204 of β_2 has been shown to be involved in hydrogen bond formation with catechol ring of catcholamines [33]. Our docking studies did not distinguish between two different orientations of ring. Agonist binding conformations were chosen on the following basis: (i) Conformation with lower energy and higher clustering. (ii) On the basis of previous mutagenesis studies [34]. Meta-hydroxyl of catechol shows interactions with Cys201 and Ser200 (corresponding Ser188 of α_{1A} AR) of α_{2A}-AR, para-hydroxy group of catechol shows interaction with Ser204 (corresponding Ser192 of α_{1A}-AR) of α_{2A}-AR. Phenylephrine (missing para-hydroxy), α_{1A} adrenergic full agonist, shows interaction of meta-hydroxy with both Ser188 and Ser192 of TM V and this concludes that only two serine residues in TMV are necessary in activation process of alpha adrenergic receptor. In α_{1A}, only two residues of TMV are involved in interaction with catechol hydroxyl group while α_{2A} shows involvement of three residues of TMV. But this could not lead to higher affinity for endogenous ligands; Indeed α_{1A} shows higher affinity than α_{2A} for endogenous ligand. This could be explained by the orientation and packing of the catechol rings in the active site. Met292 of α_{1A} makes energetically more favourable hydrophobic interactions with the catechol ring and involved in orienting ring favourable for activation.

The β-OH of (-)-NE enantiomer forms hydrogen bond with the Asp106 and Asp113 of α_{1A} and α_{2A} respectively, whereas β-OH of (+)-NE enantiomer is not involved in such type of favourable interactions and this missing interaction could be important in stereoselectivity and receptor activation. In our previous communication when rhodopsin crystal structure template was used to study interactions of the enantiomers of epinephrine similar differences between two forms were noted [35]. The chiral β-OH of both enantiomers of epinephrine was thought to involve in interaction with Asp of TM III. Ligand binding and receptor activation is a dynamic process, it could be hypothesized that the landing of ligand to the receptor occurs by stronger interactions i.e. binding of positively charged amino group and β-OH of catecholamines to aspartate of TM III. The β-OH of (+)-NE may not sustain the initial interactions during the global conformational change of receptor. Thus as compared to that of the (-)-NE, the missing β-OH interaction with Asp106 and missing meta- OH of catechole interaction with Cys201 of α_{1A}-AR makes (+)-NE less effective for receptor activation. The orientation of catechol ring differs in both enantiomers; this could explain why there is more stereoselectivity at α_{1A}-AR as compared to α_{2A}-AR. This is consistent with the K_i value of two enantiomers [36]. The binding affinity of dopamine (no β-OH group) is also lower than (-)-NE enantiomer. The β-OH interaction with aspartate of TM III is missing in dopamine. The order of pharmacological potency of these catecholamines on many organs is (-)-NE > (+)-NE = dopamine. As predicted by Easson and Stedman hypothesis that (+)-NE acts like dopamine [37,38], the β-OH interaction is missing in both cases. In α_{1A} AR, Met292, Phe288 and Phe289 of TM VI makes one side of hydrophobic pocket, which are involved in hydrophobic interactions with the catechol ring of agonist. Val107 (TM III) makes other side of hydrophobic pocket for interaction of ring. To determine optimal binding and receptor activation, details of ligand binding pocket at atomic level is needed. The resulting difference at atomic level of ligand binding pocket of α_{1A}- and α_{1A}-AR will provide rational basis for the design of better α_{1A}- and α_{1A}-AR subtype selective agonists.

Acknowledgments

This work was supported in part by an allocation of computing time from the Ohio Supercomputer Center. Early pharmacological work in part was supported by United States Public Health Service Grant from the National Institutes of Health, Grant Number GM 17859.

References

[1] Farah MB, Adler GE, Langer SZ. Possible physiological significance of the initial step in the catabolism of noradrenaline in the central nervous system of the rat. *Naunyn-Schmiedeberg's Arch. Pharmacol.* 1977; 297:119-131.

[2] Starke K, taube MD, Borowski E. Presynaptic receptor system in catecholaminergic transmission. *Biochem. Pharmacol.* 1977; 26:259-268.

[3] Langer SG. Sixth Gaddum Memorial Lecture: Presynaptic receptors and their role in the regulation of transmitter release. *Br. J. Pharmac.* 1977; 60:481-497.

[4] Westfall TC. Local regulation of adrenergic neurotransmission. *Physiol. Rev.* 1977; 57:659-728.

[5] Cherezov V, Rosenbaum DM, Hanson A, Rasmussen SG, Thian FS, Kobilka T S, Choi HJ, Kuhn P, Weis WI, Kobilka BK, Stevens RC. High-resolution crystal structure of an engineered human beta2-adrenergic G protein-coupled receptor. *Science* 2007; 318:1258-1265.

[6] Kobilka BK. Adrenergic receptor structure and activation. *Pharmaco.l Commun.* 1995; 6:119-124.
[7] Beck-Sickinger AG, Structural characterization and binding sites of G-protein-coupled receptors. *Drug Discovery Today*. 1996; 1:502-513.
[8] Gether U, Lin S, Ghanouni P, Ballesteros JA, Weinstein H and Kobilka BK Agonists induce conformational changes in transmembrane domains III and VI of the beta2 adrenoceptor. *EMBO J* 1997; 16:6737-6747.
[9] Unger VM, Hargrave PA, Baldwin JM and Schertler GF. Arrangement of rhodopsin transmembrane alpha-helices. *Nature (Lond)* 1997; 389:203-206.
[10] Gether U and Kobilka BK. G protein-coupled receptors. II. Mechanism of agonist activation. *J. Biol. Chem.* 1998; 273:17979-17982.
[11] Marjamki A, Frang H, Pihlavisto M, Hoffrn A-M, Salminen T, Johnson MS, Kallio J, Javitch JA and Scheinin M. Chloroethylclonidine and 2-aminoethyl methanethiosulfonate recognize two different conformations of the human alpha 2A-adrenergic receptor. *J. Biol. Chem.* 1999; 274:21867-21872.
[12] Salminen T, Varis M, Nyrnen T, Pihlavisto M, Hoffrn A-M, Lnnberg T, Marjamki A, Frang H, Savola J-M, Scheinin M and Johnson MS. Three dimensional models of alpha 2A-adrenergic receptor complexes provide a structural explanation for ligand binding. *J. Biol. Chem.* 1999; 274: 23405-23413.
[13] Strake K, Montel H. Influence of drug with affinity for α-adrenoceptors on noradrenaline release by potassium, tyramine and dimethylpiperazonium. *Eur. J. Pharmacol.* 1974; 27:273-280.
[14] Snedecor GW, Cochran WG. Ststistical methods. 6th ed., The Iowa State University Press, Ames, 1969.
[15] Fiser A, Do RK, Sali A. Modeling of loops in protein structures, *Protein Science* 2000; 9:1753-1773.
[16] AutoDock Tools, http://www.autodock.scripps.edu/resources/adt/index_html
[17] Goodsell GM, Halliday DS, Huey RS, Hart R, Belew WE, Olson RK, Olson AJ, Automated docking using a Lamarckian genetic algorithm and an empirical binding free energy function, *J. Comp. Chem.* 1998; 19:1639-1662.
[18] Ponder JW, Case DA. Force fields for protein simulations. *Adv. Prot. Chem.* 2003; 66:27-85.
[19] Starke K. Influence of extracellular noradrenaline on the stimulation-evoked secretion of noradrenaline from sympathetic nerve: evidence for an α-receptor-mediated feed-back inhibition of noradrenaline release. *Naunyn-Schmiedeberg's Arch. Pharmacol.* 1972; 275:11-23.
[20] Enero MA, Langer SZ. Inhibition by dopamine of ^{3}H-noradrenaline release elicited by nerve stimulation in the isolated cat's nictitating membrane. *Naunyn-Schmiedeberg's Arch. Pharmacol.* 1975; 289:179-203.
[21] Thoenen H, Huerliman A, Haefely W. Mode of action of imipramine and 5-(3`-methylaminopropylen)-dibenzo[α, e]cyclohepta [1,2,3] trine hydrochloride (Ro-60110, a new antidepressant drug, on peripheral adrenergic mechanisms. *J Pharmacol. Exp. Ther.* 1964; 144:405-414.

[22] Marjamki A, Pihlavisto M, Cockcroft V, Heinonen P, Savola JM and Scheinin M Chloroethylclonidine bind irreversibly to exposed cysteines in the fifth membrane-spanning domain of the human alpha 2A-adrenergic receptor. *Mol. Pharmacol.* 1998; 53:370-376.

[23] Farnebo LO, Malmfors T, ^{3}H-Noradrenaline release and mechanical response in the field stimulated mouse vas deferens. *Acta Physiol. Scand. Sppl.* 1971; 371:1-18.

[24] Kirpekar SM, Furchgott RF, Wakade AR, Prat JC. Inhibition by sympathomimetics amine of the release of norepinephrine evoked by nerve stimulation in the cat spleen. *J. Pharmcol. Exp. Ther.* 1973, 187:529-538.

[25] Vizi ES, Somogyi GT, Hadhary P, Knoll J. Effect of duration and frequency o stimulation on the presynaptic inhibition by α-adrenoceptor stimulation of the adrenergic transmission. *Naunyn-Schmiedeberg's Arch. Pharmacol.* 1973; 280: 79-91.

[26] Stjarne L. Selectivity for catecholamines of presynaptic alpha-receptors involved in feedback control of sympathetic neurotransmitter secretion in guines pig vas deferens. *Naunyn-Schmiedeberg's Arch. Pharmacol.* 1975; 288:295-303.

[27] Starke K, Montel H, Schuman HJ, Influence of cocain and phenoxybenzamine on noradrenaline uptake and release. *Naunyn-Schmiedeberg's Arch. Pharmacol.* 1971; 270:210-214.

[28] Ebero MA, Langer SZ, Rothlin RP, Stefano FJE. Role of the α-adrenoceptor in regulating noradrenaline overflow by nerve stimulation. *Br. J. Pharmac.* 1972; 44:672-688.

[29] Farah MB, Langer SZ. Protection by phentolamine against the effects of phenoxybenzamine on transmitter release elicited by nerve stimulation in the perfused cat heart. *Br. J. Pharmac.* 1974; 52:549-557.

[30] U'prichard DC, Greenberg DA, Snyder SH. Binding of radiolabeled agonist and antagonist at central nervous system alpha noradrenergic receptors. *Mol. Pharmacol.* 1977; 13:454-473.

[31] Hattan DG, Wolf HH. Effects of intrahypothalmic 6-hydroxy dopamine on central thermoregulatory response to (+)– and (-)-norepinephrine and dopamine. *Neuropharmacol.* 1977; 16:639-648.

[32] Porter JE, Hwa J and Perez DM. Activation of the α_{1b}-adrenergic receptor is initiated by disruption of an interhelical salt bridge constraint. *J. Biol. Chem.* 1996; 271:26318-28323.

[33] Strader CD, Candelore MR, Hill WS, Sigal IS and Dixon RAF. Identification of two serine residues involved in agonist activation of the β-adrenergic receptor. *J. Biol. Chem.* 1989; 264:13572-13578.

[34] Wang CD, Buck MA Fraser CM. Site-directed mutagenesis of alpha2A-adrenergic receptor: identification of amino acids involved in ligand binding and receptor activation by agonists. *Mol Pharmacol* 1991; 40:168-179.

[35] Patil PN, Li C, Kumari V, Hieble JP. Analysis of efficacy of chiral adrenergic agonists. *Chirality* 2008; 20(3-4):529-43.

[36] Nyronen T, Pihlavisto M, Peltonen JM, Hoffren AM, Varis M, Salminen T, Wurster S, Marjamaki A, Kanerva L, Katainen E, Laaksonen L, Savola JM, Scheinn M, Johnson

MS. Molecular mechanism of agonist-promoted α_{2A}-Adrenoceptor activation by norepinephrine and epinephrine. *Mol. Pharmacol.* 2001; 59:1343-1354.

[37] Easson LH, Stedman E, CLXX. Studies on the relationship between chemical constitution and physiological action. V. Molecular dissymmetry and physiological activity. *Biochem. J.* 1933; 27:1257-1266.

[38] Patil PN, Miller DD, Trendelenburg U. Molecular geometry and adrenergic drug activity. *Pharmacol. Rev.* 1974; 26:323-392.

In: Biogenic Amines
Editors: T. Farooqui, A. A. Farooqui

ISBN: 978-1-60876-625-3

Chapter XIII

TRACE AMINES AND THEIR RECEPTORS COMPRISE A NOVEL VERTEBRATE AMINERGIC SIGNALLING SYSTEM

Mark D. Berry[1,2,*], Jarrod Nickel[1] and Bruno Tomberli[1]
[1]Brandon University, Brandon, Manitoba, Canada;
[2]University of Saskatchewan, Saskatoon, Saskatchewan, Canada.

ABSTRACT

The trace amines are a class of endogenous amines that are found in the central nervous system of all species so far examined. Although these compounds (typified by 2-phenylethylamine, tryptamine and *p*-tyramine) have been established as neurotransmitters in invertebrate species, the available evidence suggests that trace amines do not act as traditional neurotransmitters in mammalian species. Trace amines are released from neurons, but not in an activity dependent manner. Further, administration of exogenous trace amines at doses approaching physiological concentrations, is not associated with changes in neuronal excitability. Rather, at physiological concentrations, trace amines regulate the responsivity of neurons to co-applied neurotransmitters. Since trace amine synthesis appears to be regulated in response to monoamine neurotransmitter receptor activation, it has been proposed that, at least in mammals, trace amines serve a control function, maintaining neuronal responsivity within defined physiological limits. Such a divergence of trace amine functioning between species is also seen in recent genetic studies of trace amine-associated receptors (TAAR). Vertebrate TAAR, although G-protein coupled, appear to have arisen independently of invertebrate TAAR. Further, mammalian TAAR appear to lack glycosylation sites required for membrane insertion. Additional evidence suggests that vertebrate TAAR are rapidly evolving, with a number of species specific variants

* Correspondence concerning this article should be addressed to: Mark D. Berry, Dept. Chemistry, Brandon University, 270, 18th St., Brandon, MB, Canada, R7A 6A9. Tel: 204-727-9775; Fax: 204-728-7346; e-mail: berrym@brandonu.ca.

identified. On this basis it has been suggested that they are involved in species specific adaptive responses. Possibly related to this, sub-types of TAAR have been shown to act as olfactory receptors, being activated in part by urinary amines. Thus, there appear to be significant differences between invertebrates and vertebrates with respect to trace amine functioning. This chapter will review the recent developments in the field of vertebrate trace amine pharmacology, and discuss the evidence for trace amines being the prototype of a new class of endogenous molecule, acting to modulate neural responsivity, rather than directly causing signal transduction.

Introduction

The trace amines are a class of endogenous compounds showing close structural similarity to the monoamine neurotransmitters dopamine, noradrenaline, adrenaline and 5-hydroxytryptamine (Figure 1) and sharing common metabolic pathways with these neurotransmitters [1]. Historically, the trace amines have been regarded as comprising the compounds 2-phenylethylamine (PE), *p*-tyramine (TA), octopamine, synephrine and tryptamine. As will be described below however, recent findings suggest that there may be additional members of this family, in particular some of the previously presumed inactive metabolites of monoamine neurotransmitters and thyroid hormone. Although found in all species so far examined [1], there is now considerable evidence that trace amines serve different functions in different species. In particular it is widely established that in invertebrates trace amines act as neurotransmitters, possibly serving the role of an invertebrate adrenergic system [2]. For vertebrates however, this appears not to be the case. Endogenous levels of trace amines are approximately 100-fold below those of the monoamine neurotransmitters in the central nervous system [1]. Further, there is no evidence of trace amines directly effecting neurotransmission at physiological levels, nor being released from nerve terminals in an activity dependent manner [1]. In 2001 a distinct family of G-protein coupled receptors was discovered, some of which were selectively activated by trace amines [3,4]. Prior to this vertebrate trace amines were regarded as little more than metabolic by-products. In the following sections, the physiology and pharmacology of trace amines and their associated receptors will be briefly reviewed. A number of reviews on trace amines have recently been published to which the reader is also referred for further details [1,5-8]. Here, we will focus primarily on some of the more recent developments in the field of vertebrate trace amine functioning.

Trace Amine-Associated Receptors

In 2001 two groups independently identified a family of novel G-protein coupled receptors some of which appeared to selectively respond to trace amines [3,4], thus providing a viable target for previously observed responses. As other members of this new receptor family were subsequently identified, a number of potential nomenclatures appeared in the literature. Recently, three proposals for a systematic naming system for this family of receptors have been provided [6,9,10]. The International Union of Pharmacology nomenclature [10] has only officially named one of the receptors, TAAR1. The nomenclatures of Lindemann [6] and Hussain [9] vary only in that Hussain includes the recently identified large expansion of TAAR seen in teleosts. This inclusion resulted in a modification in the classification of TAAR sub-classes previously proposed by Lindemann and colleagues [6] (Table 1).

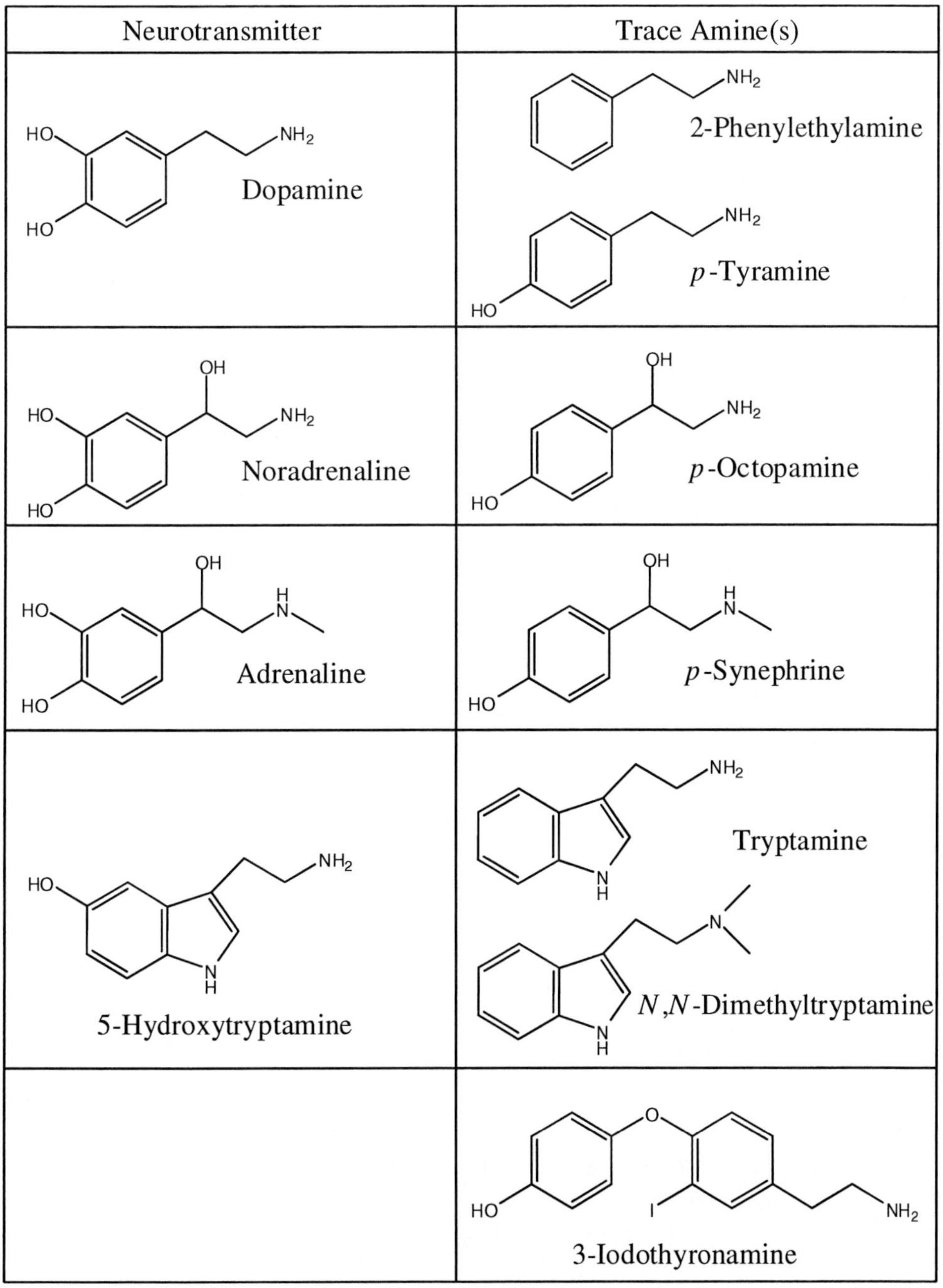

Figure 1. Structures of trace amines and the corresponding neurotransmitters.

Table 1. TAAR isoforms identified in vertebrate genomes and their sub-classifications

TAAR	Number of genes*	Sub-class (Lindemann) [6]	Sub-class (Hussain) [9]
1	1 (**TAAR1**)	1	I
2	1 (**TAAR2**)	1	II
3	1 (**TAAR3**)	1	II
4	3 (**TAAR4, TAAR4a-b**)	1	II
5	1 (**TAAR5**)	2	II
6	7 (**TAAR6, TAAR6a-f**)	3	II
7	10 (**TAAR7, TAAR7a-i**)	3	II
8	4 (**TAAR8, TAAR8a-c**)	3	II
9	7 (**TAAR9, TAAR9a-f**)	3	II
10	5 (**TAAR10, TAAR10a-d**)	Not classified	I
11	1 (**TAAR11**)	Not classified	I
12	13 (**TAAR12a-m**)	Not classified	II
13	6 (**TAAR13, TAAR13a-e**)	Not classified	II
14	12 (**TAAR14a-l**)	Not classified	III
15	2 (**TAAR15a-b**)	Not classified	III
16	7 (**TAAR16a-g**)	Not classified	III
17	3 (**TAAR17a-c**)	Not classified	III
18	11 (**TAAR18a-k**)	Not classified	III
19	22 (**TAAR19a-v**)	Not classified	III
20	31 (**TAAR20, TAAR20a-z, TAAR20a_1, TAAR20b_1, TAAR20c_1, TAAR20d_1**)	Not classified	III
21	6 (**TAAR21a-f**)	Not classified	I
22	7 (**TAAR22, TAAR22a-f**)	Not classified	III
23	16 (**TAAR23, TAAR23a-o**)	Not classified	III
24	4 (**TAAR24, TAAR24a-c**)	Not classified	III
25	12 (**TAAR25a-l**)	Not classified	III
26	28 (**TAAR26a-z, TAAR26a_1, TAAR26b_1**)	Not classified	III
27	7 (**TAAR27, TAAR27a-f**)	Not classified	I
28	8 (**TAAR28a-h**)	Not classified	III

*Number of genes identifies the total number of genes, including pseudogenes, identified in all vertebrate species so far studied. The proposed nomenclature [9] for these isoforms is given in parentheses. Note that not all isoforms of a given receptor are necessarily present in all species, while some of the isoforms may be pseudogenes in some species but remain functional in others. For example, with TAAR7 there is a single TAAR7 pseudogene in the human and opossum genomes which is distinct from the TAAR7a-i variants found in rat and mouse genomes. TAAR7c is a pseudogene in mice, while TAAR7f and TAAR7i are pseudogenes in rat. There are no functional or pseudogene variants of TAAR7g-i in the mouse genome.

Consistent with the apparent divergence of function between invertebrate and vertebrate trace amines, molecular phylogenetic analysis has revealed that vertebrate TAAR arose independently of invertebrate trace amine receptors [11]. Initially it was suggested that within mammalian and avian species there were three broad sub-classes of TAAR [6,12,13], with two additional sub-classes identified when teleost fish and amphibians were also included in phylogenetic analysis [12]. One of these additional sub-classes was only identified in teleosts, while the other was present in both teleosts and amphibians. In the amphibian and one avian species studied, there also appeared to be a lack of orthologs of class 3 TAAR [12,14]. A more comprehensive phylogenetic analysis recently published consolidated TAAR into 3 sub-classes, the third of which is restricted to teleosts [9]. This teleost specific sub-class has not been studied further, and will not be discussed in detail in this chapter. Here we will use this sub-classification scheme proposed by Hussain [9] since it is based on the most comprehensive phylogenetic analysis yet conducted.

Table 2. TAAR gene numbers in vertebrate species

	Functional TAAR Genes	Pseudogenes
Human	5	3 [11, 12]
	6	3 [6, 9]
Chimpanzee	3	6 [6]
Rat	17	2 [6, 9]
	17	1 [11]
Mouse	15	1 [6, 9, 12]
	15	2 [11]
Opossum	22	0 [16]
	22	3 [12]
	19	0 [9]
Dog	2	2 [16]
Cow	17	9 [16]
	13	0 [9]
Chicken	3	2 [12]
	3	0 [9, 14, 16]
Platypus	4	1 [16]
Frog	2	1 [9, 16]
	6	1 [12]
Zebrafish	112	4 [9]
	109	10 [12]
	57	40 [15]
Medaka	25	7 [12]
	25	1 [9]
Stickleback	49	15 [12]
	48	0 [9]
Fugu	18	0 [9]
	13	6 [12]
	8	7 [15]
Pufferfish	17	NR [17]
Tetraodon	18	0 [9]
Elephant shark	2	0 [9]

NR = not reported.

Although we will focus here on mammalian TAAR, which have been more thoroughly studied, there are interesting differences between mammalian and non-mammalian vertebrate

TAAR, which warrant brief mention. With respect to mammalian species, all so far examined in detail appear to express at least one functional receptor from each of sub-classes I and II [6,7,9]. Interestingly there is a much greater divergence of TAAR between closely related species, than is seen with other G-protein coupled receptor families, with a number of pseudogenes present in various species (Table 2) [6,9,12-17]. This has formed part of the basis of suggestions that TAAR are a rapidly evolving family of receptors, that may be involved in species-specific adaptive responses [6,9,15]. As will be detailed later, recent functional studies also appear consistent with this hypothesis for at least some of the TAARs.

The teleost specific sub-class III shows a very pronounced and remarkable expansion of TAAR isoforms [9,12], with upwards of 75% of teleost TAAR belonging to sub-class III [9]. For example, the zebrafish genome contains at least 97 [15] and possibly up to 119 [12] TAAR genes, of which between 57 and 112 [9,12,15] are thought to be functional; there are 48-64 TAAR genes in the stickleback genome, of which at least 48 appear to be functional [9,12]. This compares to the 15 functional genes plus one pseudogene seen in the mouse genome [6,9,12]. The opossum genome contains the greatest number of genes of any mammalian species so far studied with 19-22 functional TAAR and 3 possible pseudogenes [9,12,16]. The human genome consists of 6 functional TAAR with an additional 3 pseudogenes [6,9], although others only reported 5 functional TAAR [11,12]. Many of the additional TAAR genes found in teleosts appear to be species specific [9,17], and it has been hypothesized that these may form the basis of species-specific chemical communication, whether at the levels of reproductive pheromones or other social interactions [12,17]. Further, while TAAR are tightly clustered within an individual chromosome for mammalian and avian species [6,9,12,14], they are widely dispersed across multiple chromosomes in teleosts [9,12]. In addition, teleost sub-class III TAAR have seen multiple intron insertion and deletion events [9,12]. The complete coding sequence of all other TAAR is contained within a single exon with the exception of TAAR2, the gene for which contains one intron [6,12].

Sub-class I TAAR

Sub-class I TAAR consist of TAAR1, 10, 11, 21 and 27 [9,12] and includes one of only two receptors (TAAR1) shown to respond to classical trace amines [3,4,6,18]. Consequently TAAR1 is by far the most fully characterised of the TAAR. Earlier classifications based only on TAAR1-9 included TAAR2-4 in the same sub-family as TAAR1 [6], but these have subsequently been moved into sub-class II [9].

TAAR1

Functional TAAR1 have been identified in all mammalian, avian and amphibian species so far examined [6,9,11,13]. Interestingly, while TAAR1 orthologs were found in the cartilaginous elephant shark and teleost zebrafish, the zebrafish gene was reported to be a pseudogene [9], and no TAAR1 orthologs have been identified in more evolutionarily recent teleost species [9,12]. This suggests that TAAR1 was present in the common ancestor of jawed vertebrates but has subsequently been lost during the evolution of teleost species. Consistent with this, the lamprey (a jawless vertebrate) genome was reported to contain no

TAAR genes [9]. TAAR1 appears to be coupled to the $G_{\alpha s}$ G-protein, with a stimulation of adenylate cyclase and accumulation of cAMP observed following receptor activation [3,4,6,19,20]. Interestingly TAAR1 does not appear to be membrane inserted in standard expression systems, with the expressed receptor remaining intracellular [3,4,6]. Although two N-terminal N-glycosylation sites have been identified [6], the lack of membrane insertion may relate to the absence of an additional consensus glycosylation site in the N-terminal region of the full length TAAR1 [21]. Thus, despite containing 7 putative transmembrane domains, TAAR1 appears to be primarily expressed in the cytosol.

PE and TA are the most potent of the traditional trace amines as agonists at TAAR1, showing sub-micromolar EC_{50} values [3-6,22,23]. It should be noted that affinity at TAAR1 shows species dependence, human and mouse TAAR1 showing higher affinity for PE than TA, whereas rat TAAR1 shows a higher affinity for TA [3,6,22,23]. Species dependence is also seen with the affinity of tryptamine, with considerably lower affinity seen at human TAAR1 than either the rat or mouse TAAR1 [3,6,22,23]. Octopamine, synephrine and dopamine generally show 50-100 fold lower affinity, while 5-HT has an approximately 1000-fold lower affinity [3,6,22,23]. Interestingly, when N-terminal glycosylation sites were added to allow membrane insertion, the affinity of octopamine and tryptamine for human TAAR1 was increased [21] compared to that seen in previous studies [3,6,23]. The affinity of TA however remained unchanged [3,6,21,23]. Whether this relates to the ability of ligands to cross cell membranes and interact with the intracellular receptor requires confirmation. It was previously reported that PE and TA release was dependent solely on tissue levels, and was not activity dependent [24,25]. This was taken as indicative of PE and TA crossing cell membranes by simple diffusion. Indeed PE and TA are the most lipophilic of the biogenic amines [26]. We have recently begun modelling the energetics of protonated PE (the physiologically relevant form) crossing sphingomyelin lipid bilayers. Our preliminary data suggests that relatively small amounts of PE will diffuse across this lipid bilayer [27]. If confirmed this low level of penetrance raises the possibility that transport processes other than simple diffusion, presumably some form of facilitated diffusion such as trace amine 'leak' channels, may be involved.

A number of other endogenous compounds have also been shown to interact with high affinity with TAAR1. The O-methyl metabolites of dopamine, noradrenaline and adrenaline have all been shown to activate TAAR1 at low to sub-micromolar concentrations [4,22]. More recently TAAR1 has been shown to be potently activated by metabolites of thyroid hormone [20,28,29]. In particular, 3-iodothyronamine (Figure 1) has been shown to activate TAAR1 with an EC_{50} value of approximately 10-30nM [28,29], 10-fold more potent than either PE or TA. Further, the endogenous hallucinogen, *N,N*-dimethyltryptamine (DMT) (Figure 1) fully activates TAAR1 at a concentration of 1μM [4] which compares favourably with the recently described affinity of DMT for the σ1-opioid receptor [30]. Thus, there is good functional evidence for the inclusion of such endogenous amines in the class of trace amines. As will be described later, a number of psychotomimetic drugs and drugs of abuse also show a high affinity for TAAR1.

TAAR1 mRNA and protein have been localised throughout the brain in mice [3,31], rats [4], rhesus monkey [19,32] and human [3], primarily associated with catecholaminergic and serotonergic cell bodies and projection areas. In addition TAAR1 has been localised to a number of peripheral tissues including heart [33], pancreas [4,34], leucocytes [35,36], gastrointestinal tract [3,37], liver, kidney, lung, spleen, prostate gland, skeletal muscle, and stomach [3]. TAAR1 is the only TAAR which appears not to function as an olfactory receptor in mammalian species [18] or in teleosts [9]. However, it was recently reported that adult, but not larval, *Xenopus laevis* express TAAR1 in both the olfactory organ and brain [38].

A number of non-endogenous compounds have also been shown to interact potently with TAAR1. Of particular note is the agonist activity at TAAR1 of a number of drugs of abuse. Both d- and l- isomers of amphetamine and methamphetamine have been shown to activate TAAR1 with EC_{50} values in the low to sub-micromolar region [4,22,39,40]. Importantly, this concentration corresponds well with plasma levels seen following acute administration in humans [8]. Similarly, 3,4-methylenedioxymethamphetamine (MDMA; 'ecstasy') [4,21], dihydroergotamine and LSD [4] are all agonists at TAAR1 with low micromolar potency. Thus, TAAR1 provides a novel site through which some of the effects of such drugs of abuse may be manifest. Given the observed *in vivo* effects of physiological concentrations of trace amines (see later) it is interesting to note that drugs of abuse have been reported to sensitize noradrenergic and serotonergic neurons independent of dopamine releasing effects [41]. TAAR would seem an attractive putative site for the mediation of such effects. In addition, this range of ligands has allowed the determination of receptor residues involved in ligand interactions at TAAR1 [29,42], and the subsequent site-directed synthesis of increased potency agonists, based on both PE and 3-iodothyronamine [29]. This group has also described the first putative antagonists of TAAR1 [29].

Trace amines have long been implicated in a number of neuropsychiatric disorders [43]. This again received renewed interest following the discovery of TAAR. In humans all TAAR are located within a narrow region of chromosome 6q23 [3,4,6]. This is close to a well documented susceptibility locus for schizophrenia [44-48], which maps particularly close to the TAAR6 gene [6,7]. TAAR1 knockout mice have recently been developed [31,49] and proposed as a novel model for schizophrenia [49]. Such animals are largely normal in appearance and behaviour but show deficits in pre-pulse inhibition and increased sensitivity to d-amphetamine administration [8,31,49]. Such effects are regarded as indicative of an animal model of schizophrenia. There have however, been no reports directly linking TAAR1 to schizophrenia. Putative metabolites of the anti-arrhythmia drug amiodarone were recently reported to be TAAR1 agonists [50], raising the possibility that TAAR1 may be involved in either the therapeutic or adverse effects of amiodarone. However, it has subsequently been reported that in contrast to rodent TAAR1, human TAAR1 is not activated by amiodarone metabolites [50a].

TAAR10, 11, 21 and 27

With the exception of one study identifying the presence of TAAR genes in various species, the other members of sub-class I TAAR have not been studied. TAAR10, TAAR10a-d and TAAR11 gene expression have been identified in zebrafish, where they were all expressed in distinct neurons of the olfactory epithelium, with the exception of TAAR10c which was only identified in embryos [9]. The TAAR10 isoform may be a pseudogene in zebrafish [9]. TAAR21a-c and TAAR27 were identified in the stickleback genome, while the medaka and fugu genomes contained TAAR21a-f [9]. The fugu genome also encoded TAAR27. In contrast the tetraodon genome contained a number of TAAR27 isoforms (a-f) along with TAAR21a-c [9].

Sub-class II TAAR

Sub-class II TAAR consist of the remaining mammalian TAAR (2-9) along with TAAR12 and 13 [9]. Besides TAAR1, TAAR4 is the only other receptor at which cognate ligands have been conclusively demonstrated. A variety of species specific paralogs of TAAR7 and TAAR8 have been identified [6]. In humans, there is no functional TAAR7 gene, and a single functional TAAR8 [6]. Interestingly TAAR7, TAAR8 and TAAR9 are all pseudogenes in the chimpanzee [6].

TAAR2-4

Neither TAAR2 nor TAAR3 have been studied in detail. While both are functionally expressed in rat, mouse and cow [3,6,9,12,13,18,36,51], TAAR3 is a pseudogene in human and non-human primates [6,9]. TAAR2 is also a pseudogene in the chimpanzee, but appears to be functional in humans [6,9]. The endogenous ligands at these two receptors are not known. A nonsense mutation in TAAR2 showing higher prevalence in patients with schizophrenia has been reported [52], although no replication studies appear to be present. A functional TAAR2 ortholog has also been identified in chicken [9,14] and frog [38]. In contrast, Hashiguchi and Nishida [12] failed to identify a TAAR2 gene in the frog genome. Similar to the situation with TAAR1, TAAR2 was also identified in the elephant shark [9] suggesting this is the evolutionary origin of other sub-class II TAAR in jawed vertebrates.

Both TAAR2 and TAAR3 are expressed in distinct sub-sets of olfactory neurons in the mouse olfactory epithelium, co-expressed with $G_{\alpha olf}$ [12,18] and coupled to an accumulation of cAMP. TAAR2 is also expressed in low levels in olfactory neurons of the Grueneberg ganglion of embryonic, but not adult mice [51]. TAAR2 expression was reported in the olfactory organ of both larval and adult *Xenopus laevis*, while expression in the brain was only seen in adults [38]. TAAR2 and 3 expression have also been reported in mouse cardiac tissue [33] and B-cells [36]. TAAR3 was shown to be activated strongly by several primary amines, including isoamylamine, which is enriched in male versus female mouse urine, and has been suggested to act as a pheromone [18].

As stated, TAAR4 is the only other receptor at which a cognate ligand has been conclusively demonstrated. Both rat and mouse TAAR4 have been shown to respond to PE [3,18] at low micromolar concentrations and tryptamine at approximately ten-fold higher concentrations [3]. As with TAAR1, TAAR4 appears to be coupled to cAMP accumulation via $G_{\alpha s}$ [3]. The human ortholog of TAAR4 is a pseudogene, previously identified as the $5HT_4$ pseudogene [3,6]. A single, functional, TAAR4 gene has been identified in the cow genome [9]. Two isoforms of TAAR4 (TAAR4a and 4b) were reported to be present in the clawed frog (*Xenopus tropicalis*) genome [9], while Hashiguchi and Nishida [12] also identified multiple TAAR4 isoforms in the frog genome. Similar to TAAR2, TAAR4 has been shown to be expressed in distinct olfactory neurons of the olfactory epithelium [12,18], in embryonic but not adult olfactory neurons of the Grueneberg ganglion [51], in cardiac tissue [33] and B-cells [36] of mice. Beyond expression in the olfactory system, the tissue distribution of TAAR2-4 has not been systematically determined.

TAAR5

TAAR5 was previously identified as the human PNR (Putative Neurotransmitter Receptor) gene [3,6]. In the initial description of the identification of this gene, RNA transcripts were identified primarily in skeletal muscle of humans, with lower levels found in the amygdala, hippocampus, caudate nucleus, thalamus and hypothalamus [53]. A weak signal was also identified in the substantia nigra [53]. Expression in the kidney has also been reported [3]. Thus, the distribution of TAAR5 in the brain has similarities to that of TAAR1. Although expression levels in mouse olfactory epithelium were lower than other TAAR, mouse TAAR5 was shown to be activated by sexually mature male mouse urine [18], suggesting that it may function as an olfactory receptor. Specifically, mouse TAAR5 appears to respond to tertiary amines (although not primary, secondary or quarternary), including trimethylamine [18] which is enriched in sexually mature, male mouse urine. In addition, a second tertiary amine, *N*-methylpiperidine, also activated TAAR5 [18]. In this respect it would be interesting to determine the affinity of the endogenous tertiary amine, *N,N*-dimethyltryptamine for TAAR5. TAAR5 expression has also been observed in distinct neurons of embryonic although not adult Grueneberg ganglion [51]. Functional TAAR5 has also been reported in the rat [6,9], opossum [9,12], chicken [9,12,14] and bovine [9] genomes.

TAAR6

Human TAAR6 expression has been detected in the amygdala, hippocampus and kidney [3]. Particular interest has been paid to human TAAR6 as the SCZD5 schizophrenia susceptibility locus [44] was reported to correlate specifically with polymorphisms of TAAR6 [54]. Confirmatory linkages between TAAR6 and schizophrenia and possibly bipolar disorder have subsequently been reported [55-59], although others have reported marginal or no linkage [60-66]. The lack of a consistent replication of TAAR6 linkage to schizophrenia may be a function of one or more factors including disease heterogeneity, ethnic variation of populations studied and the apparent high mutation (rapidly evolving) rate of TAAR [66].

TAAR6 has been reported to be prominently expressed in distinct neurons within the mouse olfactory epithelium [18,51] and Grueneberg ganglion [51]. Although TAAR6 expression could be detected in both adult and embryonic Grueneberg ganglion cells, there was a pronounced reduction in the number of cells expressing TAAR6 during development [51]. Interestingly olfactory neuron dysfunction was recently reported to occur in schizophrenia [67]. TAAR6 would appear to provide a possible unifying explanation for the association of olfactory dysfunction with schizophrenia, although this requires specific investigation. Functional TAAR6 genes have been identified in a number of other mammalian species, with 6 isoforms present in the opossum genome, two in bovine genome and a single TAAR6 gene in the rat [9].

TAAR7

As previously stated, while TAAR7 appears to represent the major site of TAAR gene expansion in mammals, there is no functional human or chimpanzee variant [6,9]. It has been proposed that this may represent the decreased importance of olfactory cues in primates [12]. While there is a single TAAR7 pseudogene in the human and chimpanzee genomes, the rat genome contains 7 functional TAAR7 isoforms with a further 2 pseudogenes, and the mouse genome has 5 functional TAAR7 variants with an additional pseudogene [6,9,12]. Interestingly several of these isoforms appear to be species specific, with no murine form of the rat TAAR7g, TAAR7h or TAAR7i (pseudogene) variants present [6,9]. Further while TAAR7f is functional in mouse, it is a pseudogene in rat, while the reverse applies to the TAAR7c isoform [6,9]. In addition, the putative ligand binding properties of the TAAR7 isoforms show pronounced differences between rat and mouse [6]. The human and chimpanzee TAAR7 pseudogenes appear to be most closely related to the rat TAAR7h [6]. Further, while other primate pseudogenes are relatively well conserved in comparison to their rodent orthologs, the TAAR7 pseudogene has largely degenerated [6]. Again, this would appear consistent with a decreased reliance on olfactory behavioural cues in primates in comparison to other species. Interestingly, the opossum genome, which contains an expanded TAAR6 gene repertoire, does not appear to contain either functional or pseudogene TAAR7 variants [9]. TAAR7a-c genes were identified in the cow genome [9].

Hashiguchi and Nishida [12] reported that all mouse TAAR7 variants were expressed in the olfactory epithelium. Probes for TAAR7d, 7e and 7f were shown to hybridize to distinct neuronal populations in the olfactory epithelium [18], although as discussed by the authors the ability of these probes to maintain specificity for the closely related TAAR7 variants is questionable. This expression co-localised with $G_{\alpha olf}$ expression, again suggesting that these receptors function as olfactory receptors via coupling to cAMP accumulation [18]. Expression of TAAR7a and TAAR7d, but not other TAAR7 isoforms, has also been reported to occur in distinct populations of cells in the Grueneberg ganglion from both adult and embryonic mice [51]. As seen before, there was a pronounced decrease in expression during development to adulthood. The TAAR7f mouse isoform was shown to be activated by *N*-methylpiperidine, although not by a second tertiary amine, trimethylamine [18].

TAAR8

Similar to TAAR7, isoforms of the TAAR8 receptor have been identified in rat, mouse and cow genomes [6,9,12]. Both rat and mouse genomes contain three functional TAAR8 genes (TAAR8a, TAAR8b and TAAR8c) [6,9]. A single functional TAAR8 gene is present in the human genome, although similar to the situation with TAAR2, the chimpanzee TAAR8 is a pseudogene [6]. As seen with TAAR7, the opossum genome does not contain any TAAR8 genes [9]. TAAR8a and 8b were identified in the cow genome [9]. Alternative start codons for the rat isoforms were reported with only the codon leading to the shorter transcript conserved in mice and humans [6]. The human TAAR8 was suggested to be most similar to mouse TAAR8b and rat TAAR8a [6], suggesting that it is a distinct isoform.

While TAAR8 expression was observed in the olfactory epithelium [12,18], this was at a lower level than seen with other TAAR [18]. No expression of any TAAR8 isoform was detected in post-natal mouse Grueneberg ganglion [51]. TAAR8a expression was found to be the most abundant TAAR in rat cardiac tissue [33]. TAAR8 mRNAs were also found to be expressed in mouse B-cells [36]. Human TAAR8 mRNA was detected in the kidney [3], amygdala [3,5] and hippocampus [5]. There have been no ligands directly demonstrated to interact with any TAAR8 isoform at present, although it was suggested that the cardiac effects of 3-iodothyronamine were a function of an interaction with TAAR8a [68].

TAAR9

A single functional TAAR9 gene has been identified in mouse, rat, cow and human genomes [6,7,9], although one study failed to identify a human TAAR9 gene [12]. The opossum genome contains 7 TAAR9 genes (TAAR9, TAAR9a-f) [9]. The chimpanzee TAAR9 gene however, is a pseudogene [6]. Thus, the human genome contains 3 sub-class II TAAR genes (TAAR2, 8 and 9) that are not functional in the chimpanzee. On this basis it has been suggested that these may represent loci for diseases that uniquely affect humans [7], possibly including cardiovascular disease and the major psychoses [69]. A single nucleotide mutation of TAAR9 resulting in premature termination and a presumed non-functional receptor has been identified in humans [70], though it is not associated with a disease phenotype, and the prevalence of the mutation was found to be the same among control, attention deficit disorder and bipolar populations.

In humans TAAR9 was reported to be located particularly to the pituitary gland, although its expression could not be detected in a variety of other brain regions [70]. Peripherally, TAAR9 expression was only identified in skeletal muscle by one group [70] and only in kidney and no brain regions by another [3]. Nelson and co-workers [36] reported detection of TAAR9 mRNA in mouse B-cells. As with other sub-class II TAAR, TAAR9 was found in distinct neurons in the mouse olfactory epithelium where expression co-localises with $G_{\alpha olf}$ expression [18]. No expression of TAAR9 was found in cells of the Grueneberg ganglion [51]. At least one zebrafish TAAR9 ortholog has been reported, with expression predominant in the olfactory epithelium [18], although this was not detected by others [9,12]. As with TAAR8, no ligands for TAAR9 have yet been identified.

TAAR12 and 13

Beyond identification of their genes in zebrafish [9], there have been no studies concerning TAAR12 and 13. Thirteen TAAR12 isoforms were identified, all of which were expressed in the olfactory epithelium and brain with the exception of TAAR12i, TAAR12l and TAAR12m, which were only detected in the olfactory epithelium [9]. For TAAR13, 5 isoforms were identified which were restricted to expression in the brain, with the exception of TAAR13d which was only found in embryos [9]. Zebrafish TAAR genes which appear consistent with sub-class II TAAR were also reported by Hashiguchi and Nishida [12]. It seems likely that these correlate with the subsequent TAAR12 and 13 designations. At least one of these was expressed in the olfactory organ [12].

Sub-class III TAAR

Sub-class III TAAR have only been identified in teleost species and comprise approximately 75% of the teleost TAAR repertoire, consisting of TAAR14-20, TAAR22-26 and TAAR28 [9]. It has been estimated that 82% of identified, putatively functional, teleost TAAR are species specific [17]. This species specific expression appears to represent a positive selection pressure and has been proposed to be a function of olfactory organ expression, possibly related to pheromone detection [9,17]. Further, as previously described, this sub-class has been associated with a number of intron insertion and deletion events [9,12].

TAAR14-20 appear to be zebrafish specific, with a number of isoforms of each identified (Table 1) [9]. All were found to be expressed in the olfactory epithelium, with the exception of TAAR14a and TAAR15b which were only detected in embryos, and TAAR19u and TAAR19v which were only detected in brain [9]. Many of the zebrafish sub-class III TAAR were detected in both the olfactory epithelium and the brain [9].

At least one TAAR22 isoform has been identified in all teleosts with the exception of zebrafish, with again available evidence suggesting some of these variants may be species specific [9]. The medaka genome contains 15 TAAR23 and 3 TAAR24 genes [9], but only a single gene for each was identified in stickleback and none in either fugu or tetraodon [9]. A large number of TAAR25 (12) and TAAR26 (28) genes have been identified in the stickleback but in no other species studied [9]. TAAR28 genes were identified in both fugu (8 genes) and tetraodon (3 genes) but no other species [9]. In the stickleback a number of the sub-class III TAAR were found to be expressed in the olfactory epithelium and lips, with fewer forms expressed in the brain [12].

FUNCTIONAL RESPONSES

There are few studies that have conclusively shown physiological responses that are mediated by one or more TAAR, and only TAAR1 has been directly studied. The following sections will briefly outline responses to endogenous TAAR ligands that appear to be physiologically relevant, and where possible indicate which TAAR may be involved.

Trace Amine Modulation of Neuronal Activity

At pharmacologic concentrations, trace amines, in particular PE and TA, have long been recognised to have amphetamine-like, indirect-sympathomimetic effects [1]. As such, at supra-physiologic concentrations, displacement of monoamine neurotransmitters from vesicular stores, with subsequent neurotransmitter depletion has been widely demonstrated [71-74]. Indeed, historically, PE was often regarded as an endogenous amphetamine [75]. At lower, more physiologic concentrations, direct effects of trace amines on neuronal activity have not been observed. However, at these lower concentrations trace amines have been shown to modulate the responsivity of neurons to co-applied neurotransmitters [1].

Selectivity of effects were present, with individual trace amines only affecting responsivity to particular neurotransmitters (Table 3). Coupled with evidence of trace amine synthesis being regulated in response to monoaminergic tone; increased monoaminergic activity leading to decreased trace amine synthesis and vice versa [1,76], this led to the hypothesis that trace amines may function as an intrinsic regulatory mechanism, maintaining basal monoaminergic tone within defined parameters [1].

The above effects appear to be mediated at post-synaptic sites, although the exact mechanism and potential role of TAAR has yet to be determined. Lesioning of pre-synaptic terminals does not prevent the effects of exogenously administered PE [77,78], while responses to the dopamine agonists PPHT and apomorphine, neither of which is a substrate for dopamine transporters, were also potentiated [79]. It seems reasonable to suggest that these likely involve an interaction of PE with either TAAR1 or TAAR4. It should also be noted that apomorphine has subsequently been shown to have agonist activity at TAAR1 [4], thus complicating interpretation of these earlier studies.

Table 3. Interactions of trace amines with central neurotransmitters

	PE	TA	Tryptamine	Octopamine
Dopamine	+ [79, 87-89]	+ [89, 90]		- [91]
Noradrenaline	+ [77, 78, 92]	+ [89]		+ [91]
5-HT	- [92]	- [90]	+/-[a] [93, 94]	- [91]
Acetylcholine	- [78, 92]		- [94]	
GABA	- [77, 87, 92] + [95]	- [89] + [95]		
Glutamate	- [87, 92]	- [90]		

+ = Neurotransmitter response potentiated, - = No effect. [a]Excitatory 5-HT responses were either unaffected or converted to inhibitory responses in the presence of tryptamine, inhibitory responses were potentiated. Blank cells indicate that the interaction has not been studied. There are no studies at present examining the effects of synephrine or 3-iodothyronamine on neurotransmitter responses.

TAAR1 knock-out mice have also been reported to show dopaminergic hyperactivity [31], consistent with TAAR1 being involved in the normal regulation of dopaminergic activity. Pre-synaptic TAAR1 mediated modulation of dopaminergic activity has also recently been described [32,80-82]. Such effects involve alterations in dopamine transporter activity. Here TAAR1 receptor activation was seen to result in an inhibition of dopamine uptake via the transporter, with a simultaneous increase in transporter mediated efflux of dopamine [32,80,82]. Such effects would increase synaptic concentrations of dopamine, presumably increasing the duration of dopamine receptor activation. It should be noted that these effects are distinct from the indirect sympathomimetic effects that have been well documented to occur with high concentrations of PE. The endogenous TAAR1 ligand 3-iodothyronamine was also previously shown to modulate dopamine transporter activity [83], although the potency for these effects appears lower than those for 3-iodothyronamine interaction with TAAR1. This is reminiscent of the effects seen with PE, where interaction with TAAR1 occurs at considerably lower concentrations than those required for direct interaction with transporters. TAAR1 ligands may therefore affect monoamine neurotransmitter transporter function at low doses through TAAR1 mediated effects and at high doses by directly interacting with the transporters. There is thus evidence for trace amines potentiating dopaminergic responses through both pre- and post-synaptic mechanisms. An intriguing possibility is that these effects are mediated by distinct TAAR. Similar TAAR1 mediated effects were also observed at noradrenaline [32,80,82,83] and 5-HT [32,80,82] transporters.

Non-neuronal Effects

Pronounced cardiac effects have been reported to be caused by 3-iodothyronamine. In the presence of 3-iodothyronamine, both calcium and potassium currents are affected, resulting in independent negative inotropic and chronotropic effects [20,33,68,84]. Although TAAR1 has been identified in rat heart [33], considerably higher concentrations of 3-iodothyronamine are required for cardiac effects, than those shown to interact with TAAR1 [33,68,84]. Similar negative inotropic effects, but not chronotropic effects, were seen with PE, another TAAR1 ligand [68]. This suggests that the cardiac effects of 3-iodothyronamine are not mediated through TAAR1. Consistent with this, the effects were not associated with an increase in cAMP accumulation, nor affected by protein kinase A inhibition [33]. Rather, the effects of 3-iodothyronamine were sensitive to tyrosine kinase inhibition [33]. On the basis of its preponderance in cardiac tissue, it was suggested that TAAR8a may be the receptor mediating these effects [68]. However there has been no direct demonstration of PE nor 3-iodothyronamine interacting with this receptor sub-type, nor of TAAR8a being coupled to tyrosine kinase activation.

Low dose (4nmol/kg) administration of 3-iodothyronamine has also been shown to result in an increase in food intake [85]. The administered dose was suggested to be physiologically relevant to the binding affinity of 3-iodothyronamine at TAAR1. The authors suggested that the observed effects may reflect a secondary increase in neuropeptide Y release in the hypothalamus [85]. Higher dose administration has also been reported to cause hypothermic responses [20], although the relationship of this to TAAR remains to be conclusively demonstrated. Finally a number of TAAR1/TAAR4 ligands have been reported to cause vasoconstriction [86,87]. The concentrations required however are significantly above those required for TAAR1 or TAAR4 activation. Thus such effects either represent actions at other TAAR, or are non-specific in nature.

Conclusion

The discovery of a family of G-protein coupled receptors, some of which selectively respond to trace amines, has caused a resurgence of interest in the potential physiological roles of trace amines. At least 3 distinct sub-classes are present, although little is known about many of the TAAR. Only TAAR1 and TAAR4 have been shown to respond to classical trace amines, with 3-iodothyronamine more recently identified as a further endogenous ligand. TAAR are widely distributed in the body of many species, appearing to be involved in the regulation of monoaminergic neurotransmission, while also functioning as a distinct class of olfactory receptors. In this regard, TAAR3 and TAAR5 have been shown to be activated by urinary amines which may function as pheromones. Endogenous ligands at other TAAR have yet to be determined.

Acknowledgments

The authors thank Mr. Norm Finlay for expert technical assistance and Dr. Patrick Jackson for constructive feedback during manuscript preparation. The financial support of Canada Foundation for Innovation, Manitoba Research Infrastructure Fund, National Sciences and Engineering Research Council and Brandon University are gratefully acknowledged.

References

[1] Berry MD, Mammalian central nervous system trace amines. Pharmacologic amphetamines, physiologic neuromodulators. *J Neurochem.* 2004;90: 257-271.

[2] Roeder T, Tyramine and octopamine: ruling behavior and metabolism. *Annu Rev Entomol.* 2005;50: 447-477.

[3] Borowsky B, N Adham, KA Jones, R Raddatz, R Artymyshyn, KL Ogozalek, MM Durkin, PP Lakhlani, JA Bonini, S Pathirana, N Boyle, X Pu, E Kouranova, H Lichtblau, FY Ochoa, TA Branchek, and C Gerald, Trace amines: identification of a family of mammalian G protein-coupled receptors. *Proc Natl Acad Sci U S A.* 2001;98: 8966-8971.

[4] Bunzow JR, MS Sonders, S Arttamangkul, LM Harrison, G Zhang, DI Quigley, T Darland, KL Suchland, S Pasumamula, JL Kennedy, SB Olson, RE Magenis, SG Amara, and DK Grandy, Amphetamine, 3,4-methylenedioxymethamphetamine, lysergic acid diethylamide, and metabolites of the catecholamine neurotransmitters are agonists of a rat trace amine receptor. *Mol Pharmacol.* 2001;60: 1181-1188.

[5] Lewin AH, Receptors of mammalian trace amines. *Aaps J.* 2006;8: E138-145.

[6] Lindemann L, M Ebeling, NA Kratochwil, JR Bunzow, DK Grandy, and MC Hoener, Trace amine-associated receptors form structurally and functionally distinct subfamilies of novel G protein-coupled receptors. *Genomics.* 2005;85: 372-385.

[7] Lindemann L and MC Hoener, A renaissance in trace amines inspired by a novel GPCR family. *Trends Pharmacol Sci.* 2005;26: 274-281.

[8] Sotnikova TD, MG Caron, and RR Gainetdinov, Trace amine associated receptors (TAARs) as emerging therapeutic targets. *Mol Pharmacol.* 2009;76: 229-235.

[9] Hussain A, LR Saraiva, and SI Korsching, Positive Darwinian selection and the birth of an olfactory receptor clade in teleosts. *Proc Natl Acad Sci U S A.* 2009;106: 4313-4318.

[10] Maguire JJ, WA Parker, SM Foord, TI Bonner, RR Neubig, and AP Davenport, International Union of Pharmacology. LXXII. Recommendations for trace amine receptor nomenclature. *Pharmacol Rev*. 2009;61: 1-8.

[11] Gloriam DE, TK Bjarnadottir, HB Schioth, and R Fredriksson, High species variation within the repertoire of trace amine receptors. *Ann N Y Acad Sci*. 2005;1040: 323-327.

[12] Hashiguchi Y and M Nishida, Evolution of trace amine associated receptor (TAAR) gene family in vertebrates: lineage-specific expansions and degradations of a second class of vertebrate chemosensory receptors expressed in the olfactory epithelium. *Mol Biol Evol*. 2007;24: 2099-2107.

[13] Gloriam DE, R Fredriksson, and HB Schioth, The G protein-coupled receptor subset of the rat genome. *BMC Genomics*. 2007;8: 338-354.

[14] Mueller JC, S Steiger, AE Fidler, and B Kempenaers, Biogenic Trace Amine-Associated Receptors (TAARS) are encoded in avian genomes: evidence and possible implications. *J Hered*. 2008;99: 174-176.

[15] Gloriam DE, TK Bjarnadottir, YL Yan, JH Postlethwait, HB Schioth, and R Fredriksson, The repertoire of trace amine G-protein-coupled receptors: large expansion in zebrafish. *Mol Phylogenet Evol*. 2005;35: 470-482.

[16] Grus WE, P Shi, and J Zhang, Largest vertebrate vomeronasal type 1 receptor gene repertoire in the semiaquatic platypus. *Mol Biol Evol*. 2007;24: 2153-2157.

[17] Hashiguchi Y, Y Furuta, and M Nishida, Evolutionary patterns and selective pressures of odorant/pheromone receptor gene families in teleost fishes. *PLoS ONE*. 2008;3: e4083.

[18] Liberles SD and LB Buck, A second class of chemosensory receptors in the olfactory epithelium. *Nature*. 2006;442: 645-650.

[19] Miller GM, CD Verrico, A Jassen, M Konar, H Yang, H Panas, M Bahn, R Johnson, and BK Madras, Primate trace amine receptor 1 modulation by the dopamine transporter. *J Pharmacol Exp Ther*. 2005;313: 983-994.

[20] Scanlan TS, KL Suchland, ME Hart, G Chiellini, Y Huang, PJ Kruzich, S Frascarelli, DA Crossley, JR Bunzow, S Ronca-Testoni, ET Lin, D Hatton, R Zucchi, and DK Grandy, 3-Iodothyronamine is an endogenous and rapid-acting derivative of thyroid hormone. *Nat Med*. 2004;10: 638-642.

[21] Barak LS, A Salahpour, X Zhang, B Masri, TD Sotnikova, AJ Ramsey, JD Violin, RJ Lefkowitz, MG Caron, and RR Gainetdinov, Pharmacological characterization of membrane-expressed human trace amine-associated receptor 1 (TAAR1) by a bioluminescence resonance energy transfer cAMP biosensor. *Mol Pharmacol*. 2008;74: 585-594.

[22] Wainscott DB, SP Little, T Yin, Y Tu, VP Rocco, JX He, and DL Nelson, Pharmacologic characterization of the cloned human trace amine-associated receptor1 (TAAR1) and evidence for species differences with the rat TAAR1. *J Pharmacol Exp Ther*. 2007;320: 475-485.

[23] Navarro HA, BP Gilmour, and AH Lewin, A rapid functional assay for the human trace amine-associated receptor 1 based on the mobilization of internal calcium. *J Biomol Screen*. 2006;11: 688-693.

[24] Dyck LE, Release of some endogenous trace amines from rat striatal slices in the presence and absence of a monoamine oxidase inhibitor. In Boulton AA, AV Juorio, and RG Downer eds. *Trace Amines: Comparative and Clinical Neurobiology*. Humana Press, Inc.; 1988: 223-237.

[25] Henry DP, WL Russell, JA Clemens, and LA Plebus, Phenylethylamine and *p*-tyramine in the extracellular space of the rat brain: quantification using a new radioenzymatic assay and in situ microdialysis. In Boulton AA, AV Juorio, and RG Downer eds. *Trace Amines: Comparative and Clinical Neurobiology*. Clifton, New Jersey: Humana Press, Inc.; 1988: 239-250.

[26] Boulton AA, AV Juorio, and IA Paterson, Phenylethylamine in the CNS: effects of monoamine oxidase inhibiting drugs, deuterium substitution and lesions and its role in the neuromodulation of catecholaminergic neurotransmission. *J Neural Transm Suppl*. 1990;29: 119-129.

[27] Tomberli BL, J Nickel, and MD Berry, Simulation studies of trace amines passing through neuronal membranes. *Biophys. J.* 2009;96: 41a.

[28] Hart ME, KL Suchland, M Miyakawa, JR Bunzow, DK Grandy, and TS Scanlan, Trace amine-associated receptor agonists: synthesis and evaluation of thyronamines and related analogues. *J Med Chem*. 2006;49: 1101-1112.

[29] Tan ES, JC Naylor, ES Groban, JR Bunzow, MP Jacobson, DK Grandy, and TS Scanlan, The molecular basis of species-specific ligand activation of trace amine-associated receptor 1 (TAAR(1)). *ACS Chem Biol*. 2009;4: 209-220.

[30] Su TP, T Hayashi, and DB Vaupel, When the endogenous hallucinogenic trace amine N,N-dimethyltryptamine meets the sigma-1 receptor. *Sci Signal*. 2009;2: pe12.

[31] Lindemann L, CA Meyer, K Jeanneau, A Bradaia, L Ozmen, H Bluethmann, B Bettler, JG Wettstein, E Borroni, JL Moreau, and MC Hoener, Trace amine-associated receptor 1 modulates dopaminergic activity. *J Pharmacol Exp Ther*. 2008;324: 948-956.

[32] Xie Z and GM Miller, Trace amine-associated receptor 1 is a modulator of the dopamine transporter. *J Pharmacol Exp Ther*. 2007;321: 128-136.

[33] Chiellini G, S Frascarelli, S Ghelardoni, V Carnicelli, SC Tobias, A DeBarber, S Brogioni, S Ronca-Testoni, E Cerbai, DK Grandy, TS Scanlan, and R Zucchi, Cardiac effects of 3-iodothyronamine: a new aminergic system modulating cardiac function. *Faseb J*. 2007;21: 1597-1608.

[34] Regard JB, IT Sato, and SR Coughlin, Anatomical profiling of G protein-coupled receptor expression. *Cell*. 2008;135: 561-571.

[35] D'Andrea G, S Terrazzino, D Fortin, A Farruggio, L Rinaldi, and A Leon, HPLC electrochemical detection of trace amines in human plasma and platelets and expression of mRNA transcripts of trace amine receptors in circulating leukocytes. *Neurosci Lett*. 2003;346: 89-92.

[36] Nelson DA, MD Tolbert, SJ Singh, and KL Bost, Expression of neuronal trace amine-associated receptor (Taar) mRNAs in leukocytes. *J Neuroimmunol*. 2007;192: 21-30.

[37] Kidd M, IM Modlin, BI Gustafsson, I Drozdov, O Hauso, and R Pfragner, Luminal regulation of normal and neoplastic human EC cell serotonin release is mediated by bile salts, amines, tastants, and olfactants. *Am J Physiol Gastrointest Liver Physiol*. 2008;295: G260-272.

[38] Gliem S, D Schild, and I Manzini, Highly specific responses to amine odorants of individual olfactory receptor neurons in situ. *Eur J Neurosci*. 2009;29: 2315-2326.

[39] Reese EA, JR Bunzow, S Arttamangkul, MS Sonders, and DK Grandy, Trace amine-associated receptor 1 displays species-dependent stereoselectivity for isomers of methamphetamine, amphetamine, and para-hydroxyamphetamine. *J Pharmacol Exp Ther*. 2007;321: 178-186.

[40] Xie Z and GM Miller, A receptor mechanism for methamphetamine action in dopamine transporter regulation in brain. *J Pharmacol Exp Ther*. 2009;330: 316-325.

[41] Lanteri C, L Salomon, Y Torrens, J Glowinski, and JP Tassin, Drugs of abuse specifically sensitize noradrenergic and serotonergic neurons via a non-dopaminergic mechanism. *Neuropsychopharmacology*. 2008;33: 1724-1734.

[42] Lewin AH, HA Navarro, and SW Mascarella, Structure-activity correlations for beta-phenethylamines at human trace amine receptor 1. *Bioorg Med Chem*. 2008;16: 7415-7423.

[43] Berry MD, The potential of trace amines and their receptors for treating neurological and psychiatric diseases. *Rev Recent Clin Trials*. 2007;2: 3-19.

[44] Cao Q, M Martinez, J Zhang, AR Sanders, JA Badner, A Cravchik, CJ Markey, E Beshah, JJ Guroff, ME Maxwell, DM Kazuba, R Whiten, LR Goldin, ES Gershon, and PV Gejman, Suggestive evidence for a schizophrenia susceptibility locus on chromosome 6q and a confirmation in an independent series of pedigrees. *Genomics*. 1997;43: 1-8.

[45] Kaufmann CA, B Suarez, D Malaspina, J Pepple, D Svrakic, PD Markel, J Meyer, CT Zambuto, K Schmitt, TC Matise, JM Harkavy Friedman, C Hampe, H Lee, D Shore, D Wynne, SV Faraone, MT Tsuang, and CR Cloninger, NIMH Genetics Initiative Millenium Schizophrenia Consortium: linkage analysis of African-American pedigrees. *Am J Med Genet*. 1998;81: 282-289.

[46] Levinson DF, P Holmans, RE Straub, MJ Owen, DB Wildenauer, PV Gejman, AE Pulver, C Laurent, KS Kendler, D Walsh, N Norton, NM Williams, SG Schwab, B Lerer, BJ Mowry, AR Sanders, SE Antonarakis, JL Blouin, JF DeLeuze, and J Mallet, Multicenter linkage study of schizophrenia candidate regions on chromosomes 5q, 6q, 10p, and 13q: schizophrenia linkage collaborative group III. *Am J Hum Genet*. 2000;67: 652-663.

[47] Lewis CM, DF Levinson, LH Wise, LE DeLisi, RE Straub, I Hovatta, NM Williams, SG Schwab, AE Pulver, SV Faraone, LM Brzustowicz, CA Kaufmann, DL Garver, HM Gurling, E Lindholm, H Coon, HW Moises, W Byerley, SH Shaw, A Mesen, R Sherrington, FA O'Neill, D Walsh, KS Kendler, J Ekelund, T Paunio, J Lonnqvist, L Peltonen, MC O'Donovan, MJ Owen, DB Wildenauer, W Maier, G Nestadt, JL Blouin, SE Antonarakis, BJ Mowry, JM Silverman, RR Crowe, CR Cloninger, MT Tsuang, D Malaspina, JM Harkavy-Friedman, DM Svrakic, AS Bassett, J Holcomb, G Kalsi, A McQuillin, J Brynjolfson, T Sigmundsson, H Petursson, E Jazin, T Zoega, and T Helgason, Genome scan meta-analysis of schizophrenia and bipolar disorder, part II: Schizophrenia. *Am J Hum Genet*. 2003;73: 34-48.

[48] Lindholm E, B Ekholm, S Shaw, P Jalonen, G Johansson, U Pettersson, R Sherrington, R Adolfsson, and E Jazin, A schizophrenia-susceptibility locus at 6q25, in one of the world's largest reported pedigrees. *Am J Hum Genet*. 2001;69: 96-105.

[49] Wolinsky TD, CJ Swanson, KE Smith, H Zhong, B Borowsky, P Seeman, T Branchek, and CP Gerald, The Trace Amine 1 receptor knockout mouse: an animal model with relevance to schizophrenia. *Genes Brain Behav*. 2007;6: 628-639.

[50] Snead AN, M Miyakawa, ES Tan, and TS Scanlan, Trace amine-associated receptor 1 (TAAR1) is activated by amiodarone metabolites. *Bioorg Med Chem Lett*. 2008;18: 5920-5922.

[50a]Lewin AH, HA Navarro, and BP Gilmour, Amiodarone and its putative metabolites fail to activate wild type *h*TAAR1. *Bioorg Med Chem Lett*. 2009;19: 5913-5914.

[51] Fleischer J, K Schwarzenbacher, and H Breer, Expression of trace amine-associated receptors in the Grueneberg ganglion. *Chem Senses*. 2007;32: 623-631.

[52] Bly M, Examination of the trace amine-associated receptor 2 (TAAR2). *Schizophr Res*. 2005;80: 367-368.

[53] Zeng Z, P Fan, E Rand, H Kyaw, K Su, V Madike, KC Carter, and Y Li, Cloning of a putative human neurotransmitter receptor expressed in skeletal muscle and brain. *Biochem Biophys Res Commun*. 1998;242: 575-578.

[54] Duan J, M Martinez, AR Sanders, C Hou, N Saitou, T Kitano, BJ Mowry, RR Crowe, JM Silverman, DF Levinson, and PV Gejman, Polymorphisms in the trace amine receptor 4 (TRAR4) gene on chromosome 6q23.2 are associated with susceptibility to schizophrenia. *Am J Hum Genet*. 2004;75: 624-638.

[55] Abou Jamra R, I Sircar, T Becker, Y Freudenberg-Hua, S Ohlraun, J Freudenberg, F Brockschmidt, TG Schulze, M Gross, F Spira, M Deschner, C Schmal, W Maier, P Propping, M Rietschel, S Cichon, MM Nothen, and J Schumacher, A family-based and case-control association study of trace amine receptor genes on chromosome 6q23 in bipolar affective disorder. *Mol Psychiatry*. 2005;10: 618-620.

[56] Liu C, J Shi, JA Badner, H Zou, Y Qian, and ES Gershon, No association of trace amine receptor genes with bipolar disorder. *Mol Psychiatry*. 2007;12: 979-981.

[57] Pae CU, A Drago, JJ Kim, AA Patkar, TY Jun, C Lee, L Mandelli, D De Ronchi, IH Paik, and A Serretti, TAAR6 variation effect on clinic presentation and outcome in a sample of schizophrenic in-patients: an open label study. *Eur Psychiatry*. 2008;23: 390-395.

[58] Pae CU, HS Yu, D Amann, JJ Kim, CU Lee, SJ Lee, TY Jun, C Lee, IH Paik, AA Patkar, and B Lerer, Association of the trace amine associated receptor 6 (TAAR6) gene with schizophrenia and bipolar disorder in a Korean case control sample. *J Psychiatr Res*. 2008;42: 35-40.

[59] Vladimirov V, DL Thiselton, PH Kuo, J McClay, A Fanous, B Wormley, J Vittum, R Ribble, B Moher, E van den Oord, FA O'Neill, D Walsh, KS Kendler, and BP Riley, A region of 35 kb containing the trace amine associated receptor 6 (TAAR6) gene is associated with schizophrenia in the Irish study of high-density schizophrenia families. *Mol Psychiatry*. 2007;12: 842-853.

[60] Amann D, N Avidan, K Kanyas, Y Kohn, A Hamdan, E Ben-Asher, F Macciardi, JS Beckmann, D Lancet, and B Lerer, The trace amine receptor 4 gene is not associated

with schizophrenia in a sample linked to chromosome 6q23. *Mol Psychiatry*. 2006;11: 119-121.

[61] Duan S, J Du, Y Xu, Q Xing, H Wang, S Wu, Q Chen, X Li, X Li, J Shen, G Feng, and L He, Failure to find association between TRAR4 and schizophrenia in the Chinese Han population. *J Neural Transm*. 2006;113: 381-385.

[62] Ikeda M, N Iwata, T Suzuki, T Kitajima, Y Yamanouchi, Y Kinoshita, T Inada, and N Ozaki, No association of haplotype-tagging SNPs in TRAR4 with schizophrenia in Japanese patients. *Schizophr Res*. 2005;78: 127-130.

[63] Ludewick HP, SG Schwab, M Albus, B Lerer, W Maier, M Trixler, and DB Wildenauer, No support for an association with TAAR6 and schizophrenia in a linked population of European ancestry. *Psychiatr Genet*. 2008;18: 208-210.

[64] Sanders AR, J Duan, DF Levinson, J Shi, D He, C Hou, GJ Burrell, JP Rice, DA Nertney, A Olincy, P Rozic, S Vinogradov, NG Buccola, BJ Mowry, R Freedman, F Amin, DW Black, JM Silverman, WF Byerley, RR Crowe, CR Cloninger, M Martinez, and PV Gejman, No significant association of 14 candidate genes with schizophrenia in a large European ancestry sample: implications for psychiatric genetics. *Am J Psychiatry*. 2008;165: 497-506.

[65] Venken T, M Alaerts, R Adolfsson, CV Broeckhoven, and J Del-Favero, No association of the trace amine-associated receptor 6 with bipolar disorder in a northern Swedish population. *Psychiatr Genet*. 2006;16: 1-2.

[66] Vladimirov VI, BS Maher, B Wormley, FA O'Neill, D Walsh, KS Kendler, and BP Riley, The trace amine associated receptor (TAAR6) gene is not associated with schizophrenia in the Irish Case-Control Study of Schizophrenia (ICCSS) sample. *Schizophr Res*. 2009;107: 249-254.

[67] Turetsky BI, CG Hahn, SE Arnold, and PJ Moberg, Olfactory receptor neuron dysfunction in schizophrenia. *Neuropsychopharmacology*. 2009;34: 767-774.

[68] Frascarelli S, S Ghelardoni, G Chiellini, R Vargiu, S Ronca-Testoni, TS Scanlan, DK Grandy, and R Zucchi, Cardiac effects of trace amines: pharmacological characterization of trace amine-associated receptors. *Eur J Pharmacol*. 2008;587: 231-236.

[69] Olson MV and A Varki, Sequencing the chimpanzee genome: insights into human evolution and disease. *Nat Rev Genet*. 2003;4: 20-28.

[70] Vanti WB, P Muglia, T Nguyen, R Cheng, JL Kennedy, SR George, and BF O'Dowd, Discovery of a null mutation in a human trace amine receptor gene. *Genomics*. 2003;82: 531-536.

[71] Barroso N and M Rodriguez, Action of β-phenylethylamine and related amines on nigrostriatal dopamine neurotransmission. *Eur J Pharmacol*. 1996;297: 195-203.

[72] Fuxe K, H Grobecker, and J Jonsson, The effect of β-phenylethylamine on central and peripheral monoamine-containing neurons. *Eur J Pharmacol*. 1967;2: 202-207.

[73] Parker EM and LX Cubeddu, Comparative effects of amphetamine, phenylethylamine and related drugs on dopamine efflux, dopamine uptake and mazindol binding. *J Pharmacol Exp Ther*. 1988;245: 199-210.

[74] Raiteri M, R Del Carmine, A Bertollini, and G Levi, Effect of sympathomimetic amines on the synaptosomal transport of noradrenaline, dopamine and 5-hydroxytryptamine. *Eur J Pharmacol.* 1977;41: 133-143.

[75] Janssen PA, JE Leysen, AA Megens, and FH Awouters, Does phenylethylamine act as an endogenous amphetamine in some patients? *Int J Neuropsychopharmacol.* 1999;2: 229-240.

[76] Berry MD, AV Juorio, XM Li, and AA Boulton, Aromatic L-amino acid decarboxylase: a neglected and misunderstood enzyme. *Neurochem Res.* 1996;21: 1075-1087.

[77] Paterson IA, The potentiation of cortical neurone responses to noradrenaline by β-phenylethylamine: effects of lesions of the locus coeruleus. *Neurosci Lett.* 1988;87: 139-144.

[78] Paterson IA, The potentiation of cortical neuron responses to noradrenaline by 2-phenylethylamine is independent of endogenous noradrenaline. *Neurochem Res.* 1993;18: 1329-1336.

[79] Paterson IA, AV Juorio, MD Berry, and MY Zhu, Inhibition of monoamine oxidase-B by (-)-deprenyl potentiates neuronal responses to dopamine agonists but does not inhibit dopamine catabolism in the rat striatum. *J Pharmacol Exp Ther.* 1991;258: 1019-1026.

[80] Xie Z and GM Miller, Beta-phenylethylamine alters monoamine transporter function via trace amine-associated receptor 1: implication for modulatory roles of trace amines in brain. *J Pharmacol Exp Ther.* 2008;325: 617-628.

[81] Xie Z and GM Miller, Trace Amine-Associated Receptor 1 as a monoaminergic modulator in brain. *Biochem Pharmacol.* 2009;78: 1095-1104.

[82] Xie Z, SV Westmoreland, and GM Miller, Modulation of monoamine transporters by common biogenic amines via trace amine-associated receptor 1 and monoamine autoreceptors in human embryonic kidney 293 cells and brain synaptosomes. *J Pharmacol Exp Ther.* 2008;325: 629-640.

[83] Snead AN, MS Santos, RP Seal, M Miyakawa, RH Edwards, and TS Scanlan, Thyronamines inhibit plasma membrane and vesicular monoamine transport. *ACS Chem Biol.* 2007;2: 390-398.

[84] Ghelardoni S, S Suffredini, S Frascarelli, S Brogioni, G Chiellini, S Ronca-Testoni, DK Grandy, TS Scanlan, E Cerbai, and R Zucchi, Modulation of cardiac ionic homeostasis by 3-iodothyronamine. *J Cell Mol Med.* 2009;Postprint;10.1111/j.1582-4934.2009.00728.x.

[85] Dhillo WS, GA Bewick, NE White, JV Gardiner, EL Thompson, A Bataveljic, KG Murphy, D Roy, NA Patel, JN Scutt, A Armstrong, MA Ghatei, and SR Bloom, The thyroid hormone derivative 3-iodothyronamine increases food intake in rodents. *Diabetes Obes Metab.* 2009;11: 251-260.

[86] Broadley KJ, M Akhtar Anwar, AA Herbert, M Fehler, EM Jones, WE Davies, EJ Kidd, and WR Ford, Effects of dietary amines on the gut and its vasculature. *Br J Nutr.* 2009;101: 1645-1652.

[87] Herbert AA, EJ Kidd, and KJ Broadley, Dietary trace amine-dependent vasoconstriction in porcine coronary artery. *Br J Pharmacol.* 2008;155: 525-534.

[88] Berry MD, E Scarr, MY Zhu, IA Paterson, and AV Juorio, The effects of administration of monoamine oxidase-B inhibitors on rat striatal neurone responses to dopamine. *Br J Pharmacol.* 1994;113: 1159-1166.

[89] Harris J, S Trivedi, and BL Ramakrishna, A contribution to the neuromodulatory/neurotransmitter role of trace amines. In Boulton AA, AV Juorio, and RG Downer eds. *Trace Amines: Comparative and Clinical Neurobiology*. Clifton, New Jersey: Humana Press, Inc.; 1988: 213-221.

[90] Jones RS and AA Boulton, Interactions between *p*-tyramine, *m*-tyramine, or β-phenylethylamine and dopamine on single neurones in the cortex and caudate nucleus of the rat. *Can J Physiol Pharmacol.* 1980;58: 222-227.

[91] Jones RSG, Trace biogenic amines:a possible functional role in the CNS. *Trends Pharmacol Sci.* 1983;5: 426-429.

[92] Jones RS, Noradrenaline-octopamine interactions on cortical neurones in the rat. *Eur J Pharmacol.* 1982;77: 159-162.

[93] Paterson IA and AA Boulton, β-Phenylethylamine enhances single cortical neurone responses to noradrenaline in the rat. *Brain Res Bull.* 1988;20: 173-177.

[94] Jones RS, Tryptamine modifies cortical neurone responses evoked by stimulation of nucleus raphe medianus. *Brain Res Bull.* 1982;8: 435-437.

[95] Jones RS and AA Boulton, Tryptamine and 5-hydroxytryptamine: actions and interactions of cortical neurones in the rat. *Life Sci.* 1980;27: 1849-1856.

[96] Berretta N, M Giustizieri, G Bernardi, and NB Mercuri, Trace amines reduce GABA(B) receptor-mediated presynaptic inhibition at GABAergic synapses of the rat substantia nigra pars compacta. *Brain Res.* 2005;1062: 175-178.

II. Biogenic Amines in Invertebrate System

In: Biogenic Amines
Editors: T. Farooqui, A. A. Farooqui

ISBN: 978-1-60876-625-3

Chapter XIV

QUANTIFICATION OF OCTOPAMINE IN INVERTEBRATE AND VERTEBRATE NERVOUS SYSTEMS

Tahira Farooqui*
The Ohio State University, Columbus, OH, 43210, USA.

ABSTRACT

Octopamine belongs to a group of compounds known as biogenic amines. It is found in the nervous system of both invertebrates and vertebrates. Octopamine is present in higher concentrations in invertebrates, whereas it occurs as a trace amine in vertebrates. Octopamine acts as a neurotransmitter, neuromodulator and neurohormone in invertebrates, therefore, has a prominent role in influencing multiple physiological events. However, its physiological role in vertebrates is not fully understood. The unusually higher levels of octopamine observed in various pathological conditions suggest that octopamine may have some pathophysiological role in humans. Therefore, it is important to select a reliable, sensitive and rapid procedure for the quantification of octopamine, especially in vertebrate system, due to its low endogenous concentration and rapid turnover. Octopamine has been identified and quantified using a wide variety of analytical procedures. Some assays lack necessary specificity because they can not distinguish between the positional isomers of octopamine. The purpose of this review is to collate the information on available octopamine estimation procedures and point out advantages and drawbacks of each method including factors that may affect octopamine levels in tissues and biological fluids. The author hopes that this overview will facilitate the selection of a most suitable procedure for the routine analysis of octopamine in tissues and biological fluids of invertebrates and vertebrates.

* Correspondence concerning this article should be addressed to: Tahira Farooqui, Department of Entomology/ Center for Molecular Neurobiology, The Ohio State University, Columbus, OH 43210-1220, Telephone: (614) 783-4369, Email: farooqui.2@osu.edu.

Keywords: Octopamine; biogenic amine; sympathomimetic amine; trace amine; chiral molecule.

1. Introduction

Octopamine - a monohydroxylated analog of norepinephrine - is a biogenic amine, first discovered by Erspamer in the salivary gland of a mollusc *Octopus*, by a two dimensional paper chromatography technique [1]. It is a major neurotransmitter in invertebrates, where it is present in high concentration in the central and peripheral nervous systems and in various other tissues [1-4]. Octopamine occurs in three positional isomeric forms, *para*, *meta* and *ortho* (Figure 1). Each isomeric form occurs as (+) or (-) enantiomers. The naturally occurring isomer of octopamine is *p*-octopamine in many invertebrate nervous systems including insects. The (-)-enantiomer of *p*-octopamine is the naturally occurring form of octopamine in honeybees [5]. The (-)-enantiomer of *p*-octopamine is over 200 times more potent than (+)-enantiomer in stimulating adenylyl cyclase (AC) activity in cockroach brain [6]. In vertebrates, *p*- and *m*- isomers of octopamine are present at trace levels in the mammalian brain [7]. However, *m*-isomer of octopamine is more potent than the *p*-isomer with higher potency for the (-)-enantiomer than the (+)-enantiomer for stimulating adrenergic receptors [8-9]. *o*- Isomer of octopamine is found in rat urine, following monoamine oxidase (MAO) inhibition [10]. Its major urinary metabolite (*o*-hydroxymandelic acid) is detected in both rat [10] and human urine [11]. However, *o*- octopamine is not found in any organ even after MAO inhibition, suggesting it occurs below the lower limits of detection [7].

Octopamine is present in high amounts in invertebrates, whereas it's chemically identical form, norepinephrine, is found predominantly in vertebrates [2-4,9,12-13]. Under normal conditions, octopamine is present as a trace amine in the mammalian brain and biological fluids. However, it occurs in unusually high concentrations in certain diseases [14] and in patients treated with monoamine oxidase inhibitors (MAOIs), suggesting that it may have some pathophysiological role. Due to low concentration and rapid turnover of octopamine, its role as a neurotransmitter in mammalian systems has not been thoroughly examined. Octopamine quantification in the biological fluids may be affected by the presence of interfering substances, and also due to many methodological factors. Using a non-suitable quantification method, the obtained value of octopamine may either be much lower or much higher than the expected value. Therefore, it is important to use a suitable quantification procedure that can distinguish between three positional isomers of octopamine and separate octopamine from other amines, allowing its correct and rapid determination in tissues and biological fluids.

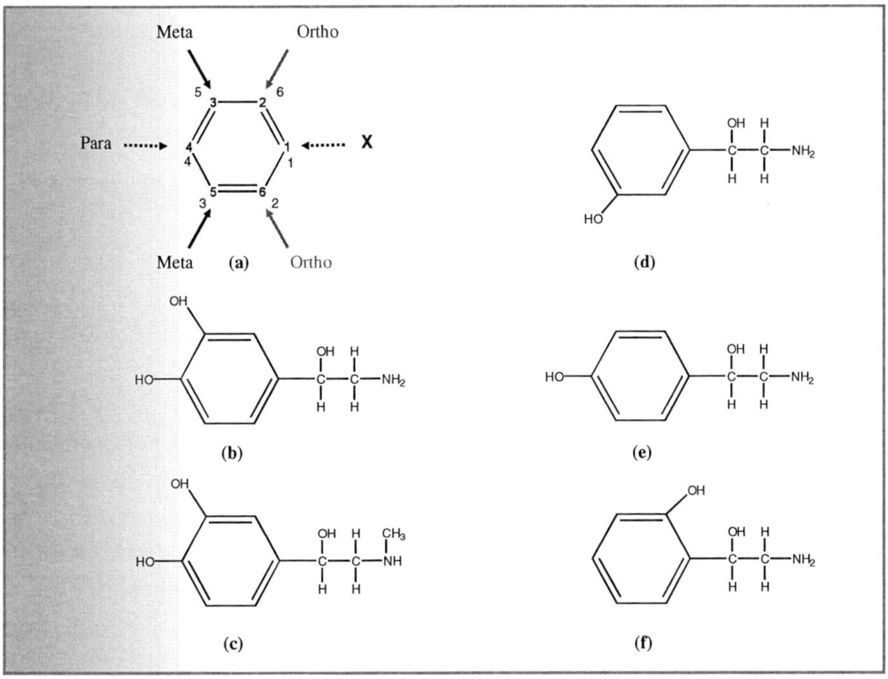

Figure. 1. Chemical structures of norepinephrine, epinephrine, and octopamine and its positional isomers. (a) *Ortho*, *meta* and *para* configuration, (b) norepinephrine, (c) epinephrine, (d) *p*-octopamine, (e) *m*-octopamine, and (f) *o*-octopamine. *Ortho* or *o*-disubstituted benzene has two substituents in 1,2 relationship on the ring, whereas *meta* or *m*-disubstituted benzene has two substituents in 1,3 relationship on the ring. *Para* or *p*- disubstituted benzene has two substituents in 1,4 relationship on the ring.

2. Importance of Octopamine in Invertebrates

Among invertebrates, octopamine is a well-established neuromodulator that mediates a diverse range of physiological processes. Octopamine has been reported to modulate the foraging preference in honeybees [15-16], division of labor in honeybees [17], mobilization of lipids and carbohydrates in crickets [18], sting response in honeybees [19], ovarian muscle contraction in flies [20], aggression in crickets and flies [21-22], arousal in locusts [23], learning and memory in honeybees and flies [24-26], development of ethanol tolerance in flies [28], sensory systems in insects [29], and photoreceptors response in horseshoe crabs [30]. Collective evidence suggests that octopamine plays a major neuromodulatory role in invertebrate behaviors.

In invertebrates, octopamine specifically interacts with octopamine receptors that belong to the superfamily of G protein-coupled receptors (GPCR). Based on information from newly cloned *D*rosophila octopamine receptors, a new classification scheme for octopamine receptors has been proposed [32]. According to this classification octopamine receptors can be classified into three classes: (1) α-adrenergic-like octopamine receptors (OctαRs), (2) β-adrenergic-like octopamine receptors (OctβRs) and (3) octopamine/tyramine or tyraminergic receptors (OctRs/TyrRs or TyrRs). The activation of OctαRs increases levels of intracellular calcium and intracellular cAMP, whereas activation of OctβRs increases levels of intracellular cAMP without affecting intracellular calcium. Both groups have a higher affinity for octopamine than tyramine. The interaction of tyramine with OctRs/tyrRs or TyrRs inhibits adenylyl cyclase activity via the inhibitory G_i protein, causing a decrease in intracellular cAMP level [32]. A definitive designation for this class as "dual octopamine/tyramine or specific tyraminergic" remains to be elucidated.

3. Importance of Octopamine in Mammals

Octopamine is present at very low concentration in the central and sympathetic nervous systems of mammals [13,31]. The physiological role of octopamine has not been established in mammals. It is called as false neurotransmitter in mammals because it exerts its effects at millimolar range. Octopamine, a sympathomimetic amine, is localized in nerves in peripheral tissues and brain where it seems to coexist with other catecholamines [2]. Octopamine is synthesized from tyramine and co-released from nerves with norepinephrine. Therefore, many of its effects may be indirect, possibly due to some interference with adrenergic pathway. Octopamine has been reported to interact with α- and β-adrenergic receptor subtypes in mammals [33-34]. Like norepinephrine, octopamine is rapidly metabolized by monoamine oxidase (MAO). Following treatment with MAO inhibitors, octopamine concentration has been reported to increase in the mammalian brain [35]. Octopamine influences several functions, including lipolysis, glucose uptake, locomotor activity, blood pressure, and anti-depressive effect in mammals [36-39]. Octopamine is found in relatively higher concentration in humans only under abnormal conditions such as neuroblastoma [40,41], renal disease [42], hepatic encephalopathy [43], migraine and cluster headaches [44], phenylketonuria [45], and in patients treated with monoamine oxidase inhibitors [35]. Therefore, it is assumed that octopamine may have some pathophysiological role in humans [14]. Octopamine has very low affinity for human adrenergic receptors that makes its physiological relevance unlikely. Furthermore, the occurrence of octopamine receptors in humans has not yet been established. In mammals, octopamine is present at generally low levels and has a rapid turnover rate. Therefore, the selection of a sensitive, reliable and rapid procedure is particularly necessary for its quantification in the clinical samples.

4. Methods for the Quantification of Octopamine

In last four decades, a number of methods, including radioenzymic assay, high performance liquid chromatography with electrochemical and fluorimetric detection, gas chromatographic-mass spectrometry, and capillary electrophoresis, have been developed for the quantification of octopamine. Several modifications, such as changes in the composition of the mobile phase, addition and/or alteration in the separation step, inactivation of potent endogenous enzyme inhibitors and removal of other interfering electroactive compounds, have been made in original procedures to increase the sensitivity of octopamine assay in invertebrate and vertebrate samples. Few modifications in these methods have also been employed in clinical laboratories to analyze octopamine concentration in human body fluids. A critical discussion on currently available estimation procedures will help researchers in selecting the "most practical and effective" technique(s) for correct quantification of octopamine in tissues and biological fluids.

4.1. Quantification of Octopamine by Radioenzymic Assay

The original radioenzymic assay used for the determination of octopamine is based on the *N*-methylation of the β-hydoxylated amines (such as octopamine and phenylethanolamine to their *N*-methyl derivatives) by the enzyme phenylethanolamine-*N*-methyltransferase (PNMT), using radioactively labeled *S*-adenosylmethionine as a methyl donor in the reaction mixture (Figure 2) [46]. In this assay, labeled methyl derivative of octopamine (N-methyloctopamine, synephrine) is extracted in a mixture of toluene-isoamyl alcohol (3:2, v/v), whereas labeled methyl derivative of phenylethanolamine (N-methyl phenylethanolamine) is extracted into 3% isoamyl alcohol in toluene. After extraction, radioactive samples are quantified using liquid scintillation counter and identified by thin layer chromatography on chromagram sheets of silica gel, using solvent containing butanol saturated with 1N HCl and a mixture of isopropyl alcohol-ammonium hydroxide-water (80:10:19, v/v). The sheets are developed, stained, and the radioactivity is determined in the marked sections. The LOD for octopamine by this procedure is 0.05 ng of octopamine per 0.5 ml sample volume (6.53 pM) [46].

Some investigators have expressed doubts about the specificity of extraction procedures used in the original assay [35,47], whilst others have shown concerns because original assay does not allow the simultaneous determination of *m*- and *p*- isomers of octopamine [48]. Therefore, additional modifications such as further purification of co-crystallization from methanol solution by adding ethyl acetate after enzymic methylation of octopamine and selective isolation of labeled synephrine using unlabeled synephrine as a carrier have been made in this assay [47]. The radioactivity is determined by liquid scintillation counting, and the identity of radioactive product is checked by chromatography of the thrice crystallized product on TLC plates of silica gel using three solvents: (1) *n*-butanol saturated with 1M HCl, (2) isopropanol-ammonia-water (80:19:10, v/v), and (3) *n*-butanol-acetic acid-water (4:1:1, v/v) [49]. The advantage of crystallization is to avoid high and erratic radioactivity values that have been observed in the hemolymph of the locusts *Schistocerca americana gregaria* in the absence of this step [47]. Other modification accounts for improving the sensitivity of the assay as well as allowing simultaneous determination of *m*- and *p*-octopamine isomers and phenylethanolamines that are found in rat salivary glands [48]. The modifications such as dansylation of amines after stopping the enzymic reaction, extraction of these amines in benzene, and purification by TLC gel plates using three solvent systems: (1) chloroform-butylacetate (5:2, v/v), (2) toluene-triethylamine-methanol (10:1:0.5, v/v), and (3) ethylacetate-cyclohexane (35:25, v/v) give reproducible results in a variety of insect tissues [48,50]. The disadvantage of original assay is its nonspecificity. PNMT can accept all three isomers of octopamine as substrates, but all isomers are identified and quantitated as *p*-octopamine. Furthermore, presence of other alternative substrates (such as norepinephrine, normetanephrine, and phenylethaolamine) in tissues may result in lack of specificity (Table 1). In addition, the solvent extraction using *n*-butanol saturated with 1 M HCl or acetone 35% aqueous ammonia, 99:1 (v/v) gives poor results. Saavedra's modified assay gives LOD for octopamine as low as 50 pg (0.0003265 pmol) and allows simultaneous determination of phenylethanolamine in the same tissue sample [51].

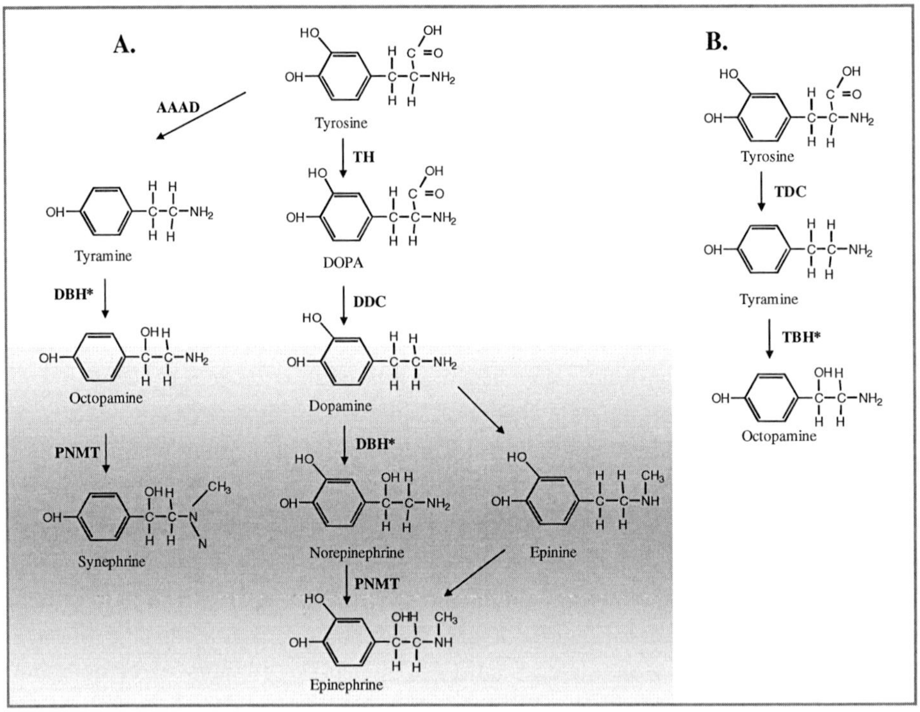

Figure. 2. Biosynthesis of catecholamines: (A) in vertebrates and (B) in invertebrates. TH, Tyramine hydroxylase; DDC, Dopamine decarboxylase; DBH, Dopamine-β-hydroxylase; PNMT, Phenyl-ethanolamine-N-methyltransferase; AAAD, Aromatic amino acid decarboxylase; TDC, Tyramine decarboxylase; TBH, Tyramine-β-hydroxylase. *Both enzymes DBH and TBH recognize either tyramine or dopamine as substrates but have different affinities. Both enzymes have similar properties, and catalyze the hydroxylation step, suggesting that they are functionally homologous [132].

Using modified radioenzymic assay, different assay sensitivities in various insect systems, such as 0.05-0.06 pmol in the lantern and abdominal nerve cord of fire fly *Photuris versicolor* larvae [52], 0.1 pmol in locust *Schistocerca gregaria* extensor-tibiae muscles [53], and 0.1-2.5 pmol in moth *Manduca sexta* brain during metamorphosis [54], have been recorded. This assay can measure pg quantities of monoamines, and allows determination in very small volume of body fluids in brain perfusates, individual brain nuclei, and individual cells of some simple animals [55]. The modified radioenzymic assay allows the simultaneous determination of *m*- and *p*-isomers of octopamine and phenylethanol amines levels in tissues [48]. It is used to determine the concentration of free and conjugated normetanephrine and octopamine excretion in humans [56], as well as octopamine metabolites in the urine of healthy individuals and neuroblastoma patients [41]. This assay allows measurement of the labeled methyl derivative of octopamine, but without incorporation of rigourous chromatographic purification step it results in artificially high amine levels (Table 1).

Table 1. Advantages and disadvantages of methods used for octopamine determination

Assay	Advantages	Disadvantages
Radioenzymic (Original assay)	More sensitive than non- colorimetric and bioassays -Small sample, no sample pretreatment	- The solvent extraction procedures used in this assay give poor results - Determines only one compound per sample -Does not allow simultaneous determination of *m*- and *p*- isomers -Absence of any rigorous purification results in artificially high octopamine levels
Radioenzymic (Modified assay)	- Allows simultaneous determination of *m*- and *p*- octopamine isomers and phenylethanolamines -Sensitive and selective	-Intra- and interlaboratory variability, assay complexity, and procedure time are more than chromatographic methods -Not suitable in plasma
HPLC-ECD	-Stable, efficient, rapid, reproducible, and easy sample preparation, but not suited for plasma - High applicability -It can simultaneously determine several monoamines/sample	-Derivatization in the absence of a reducing agent causes problems with stability and reproducibility -Protein precipitation of the sample preparation in perchloric acid may interfere with detection -Adsorption of the oxidation or reduction products on the electrode may cause contamination
HPLC-FD	- Simultaneous analyses of tyramine, octopamine phenethylamine, and phenylethanolamine -Suited for plasma - As sensitive and selective as HPLC-ECD -Pre-column derivatization enhances fluorescence detection -Gradient elution with μ Bondapak /phenyl column shortens analysis time -Suitable for routine analysis of biological samples in healthy humans, patients, and animals	-Affected by many interfering substances method has been applied to the routine
CE	-High degree of resolution, high separation efficiency, rapid analysis, requires small sample volumes (femtoliter to picoliter) -Allows chiral analysis -Better than HPLC - Particularly well-suited to perform cellular and subcellular analysis	-Joule heat generation requires an effective cooling system -Non-specific and incomplete derivatization may limit application of this method

Table 1 (Continued)

Assay	Advantages	Disadvantages
Borate-MEKC-EC	-Sensitive, selective, and reproducible separation; short run time -Straight identification of biogenic amines and their metabolites -Surfactant improves separation of similarly charged and sized biogenic amines and metabolites -Allows quantitative analysis of metabolites in conjunction with biogenic amine precursors in a single separation	-Peak shifting upon exposure to biological solutions -Surfactant can decrease or reverse ESF
CE- optical fiber LED-IF	-Extremely sensitive, low running cost, and short preparation time (< 5 min) -It can be applied to single cell cytoplasm -Sample volume as low as 270 fL	-The derivatization between octopamine and NDA is affected by endogenous components
GC-MS	-Sensitive -It can distinguish among compounds that are modified by the inclusion of isotopes	-Residual impurities may result in an unexpected peak -Background noise in the MS if GC does not completely separate the components
GC-NICIMS	-Highly sensitive, accurate, and specific	-Expansive
GC-ECD	-Simple, sensitive, accurate and specific -Economically good alternative for laboratories that only have gas chromatographic instrumentation available -Resulting derivative is highly responsive to ECD at subpmole level. GC-ECD quantification is confirmed by GCMS	-Alcohol and amine groups can simultaneously react with PFPA
Autoradiographic	-Sensitive, very high spatial accuracy -Easily quantifiable -Selective labeling of fibers	-Extremely long developing time, expansive -Lacks ultrastructural resolution
Immunostaining	-Specific and repeatable from one animal to next -Direct measurement of octopaminergic neurons and determination of the origin of fibers	-Minor variation in intensity between two preparations in the same taxon -Possibility of octopamine degradation before fixation -Cellular projections may look fine, but may not uniformly contain octopamine -Some of the labeled neurons may be tyraminergic

Gas chromatography-mass spectroscopy (GC-MS); high-performance liquid chromatography coupled to electrochemical detection (HPLC-ECD); high-performance liquid chromatography florometric detection (HPLC-MCD); borate micellar electrokinetic capillary chromatography (Borate ME-KCEC); capillary electrophoresis (CE); capillary electrophoresis optical fiber LED-induced fluorescence detection (LED-IF); Naphthalene-2,3-dicarboxaldehyde (NDA); gas chromatography (GC); capillary gas chromatography with electron capture detector (GC-ECD); pentafluoropropionic anhydride (PFPA); electroosmotic flow (EOF).

Another radioenzymic assay is based on determination of dopamine-β-hydroxylase (DBH) activity in vertebrates, and tyramine-β-hydroxylase (TBH) in invertebrates. Both enzymes are functionally homologous. They require O_2 and ascorbic acid as cosubstrates, bind copper, and catalyze the β-hydroxylation reaction. DBH uses a phenyl-substituted alkylamine (dopamine or tyramine) as a substrate (Figure 2). DBH activity has been detected by converting [^{14}C]tyramine to [^{14}C]octopamine, oxidizing with periodate to form [^{14}C]*p*-hydroxybenzaldehyde, separating by solvent extraction into ether, and then counting the radioactivity [57]. It is necessary to inactivate potent endogenous inhibitors of DBH that are present in the tissue homogenates. Otherwise, these inhibitors interact with the copper (Cu^{2+}) present in DBH [57-58]. DBH assay allows LOD as low as 7.5 pmol of the product [57], whereas using radioactively labeled *S*-adenosylmethionine, Henry et al. [58] have reported the LOD value of 33 pmol. The further modified radioenzymic two dimensional TLC-based assay uses purified PNMT with optimal incubation conditions [59]. DBH catalyzes the hydroxylation of *p*-tyramine to form *p*-octopamine, which is then methylated to synephrine by PNMT using [^{3}H]*S*-adenosylmethionine as the methyl donor (Figure 2). This assay simultaneously detects as low as 2 pg of *p*-octopamine and *p*-tyramine. It allows less than 0.2% contamination from other structurally similar amines, suggesting improved specificity for *p*-octopamine and *p*-tyramine [59]. TBH assay relies on the conversion of radioactive tyramine to octopamine (Figure 2), which is then methylated to synephrine by PNMT [54]. Collectively, the sensitivity of radioenzymic assays varies depending on the buffer, sample type (source, tissue, biological fluid), maximal concentration of substrate, and enzyme level in the reaction mixture.

4.2. Quantification of Octopamine by High Performance Liquid Chromatography

The high Performance Liquid Chromatography (HPLC) is basically a highly improved form of column chromatography in which analyte is forced through a column packed with chromatographic material (stationary phase) by pumping the solvent(s) (mobile phase) under high pressure. A very small particle size of the stationary phase increases the surface area, allowing greater separation. However, it decreases the flow rate of the mobile phase therefore high pressure is required to increase the flow rate of solvent through the column. After separation of the chemical mixture, a means of detection system is required to detect the amount of each of the chemical constituents in the original mixture. A detector shows the retention times of the molecules, which may vary depending on the interactions between the stationary phase, molecules to be analyzed, and the solvent(s) used. There are many types of detectors, such as spectrophotometeric, refractometeric, solute transport detectors, radioactive, and electrochemical, which can be employed. Here, I discuss two most popular detection systems used with HPLC for octopamine determination: (1) electrochemical detection, and (2) fluorometric detection.

4.2.1. HPLC with Electrochemical Detection (HPLC-ECD)

The HPLC-ECD is based on the measurement of current resulting from the flow of electrons produced by the oxidation or reduction of electrochemically active substances on the surface of an electrode [60]. Catecholamines can be oxidized to corresponding quinones by the loss of two electrons in the detector electrode, therefore, they provide the basis for their detection by electrochemical detection. However, other compounds that do not undergo oxidation unless large electric potentials are applied can not be detected by this detection system. Although, biogenic amines are generally present at low concentrations in animal tissues and fluids, but can be sensed by the electrochemical cell of the detector. HPLC-ECD allows the simultaneous estimation of a variety of biogenic amines (including octopamine) and their metabolites in tissues and biological samples (Table 1), in both man and animals [61-64]. Due to the presence of interfering electroactive compounds in biological fluids, this procedure is considered to be unsuitable for detecting and quantifying octopamine in the plasma [61]. The separation of catecholamines and their metabolites has been demonstrated by reverse-phase chromatography with aqueous isocratic eluents where mixtures of both acidic and basic substances can be separated by a single octadecyl-silica column [65]. The retention is governed by hydrophobic interactions between the nonpolar moiety of the solute and the stationary phase. The eluent at the right pH allows the separation of both basic and acidic substances. Due to the high column stability and efficiency, reversed phase liquid chromatography is assumed to have an ability to separate a wide range of biogenic amines in clinical samples [65].

In invertebrates, HPLC-ECD is used to measure N-acetyltransferase activity against octopamine, dopamine and 5-hydroxytryptamine as well as for perfoming the bi-substrate kinetics of this enzyme in the cerebral ganglion of the American cockroach *Periplaneta Americana* [66]. In addition, HPLC-ECD is used to determine octopamine levels in various insect species, including locust *Locusta migratoria* [66], cockroach *Periplaneta Americana* [67-68], and fruit fly *Drosophila melanogaster* [69]. Robinson's group has used HPLC-ECD to determine behavior-related changes and age-related changes in the levels of various biogenic amines, such as octopamine, dopamine and serotonin, in the antennal lobes and mushroom bodies of individual honeybee *Apis mellifera* brains [70-71]. The mobile phase in this analysis contains 15% methanol, 15% acetonitrile (a chemical modifier), 1.5 mmol. l^{-1} sodium dodecyl sulfate, 75 mmol.l^{-1}, sodium phosphate monobasic, and trace amounts of triethylamine and EDTA [70]. The LOD's for octopamine, dopamine and serotonin are 7.5 pg, 15 pg and 15 pg, respectively [70]. The use of perchloric acid for precipitating proteins in the sample preparation has become a contravercial issue due to its interference with biogenic amine detection, giving high background [70,72]. Peak identification in HPLC is determined on the basis of changes in retention times or peak height as a function of systematic changes in chromatographic conditions (such as pH, percentage organic compounds and applied channel voltage) [73]. Octopamine is considered as 'fight or flight' hormone in insects, and its level in haemolymph increases during stress. Adamo et al. measured the levels of octopamine in the haemolymph of field crickets *Gryllus bimaculatus* after performance of several behaviors, showing increase in octopamine levels in haemolymph during aggressive behavior (from 0.03 to 0.16 pM), after flying behavior (from 0.03 to 0.29 pM), but did not observe an increase after an escape run [74].

In *Drosophila* brain, previously reported biogenic amines levels are highly contradictory, which may be either due to type of detection method used or variability in the preparation of the sample or composition of mobile phase. Some groups have utilized a mobile phase containing a phosphate buffer pH 2.5-3.0, 1-2 mM octanesulfonic acid (an ion-pairing agent), and 6-10% of acetonitrile. It is known that higher pH (such as pH 4.5) facilitates the detection of octopamine and tyramine [69,75], but phosphate buffer does not provide adequate buffering capacity at this pH. Therefore, increasing the pH of the mobile phase (from pH 3.0 – pH 4.5), using a citrate acetate buffer mobile phase for the remaining separation, and applying a high voltage (+850 mV) helps in the detection of octopamine and tyramine [76]. The LOD for octopamine by the improved method is over a range of 5-10 pg (0.00003265-0.0000653 pmol). The improved HPLC-ECD involves minimal sample preparation and can reproducibly detect octopamine.

4.2.2. High Performance Liquid Chromatography with Fluorometric Detection

The fluorometric detection is based on the phenomenon in which light from the ultraviolet spectrum is absorbed by a substance, and the fluorescence is emitted in the visible spectrum. Biogenic amines can absorb light energy that can be dissipated by the fluorescence emission. The number of photons in light emission is proportional to the number of molecules involved, referring to the concentration of fluorescent substances in the sample. The HPLC separation combined with the enhanced sensitivity of fluorescence detection allows LOD less than 100 pg (< 0.000653 pmol) for octopamine [77]. The increased sensitivity is in part due to pre-column derivatization, which facilitates extraction and improves detection of nanogram amounts. The derivatization reagents such as fluorescamine and *o*-phthalaldehyde are used to enhance natural fluorescence of biogenic amines. Fluorescamine rapidly reacts with primary amines, whereas *o*-phthalaldehyde requires a reducing agent for forming a strong fluorescence compound with primary amines [78-80]. Biogenic amines, including octopamine, tyramine, dopamine, norepinephrine, and 3,4-dihydroxybenzylamine, react readily with naphthalene- 2,3-dicarboxaldehyde in the presence of cyanide ion under mild conditions, resulting in highly fluorescent cyanobenz[*f*]isoindole (CBI) products [81]. The high-fluorescence quantum efficiency of the CBI fluorophore and the ability to excite these adducts in the visible region increase the sensitivity and selectivity of the derivatization detection. Mell and Carpenter [82] have extracted octopamine from tissue homogenates of ganglia in *Aplysia californica*, using perchloric acid and derivatized it with *o*-phthalaldehyde prior to chromatography. The separation of the fluorescent octopamine-OPT adduct from other biogenic amines is achieved by using a $^{\mu}$Bondapak C_{18} reversed-phase column and isocratic elution with a mobile phase containing methanol-(0.08 mol/liter) and acetic acid (50:50 by vol, pH 2.9). Linearity is ranged from 500 pg to 30 ng injected onto the column, and LOD for octopamine is 0.07 pmol/mg tissue [82].

The fluorometric method involving ion-exchange chromatography, derivatization with fluorescamine, solvent extraction and then separation by thin-layer chromatography can simultaneously determine levels of various biogenic amines, including octopamine, tyramine, phenethylamine and phenylethanolamine, in the rat brain [83]. The LODs by HPLC-FD method are 20 pmol for octopamine and 10 pmol for tyramine, phenethylamine and phenylethanolamine [83]. Another method measuring β-mono oxygenase activity (involving the hydroxylation of tyramine to octopamine) requires no derivatization before detection, and has LOD of 50 pmole for octopamine in human serum [84]. The separation in cation-exchange resin followed by the measurement of the native fluorescence provides a sensitive assay of dopamine β-monooxygenase. Collectively, HPLC-FD system (Table 1) provides a sensitive and selective tool for the estimation of octopamine and other biogenic amines in both man and animals.

4.3. Determination of Octopamine by Gas Chromatography Coupled with Mass Spectrometry (GC-MS)

Gas chromatography (GC) and mass spectrometry (MS) make an effective combination tool for detecting levels of bioamines. GC-MS relies not only on chromatographic retention but also on molecular mass to ascertain identity [85]. Therefore it offers a significant advantage over methods such as HPLC-ECD, which rely on retention time alone. In GC procedure, a small amount of the sample to be analyzed is drawn up into a syringe. The injector is set to a temperature higher than the components' boiling points so that components in the mixture evaporate into the gaseous phase inside the injector. A carrier gas flowing through the injector pushes the gaseous components onto the GC column, and the separation of organic compounds occurs due to differences in their partitioning behavior between the mobile phase and the stationary phase. The detector measures the time at which each component elutes from the column and the amount of that component can be determined. Many detectors, such as argon ionization detector, flame ionization detector, flame emission detector, cross section detector, thermal conductivity detector, and the electron capture detector, are commercially available for GC analysis. The comparison of retention time of analytes is the analytical power to GC that makes it very similar to other forms of chromatography (such as HPLC, TLC), but the traveling of the solution through the column in a gaseous state makes GC different from others. A derivatization-capillary GC with electron capture detection method (GC-ECD), involving a derivatization reaction between pentafluoropropionic anhydride and octopamine in buffer, is used to detect octopamine levels in the head tissue of German cockroach *Blattella germanica* [86]. GC-ECD chromatogram shows a single peak, consequently confirmed by GC-MS to be the tris-pentafluoropropionyl-octopamine. GC-ECD shows LOD for octopamine as low as 0.1 ng/gm tissue in *Blatella* [86]. It may be a good alternative for laboratories that only have GC instrument available (Table 1).

MS identifies substances by electrically charging the specimen molecules, accelerating them through a magnetic field, breaking the molecules into charged fragments and detecting different charges. Quantification by MS is based on peak areas from mass chromatogram or from selected ion monitoring. With mass chromatogram all of the masses are scanned, whereas ion monitoring records the responses from only few pre-selected masses. The GC-

MS method positively identifies the actual presence of a particular substance in a given sample, relying not only on chromatographic retention but also on molecular mass (Table 1). Negative ion chemical ionization of pentafluoropropionate derivatives of octopamine and tyramine together with the use of detureated internal standards is employed to quantitate octopamine and tyramine levels in different regions of the rat brain, with LOD of 50 pmol for octopamine [87]. The use of methane as the reagent gas, conversion of amines to their ditriofluoromethyl benzamide derivatives prior to analysis and quantification against deturiated internal standards added prior to processing the samples for negative-ion chemical ionization gas chromatography mass spectrometry (GC-NICIMS) have been shown to provide a high degree of precision in identification and quantification of biogenic amines, their putative amino acid precursors and metabolites in locusts *Schistocerca gregaria* and *Locusta migratoria*, and fruit fly *Drosophila melanogaster* [88-90,85]. Moreover, use of the chiral derivatization reagent, (-)-hepta-fluorobutyrylphenylalanyl chloride establishes that enantiomer of *p*-octopamine in the thoracic nervous system of the locust occurs in R configuration [89]. In rat, application of negative chemical ionization GC-MS to several sympathetically innervated organs has revealed the presence of both *m*-octopamine and *p*-octopamine [7]. Using GC-NICIMS method for the quantitative analysis of isomeric octopamine and synephrine in human urine has exhibited LOD of 50-100 ng/mg creatinine [91]. By combining ion exchange chromatography, capillary column gas chromatography, and electron capture negative ion chemical ionization mass spectrometry, the levels of *o*-octopamine, *m*-octopamine and *p*-octopamine in normal human urine are 0.6, 2.1 and 25 ng mg-1 creatinine, respectively [91]. After stable isotope dilution, another group has used GC-NICIMS procedure to determine levels of trace amines in plasma of healthy and hypertensive human subjects [92]. The limit of detection under optimal conditions is 5-20 pg ml^{-1} (0.032 pM -0.130 pM). However, levels of *m*- and -*p* isomers of octopamine, tyramine and synephrine detected in human plasma are low and inconsistent. This may be due to involvement of various factors, such as individual, dietary, environmental, and methodological (not using a sufficient n value), affecting the concentrations of these amines in human subjects [92].

By using an extraction-derivatization procedure, involving reactions with 3,5-di(trifluoromethyl)benzoyl chloride in the aqueous phase, followed by extraction into an organic solvent, hydrolysis of phenolic esters, and conversion of free hydroxyl groups to trimethylsilyl ethers, biogenic amines have been quantified in bovine brain [93]. The limits of detection for 3,5-di(trifluoromethyl)benzoyI chloride-trimethylsilyl derivatives by NCI-GC-MS are below the picogram level. The principal amines found in bovine retina using this method are *p*-tyramine, dopamine, and serotonin, with no evidence of *p*-octopamine [93]. Durden and Davis (1993) have used this procedure to determine the regional distribution of biogenic amines (such as phenylethylamine, *m*eta-tyramine and *p*ara-tyramine) in brain regions from a single rat as well as presence of these amines in human and dog plasma [94]. The standard curve for three amines is linear over the range 25 to 800 pg. The method has been found useful for unilateral experiments involving the rat striatum as the unaffected side as control for the experimental side. The values of human plasma study are supported by the dog plasma study as concentrations of phenylethylamine do not significantly increase after adding monoamine oxidase inhibitors in the blood, suggesting the absence of MAO-B in samples [94]. Collectively, GC-MS is considered as a sensitive and accurate quantification method (Table 1) that gives high analytical reproducibility and low biological variability.

4.4. Capillary Electrophoresis (CE)

The term capillary electrophoresis (CE) includes a variety of related techniques that rely on a high voltage electric field applied over a buffer solution held in narrow bore-capillaries to separate components of a mixture based on the differences in charge, size, and hydrophobic nature. Sample introduction is accomplished by immersing the end of the capillary into a sample vial followed by application of voltage allowing migration of molecules through the solution towards the electrode of opposite charge. Therefore, based on the charge, each molecule moves through at different speeds in the electric field. A photocathode measures the absorbency of molecules as they pass through the solution, and process is analyzed by a computer. CE has several advantages and disadvantages (Table 1). Due to high voltage, CE produces very efficient separation and reduces separation time. It is a technically a simple and sensitive procedure that consumes relatively small sample volumes (femtoliter to picoliter). It is an ideal method for the chiral analysis of compounds due to the high separation power of the technique [95]. The major drawbacks of CE include: (1) no effect on electrically neutral particles, and (2) small capillaries efficiently dissipate the heat, which is produced due to high electric field. A variety of detection systems, including electrochemical, laser-induced fluorescence and mass spectrometric, have been used in conjunction with CE.

The micellar electrokinetic capillary chromatography coupled to amperometric electrochemical (MEKC-EC) detection has been used to separate 14 biogenic amines in a single separation from head homogenates of *Drosophila melanogaster* [96]. The optimal conditions in CE include use of 25 mM borate buffer containing 50 mM sodium dodecyl sulfate and 2% 1-propanol (pH 9.5), which are responsible for improving the separation of several biogenic amines such as dopamine, epinephrine, norepinephrine, octopamine, L-3, 4-di-hydroxyphenylalanine, tyramine, and serotonin as well as their metabolites including 5-hydroxyindol-acetic acid, 3,4-dihydroxyphenylacetic acid, homovanillic acid, and 3-methoxytyramine in addition to *N*-acetylated metabolites including *N*-acetyldopamine, *N*-acetyloctopamine and *N*-acetylserotonin [96]. Previous investigators have used buffer containing 10 mM TES, 30 mM SDS and 2% 1-propanol pH 7.4 for the separation of biogenic amines [97], resulting in an incomplete separation of similarly charged and sized biogenic amines and metabolites. However, the modified separation procedure is more sensitive and selective in detecting changes in tyramine, octopamine, and N-acetyloctopamine, which are present in the head homogenate of fruit fly [96]. This suggests that the right composition of the separation buffer, surfactant, pH and additive are important factors for enhancing separation efficiency and selectivity of bioamines. MEKC-EC procedure has tremendous potential for understanding the physiological processes and underlying mechanisms mediated by biogenic amines. This procedure provides a complete quantitative analysis of all metabolites in conjunction with precursors in a single run. The

LOD for octopamine by borate-MEKC-EC procedure is 3.4 amol, which makes it attractive for volume limited sample analysis [96].

CE separation along with optical fiber light emitting diode (LED)-induced fluorescence detector has been used in determining octopamine levels in human plasma [98]. The selection of LED is based according to the excitation and emission spectra of analyte. The pre-column derivatization of octopamine with naphthalene-2,3-dicarboxaldehyde (NDA) forms NDA-labeled octopamine, a high fluorescent compound. Optimal separation is obtained by using a background electrolyte solution (consisting 25 mM sodium borate at pH 9.2) with applied voltage (20 kV) to achieve high sensitivity detection with LED as the excitation source. CE-LED has LOD for octopamine as low as 5.0 nM [98]. The measured concentrations of octopamine by CE-LED in human plasma are in the range of 0.95×10^{-8} to 2.41×10^{-8} M [98]. The concentration range is in accordance with previous values obtained by the HPLC method [44]. Due to sufficient sensitivity, short separation time (less than 5 min), and low cost, CE-LED method can be considered as a good alternative to existing HPLC for measuring octopamine levels in biological fluids (Table 1). CE with laser-induced fluorescence detection (CELINF) system uses a photomultiplier tube-based (PMT) multichannel detection system with a 224-nm HeAg laser as its excitation source, enabling analysis of indolamines and catecholamines (Table 1). CELINF is used to analyze biogenic amines levels in a single neuron in the brain of pond snail *Lymnaea stagnalis* [99]. In PMT-based detection system, the LOD for octopamine is 11 nM [99]. Collectively, it is suggested that adequate sensitivity of CELINF may allow the analysis of octopamine and other neurotransmitters even in the subcellular compartments of individual neurons.

4.5. Determination of Octopamine by Autoradiographic Localization

Autoradiography is a process by which radioactive molecules within the cell make their location known by exposing to a photographic emulsion forming a pattern on the photographic films corresponding to their location within the cell. The autoradiographic method is used to locate the sites of octopamine synthesis in invertebrates [100]. Octopamine is synthesized from both tyrosine and tyramine in the lateral, median, and ventral eyes of *Limulus.* It is located in efferent fibers, which innervate ventral photoreceptors and lateral eye ommatidia by using light microscopic and electron microscopic autoradiography. In this procedure, signal can be evaluated quantitatively. Signal can be compared in several structures at the same time. Autoradiography is sensitive, but compared to immunostaining techniques it shows less microscopic resolution (Table 1). Octopamine modulates basic retinal processes, such as photoreceptor sensitivity, photomechanical movements, and photoreceptive membrane turnover [100]. The efferent fibers, which project to ventral and lateral eyes from the central nervous system can be selectively labeled during *in vitro* incubations with ^{3}H-tyramine. Since in the ventral eye, ~ 95% of the efferent fibers are labeled, therefore, the selective labeling of efferent fibers during incubation with ^{3}H-tyramine is used as an approach to evaluate the number and distribution of efferent fibers within the ventral eye [101]. Thus, octopamine may mediate the biochemistry and physiology of ventral photoreceptor cells in *Limulus* retinal efferent fibers. Many of theses effects may be due to circadian efferent innervation to the lateral eye of *Limulus.*

Autoradiographic techniques normally require β emittors such as ^{14}C, ^{3}H or ^{35}S. However, autoradiography can also be performed by using a radionuclide (such as ^{99m}TC), which emit many low energy electrons (gammas, x-ray, and monoenergetic) while decaying [102]. The detector is gas chamber, which allows direct quantification of images and provides a clear advantage of autoradiographic film by improving image resolution, efficiency, linearity, high dynamic range and low noise levels [102]. Collectively, by using autoradiography technique, one can detect presence of radioactive isotope in tissues by covering tissue sections with photographic emulsion, followed by subsequent development and fixation, and finally the radiated silver resulting into black grains [100]. Autoradiography technique is suitable for studying sites for octopamine metabolism. It can also be used to visualize and quantitate radiolabeled octopamine binding proteins in tissue homogenate (not discussed in this review).

4.6. Determination of Octopaminergic Neurons by Immunostaining Techniques

Immunostaining techniques detect specific antibody-antigen interactions after tagging the antibody with a visible label. The most commonly used immunostaining labels for octopaminergic neuronal distribution are fluorescence and enzyme mediated detection. Immunocytochemistry is performed on samples of intact cells that have had most, if not all of their surrounding extracellular matrix removed, whereas immunohistochemistry is performed on sections of tissue in which each cell is surrounded by the tissue architecture. The binding of antibody with a cellular antigen results in a specific stain that can be readily visualized with a microscope. The control preparations in which the primary antibody is omitted should not reveal any immunoreactive cell bodies or fibers. Fluorescent images are viewed with a laser-scanning, confocal microscope equipped with lasers and appropriate filters. In immunohistochemistry, the staining intensity is closely related to the degree of fixation and affected by under- or over-fixation and different immersion/perfusion methods.

Octopamine is found ubiquitously in invertebrates including insects, modulating a wide spectrum of behaviors. Immunostaining procedures using polyclonal or monoclonal antibodies have been used to examine the distribution of octopaminergic neurons in many brain regions in several insect species, including hawkmoth *Manduca sexta* [103], honeybee *Apis mellifera* [104-105], cockroach *Periplaneta americana* [106-108], fruit fly *Drosophila melanogaster* [109-110], blow fly *Phaenicia sericata* [111], and locust *Locusta migratoria* and *Schistocerca gregaria* [112-113]. One group has elegantly combined HPLC system with immunocytochemistry (HPLC to detect octopamine levels and immunocytochemical approach to observe the distribution of octopamine-containing nerve cells) in the alimentary tract of the earthworm *Eisenia fetida*, suggesting the occurrence of octopamine-containing nerve cells in invertebrate peripheral nervous system [114]. The immunocytochemistry has been helpful in resolving the neural pathways that mediate adaptive behavioral changes at a single cell level in honeybee [105]. Furthermore, octopamine immunoreactivity in the adult

Drosophila CNS reveals many additional neurons compared to the larval CNS, suggesting that at least a subset of adult octopamine neurons differentiate during metamorphosis [109-110].

The use of immunostaining techniques allow to compare possible morphological and anatomical similarities and dissimilarities in the number of octopaminergic neurons, their branching patterns and projections among different taxa and species. Immunostaining techniques are specific and repeatable from one animal to the next. However, fixation time, type of fixative, dehydration, embedding, and antibody conentration (Table 1) may cause minor variation in signal intensity and/or resolution between two preparations in the same taxon [104]. The variation in signal intensity of octopamine also depends on the difference in the age of individual insect [70-71]. Collective evidence supports the view that identification of octopamine-immunoreactive neurons in invertebrate brains sets the stage to examine the sequence of appearance of octopamine-immunoreactive neurons and to correlate the appearance of these neurons with the functional role during development.

In vertebrates, trace amines (including *m*-octopamine, *p*-octopamine, *m*- tyramine, *p*-tyramine, and 2-phenylethylamine) are present at least two orders of magnitude below those of neurotransmitters, such as dopamine, norepinephrine and serotonin [115]. Catecholamines differ from each other by the presence of one or two chemical groups placed either on the ring or on the side chain (Figure 1). A slight structural difference such as hydroxyl or carboxylic group may lead to a similar, different and specific immune response [116]. Paterson et al. (1990) have reported that low levels of trace amines (including octopamine) require an antibody cross-reactivity of greater than 100,000:1 with respect to other classical monoamine neurotransmitters [117]. Thus, due to high cross-reactivity of octopamine antiserum, trace amounts of octopamine and uncertainty of octopamine receptors in the vertebrate brain, the immunostaining profiles of octopamine may not be specific and valid.

5. Quantification of Radionuclides in Experimental Animals and Humans

The positron emission tomography (PET) and single photon emission computed tomography (SPECT) are i*n vivo* nondestructive imaging methods that allow quantification of radionuclides in small regions of large experimental animal and humans. By using these imaging techniques, a drug can be injected into the subject's body, and its uptake, the tissues in which it concentrates, receptor occupancy, and its elimination can be easily monitored far more quickly than the previously used techniques [118]. In PET, the subject receives a short half-lived radiopharmaceutical that discharges positrons from wherever they are used in the body. When these positrons encounter electrons within the body, it results in production of gamma rays. PET enables to see the location of the metabolic process. The most commonly used radionuclides in PET are ^{15}O, ^{13}N, ^{11}C and ^{18}F, which are short-lived. SPECT is another related technique in which most commonly used radionuclides are ^{99m}Tc, ^{123}I, ^{67}G, ^{201}TI, which are heavier and longer-lived radioactive atoms that emit high-energy photons called gamma rays, instead of positrons. Long half-lives of these compounds can limit the number of times a patient can be scanned. PET has advantage over SPECT for being more sensitive in quantification and topographic isolation because of the positron decay mode. However, the longer life of radionuclide in SPECT (such as 13 hr for ^{123}I, 6 hr for ^{99m}Tc) gives more flexibility in imaging studies. SPECT is used for many of the same applications as PET, but is less expensive than PET, and the resulting picture is usually less sharp than a PET image. These techniques are relatively safe, and have been used in patients suffering from cancer, neurodegenerative and neuropsychiatry disorders and heart diseases [118-122], but because of the high cost only few hospitals and laboratories are capable of maintaining such systems.

A radioligand, 1*R*-[^{11}C]Phenylephrine ([^{11}C]PHEN), has been developed for PET imaging of the sympathetic nervous system. It is a catecholamine analog labeled with ^{11}C, a positron emitter with half life of 20.4 min [123]. The chiral synthesis of PHEN is achieved by direct methylation of (-)-*m*-octopamine with either CH_3I or $CF_3SO_3^{11}$ [123]. [^{11}C]PHEN is a radiolabeled analog of norepinephrine. Collective evidence suggests that probes with structures closely related to norepinephrine (such as octopamine) can be labeled with ^{11}C, a positron emitter, which can then be used to assess octopamine metabolism in various human diseases.

6. Advantages and Disadvantages of Various Procedures

Advances in methodology have resulted in the enhancement of the analytical sensitivity and selectivity of the assay procedure for estimating catecholamines and their metabolites in invertebrate and vertebrate systems. Early estimates of these amines in tissues and biological fluids have been done by using colorimetric and bioassays. These assays are relatively inexpensive but they lack sensitivity and specificity [124]. Followed by colorimetric and bioassays, the most commonly used methods to estimate catechecholamines have been the radioenzymic methods [46-59], which have gained limited use due to the technical disadvantages including the complexity and length of the procedures. The original radioenzymic method of Molinoff et al. (1969) to measure *p*-octopamine has now been found to be nonspecific because PNMT can utilize all three positional isomers of octopamine and the corresponding isomeric synephrines as its substrates [125,7]. The modification in the original radioenzymic assay, using dansylation of the crude extract followed by three successive TLC separations, is responsible for the discovery of *m*-octopamine simultaneously present with *p*-octopamine in rat salivary gland [50], and following MAO inhibition in the rat brain [48]. The modified radioenzymic method can determine both *m*- and *p*-octopamine in vertebrates but does not take into account the possibility of the presence of *o*-octopamine and *o*-, *m*-and *p*-synephrine [7]. *O*-synephrine has never been detected in any biological tissue or fluid [7]. However, the presence of *o*-octopamine has been reported in rat urine after MAO inhibition [126]. Another disadvantage of radioenzymic assays is their unsuitability for the measurement of octopamine due to the presence of endogenous inhibitors of enzymes (such as DBH and PNMT) in the biological fluids, altering the levels of octopamine and synephrine, respectively. Furthermore, the cost of chemicals in radioenzymic assays (e.g. radioactively labeled *S*-adenosylmethionine, solvents, and enzyme) is also quite high compared to the materials used for HPLC.

The estimation of octopamine and other catecholamines using HPLC methods coupled with electrochemical detection [60-76] or fluorometric detection [77-84] involving pre-column derivatization or post- column derivatization, is quite promising. HPLC methods are useful tools for measuring plasma catecholamines because unlike radioenzymic assays chromatographic conditions can be altered to eliminate endogenous or exogenous interferences. HPLC relies only on the chromatographic retention, whereas GC-MS permits unequivocal identification because it is based on chromatographic retention and molecular mass. Therefore GC-MS has advantage over HPLC, but it is not used in the routine clinical setting because of the complexity of the procedure [85-96]. CE is a novel separation procedure that enables rapid separations with high efficiency, chiral selectivity, and compatibility with small sample volumes [96-98]. It is sensitive enough to determine octopamine levels in human plasma from healthy subjects [98]. At present, among all available analytical techniques to determine octopamine, CELINF is the only technique that can allow detection of octopamine and other catecholamines in single neurons of the pond snail brain [99].

Invertebrate preparations are richer sources of octopamine, and unique octopaminergic neurons exist in invertebrate preparations. Therefore, immunostaining techniques have been routinely used in invertebrate brains. However, in mammalian systems, it is still uncertain whether unique octopaminergic neurons exist or octopamine is found only in neurons that utilize norepinephrine as their transmitter substance. Therefore, use of antibody against octopamine may not allow the labeling of the distribution of immunoreactive profiles of octopaminergic neurons in the vertebrate brain. Another point that has to be emphasized here is the chiral nature of octopamine. Studies may get complicated depending on the presence of enantiomer *in vivo*. For example, in insects, during functional studies most investigators have simply used DL-*p*-octopamine for all their studies, which could give misleading information on the relative potency of octopamine vs. other biogenic amines such as tyramine. Therefore, it is important to evaluate whether octopamine is only present as *p*-isomer in all insect species or the presence of isomeric form(s) varies from one taxon to the other. Furthhermore, for future biochemical and behavioral studies in invertebrates, instead of just evaluating the effect of the racemic compound (DL-*p*-octopamine), the effect of separate enantiomers (D-*p*-octopamine, L-*p*-octopamine) should be tested. In addition, DL-*m*-octopamine (racemic) and its enantiomers D-*m*-octopamine and L-*m*-octopamine should be evaluated.

Furthermore, when determining octopamine in biological fluids, it is very important to consider dietary influence on its concentration because foods enriched in aromatic amino acids or amines may certainly influence octopamine levels in biological systems [127]. Furthermore, accidental poisoning with octopaminergic pesticides [128-129] or with medicinal plant extracts containing high levels of octopamine-related drugs may contribute to alteration in octopamine levels [130]. Both tyramine and octopamine are substrate for MAO. The catabolism of these amines by MAO results in the production of toxic compounds, such as hydrogen peroxide, ammonia, and aldehyde. Therefore, increase in MAO levels may decrease octopamine concentration and increase free radical-mediated oxidative stress to the cell that is seen under several pathological diseases. Thus, patients have to be on a regular diet several days before the quantification of octopamine or its metabolites in their biological fluids. The estimated levels of octopamine and other catecholamines in body fluids of normal individuals are different from patients suffering from psychiatric and neurological diseases [131].

A considerable variation in endogenous octopamine concentration can be from from animal to animal, which may be due to differences in dietary, environmental, and genetic factors. Different levels of octopamine are present in vertebrates and invertebrates. Therefore, methods with different selectivities and sensitivities are required to quantify octopamine, particularly with respect to detecting the trace endogenous levels of octopamine in vertebrates.

7. Conclusion

Octopamine is a multifunctional naturally occurring biogenic amine, which plays a major neuromodulatory role in invertebrates. However, its physiological role remains elusive in vertebrates due to its low endogenous concentration in the central and sympathetic nervous systems and rapid turnover. Octopamine is coreleased with other catecholamines, therefore many of its effects may be indirect through its interference with other aminergic pathways. However, octopamine is present in unusually high concentration in certain diseases and in patients treated with MAOI [40,42-45], implicating some pathophysiological role. Therefore, it is important to select a sensitive and rapid procedure for the quantification of octopamine, especially in vertebrate system. In this overview, author has critically pointed out the strengths as well as drawbacks of each assay system, and provided LODs achieved by using different assay systems in both invertebrate and vertebrate systems. Collective evidence suggests that the sensitivity of each assay system varies among vertebrates and invertebrates. At present, highly sensitive assay is CE-MEKC, which can detect lowest octopamine level in *Drosophila* head. In human plasma, GC-NICIMS and CE-LED have been quite successful in detecting octopamine. In rat brain, improved radioenzymic assay detects significantly lower concentrations of *p*-octopamine and *p*-tyramine. Improved sensitivity and selectivity of these procedures indicate that these procedures will be helpful in evaluating the physiological and pathophysiological roles of octopamine and other trace amines in the mammalian nervous system.

Acknowledgments

I dedicate this review to the memory of my Ph.D advisor, late Professor Norman J. Uretsky. He will always be remembered for his dynamism and passion for biogenic amines. I thank anonymous referees who helped me in significantly improving this manuscript.

REFERENCES

[1] Erspamer V, Boretti G. Identification and characterization, by paper chromatography, of enteramine, octopamine, tyramine, histamine and allied substances in extracts of posterior salivary glands of octopoda and in other tissue extracts of vertebrates and invertebrates. *Archives Internationales de Pharmacodynamie et de Therapie.* 1951;88:296-332.

[2] Axelrod J, Saavedra JM. Octopamine. *Nature.* 1977;265:501-504.

[3] Evans PD. Octopamine. In: Kerkut GA, Gilbert LI, eds. *Comprehensive Insect Physiology, Biochemistry and Pharmacology.* Pergamon Press: Oxford; 1985: 499-529.

[4] David JC, Coulon JF. Octopamine in invertebrates and vertebrates. A review, *Progress in Neurobiology.* 1985;24:141-185.

[5] Mercer AR, Menzel, R. The effects of biogenic amines on conditioned and unconditioned responses to olfactory stimuli in the honeybee *Apis mellifera. Journal of Comparative Physiology. A.* 1982;145:363-368.

[6] Harmar, AJ, Horn, AS. Octopamine-sensitive adenylate cyclase in Cockroach brain: effects of agonists, antagonists, and guanylyl nucleotides. *Molecular Pharmacology.* 1976;13:512-520.

[7] Ibrahim KE, Couch MW, William CM, Fregly MJ, Midgley JM. *m*-Octopamine: normal occurrence with *p*-octopamine in mammalian sympathetic nerves. *Journal of Neurochemistry.* 1985;44:1862-1867.

[8] Brown CM, McGrath JC, Midgley JM, Muir AG, O'Brien JW, Thonoor CM, Williams CM, Wilson VG. Activities of octopamine and synephrine stereoisomers on α-adrenoceptors. *British Journal of Pharmacology.* 1988;93: 417-429.

[9] Airriess CN, Rudling JE, Midgley JM, Evans PD. Selective inhibition of adenylyl cyclase by octopamine via a human cloned $\alpha_{\text{-}2A}$-adrenoceptor. *British Journal of Pharmacology.* 1997;122:191-198.

[10] James MI, Midgley JM, Williams CM. The metabolism and biosynthesis of (+/-)-*o-octopamine and (+/-)-o-synephrine in the rat. Journal of Pharmacy and Pharmacology.* 1983;35:559-565.

[11] Midgley JM, Couch MW, Crowley JR, Williams C M. Identification and quantitative determination of *o*- and *m*-hydroxymandelic acid in human urine. Biomedical Mass Spectrometry. 1979; 6:485-490.

[12] Orchard I. Octopamine in insects: neurotransmitter, neurohormone, and neuromodulator. *Canadian Journal of Zoology.*1982; 60:659-669.

[13] Roeder T. Octopamine in invertebrates. *Progress in Neurobiology.* 1999;59:533-561.

[14] Borowsky B, Adham N, Jones KA, Raddatz R, Artymyshyn R, Ogozalek, KL, Durkin MM, Lakhlani PP, Bonini JA, Pathirana S, Boyle N, Pu X, Kouranova E, Lichtblau H, Ochoa FY, Branchek TA, Gerald C. Trace amines: identification of a family of mammalian G protein-coupled receptors. *Proceedings of the National Academy of Sciences USA.* 2001;98:8966-8971.

[15] Barron AB, Schulz DJ, Robinson GE. Octopamine modulates responsiveness to foraging-related stimuli in honey bees *Apis mellifera. Journal of Comparative Physiology. A.* 2002;188:603-610.

[16] Giray T, Galindo-Cardona A, Oskay D. Octopamine influences honey bee foraging preference. *Journal of Insect Physiology.* 2007;53:691-698.

[17] Schulz DJ, Barron AB, Robinson G. A role for octopamine in honeybee division of labor. *Brain Behavior and Evolution.* 2002;60:350-359.

[18] Fields PE, Woodring JP. Octopamine mobilization of lipids and carbohydrates in the house cricket, *Acheta domesticus. Journal of Insect Physiology.* 1991;37: 193-199.

[19] Burrell, BD, Smith BH. Modulation of the honeybee (*Apis mellifera*) sting response by octopamine. *Journal of Insect Physiology.* 1995;41:671-680.

[20] Monastirioti M. Distinct octopamine cell population residing in the CNS abdominal ganglion controls ovulation in *Drosophila melanogaster. Developmental Biology.* 2003;264:8-49.

[21] Stevenson PA, Dyakonova V, Rillich J, Schildberger K. Octopamine and experience-dependent modulation of aggression in crickets. *Journal of Neuroscience.* 2005;25:1431–1441.

[22] Dierick HA. Fly fighting: octopamine modulates aggression. *Current Biology.* 2008;18:R161-R163.

[23] Bacon JP, Thompson KS, Stern M. Identified octopaminergic neurons provide an arousal mechanism in the locust brain. *Journal of Neurophysiology.* 1995;74: 2739-2743.

[24] Hammer M, Menzel R. Multiple sites of associative odor learning as revealed by local brain microinjections of octopamine in honeybees. *Learning and Memory.* 1998;5:146-156.

[25] Farooqui T, Robinson K, Vaessin H, Smith BH. Modulation of early olfactory processing by an octopaminergic reinforcement pathway in the honeybee, *Journal of Neuroscience.* 2003;23:5370-5380.

[26] Schwaerzel M, Monastirioti M, Scholz H, Friggi-Grelin F, Birman S, Heisenberg M. Dopamine and octopamine differentiate between aversive and appetitive olfactory memories in *Drosophila. Journal of Neuroscience.* 2003;23:10495-10502.

[27] Schroll C, Riemensperger T, Bucher D, Ehmer J, Völler T, Erbquth K, Gerber B, Hendel T, Naqel G, Buchner E, Fiala A. Light-induced activation of distinct modulatory neurons triggers appetitive or aversive learning in *Drosophila* larvae. *Current Biology.* 2006;16:1741-1747.

[28] Scholz H, Franz M, Heberlein U. The hangover gene defines a stress pathway required for ethanol tolerance development. *Nature.* 2005;436-845-847.

[29] Farooqui T. Octopamine-mediated neuromodulation of insect senses. *Neurochemical Research.* 2007;32:1511-1529.

[30] Lim CC, Wasserman GS. Categorical and prolonged potentials are evoked when brief, intermediate-intensity flashes stimulate horseshoe crab lateral eye photoreceptors during octopamine neuromodulation. *Biological Signals and Receptors.* 2001;10:399-415.

[31] Roeder T. Tyramine and octopamine: Ruling behavior and metabolism. *Annual Review of Entomology.* 2005;50:447-477.

[32] Evans PD, Maqueira B. Insect octopamine receptors: a new classification scheme based on studies of cloned *Drosophila* G-protein coupled receptors. *Invertebrate Neuroscience*. 2005;5:111-118.
[33] Rudling JE, Richardson J, Evans PD. A comparison of agonist-specific coupling of cloned human α_2-adrenoceptor subtypes. *British Journal of Pharmacology*.2000;131:933 -941.
[34] Carpêné C, Galitzky J, Fontana E, Atgieâ C, Lafontan M, Berlan M. Selective activation of β3-adrenoceptors by octopamine: comparative studies in mammalian fat cells. *Naunyn-Schmiedebergs Archives of Pharmacology*. 1999;359:310-321.
[35] Harmar AJ, Horn AS. Octopamine in mammalian brain: rapid post mortem increase and effects of drugs. *Journal of Neurochemistry*. 1976;26:987-993.
[36] V. Visentin, N. Morin, E. Fontana, D. Prevot, J. Boucher, I. Castan, P. Valet, D. Grujic, C. Carpene, Dual action of octopamine on glucose transport into adipocytes: inhibition via β_3-adrenoceptor activation and stimulation via oxidation by amine oxidases. *Journal of Pharmacology and Experimental Therapeutics*. 2001;299:96-104.
[37] Delbarre B, Delbarre G, Casset-Senon D, Sestillange P. Effects of drugs interfering with the metabolism of octopamine on blood pressure of rats. *Comparative Biochemistry and Physiology. C*. 1982;72:153-157.
[38] Bulach C, Doaré L, Massari B, Simon P. Antidepressive effects of 3 endogenous monoamines: psychopharmacologic profiles of noradrenaline, octopamine and phenethylamine. *Journal of Pharmacology*. 1984;15:1-15.
[39] Delacour J, Guenaire C. Octopamine and locomotor activity of rats. *Psychopharmacology*. 1983;80:50-52.
[40] Ibrahim KE, Couch MW, Williams CM. Increased urinary excretion of *m*-octopamine in neuroblastoma. *Biochemical Medicine*. 1985;33:53-59.
[41] Couch MW, Greer DM, Williams CM. Excretion of octopamine metabolites in neuroblastoma. *Clinica Chimica ACTA*. 1986;158:109-114.
[42] Kinniburgh DW, Boyd ND. Determination of plasma octopamine and its level in renal disease. *Clinical Biochemistry*. 1979;12:27-32.
[43] Nespoli A, Bevilacqua G, Staudacher C, Rossi N, Salerno F, Castelli, MR. Pathogenesis of hepatic encephalopathy and hyperdynamic syndrome in cirrhosis. Role of false neurotransmitters. *Archives of Surgery*. 1981; 116:1129-1138.
[44] D'Andrea G, Terrazzino S, Leon A, Fortin D, Perini F, Granella F, Bussone G. Elevated levels of circulating trace amines in primary headaches. *Neurology*. 2004;62:1701-1705.
[45] Premont RT, Gainetdinov RR, Caron MG. Following the trace of elusive amines, *Proceedings of the National Academy of Sciences USA*. 2001;98:9474-9475.
[46] Molinoff PB, Landsberg L, Axelrod J. An enzymatic assay for octopamine and other β-hydroxylated phenylethylamines, *Journal of Pharmacology and Experimental Therapeutics*. 1969;170:253-261.
[47] Goosey MW, Candy DJ. The d-octopamine content of the haemolymph of the locust, *Schistocerca americana gregaria* and its elevation during flight. *Insect Biochemistry*. 1980;10:393-397.

[48] Danielson TJ, Boulton AA, Robertson HA. *m*-Octopamine, *p*-octopamine and phenylethanolamine in rat brain: a sensitive, specific assay and the effects of some drugs. *Journal of Neurochemistry*. 1977;29:1131-1135.
[49] Evans PD. Octopamine distribution in the insect nervous system. *Journal of Neurochemistry*. 1978;30:1009-1013.
[50] Robertson HA, David JC, Danielson TJ. Effects of denervation of the levels of *p*-octopamine *m*-octopamine (norphenylephrine) and phenylethanolamine in rat salivary gland. *Journal of Neurochemistry*. 1977;29:1137-1139.
[51] Saavedra JM. Enzymatic-isotopic method for octopamine at the picogram level. *Analytical Biochemistry*. 1974;59:628-633.
[52] Christensen TA, Sherman TG, McCaman RE, Carson AD. Presence of octopamine in firefly photomotor neurons. Neuroscience, 1983;9:183-189.
[53] Howell KM, Evans PD. The characterization of presynaptic octopamine receptors modulating octopamine release from an identified neurone in the locust. *Journal of Experimental Biology*. 1998;201:2053-2060.
[54] Lehman HK, Murgiuc CM, Hildebrand JG. Characterization and developmental regulation of tyramine-β-hydroxylase in the CNS of the moth, *Manduca sexta. Insect Biochemistry and Molecular Biology*. 2000;30:377-386.
[55] Philips SR. Radioenzymatic analysis of neurotransmitters, *Life Sciences*. 1987;41:877-880.
[56] Kobayashi K, Foti A, Dequattro V, Kolloch R, Miano L. A radioenzymatic assay for free and conjugated normetanephrine and octopamine excretion in man. *Clinica Chimica ACTA*. 1980;107:163-173.
[57] Wise CD. A sensitive assay for dopamine β-hydroxylase. *Journal of Neurochemistry*. 1976;27:883-888.
[58] Henry DP, Johnson DG, Starman BJ, Williams RH. Kinetic characterization of rat serum dopamine-β-hydroxylase using a simplified radioenzymatic assay. Life Sciences. 1975;17:1179-1186.
[59] Van Huysse JW, Bowsher RR, Willis LR, Henry DP. A radioenzymatic assay for *para*-tyramine and *para*-octopamine. *Biochemical Medicine and Metabolic Biology*. 1988;40:50-60.
[60] Hashimoto H, Maruyama Y. High performance liquid chromatography with electrochemical detection. In: Parvez S, Nagatsu T, Nagatsu I, Pervez H, eds. *Methods in Biogenic Amine Research.* Elsevier Science Publishers: B.V.; 1983: 35-74.
[61] Kissinger PT, Riggin RM, Alcorn RL, Rau L.D. Estimation of catecholamines in urine by high performance liquid chromatography with electrochemical detection, *Biochemical Medicine*. 1975;13:299-306.
[62] Auger J, Boulay R, Jaillais B, Delion-Vancassel S. Analysis of biogenic amines by solid-phase microextraction and high-performance liquid chromatography with electrochemical detection. *Journal of Chromatography*. 2000;870:395–403.
[63] Cheng FC, Kuo JS. High-performance liquid chromatographic analysis with electrochemical detection of biogenic amines using microbore columns. *Journal of Chromatography.* 1995;665:1-13.

[64] Duine JM, Floch F, Cann-Moisan C, Mialon P, Caroff J. Simultaneous measurement of monoamines, their metabolites and 2,3- and 2,5-dihydroxy- benzoates by high-performance liquid chromatography with electrochemical detection. Application to rat brain dialysates. *Journal of Chromatography. B.* 1998;716:350–353

[65] Molnar I, Horvath C. Reverse-phase chromatography of polar biological substances: separation of catechol compounds by high-performance liquid chromatography. *Clinical Chemistry*. 1976;22:1497-1502.

[66] Martin RJ, Downer RG. Microassay for N-acetyltransferase activity using high-performance liquid chromatography with electrochemical detection. *Journal of Chromatography*. 1989;487:287-293.

[67] Clark J, Lange AB. Octopamine modulates spermathecal muscle contractions in *Locusta migratoria. Journal of Comparative Physiology. A.* 2003;189:105-114.

[68] Downer RGH, Bailey BA, Gole JWD, Martin RJ, Orr GL. Estimation of octopamine and its role in the release of a hypertrehalosemic factor in the American cockroach, *Periplaneta Americana.* In: Borkovec AB, Kelley TJ, eds. *Insect Neurochemistry and Neurophysiology*. Plenum Press: New York; 1983: 349-352.

[69] Monastirioti M, Linn CE Jr., White K. Characterization of *Drosophila* tyramine β-hydroxylase gene and isolation of mutant flies lacking octopamine. *Journal of Neuroscience*. 1996;16:3900-3911.

[70] Schulz DJ, Robinson GE. Biogenic amines and division of labor in honey bee colonies: behaviorally related changes in the antennal lobes and age-related changes in the mushroom bodies. *Journal of Comparative Physiology. A.*1999; 184:481-488.

[71] Schulz DJ, Barron AB, Robinson GE. A role of octopamine in honeybee division of labor. *Brain Behavior and Evolution*. 2002;60:350-359.

[72] Linn CE, Poole KR, Wu WQ, Roelofs WL. Circadian changes in melatonin in the nervous system and hemolymph of the cabbage looper moth, *Trichoplusia ni. Journal of Comparative Physiology. A.* 1995;176:761-771.

[73] Linn CE, Poole KR, Roelofs WL. Studies on biogenic amines and their metabolites in nervous tissue and haemolymph of adult male cabbage looper moths. I. Quantitation of photoperiod changes. *Comparative Biochemistry and Physiology. C.* 1994;108:73–85.

[74] Adamo SA, Linn CE, Hoy RR. The role of neurohormonal octopamine during 'fight or flight' behaviour in the field cricket *Gryllus Bimaculatus. Journal of Experimental Biology*. 1995;198:1691-1700.

[75] Seiler N, Knodgen B. High-performance liquid chromatographic procedure for the simultaneous determination of the natural polyamines and their monoacetyl derivatives. *Journal of Chromatography*. 1980;221:227-235.

[76] Hardie SL, Hirsh J. An improved method for the separation and detection of biogenic amines in adult *Drosophila* brain extracts by high performance liquid chromatography. *Journal of Neuroscience Methods*. 2006;153:243-249.

[77] Davis TP, Gehrke CW, Gehrke CW Jr., Cunningham TD, Kuo KC, Gerhardt KO, Johnson HD, Williams CH. High-performance liquid-chromatographic separation and fluorescence measurement of biogenic amines in plasma, urine, and tissue. *Clinical Chemistry*. 1978;24:1317-1324.

[78] De Bernardo S, Weigele M, Toome V, Manhart K, Leimgruber W, Bohlen P, Stein S, Udenfriend S. Studies on the reaction of fluorescamine with primary amines. *Archives of Biochemistry and Biophysics*. 1974;163:390-399.

[79] Maickel RP, Miller FP. Flourescent products formed by reaction of indole derivative and *o*-phthalaldehyde. *Analytical Chemistry*. 1966;38:1937-1938.

[80] Simons SS, Johnson DF. The structure of the flourescent adduct formed in the reaction of *o*-phthalaldehyde and thioles with amines. *Journal of the American Chemical Society*. 1976; 98: 7098-70994.

[81] Kawasaki T, Higuchi T, Imai K. Determination of dopamine, norepinephrine, and related trace amines by prechromatographic derivatization with naphthalene-2,3-dicarboxaldehyde. *Analytical Biochemistry*. 1989;180:279-285.

[82] Mell LD Jr, Carpenter DO. Fluorometric determination of octopamine in tissue homogenates by high-performance liquid chromatography. *Neurochemical Research*. 1980;5:1089-1096.

[83] Kamata S, Imura K, Okada A, Kawashima Y, Yamatodani , Y A, Watanabe T, Wada H. Simultaneous analyses of phenethylamine, phenylethanolamine, tyramine and octopamine in rat brain using fluorescamine. *Journal of Chromatography*. 1982;231:291-299.

[84] Flatmark T, Skotland T, Ljones T, Ingebretsen OC. Fluorimetric detection of octopamine in high-performance liquid chromatography and its application to the assay of dopamine β-monooxygenase in human serum. *Journal of Chromatography*. 1978;146:433-438.

[85] Watson DG, Zhou P, Midgley JM, Milligan CD, Kaiser K. The determination of biogenic amines in four strains of the fruit *fly Drosophila melanogaster. Journal of Pharmaceutical and Biomedical Analysis*. 1993;11:1145-1149

[86] Pan C, Li W, Wang Y, Jiang S. Octopamine levels in *Blattella Germanica* L. Tissues by capillary gas chromatography with electron capture detection. *International Journal of Molecular Sciences*. 2005;6:188-197.

[87] Duffield PH, Dougan DF, Wade DN, Duffield AM. A chemical ionization gas chromatographic mass spectrometric assay for octopamine and tyramine in rat brain. Biomedical Mass Spectrometry. 1981;8:170-173.

[88] Macfarlane RG, Midgley JM, Watson DG, Evans PD. Identification and quantitation of phenylalanine, tyrosine and dihydroxyphenylalanine in the thoracic nervous system of the locust, *Schistocerca gregaria*, by gas chromatography–negative-ion chemical ionization mass spectrometry. *Journal of Chromatography*. 1990;532:1-25.

[89] Macfarlane RG, Midgley JM, Watson DG, Evans PD. Biogenic amines: their occurrence, biosynthesis and metabolism in the locust, *Schistocerca gregaria*, by gas chromatography–negative ion chemical ionisation mass spectrometry. *Journal of Chromatography*. 1991;562:585-98.

[90] Stern M. Octopamine in the locust brain: cellular distribution and functional significance in an arousal mechanism. *Microscopy Research and Technique*. 1999;45:135-141.

[91] Ibrahim KE, Couch MW, Williams CM, Budd MB, Yost RA, Midgley JM. Quantitative measurement of octopamines and synephrines in urine using capillary

column gas chromatography negative ion chemical ionization mass spectrometry. *Analytical Chemistry*. 1984;56:1695-1699.

[92] Andrew R, Best SA, Watson DG, Midgley JM, Reid JL, Squire IB. Analysis of biogenic amines in plasma of hypertensive patients and a control group. *Neurochemical Research*. 1993;18:1179-1182.

[93] Macfarlane RG, Macleod SC, Midgley JM, Watson DG. Analysis of biogenic amines in bovine retina by gas chromatography–negative ion chemical ionisation mass spectrometry. *Journal of Neurochemistry*. 1989;53:1731-1736.

[94] Durden DA, Davis BA. Determination of regional distributions of phenylethylamine and meta-and para-tyramine in rat brain regions and presence in human and dog plasma by an ultra-sensitive negative chemical ion gas chromatography-mass spectrometric (NCI-GC-MS) method. *Neurochemical Research*. 1993;18:995-1002.

[95] Scriba GK. Recent advances in enantioseparations of peptides by capillary electrophoresis. *Electrophoresis*. 2003;24:4063-4077.

[96] Paxon TL, Powell PR, Lee HG, Han KA, Ewing AG. Microcolumn separation of amine metabolites in the fruit fly. *Analytical Chemistry*. 2005;77:5349-5355.

[97] Ream PJ, Suljak SW, Ewing AG, Han KA. Micellar electrokinetic capillary chromatography-electrochemical detection for analysis of biogenic amines in *Drosophila melanogaster. Analytical Chemistry*. 2003;75:3972-3978.

[98] Yu Q, Zhao S, Ye F, Li S. Determination of octopamine in human plasma by capillary electrophoresis with optical fiber light-emitting diode-induced fluorescence detection. *Analytical Chemistry*. 2007;369:187-191.

[99] Lapainis T, Scanlan C, Rubakhin SS, Sweedler JV. A multichannel native fluorescence detection system for capillary electrophoretic analysis of neurotransmitters in single neurons. *Analytical and Bioanalytical Chemistry*. 2007; 387:97-105.

[100] Batelle BA, Evans JA, Chamberlain SC. Efferent fibers to Limulus eyes synthesize and release octopamine. *Science*. 1982;216:1250-1252.

[101] Evans JA, Chamberlain SC, Battelle BA. Autoradiographic localization of newly synthesized octopamine to retinal efferents in the Limulus visual system. *Journal of Comparative Neurology*. 1983;219:369-383.

[102] Barthe N, Coulon P, Hennion C, Ducassou D, Basse-Cathalinat B, Charpak G. Optimization of a new scintillation gas detector used to localize electrons emitted by 99mTc. *Journal of Nuclear Medicine*. 1999; 40:868-875.

[103] Dacks AM, Christensen TA, Agricola HJ, Wollweber L, Hildebrand J.G. Octopamine-immunoreactive neurons in the brain and subesophageal ganglion of the hawkmoth *Manduca sexta. Journal of Comparative Neurology*. 2005;488: 255-268.

[104] Kreissl S, Eichmuller S, Bicker G, Rapus J, Eckert M. Octopamine-like immunoreactivity in the brain and subesophageal ganglion of the honeybee. *Journal of Comparative Neurology*. 1994;348:583-595.

[105] Bicker G. Biogenic amines in the brain of the honeybee: cellular distribution, development, and behavioral functions. *Microscopy Research and Technique*. 1999;44:166-178.

[106] Eckert M, Rapus J, Nűrberger A, Penzlin H. A new specific antibody reveals octopamine-like immunoreactivity in cockroach ventral nerve cord. *Journal of Comparative Neurology*. 1992;322:1-15.

[107] Sinakevitch I, Niwa M, Strausfeld NJ. Octopamine-like immunoreactivity in the honey bee and cockroach: comparable organization in the brain and subesophageal ganglion. *Journal of Comparative Neurology.* 2005;488:233-254.

[108] Sinakevitch IG, Geffard M, Pelhate M, Lapied B. Octopamine-like immunoreactivity in the DUM neurons, innervating the accessory gland of the male cockroach *Periplaneta americana* (L). *Cell and Tissue Research.* 1994;276: 15-21.

[109] Python F, Stocker RF. Immunoreactivity against choline acetyltransferase, γ-aminobutyric acid, histamine, octopamine, and serotonin in the larval chemosensory system of *Dosophila melanogaster. Journal of Comparative Neurology.* 2002;453:157-167.

[110] Monastirioti M, Gorczyca M, Rapus J, Eckert M, White K, Budnik V. Octopamine immunoreactivity in the fruit fly *Drosophila melanogaster. Journal of Comparative Neurology*. 1995;356:275-287.

[111] Sinakevitch I, Strausfeld NJ. Comparison of octopamine-like immunoreactivity in the brains of the fruit fly and blow fly. *Journal of Comparative Neurology.*2006; 494:460-475.

[112] Konings PN, Vullings HG, Geffard M, Buijs RM, Diederen JH, Jansen WF. Immunocytochemical demonstration of octopamine-immunoreactive cells in the nervous system of *Locusta migratoria* and *Schistocerca. Cell and Tissue Research.* 1988;251:371-379.

[113] Kononenko NL, Wolfenberg H, Pflüger H-J. Tyramine as an independent transmitter and a precursor of octopamine in the locust central nervous system: An immunocytochemical study. *Journal of Comparative Neurology.* 2008;512: 433-452.

[114] Csoknva M, Barna J, Bànvölgyi T, Hiripi L, Eckert M, Hámori J, Elekes K. Octopamine-containing neurons in the alimentary tract of the earthworm *Eisenia fetida. Brain Research.* 1997;778:414-417.

[115] Berry MD. The potential of trace amines and their receptors for treating neurological and psychiatric diseases. Reviews on Recent Clinical Trials. 2007;2:3-19.

[116] Mons N, Geffard M. Specific antisera against the catecholamines: L-3,4-dihydroxyphenylalanine, dopamine, noradrenaline, and octopamine tested by an enzyme-linked immunosorbent assay. *Journal of Neurochemistry.* 1987;48:1826-1833.

[117] Paterson IA, Juorio AV, Boulton AA. 2-Phenylethylamine: a modulator of catecholamine transmission in the mammalian central nervous system? Journal of Neurochemistry. 1990;55:1827-1837.

[118] Wong DF, Grunder G, Brasic GR. Brain imaging research: does the science serve clinical practice? *International Review of Psychiatry*. 2007;19:541-558.

[119] van Dalen JA, Vogel WV, Corstens FH, Oyen WJ. Multi-modality nuclear medicine imaging: artefacts, pitfalls and recommendations. *Cancer Imaging*. 2007;7:77-83.

[120] Thobois S, Jahanshahi M, Pinto S, Frackowiak R, Limousin-Dowsey P. PET and SPECT functional imaging studies in Parkinsonian syndromes: from the lesion to its consequences. *Neuroimage*. 2004;23:1-16.

[121] Arahata Y, Kato T, Ito K. Molecular imaging in Parkinson's disease. *Nippon Rinsho. Japanese Journal of Clincal Medicine*. 2007;365:327-331.

[122] Miller TD, DiCarli MF. Nuclear cardiac imaging for the assessment of coronary artery disease in the elderly. *American Journal of Geriatric* Cardiology. 2007;16: 355-362.

[123] Del Rosario RB, Jung YW, Caraher J, Chakraborty PK, Wieland DM. Synthesis and preliminary evaluation of [11C]-(-)-phenylephrine as a functional heart neuronal PET agent. *Nuclear Medicine and Biology*. 1996;23:611-616.

[124] Vane RJ. The estimation of catecholamines by biological assay. *Pharmacological Reviews*. 1966;18:317-324.

[125] Fuller RW, Hemrick-Luecke SK, Midgley JM. Comparison of o-octopamine and other substituted phenylethanolamines as substrates for norepinephrine N-methyltransferase. *Research Communications in Chemical Pathology and Pharmacology*. 1981;33:207-213.

[126] James MI, Midgley JM, Williams CM. The metabolism and biosynthesis of (±)-*o*-octopamine and (±)-*o*-synephrine in the rat. *Journal of Pharmacy and Pharmacology*. 1983;35:559-565.

[127] Coutts RT, Baker GB, Pasutto FM. Foodstuffs as sources of psychoactive amines and their precursors:content significance and identification. *Advances in Drug Research*. 1986;15:169-232.

[128] Evans PD, Gee JD. Action of formamidine pesticides on octopamine receptors. *Nature,* 1980;287:60-62.

[129] Costa LG, Olibet G, Murphy SD. α_2-adrenoceptors as a target for formamidine pesticides: in vitro and in vivo studies in mice. *Toxicology and Applied Pharmacology*. 1988;93:319-328.

[130] Calapai G, Firenzuoli F, Saitta A, Squadrito F, Arlotta MR, Costantino G, Inferrer AG. Antiobesity and cardiovascular toxic effects of *Citrus aurantium* extracts in the rat: a priliminary report. *Fitoterapia.*1999;70: 586-592.

[131] Davis BA. Biogenic amines and their metabolites in body fluids of normal, psychiatric and neurological subjects. *Journal of Chromatography*. 1989;466: 89-218.

[132] Wallace BG. The biosynthesis of octopamine-characterization of a lobster tyramine β-hydroxylase. *Journal of Neurochemistry*. 1976;26:761-770.

In: Biogenic Amines
Editors: T. Farooqui, A. A. Farooqui

ISBN: 978-1-60876-625-3

Chapter XV

Pharmacology of Invertebrate Octopamine and Tyramine Receptors

*Thomas Roeder**
University of Kiel, Zoophysiology II, Kiel, Germany.

Abstract

The monoamines octopamine and tyramine are the invertebrate counterparts of our adrenergic transmitters. Although tyramine is the biological precursor of octopamine, both compounds are independent neurotransmitters, acting mostly through G-protein coupled receptors. Especially octopamine modulates countless behaviors, as well as the performance of sense organs and of other peripheral tissues. As these two monoamines are the only biogenic amines whose physiological significance is restricted to invertebrates, the attention of pharmacologists was focused on the corresponding receptors, which are still believed to represent promising targets for new insecticides. This assumption is especially valid for octopamine receptors, where a number of highly active receptor agonists with potent insecticidal activities have been identified. Future research based on these *leitstructures* should yield a number of novel insecticides acting on these highly relevant receptors.

Introduction

It is common knowledge that adrenergic signaling dictates various aspects of our life. The corresponding transmitters/hormones, adrenaline and noradrenaline are unique to vertebrates. In invertebrates (including insects) where both compounds have no physiological

* Correspondence concerning this article should be addressed to: Prof. Dr. Thomas Roeder, Christian-Albrechts University Kiel, Zoophysiology, Am Botanischen Garten 1-9, 24098 Kiel, Germany. Email: troeder@zoologie.uni-kiel.de; Phone: 0049-431-880-4181

relevance, their role is taken by the two monoamines octopamine and tyramine [1,2]. They are structurally related to adrenaline and noradrenaline and have very similar physiological roles, pointing to a very early evolutionary origin of the adrenergic/octopaminergic system. Octopamine is found in a huge variety of different invertebrates. As tyramine is its biological precursor, it is also found in these tissues. In vertebrates, only trace amounts of these monoamines are found leading to their categorization as trace amines.

A wealth of information regarding the physiological significance of especially OA has been generated over the last few decades. It became apparent that this compound regulates a plethora of behaviors, sensory inputs and outputs by peripheral organs [1,2]. Naming for octopamine was its discovery in the salivary glands of *Octopus vulgaris* [3]. Ironically, this is one of the very few sources of octopamine, where we still have no information regarding its normal physiological role.

Octopaminergic action appears to be among the most relevant modulatory inputs of invertebrate neuronal and peripheral tissues. Apparently, one major role of octopamine is to orchestrate complex sets of behaviors, thus enabling to switch from one behavioral state to the other, simply by inducing secretion of this hormone/transmitter.

Both, octopamine and tyramine take a prominent position because they are the only non-peptide transmitters/hormones that are restricted in their occurrence to invertebrates. The lack of corresponding receptors in vertebrates promises that tailored compounds are active in invertebrates but almost inactive in vertebrates, thus having the potential to show negligible vertebrate toxicity. This, together with other features focused the interest of numerous researchers on topics related with octopamine and tyramine, leading to a plethora of original contributions and reviews dealing with this issue. The main emphasis of this article is to focus onto the pharmacology of these highly relevant receptors.

Physiological Role of Octopamine and Tyramine

The overwhelming similarities between octopaminergic and adrenergic signaling systems imply that both systems can be traced back to a common ancestor that lived prior to the split of bilaterian animals into deutero- and protostomes. Apparently, all deuterostomes use adrenaline/noradrenaline, whereas the protostomian lineage (the invertebrates) uses octopamine and tyramine. Similarities between the octopaminergic/tyraminergic and the adrenergic system are not restricted to the structure of these compounds. Apparently, all major aspects of the corresponding signaling cascades are conserved. This comprises the structure and pharmacology of the receptors as well as the behaviors and organs that are modulated. Octopamine and tyramine, just like adrenaline and noradrenaline, are hormones that are secreted in times of stress with the goal to prepare the organism for situations such as the highly demanding fight or flight response [2]. Stressful stimuli, such as the encounter with predators or rivals, induce this metabolic and behavioral adaptation, leading to mobilization of energy stores, increased muscle performance, increased sensory perception and a matched behavior. Although this reaction appears to be simple, it requires an orchestrated reaction of the entire organism. In the central nervous system, the general level of arousal is enhanced leading to an increased tendency to move. To match this generally increased motivation for activity, a whole battery of peripheral organs is modulated, meaning that muscle performance; energy supply or heart rates are increased concurrently. This complex type of reaction can be seen in both, vertebrates and insects, pointing to a very ancient origin of even this behavioral trait. In addition to the modulatory effects that are associated with the fight or flight response or with a general stress response, other aspects of the animal's biology are also modulated. Among the first behaviors that had been associated with octopamine action are rhythmic behaviors such as walking of flying. Apparently their initiation and maintenance depends on this neurohormone, giving octopamine a central role for the execution of behaviors [4]. Surprisingly, almost all sense organs of insects and presumably also of other invertebrates are modulated by octopamine action [5]. As a general rule of thumb, octopamine appears to increase the sensitivity of these sense organs. This reaction may help to increase the attention in times of stress. In addition, countless peripheral tissues are modulated if confronted with octopamine. Most intriguingly, the performance of skeletal muscles is increased, meaning that they show a higher output, presumably also to enable an adequate reaction to altered situations. Regarding the physiological significance of tyramine in invertebrates, our knowledge is restricted to very few examples. In *Drosophila*, octopamine and tyramine appear to have opposite effects on some aspects of larval movement, which may represent an example for the interconnected architecture of both systems [6]. Dissection of octopaminergic and tyraminergic activities was most successful in the nematode *C. elegans*. In these animals, tyramine appears to be more important than octopamine and some very distinct and well-characterized behaviors can be attributed to tyramine action only [7].

Compared with the plethora of studies dealing with octopamine action that used classical physiological and pharmacological approaches, only a minority of them has been confirmed using genetic tools. In both model organisms *Drosophila* and *C. elegans*, mutants defective in the synthesis of tyramine, thus being devoid of both tyramine and octopamine, or in the synthesis from tyramine to octopamine, thus lacking octopamine only are available. All mutant flies and worms are surprisingly healthy, showing no very obvious phenotypical peculiarities [7,8]. In *Drosophila*, a female sterility can be observed, depending on the inability to lay eggs, which is caused by the lack of octopamine-induced contractions of the oviduct muscles[8]. Nevertheless, a number of different behaviors have been shown to particularly depend on octopaminergic signaling. Some of these results are in line with earlier findings. As already mentioned, octopamine is of vital importance for the initiation and maintenance of flight behaviors [4]. Recently, using *Drosophila* mutants defective in the synthesis of octopamine, Brembs and colleagues showed that these flies have a significantly reduced tendency to initiate flight sequences and that the duration of these flight sequences is also severely decreased [9]. Another highly relevant behavior that depends on octopamine is learning and memory. Octopamine appears to function as the neuronal correlate of sugar in associative learning paradigms [10]. It is long known that aggression is also influenced by octopamine, which appears to be necessary to develop a certain degree of aggressiveness [11]. Unexpectedly, even the regulation of wake/sleep cycles depend on octopaminergic signaling, meaning that flies devoid of this transmitter have significantly shortened wake phases. Our knowledge regarding the physiological role of octopamine and tyramine in the second invertebrate model, the nematode *C. elegans* is even more restricted. As already mentioned, tyramine appears to be more relevant than octopamine. Transforming stress signals into a meaningful physiological response seems to be primary role of both monoamines. Inhibition of egg laying and increase in movement velocity are among their most conspicuous effects [7,12]. A major contribution of this model was the unequivocal distinction between octopamine and tyramine action [7,13]. Surprisingly, the plethora of reports dealing with the role of octopamine in peripheral tissues still awaits verification in these knockout systems.

RECEPTORS FOR OCTOPAMINE AND TYRAMINE

Both, octopamine and tyramine act through sets of specific receptors. Most of them are belong to the family of G-protein coupled receptors. Very recently, tyramine-gated chloride channels were identified in *C. elegans* and they have been associated with some very peculiar tyramine dependent behaviors [14, 15]. Regarding their pharmacology, we still have no valid information, meaning that the only well characterized agonist is tyramine itself. In the following parts of the manuscript, I will focus on the G-protein coupled receptors only. Reflecting the similarities on the physiological field, octopamine and tyramine receptors share structural features with the corresponding adrenaline and noradrenaline receptors, the α- and β-adrenergic ones. Octopamine receptors, and to a much lesser degree tyramine receptors were detected on almost every peripheral organ in a huge number of different insect species. Insect neuropharmacologists were engaged for more than two decades to study if and how organs and tissues from all available insects react if confronted with octopamine [1,16]. This extensive set of studies left us with the knowledge that almost every organ reacts to octopamine and therefore contains octopamine receptors. Unfortunately, the pharmacological studies performed with different tissues from different insect species are hardly comparable. In addition, the knowledge that the identification of a new receptor type attains more interest than the characterization of homologous receptors from a different species seduced various researchers to state that "their" octopamine receptor is new. Using the locust and no other species, a total number of 4 different octopamine receptors were identified by their different pharmacological features [17-21], which nicely matches with the identification of four different octopamine receptor genes in the fly [22,23]. Most actions of octopamine in the nervous system appear to be mediated via the "neuronal" type octopamine receptor [20], although a direct assignment to one of the cloned receptors has not been made yet. A number of different tyramine receptors have been identified in the different model organisms and gave us an overview of almost the entire set of octopamine and tyramine receptors present in invertebrates.

PHYLOGENETIC ANALYSIS

As already mentioned, except one tyramine gated chloride channel, all other receptors for octopamine and tyramine belong to the large family of G-protein coupled receptors for biogenic amines. Currently, we have four different octopamine receptors in insects (only one in nematodes) and three different tyramine receptors, both in insects and nematodes. Within the large and diverse group of bioamine receptors, all or most of them are associated with the receptors for adrenaline and noradrenaline, namely the α- and β-adrenergic ones. As a cautionary note, it has to be mentioned that the overall similarities between the amino acid sequences of the corresponding receptors are often relatively low, making the outcomes of phylogenetic analyses variable and not always reliable.

The only well characterized octopamine receptor, DmOAR1 (OAMB) [24] clusters together with other OARs from insects and crustaceans, the type 1 receptor of the snail *Lymnea stagnalis* [25], and the ser-3 receptor of the nematode *Caenorhabditis elegans*. Interestingly, this entire group is the sister-group of the vertebrate alpha1-adrenergic receptors. Obviously, only one receptor from each of these invertebrate species constitutes this group of "alpha1-adrenergic" like octopamine receptors.

As for the octopamine receptors, the phylogenetic positions of tyramine receptors are not as clear-cut as expected. The first group of tyramine receptors cluster into a totally different group, unrelated to the "classical" octopamine receptors mentioned above. This group consists of the first tyramine receptors from *Drosophila* (DmTyrR) and other insects as well as the tyramine receptor from *C. elegans* (ser-2) [26]. Common to them is that they have been misinterpreted as octopamine receptors or mixed tyramine/octopamine receptors. In *C. elegans*, a total of three different tyramine receptors have been identified [13]. The same number of tyramine receptors is known from *Drosophila* [27]. A direct assignment of the phylogenetic positions of the receptors from nematodes and insects is still missing. Apparently the different tyramine receptors belong to different branches within the phylogenetic tree of bioamine receptors.

One very peculiar octopamine receptor of the snail *Aplysia californica* was the founding member of another class of octopamine/tyramine receptors. In opposite to the first group of octopamine receptors, this one shows a substantial homology to ß-adrenergic receptors. In the fruit fly, three different receptor subtypes are part of this ß-adrenergic like receptors. OAR-2 (DmOctβ1), DmOctβ2 and DmOctβ3 are part of this group of octopamine receptors, which was the reason why they were named β-like receptors [22].

A second octopamine receptor of the snail *Lymnea stagnalis* doesn't fit into this classification scheme, thus it forms its own subgroup of receptors, which is in accordance with the very peculiar type of signaling observed for this subtype. It is directly coupled to opening of a chloride channel, which is very unusual for bioamine receptors [28]. Regarding the second-messenger coupling of octopamine and tyramine receptors almost everything is possible. The classical octopamine receptor that was characterized in pharmacological studies is coupled to activation of cAMP-synthesis. All three receptors of the β-adrenergic subtype show this coupling [23]. In addition, the only well characterized octopamine receptor, OAMB, is linked to Ca^{2+}-signalling, but the homologous receptor of the nematode (ser-3) is coupled to activation of adenylate cyclase. Regarding the tyramine receptors, the picture is less clear-cut. Some of them are linked to inhibition of adenylate cyclase, whereas others induce an increase in Ca^{2+}-levels, presumably via activation of the Gq G-protein [27].

Preparations to Study Octopamine Receptor Pharmacology

Pharmacology of octopamine receptors has been studied for almost half a century. In the early times, informative models have been used that allow quantitative octopamine receptor pharmacology without the need to clone the corresponding receptors. For this purpose extracts of brain tissues, but also complete and functional preparations were used. The first preparation that was used extensively for pharmacological studies was a brain extract of the cockroach, where agonist induced production of cAMP was the read-out [29]. cAMP determination in different tissues was the most relevant type of pharmacological readout in the early times of octopamine receptor pharmacology. This approach was supplemented by binding studies with radioactive ligands, namely with tritiated octopamine [30,31]. Later, highly specific agonists were used for this purpose [32]. Cell lines, such as the *Drosophila* S2 line or even more the SF9 cell line (derived from *Spodoptera frugiperda*) were believed to be ideally suited, as the octopamine receptor is apparently the only amine receptor expressed in the cell line (at least in the SF9 line). Unfortunately, the receptor density was not high enough to enable simple and easy to perform readouts [33]. Therefore, the lines lost most of their attractiveness. Presumably the most advanced and "intelligent" preparation was the firefly lantern. In the fire beetles, octopamine is the transmitter that induces light production of the lantern. The activity of the transmitter is mediated via activation of cAMP-production [34]. The major advantage of this preparation is the absence of other bioamine receptors in the lantern. Technically, lanterns are isolated and the material is used for a luminescence assay, meaning that only the compound of interest has to be added and the response measured in a luminometer. Thus this system provides all advantages that are characteristic for modern expression systems, where heterologously expressed receptors are coupled to activation of a reporter system [35,36]. Using this system, the group of James Nathanson identified a number of highly potent agonists for the major octopamine receptor subtype [32,34]. In addition to these assays in different states of reductionism, skeletal muscle preparations of the locust were used for different aspects of octopamine receptor pharmacology. Especially the group of Peter Evans developed a very simple physiological system to address pharmacological differences between different octopamine receptor subtypes, based on muscle preparations, where only single receptor subtypes are present [17-19]. In addition, we used the locust brain as a valuable source for the characterization (especially the pharmacological characterization) of octopamine receptors in general [31].

With the availability of cloned receptors for pharmacological studies, these old, but very informative approaches have almost completely been replaced. Currently, pharmacological studies of octopamine and tyramine receptors are performed with heterologously expressed, cloned receptors from different species. To simplify the analysis, a number of different expression systems are in use allowing quantitative pharmacological analyses based on the activity of suitable reporter systems. These reporter systems are tailored to the different signal transduction cascades activated by the receptor subtypes different. Among them the different detection methods for ligand induced changes in cAMP- or Ca^{2+}-transients are the most popular ones, in particular as they are amenable to high-throughput screening. This is especially true for Ca^{2+}-signalling, which is extensively used for this purpose. Alternatively, expressions systems are in use, which combine second messenger dependent activation of gene expression with the use of suitable reporter molecules as luciferase, GFP or soluble excreted alkaline phosphatase (SEAP) [35,36].

RECEPTOR PHARMACOLOGY

Our knowledge regarding the pharmacological characteristics of octopamine receptors is copious, whereas we know only very basic aspects about tyramine receptor pharmacology. Thus tyramine receptors are in a situation that is very similar for most invertebrate bioamine receptors; we have no, or only very few information regarding their pharmacological characteristics. One major drawback is the lack of highly potent and specific agonists for tyramine receptors in general. The major classification tool for tyramine receptors is the higher activity of tyramine compared with octopamine. The same holds true for the antagonists available. Only very few ones, belonging to the group of "old" adrenergic antagonists had been shown to be active in blocking tyraminergic neurotransmission. One of this very few compounds is yohimbine, an indol-alkaloid of the yohimbe tree. Khan and colleagues reported that members of the 5-Phenyloxazoles contain both, octopamine and tyramine receptor agonists. This observation needs further experimental support [37]. But as long as tyramine receptors are not considered to represent a valuable target for insecticides, we will not learn more about its pharmacology and no relevant *leitstructures* will be presented.

The situation is totally different for the octopamine receptor, where a wealth of structural information is available. Hundreds of different compounds are known that display high affinity properties for octopamine receptors. Most of them are agonists and only very few are antagonists. The attention of pharmacologists was torn to octopamine receptor pharmacology after it became apparent that these receptors are exclusively present in invertebrates, representing an ideal target for specific insecticides [1,16]. This imbalance between our knowledge regarding the features of agonists and antagonists might reflect the observation that octopaminergic insecticides are always octopamine receptor agonists and never antagonists. This difference can obviously be attributed to the mode of action of octopaminergic insecticides. Usually they induce a conserved and stereotype syndrome that is best characterized by the leaf walk off phenotype. Following application of these insecticides, the animals start to become "nervous", a phenotype that increases over time leading to this leaf walk off phenomenon. Insects usually feeding on leaves have a too high motor activity to stay where they are, they stop feeding and finally they walk off the leaves, presumably induced by this higher motor activity. This hyperactivity of the nervous system is apparently the major reason for death of the poisoned animals. Application of antagonists would do the opposite, meaning that treated individuals are relatively calm and behave such as those mutants devoid of this hormone, meaning almost normally.

During the time, a number of different *leitstructures* were identified that served as starting points for the development of even higher affinity compounds (Figure 1). Based on octopamine´s own structure, substances such as synephrine were characterized by their higher affinity. The first agonists with high insecticidal potential were the formamidines, with chlordimeform as the prototype [17,38]. Subsequently, additional compounds such as the phenyliminoimidazolidines or aminooxazoline derivatives were developed [21,32]. They displayed extremely high affinities for octopamine receptors with K_D-values in the sub-nanomolar range. Some of these substances display a high insecticidal potency, but only "old" octopaminergic insecticides such as amitraz or chlordimeform were used for this purpose.

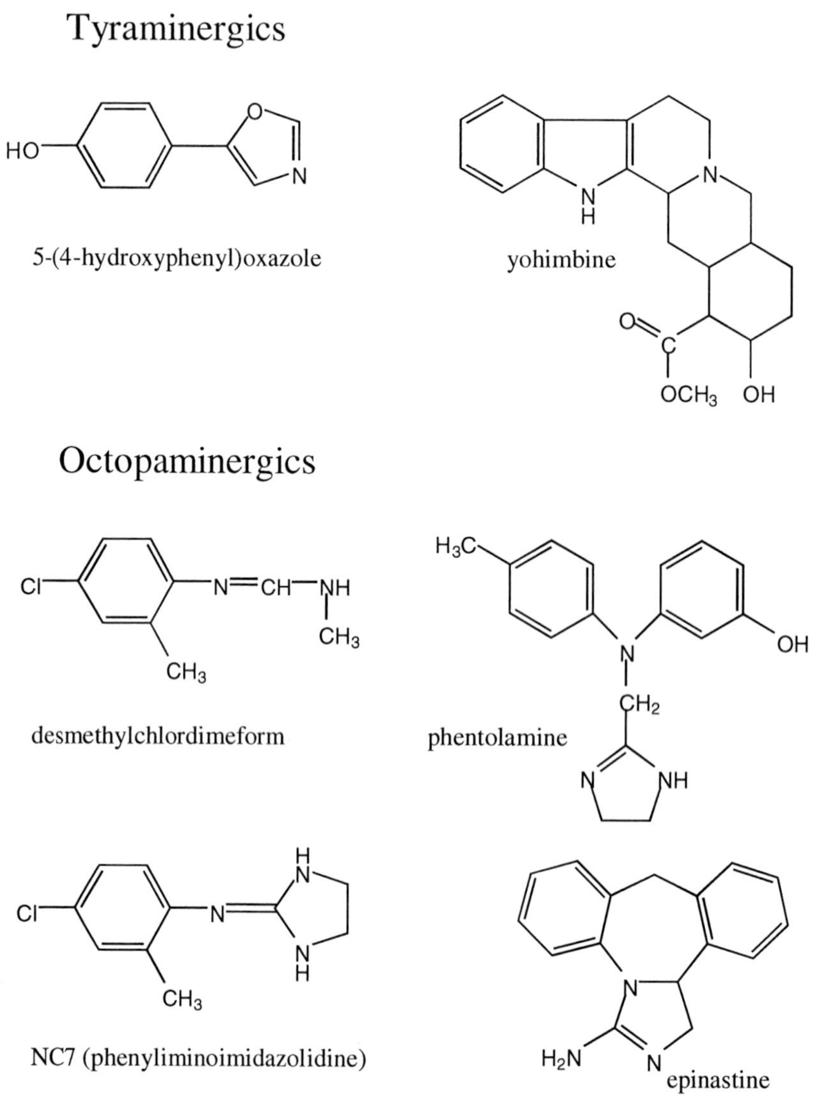

Figure 1. Tyraminergics and octopaminergics. Compounds with high affinity properties for tyramine and octopamine receptors of insects. Only very few tyraminergics are known. Shown is the entire repertoire of known tyramine agonists (5-(4-hydroxyphenyloxazole) and antagonists (yohimbine). In contrast, a plethora of octopaminergic agonists are known and available (NC7 and demethylhlordimeform). The number of valuable antagonists is smaller, shown are the two most important ones (phentolamine and epinastine).

A closer look at the structural features of highly active octopaminergic agonists shows a surprisingly high degree of similarities. Structurally, these antagonists can be traced back to octopamine's own structure. Common to them is a phenolic function, which is obviously homologous to the one of octopamine. Substitutions at some positions are beneficial, whereas others tend to be detrimental. Again, in analogy with the structure of octopamine, substitutions in para- position are mostly beneficial. The other structural group may be an equivalent to the amino-group of octopamine. Formamidines usually have this archetypical structure, but in the more advanced compounds, this group is a 5 membered heterocycle. A spacer separates these two functional groups (phenolic function and 5 membered heterocycle). This spacer is of variable length, but if it is too small or too long, the corresponding compounds have a greatly reduced affinity. In Figure 1 some highly effective octopamine receptors are displayed. Only a few high affinity compounds with slightly different structural feature have been reported. The most promising candidate structures have affinities for octopamine receptors in the subnanomolar range, which means that this feature does not have to be optimized. To complement the classical pharmacological studies, modeling approaches were undertaken to identify the active site, thus allowing to focus on only some positions within the structure o the potential insecticides [39]. The greatest challenge for the development of novel, highly valuable insecticides is to transform the high affinity properties into a high insecticidal activity. Thus, hydrophilic moieties have to be replaced by lipophilic ones, allowing an easy penetration of the insect cuticle and their ability to reach their primary target sites within the nervous system.

Unfortunately, our knowledge regarding the actions of antagonists is limited. Only very few compounds are known to inhibit octopamine's action. Among them are especially those that show high affinity properties for other bioamine receptors, especially those of vertebrates. Very well known compounds that fall into this category are phentolamine and mianserin [40]. Most of them show a limited degree of specificity for octopamine receptors as other receptor systems are also affected, sometimes at concentrations in the same range. Nevertheless, these antagonists were used to establish a valuable classification scheme for octopamine receptors. The lack of specificity for octopamine receptors impedes their use as *in vivo* pharmaceutical tools, because one never can be sure if the observed effects can be attributed to the block of octopaminergic neurotransmission. Only one antagonist combines high affinity with high specificity. It is called epinastine, and has a relatively broad pharmacological spectrum in vertebrates. In insects, it has an extremely narrow pharmacological profile, which allows very specific blockade of octopaminergic neurotransmission. This made it to an ideal tool for the dissection of octopaminergic neurotransmission *in vivo* [41], a feature that was used to elucidate some aspects of octopaminergic neurotransmission.

Do Different Octopamine Receptors from Distantly Related Species Show the Same Pharmacology?

This question of seemingly only academic importance has central importance for the development of new, octopaminergic insecticides. Due to the enormous costs associated with the development of new insecticides, species-specific acting compound will never have the chance to come to the market. Taking into account that insects are a rather diverge group of animals that occurred about 400 million years ago, it sounds likely that the pharmacological properties of octopamine receptors from different insect species are very different. Insects from different groups (such as hymenoptera, diptera or orthoptera) contain homologous octopamine receptors. Regarding the observed degree of sequence similarities, it was very astonishing to learn that the pharmacological properties of octopamine receptors from species as diverse as locusts and bees have almost identical pharmacological features. Bees, as all other hymenopteran insects, are holometabolous and locusts as orthopterans are hemimetabolous. Their last common ancestor lived more than 330 million years ago, a time comparable to the period where the ancestors of birds and mammals splitted. This implies that pharmacological studies performed on one model could be interpolated to almost every insect species [42]. From this, it follows that octopaminergic insecticides should be suitable for a broad variety of different insects, an aspect that might help to bring them back into the market.

Acknowledgments

I would like to thank the Deutsche Forschungsgemeinschaft (DFG) for continuous support (DFG Ro 1241).

References

[1] Roeder T. Octopamine in invertebrates. *Progress in Neurobiology.* 1999;59:533-561.

[2] Roeder T. Tyramine and octopamine: ruling behavior and metabolism. *Annual Review of Entomology.* 2005;50:447-477.

[3] Erspamer V. Identification of octopamine as l-p-hydroxyphenylethanolamine. *Nature.* 1952;169:375-376.

[4] Sombati S, Hoyle G. Generation of specific behaviors in a locust by local release into neuropil of the natural neuromodulator octopamine. *Journal of Neurobiology.* 1984;15:481-506.

[5] Farooqui T. Octopamine-mediated neuromodulation of insect senses. *Neurochemical Research.* 2007;32:1511-1529.

[6] Saraswati S, Fox LE, Soll DR, Wu CF. Tyramine and octopamine have opposite effects on the locomotion of Drosophila larvae. *Journal of Neurobiology.* 2004;58:425-441.

[7] Alkema MJ, Hunter-Ensor M, Ringstad N, Horvitz HR. Tyramine Functions independently of octopamine in the *Caenorhabditis elegans* nervous system. *Neuron.* 2005;46:247-260.

[8] Monastirioti M, Linn CE, White K. Characterization of Drosophila tyramine beta-hydroxylase gene and isolation of mutant flies lacking octopamine. *Journal of Neuroscience.* 1996;16:3900-3911.

[9] Brembs B, Christiansen F, Pflüger HJ, Duch C. Flight initiation and maintenance deficits in flies with genetically altered biogenic amine levels. *Journal of Neuroscience.* 2007;27:11122-11131.

[10] Schwaerzel M, Monastirioti M, Scholz H, Friggi-Grelin F, Birman S, Heisenberg M. Dopamine and octopamine differentiate between aversive and appetitive olfactory memories in Drosophila. *Journal of Neuroscience.* 2003;23:10495-10502.

[11] Hoyer SC, Eckart A, Herrel A, et al. Octopamine in male aggression of *Drosophila. Current Biology.* 2008;18:159-167.

[12] Horvitz HR, Chalfie M, Trent C, Sulston JE, Evans PD. Serotonin and octopamine in the nematode Caenorhabditis elegans. *Science.* 1982;216:1012-1014.

[13] Wragg RT, Hapiak V, Miller SB, et al. Tyramine and octopamine independently inhibit serotonin-stimulated aversive behaviors in Caenorhabditis elegans through two novel amine receptors. *Journal of Neuroscience.* 2007;27:13402-13412.

[14] Pirri JK, McPherson AD, Donnelly JL, Francis MM, Alkema MJ. A tyramine-gated chloride channel coordinates distinct motor programs of a Caenorhabditis elegans escape response. *Neuron.* 2009;62:526-538.

[15] Ringstad N, Abe N, Horvitz HR. Ligand-gated chloride channels are receptors for biogenic amines in C. elegans. *Science.* 2009;325:96-100.

[16] Roeder T, Seifert M, Kähler C, Gewecke M. Tyramine and octopamine: antagonistic modulators of behavior and metabolism. *Archives of Insect Biochemistry and Physiology.* 2003;54:1-13.

[17] Evans PD, Gee JD. Action of formamidine pesticides on octopamine receptors. *Nature.* 1980;287:60-62.

[18] Evans PD. Multiple receptor types for octopamine in the locust. *Journal of Physiology.* 1981;318:99-122.

[19] Evans PD. A modulatory octopaminergic neurone increases cyclic nucleotide levels in locust skeletal muscle. *Journal of Physiology.* 1984;348:307-324.

[20] Roeder T. A new octopamine receptor class in locust nervous tissue, the octopamine 3 (OA3) receptor. *Life Sciences.* 1992;50:21-28.

[21] Roeder T. Pharmacology of the octopamine receptor from locust central nervous tissue (OAR3). *British Journal of Pharmacology.* 1995;114:210-216.

[22] Balfanz S, Strünker T, Frings S, Baumann A. A family of octopamine [corrected] receptors that specifically induce cyclic AMP production or Ca2+ release in *Drosophila melanogaster. Journal of Neurochemistry.* 2005;93:440-451.

[23] Maqueira B, Chatwin H, Evans PD. Identification and characterization of a novel family of Drosophila beta-adrenergic-like octopamine G-protein coupled receptors. *Journal of Neurochemistry.* 2005;94:547-560.

[24] Han KA, Millar NS, Davis RL. A novel octopamine receptor with preferential expression in *Drosophila* mushroom bodies. *Journal of Neuroscience.* 1998;18:3650-3658.

[25] Gerhardt CC, Bakker RA, Piek GJ, et al. Molecular cloning and pharmacological characterization of a molluscan octopamine receptor. *Molecular Pharmacology.* 1997;51:293-300.

[26] Rex E, Komuniecki RW. Characterization of a tyramine receptor from *Caenorhabditis elegans. Journal of Neurochemistry.* 2002;82:1352-1359.

[27] Cazzamali G, Klaerke DA, Grimmelikhuijzen CJ. A new family of insect tyramine receptors. *Biochemical and Biophysical Research Communications.* 2005;338:1189-1196.

[28] Gerhardt CC, Lodder HC, Vincent M, et al. Cloning and expression of a complementary DNA encoding a molluscan octopamine receptor that couples to chloride channels in HEK293 cells. *Journal of Biological Chemistry.* 1997;272:6201-6207.

[29] Nathanson JA, Greengard P. Octopamine-sensitive adenylate cyclse: evidence for a biological role of octopamine in nervous tissue. *Science.* 1973;180:308-310.

[30] Dudai Y. High-affinity octopamine receptors revealed in Drosophila by binding or [^{3}H]octopamine. *Neurosci Letters.* 1982;28:163-167.

[31] Roeder T, Gewecke M. Octopamine receptors in locust nervous tissue. *Biochemical Pharmacology.* 1990;39:1793-1797.

[32] Nathanson JA. Phenyliminoimidazolidines. Characterization of a class of potent agonists of octopamine-sensitive adenylate cyclase and their use in understanding the pharmacology of octopamine receptors. *Molecular Pharmacology.* 1985;28:254-268.

[33] Van Poyer W, Torfs H, Poels J, et al. Phenolamine-dependent adenylyl cyclase activation in Drosophila Schneider 2 cells. *Insect Biochemistry and Molecular Biology.* 2001;31:333-338.

[34] Nathanson JA. Octopamine receptors, adenosine 3',5'-monophosphate, and neural control of firefly flashing. *Science.* 1979;203:65-68.

[35] Roeder T, Görich D, Heyden D, Gewecke M. A green-fluorescent-protein-based assay for the characterization of G-protein-coupled receptors. *Analytical Biochemistry.* 2004;332:38-45.

[36] Thomsen W, Frazer J, Unett D. Functional assays for screening GPCR targets. *Current Opinion in Biotechnology.* 2005;16:655-665.

[37] Khan MA, Nakane T, Ohta H, Ozoe Y. Positive and negative modulation of Bombyx mori adenylate cyclase by 5-phenyloxazoles: identification of octopamine and tyramine receptor agonists. *Archives of Insect Biochemistry and Physiology.* 2003;52:7-16.

[38] Gole JW, Orr GL, Downer RG. Interaction of formamidines with octopamine-sensitive adenylate cyclase receptor in the nerve cord of *Periplaneta americana* L. *Life Sciences.* 1983;32:2939-2947.

[39] Hirashima A, Eiraku T, Shigeta Y, Kuwano E. Three-dimensional pharmacophore hypotheses of octopamine/tyramine agonists which inhibit [1-^{14}C]acetate incorporation in *Plodia interpunctella*. *Bioorganic and Medicinal Chemistry.* 2003;11:95-103.

[40] Roeder T. High-affinity antagonists of the locust neuronal octopamine receptor. *European Journal of Pharmacology.* 1990;191:221-224.

[41] Roeder T, Degen J, Gewecke M. Epinastine, a highly specific antagonist of insect neuronal octopamine receptors. *European Journal of Pharmacology.* 1998;349:171-177.

[42] Degen J, Gewecke M, Roeder T. Octopamine receptors in the honey bee and locust nervous system: pharmacological similarities between homologous receptors of distantly related species. *British Journal of Pharmacology.* 2000;130:587-594.

In: Biogenic Amines
Editors: T. Farooqui, A. A. Farooqui
ISBN: 978-1-60876-625-3

Chapter XVI

The Cellular and Molecular Biology of Octopaminergic Neurons

Herman K. Lehman *

Biology Department and Program in Neuroscience Hamilton College, Clinton, NY 13323, USA.

Abstract

Biogenic amines are among the most important modulators of vertebrate and invertebrate nervous system function and behavior. The biogenic amine octopamine (OA) is one of the most widely studied neuromodulators in the arthropod nervous system and immunoreactive octopaminergic neurons have been identified in an assortment of insects. OA is responsible for an impressive array of actions; generally, OA is a modulatory agent of skeletal, visceral, and sensory systems. In addition, OA has an equally impressive array of modulatory actions within the CNS and OA has been widely implicated in the regulation of behavioral plasticity for such complex behaviors as learning and memory, aggression, flight motivation, reward, and age-related division of labor.

A change in the amount of a neuromodulator released by neurons is a key mechanism in neurochemical regulation of neural and behavioral plasticity. This implies that understanding the regulation of neuromodulator biosynthesis is an important element in a comprehensive analysis of neuromodulation. The rich background knowledge of OA in insects provides a unique opportunity to address important questions regarding the synthesis of this neuromodulator. Tyramine beta-hydroxylase (TBh) is the pivotal enzyme that that mediates OA synthesis; it is a catalyst for the hydroxylation of tyramine to form OA, and it is found exclusively in OAergic neurons. TBh shares many structural and functional similarities to the mammalian enzyme, dopamine beta-hydroxylase, and recent studies have suggested that the substrates and co-factors necessary for TBh

* Correspondence concerning this article should be addressed to: Herman K. Lehman, Ph.D., Biology Department, Hamilton College, 198 College Hill Road, Clinton, NY 13323, Tel: 315-859-4298, Fax: 315-859-4807, Email: hlehman@hamilton.edu.

function (tyramine, copper, ascorbate, and oxygen) have profound effects on TBh rate. Therefore, substrate transport and metabolism (tyrosine, tyramine, and OA), co-substrate transport and metabolism (pyridoxal 5'-phosphate, copper, and ascorbic acid), and oxygen availability may all potentially influence OA levels and in turn, the amount of OA released. To clearly understand how this potent neuromodulator regulates multiple behaviors it is vital to understand the cellular and molecular processes that underlie its synthesis. Here, I review how and where these processes of OA synthesis occur within OAergic neurons, and suggest that tyramine and Vitamin C transport play important roles in the regulation of OA synthesis.

1. Introduction

A universal characteristic of biogenic amines is their diverse distribution and function in animal nervous systems. Octopamine (OA, 4-(2-amino-1-hydroxy-ethyl)phenol) is an exemplary example of biogenic amines - it appears to be evolutionarily related to the vertebrate adrenergic system, it occurs in all triploblastic organisms investigated to date, and is one of the most abundant biogenic amine neurohumors in the insect nervous system [1]. The interactions of OA with its cognate receptors lead to an impressive array of actions; in general, OA is a modulatory agent in skeletal, visceral, and sensory systems. More specifically, the sensitivities of insect photo-, mechano- and olfactory receptors are enhanced by OA [2,3,4, reviewed by 5]. In addition to these peripheral actions, OA has an equally impressive array of modulatory actions within the CNS. OA has been widely implicated in the regulation of behavioral plasticity for such complex behaviors as olfactory processing, learning and memory, aggression, and flight motivation [reviewed by 6,7]. Recently, OA also was found to play an important role in a complex system of social behavior, age-related division of labor in the honey bee colony [8,9].

A change in the amount of OA and other neuromodulators released by neurons is a key mechanism in neurochemical regulation of neural and behavioral plasticity. This implies that understanding the regulation of neuromoduator biosynthesis is an important element in the comprehensive analysis of neuromodulation. The rich background knowledge of OA in insects, coupled with the completed genome sequence of several insects, provides a unique opportunity to address the underlying cellular and molecular control of OA synthesis in insects.

2. Octopaminergic Neurons

The identification of OAergic neurons in a wide range of invertebrates has been accomplished primarily due to the development of OA antisera [10,11,12]. Although immunoreactivity does not authenticate the identity of the antigen, insect immunoreactive octopaminergic (ir-OA) neurons have been identified in locusts (*Schistocerca gregaria* and *Locusta migratoria*), crickets (*Gryllus bimaculatus*), cockroaches (*Perplaneta americana*), moths (*Manduca sexta*), honeybees, (*Apis mellifera*), and flies (*Drosophila melanogaster* and

Phaenicia sericata) [13,14,15,16,17]. These studies show that relatively few ir-OA neurons richly innervate most regions of the insect nervous system.

Locations of OAergic neurons were originally described in studies of crickets and locusts and most of these neuronal locations appear to be well conserved in other insects [15,16,18]. Approximately 100-200 ir-OA neurons are found in insects, and these neurons are distributed among eleven neuronal clusters. Four classes of OA-ir neurons occur in ganglia of the insect ventral nerve cord and seven classes of bilaterally distributed OA-ir neurons occur in the insect brain.

The most conspicuous, well-studied class of OAergic neurons is a group of large (30-60 μm) unpaired medial neurons dorsally located in orthopteroid insects (DUM neurons) or ventrally located in Lepidoptera, Diptera, and Hymenoptera (VUM neurons). These midline neurons reside within the suboesophageal ganglia, thoracic and abdominal ganglia; DUM neurons in the thoracic and abdominal ganglia have a characteristic single neurite that projects anteriorly within the ganglia and exits the ganglia via anterior nerves in a bilaterally symmetrical fashion. These DUM neurons mediate the neuromodulatory and neurohormonal effects of OA on muscles and sensory neurons whereas DUM/VUM neurons in the suboesophageal ganglia are more diverse [19-23]. For example, eleven different VUM neuron types have been described in honeybee suboesophageal ganglia [24] and they project anteriorly, bifurcate, and distinctively innervate regions of the brain, including mushroom bodies, antennal lobes, dorsal lobes, central complex, antennal nerves, and nerves supplying the corpora cardiaca [15,16,24,25,26,27]. Three other clusters of OA-ir neurons have been described in the ventral nervous system, but much less is known about these anterior medial, ventral paired, and small paired lateral neurons [15,16,18].

The location and innervation of seven major classes of bilaterally distributed OA-ir neurons have also been well characterized in insect brain. OA-ir somata in the insect brain emanate from a few major regions of the brain (including the protocerebrum and tritocerebrum) and exist as single cells or as clusters of up to ca. 17 neurons. These OA-ir neurons range in size from 5-10 nm in *Drosophia melanogaster* to 45 nm in *Apis mellifera*, and innervate the retrocerebral complex (neurosecretory function), the inner and outer medulla, lobula and lobula plate of the optic lobes (visual processing), mushroom bodies (learning and memory) and central complex (multimodal information processing and motor behavior) [15,16,28,29,30]. A detailed study of individual *Drosophila* OAergic neurons show that these neurons are highly polarized with clearly defined and separated pre and postsynaptic regions [31]. Each neuron innervates a distinct combination of brain regions, suggesting that OAergic neurons may be organized and activated in a combinatorial fashion. Thus, each OAergic neuron appears to participate in a unique neuronal circuit, capable of modulating separate brain regions in different combinations. This notion is reminiscent of the orchestration hypothesis originally articulated by Hoyle [32].

In sum, a few clusters of ir-OAergic neurons widely innervate the calyces of mushroom bodies, antennal lobes, central complex, optic lobes and lateral protocerebrum of the insect brain, and play a role in motor control, sensory processing, sensory association, and information storage in insects. These anatomical studies also indicate that there is diversity in the number, size, and innervation patterns among ir-OA neurons. For example, OA-ir VUM neurons are among the largest neurons in the insect nervous system (up to 60 μm dia), are few in number (2-10 neurons per cluster), and have broad innervation patterns [24]. In contrast, much smaller ir-OA neurons are located in the G4 cluster (G4b; 8-10 μm) that only innervates the protocerebral bridge and ellipsoid body of the central body complex [15,16]. Moreover, the concentrations of substrates and neurotransmitters contained within OA neurons (tyrosine, tyramine, and OA) appear to be highly variable in insect nervous systems [33-39]. The levels of tyrosine (208 pmol/tissue), tyramine (0.21 pmol/tissue), and OA (2.2 pmol/tissue) in the brain of *M. sexta* larvae are in marked contrast to the levels found in the brain of another moth, *Trichoplusia ni* (tyrosine, 24.3 pmol/tissue; tyramine, 1.7 pmol/tissue; and OA, 5.3 pmol/tissue) [40,41]. Furthermore, the ratios of OA to tyramine vary from 4.4 to 10.1 in nervous system structures of *Manduca sexta* larvae and from 2.8 to 6.9 in nervous system structures of locusts [40,42]; and even within a single species, OA levels vary according to different stages of development or behavior [9,43,44]. What is the basis of this variability? One possibility is that a single key component of the biosynthetic pathway may be specifically controlled. Conversely, several or all of the biosynthetic elements may be controlled independently or in unison. The first step in sorting out of these possibilities is to identify these components of neurotransmitter synthesis. Here, I explore the cellular and molecular facets of OA synthesis to illuminate regulatory elements that may affect synthesis and may ultimately underlie the control of specific insect behaviors.

3. Octopamine Synthesis

Octopamine is synthesized in the insect nervous system from the amino acid tyrosine in two enzymatic steps. First, tyrosine decarboxylase (Tdc) catalyzes the conversion of tyrosine to tyramine and secondly, tyramine beta-hydroxylase (Tbh) catalyzes the conversion of tyramine to OA (Figure 1) [45]. Tdc activity has been detected in nervous system extracts of several insects and this activity is dependent on pyridoxal-5'-phosphate [42]. Recently, the gene necessary for tyrosine decarboxylation of tyrosine in the *Drosophila* nervous system has been identified (*DmTdc2*). DmTdc2 encodes a protein that is pyridoxal-5'-phosphate dependent and is structurally related to mammalian aromatic amino acid decarboxylase (AADC) [46,47]. Tbh activity has also been measured in several arthropod nervous systems extracts (including the American lobster, *Homarus americanus* and the moth *M. sexta*) and the conversion of tyramine to octopamine is dependent upon pH, ascorbic acid, and copper [48,49]. In 1996, Monastirioti et al. identified, cloned, and sequenced the *Drosophila* Tbh gene (*DmTbh*) and confirmed that Tbh is a copper-dependent hydroxylase belonging to a group of enzymes that include dopamine beta-hydroxylase (Dbh) and alpha-hydroxylating monooxygenase (PHM) [50]. Many subsequent studies have illuminated the importance of Tbh for the synthesis of octopamine and the modulation of many complex insect behaviors [51-55].

3.a. Structure and Function of Synthetic Enzymes

Tyrosine Decarboxylase

Tyrosine decarboxylase (Tdc) belongs to an extensive group of aromatic amino acid decarboxylases (AADC) that are dependent on pyridoxal-5'-phosphate. AADCs have been well characterized in plants and animals (including vertebrates and invertebrates), and specific AADCs have been localized in the nervous systems of many animals. Within the vertebrate central nervous system a single AADC decarboxylates 5-hydroxytryptophan, L-dopa, tyrosine, phenylalanine, tryptophan, and tyramine, [56,57]; whereas, separate AADCs are responsible for the conversion of L-dopa to dopamine and of 5-hydroxytryptophan to 5-HT (dopa decarboxyase), and tyrosine to tyramine (tyrosine decarboxylase) in invertebrate nervous systems [45,58,59,60].

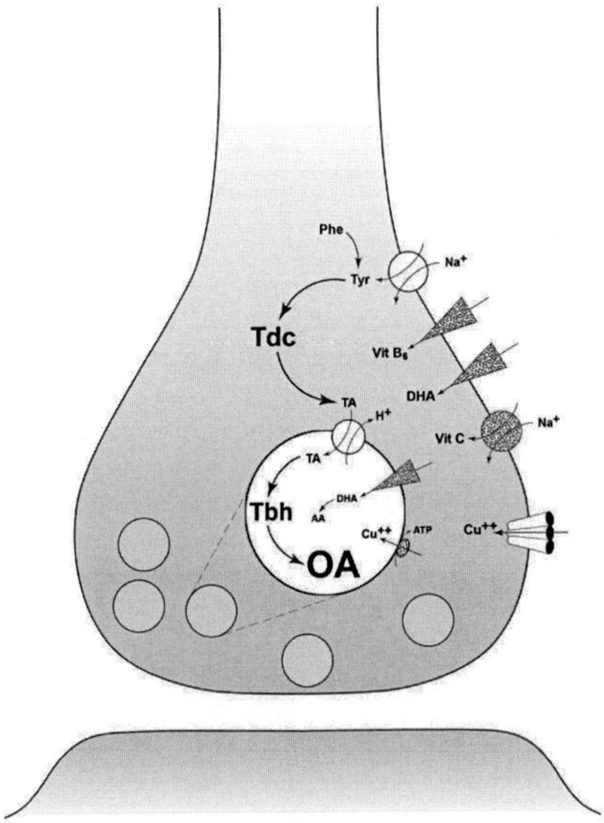

Figure 1. Schematic model of an insect OA neuron containing known and predicted proteins involved in OA biosynthesis. All locations are putative. Tdc, tyrosine decarboxylase; Tbh, tyramine beta-hydroxylase; Phe, phenylalanine; Tyr, tyrosine, Vit B6, vitamin B6; DHA, dehydroascorbic acid; Vit C, vitamin C; Cu++, copper; TA , tyramine. Transporters diagrammed as large circles represent plasma membrane, Na^+ dependent antiporters; as triangles represent facilitative transporters; and as small circles represent ATP-dependent pumps. Open transporters have been described in insects whereas stippled transporters are speculative (see text).

The biochemical characterization of Tdc from locust brain extracts revealed two binding affinity components, suggesting that multiple forms Tdc may exist ($K_{m,\ tyrosine}$=1.31 and 50.8 μM) [60]. Recently, two *Drosophila* Tdc genes were identifified (*dTdc1* and *dTdc2*). *DTdc2* encodes a 637 amino acid polypeptide with a predicted mass of 72,000 Da that is restricted to the insect nervous system and critical for octopamine synthesis [47]. DTdc2 is 75% identical to other putative insect Tdcs (*Anophiles gambiae* and *A. mellifera*), 65% identical to mammalian AADCs, and 55% identical to plant AADCs. All AADCs contain a highly conserved pyridoxal-dependent domain characteristic of group II decarboxylases and studies of plant AADCs have revealed the role of pyridoxal 5'-phosphate in the decarboxlation reaction [61,62]. Direct experimental evidence has not confirmed the insect Tdc reaction mechanism, however, the conserved sequence homology among AADCs suggests that the catalytic mechanism is similar, if not also conserved. Pyridoxal 5'-phosphate first binds to AADCs via a Schiff base linkage with the ε-amino group of a specific Lys residue. A transaldimination reaction follows in which the enzyme-pyridoxal-5'-phosphate Schiff base linkage is broken and a new Schiff base is formed between the substrate and pyridoxal-5'-phosphate. The substrate is then decarboxylated to form tyramine, and the enzyme and pyridoxal-5'-phosphate are regenerated. Thus, pyridoxal-5'-phosphate is vital for AADC and Tdc activity.

Tyramine Beta-hydroxylase (Tbh)

The *Drosophila melanogaster* Tbh gene (*DmTbh*) encodes a 660 amino acid polypeptide with a predicted mass of 76,000 Da. The protein encoded by *DmTbh* is 35-54% identical (and 54-70% similar) to other invertebrate Tbh proteins now predicted or described (including *Drosophila sp. A. mellifera, Anopheles gamiae, Tribolium castaneum, Nasonia vitripennis and Caenorhabditis elegans*), and is ca. 41% identical (and 57% similar) to the vertebrate enzyme Dbh (Figure 2) [50]. Tbh, Dbh, and PHM belong to a group of copper type II, ascorbate-dependent monooxygenases that require copper as a cofactor and uses ascorbate as an electron donor. Ascorbate acid is needed as a co-substrate to regenerate the reduced form of the enzyme following catalysis whereby ascorbate is converted to semidehydroascorbate (SHA) during the one electron reduction of two copper ions. Copper is bound to Tbh by two clusters of conserved histidine residues [63]. More specifically, Dbh and Tbh use two copper centers, Cu_M and Cu_H to hydroxylate substrates. The Cu_M site is coordinated by two histidine residues and a methionine residue and serves as the location for oxygen binding and substrate hydroxylation. The Cu_H site is coordinated by other histidine residues and it functions as an electron transfer site in the reaction mechanism [64,65]. Both of these copper centers are well conserved in all Dbh and Tbh proteins identified to date (Figure 2). In addition, conserved Cys residues are believed to form intramolecular disulfide bonds that are critical for the formation of the Dbh teritary structure; all Tbh and Dbh proteins possess twelve Cys residues in identical locations [66] (Figure 2). Moreover, substrate analogs bind to a site close to the Cu_M site and near an arginine residue in PHM; interestingly, a conserved Arg residue is present 12 amino acids downstream from the Cu_M in all Tbh proteins [67]. Finally, Dbh and Tbh have comparable substrate affinities. For example, the affinities of *Drososophia* Tbh to tyramine (apparent $K_{M,\ tyramine} = 87.6 \pm 12.7$ μM) and ascorbate (apparent $K_{M,\ ascorbate} > 16$

```
Drosophila          -----MPVMLLLLMATLLTRPLSAFSN--------------------------------- 22
Anopheles           --------------CTLLNSIFVIIVS--------------------------------- 13
Bos                 ---MQVPSPSVREAASMYGTAVAVFLV--------------------------------- 24
Caenorhabditis      -----------MRSAVALLFLLVAYCG--------------------------------- 16

Drosophila          --RLSDT---------KLHEIYLD-DKEIKLSWMVDWYKQEVLFHLQNAFN-EQHRWFYL 69
Anopheles           --DISSE---------RLHSLKLN-HDSTKLTWMVDWPKKEVLFYINNTFDNGKFKTFAI 61
Bos                 -ILVAALQGSAPAESPFPFHIPLDPEGTLELSWNISYAQETIYFQLLVRELK---AGVLF 80
Caenorhabditis      --GVVHAG----------EIVAELYHTNVTVKWHTDYERQLVDFSIWFGASTP--DVLFL 62
                                                  .    .     :               . .

Drosophila          GFSKRGGLADADICFFENQNGFFNAVTDTYTSPDGQWVRRDYQQDCEVFKMDE-----FT 124
Anopheles           GFSQRGELSRCDLCVFTSVPKLYQQVHDSYTSRKFDHIFRDTLQNCEVMYMDD-----NS 116
Bos                 GMSDRGELENADLVVLWTD-RDGAYFGDAWSDQKG-QVHLDSQQDYQLLRAQR-TPEGLY 137
Caenorhabditis      GFSDFGDTNNSDVLMYYNS---KKEIKDAYTNRDF-KITSDLQQDFQLLRKRK-----DH 113
                    *:*  *    .*  .          . *            *  *:   :

Drosophila          LAFRRKFDTCD-PLDLRLHEGTMYVVWARGETELALEDHQFALPN--------------- 168
Anopheles           VAFRRKFDTCD-PQDIVFHTGTMYIVWLRSNALLEWSNNSTIIP---------------- 159
Bos                 LLFKRPFGTCD-PNDYLIEDGTVHLVYGFLEEPLRSLESINTSG---------------- 180
Caenorhabditis      IVVRRKLTTCD-SRDYAFLPGTTQFYIAASWGSTNLVDIRDKRWV--------------- 157
                    . . * : ***   *  :  .*  .

Drosophila          ---------VTAPHEAGVKMLQLLRADKILIPET--ELDHMEITLQEAP-IPSQETTYWC 216
Anopheles           ---------KHSTKNQGVLPVQLLRADKIRIPETGQILKKLDVRLNNVS-VPAAETTYWC 209
Bos                 -------------LHTGLQRVQLLKPSIPKPALP-ADTCTMEIRAPDVL-IPGQQTTYWC 225
Caenorhabditis      ---------VDKKFGKVIEGPTDQPNIEEEPAALEKDVKVVIVNSNSPDPIPNVETTYKC 208
                                     :                                :*  :* * *

Drosophila          HVQRLEGNLR-RRHHIVQFEPLIR--TPGIVHHMEVFHCEAGEHEEIPLYN-GDCEQ--L 270
Anopheles           KIQQLDPWLTNAKHHIVQFEPIID--NEALVHHMEVFQCIAGN-AEIPTYD-GPCQN--M 263
Bos                 YVTELPDGFP--RHHIVMYEPIVTEGNEALVHHMEVFQCAAEF-RDHPHFS-GPCDSKMK 281
Caenorhabditis      IIRKMPFDTVNNMYHVVRMEPYVTPGNEHLVHHMEIFMC----RDEVEEWS-GSCNDPKK 263
                     : .:        :*:    . :   .  .**** :. *               *

Drosophila          PPRAKICSKVMVLWAMGAGTFTYPPEAGLPIGGPGFNPYVRLEVHFNNPEKQSGLVDNSG 330
Anopheles           PASGHLCSKVMALWAMGAGSFTYPREAGLPIGGKDFNPHIRLEVHFNNPRMLSGYNDSSG 323
Bos                 PQRLNFCRHVLAAWALGAKAFYYPEEAGLAFGGPGSSRFLRLEVHYHNPLVITGRRDSSG 341
Caenorhabditis      PPKSKSCSHVIAAWAMGEGPIHYPKEAGLPIGGKGKNAYVMVEIHYNNPELHKGVIDSSG 323
                    *     * :::. **        * ..* .:*. .   .  :*:*:.**    *  *.**

Drosophila          FRIKMSKTLRQYDAAVMELG-LEYTDKMAIPPGQTAFPLSGYCVADCTRAALPATG---I 386
Anopheles           MRINVVSKLRRYDAAIMELG-LEYTDKMAIPPEQLAFPLHGYCIAECSKIALPKTG---I 379
Bos                 IRLYYTAALRRFDAGIMELG-LAYTPVMAIPPQETAFVLTGYCTDKCTQLALPASG---I 397
Caenorhabditis      FQFFVTGQLRKYDAGIMELG-LIYSDANSVPPNQKAWAMNGYCPSQCTKN-LPEEG---I 378
                    .:                    :  :   :**    :   * *  .*    :   *   *

Drosophila          IIFGSQLHTHLRGVRVLTRHFRGEQELREVNRDDYYSNHFQEMRTLHYKPRVLPGDALVT 446
Anopheles           VVFGSQLHTHLRGVRVLTRHFRGKTELPILNRDDFFSHHYQEIRQLRYKPRVLPGDALVT 439
Bos                 HIFASQLHTHLTGRKVVTVLARDGRETEIVNRDNHYSPHFQEIRMLKKVVSVQPGDVLIT 457
Caenorhabditis      NIFASQLHAHLTGRKLFTSQYRSGVRIGDVNRDEHYSPHWQHLQQLRPVVKVMPGDTLVT 438
                    :... **:** *  : .   *.  .   :  *: :. .:*. : *     :  ** * .

Drosophila          TCYYNTKDDKTAALGGFSISDEMCVNYIHYYPATKLEVCKSSVSEETLENYF-------- 498
Anopheles           SCYYDTRGYNSTTLGGFAISDEMCVNYIHYYPATELEVCKSSISENSLYEYF-------- 491
Bos                 SCTYNTEDRRLATVGGFGILEEMCVNYVHYYPQTQLELCKSAVDPGFLHKYF-------- 509
Caenorhabditis      TCVYDTRKRSKVTFGGYRIVDEMCVNYIYYYPASDVEVCKSAISNSTLRAYF-------- 490
                     * *::      :  *    :***:.:: *:*   :.:*:::     *  **

Drosophila          ------IYMKRTEHQHG-VHLNGARSSNYRSIEWTQPRIDQLYTMYMQEPLSMQCNRSDG 551
Anopheles           ------LYMKNIHKQNI-TSPNGPRSENYRVIDWNQAKANELIDVYVTEPISMQCNRSNG 544
Bos                 ------RLVNRFNSEEVCTCPQASVPEQFASVPWNSFNREVLKALYGFAPISMHCNRSSA 563
Caenorhabditis      ------SERHGMDGKR-----MQISDMYSNVKDWGNGVDEEFYNVLNVGNMNMNCLKSNG 539
                                  .                  *                 .  *  .

Drosophila          TRFEGR-----SSWEGVAATPVQIRIPIHRKLCPNYNP--LWLKPLEKGDCDLLGECIY- 603
Anopheles           LRFDG------FEWENAPITSFNLAPPDQSKTCSPVKSGIGWFRSLNNGLCDNYGDCIYA 598
Bos                 VRFQG-------EWNRQPLPEIVSRLEEPTPQCPASQAQSPAGPTVLNISGGKG------ 610
Caenorhabditis      EPFEFESKDSRQSWENMARPTFVSGSFITTRDRFQCPAINDMINFE-------------- 585
```

Figure 2. Deduced amino acid sequences of invertebrate Tbh and vertebrate Dbh. Alignment using Clustal W2 of *Drosophila melanogaster*, *Anopheles gamiae, and Caenorhabditis elegans* Tbh and a Dbh protein from a mammalian species (*Bos taurus*). Putative signal sequences at the amino termini as predicted by SignalP are underlined. Conserved Cys residues critical for intermolecular disulfide bonds are highlighted in yellow. Conserved His residues that form the Cu_M site are highlighted in purple, and Cu_H site are highlighted in blue.

mM) are similar to values reported for *M. sexta* Tbh, *H. americana* Tbh, and mammalian Dbh [48,49,68-74]. Based on the high sequence similarities and reactivity studies of Tbh and Dbh, the reaction mechanisms of these enzymes were thought to be similar; however, recent studies indicate that the reaction mechanisms of Tbh and Dbh are different [68]. Unlike the mammalian enzyme Dbh, Tbh is subject to substrate inhibition. Tyramine and ascorbate competively bind to the oxidized form of Tbh - high levels of tyramine interferes with ascorbate function at the Cu_M site, resulting in the formation of an inhibitory complex, while, increasing concentrations of ascorbate leads to a greater amount of enzyme in the reduced and active form. This study suggests that changes in the levels of ascorbate and tyramine affect Tbh rate constants which may, in turn, affect OA levels in the insect nervous system (Figure 1).

3.b. Subcellular localization of Tdc and Tbh

OA is localized and released from specific neurons in a wide variety of invertebrates, but the subcellular location of OA is less well known. Hoyle et al. were first to observe large dense-cored and small clear vesicles within locust OAergic dorsal unpaired median neurons that innervate the extensor tibiae muscle [75]. Similar large dense-cored and small clear vesicles are present in the OAergic innervated firefly luminescent organ [76]. More specifically, EM immunocytochemistry was used to demonstrate that OA-ir is present in nerve processes, endings, and large cylindrical dense-core granules in the *Limulus polyphemus* nervous system [77]. Moreover, Tbh is likely located in secretory vesicles. A signal peptide is predicted to occur at the amino-terminus of all Tbh proteins identified to date by SignalP and TargetP, indicating the protein is likely to enter into the secretory pathway and be contained within vesicles [78] (Figure 2). In contrast, tyrosine decarboxylases (Tdc) are not predicted to encode signal peptides. Thus, DTdc2 is likely located cytoplasmically where it catalyses the formation of tyramine. Tyramine is then transported into synaptic vesicles and hydroxylated by vesicular Tbh to form octopamine in large dense-cored vesicles within the insect nervous system (Figure 1). This scenario is similar to the catabolic pathway of catecholamines in mammals. Norepinephrine synthesis from tyrosine involves two cytosolic enzymes (tyrosine hydroxylase and DOPA decarboxylase) and one secretory enzyme, (Dbh) in chromaffin granules of the adrenal medulla and in the large dense-cored synaptic vesicles of the symphathetic nervous system [72].

4. Cellular Transport Mechanisms

4.a. Amino Acid Substrates

Substrates for cytoplasmic Tdc and vesicular Tbh are also required to be located in separate subcellular compartments for proper OA synthesis. Tyrosine must be transported to the cellular cytoplasm, whereas tyramine is synthesized in the cytoplasm and must be transported across the vesicular plasma membrane into the lumen of synaptic vesicles. Separate and distinct transporters are necessary for transport of these amino acid substrates to their respective cellular destinations.

Tyrosine

Tyrosine is a nonessential amino acid that can be obtained either from dietary sources or can be synthesized from the essential amino acid phenylalanine. Transport of dietary tyrosine and other neutral amino acids across biological membranes occurs through nutrient amino acid transporters (NAT), a subfamily of the sodium neurotransmitter symporter family (SNF, a.k.a SLC6). The SLC6 transporter family meditates the transport of many neurotransmitters, neuromodulators, and amino acids across the plasma membrane by coupling solute transport to the co-transport of sodium and chloride down their electrochemical gradients [79]. SLC6 transporters are characterized by a predicted twelve transmembrane domain topology, and highly conserved amino acids in transmembrane domains 1, 2 and 4-8 [reviewed by 80,81]. Moreover, TM1, TM6 and TM8 appear to be involved in substrate and sodium binding [82]. Based upon these structural characteristics, the first aromatic nutrient transporter from insects was cloned and characterized [83]. This insect transporter (agNAT8) has high sequence similarity to other NATs, a high transport velocity, and high affinity (apparent $E_{0.5,\ \mathrm{phenylalanine}} = 0.194 \pm 0.015$ mM) for a narrow range of large aromatic amino acids, esp. tyrosine and phenylalanine. AgNAT8 is widely expressed in *Anopheles gambiae*, including the gut, central nervous system, and peripheral nervous system [80,83]. A structurally-related orthologue is present in *Drosophila* (DmNAT1;CG 3252); DmNAT1 belongs to the same nutrient amino acid transporter family, has a broader selectivity, and is also expressed in the salivary glands, brain, imaginal dics and posterior midgut [84].

Tyrosine can also be synthesized by phenylalanine hydroxylase (PAH), a metabolic enzyme important in phenylalanine catabolism, protein, and neurotransmitter synthesis. PAH belongs to a broad group of aromatic amino acid hydroxylases that require iron for activity and the bound iron must be in the reduced (ferrous) state for catalysis to occur. Tetrahydrobiopterin (BH_4), is necessary for the reduction of iron and is required along with molecular oxygen for hydroxylase activity [85]. *Drosophila* tryptophan-phenylalanine hydroxylase (DTPH) is a single enzyme capable of differentially hydroxylating tryptophan and phenylalanine [86,87]. The $K_{\mathrm{m,phenylalanine}}$ of DTPH-PAH activity is 44.66 μM, as compared to the $K_{\mathrm{m,\ tryptophan}}$ for DPTH-TRH activity of 3.19 μM. Interestingly, DTPH has different enzyme features with phenylalanine as a substrate; PAH activity is inhibited by tyrosine, dopamine, and serotonin, whereas TRH reactions are not subject to feedback inhibition. In addition, phosphorylation of DTPH by Protein Kinase A decreased PAH activity, increased TRH activity, and released end-product inhibition of PAH [87]. DTPH has a broad distribution with multiple roles: metabolism of phenylalanine in the fat body, the production of tyrosine in cuticle formation, and the synthesis of the neurotransmitters dopamine and serotonin. Although DTPH does not appear to be present in *Drosophila* VUM neurons, its presence in other OAergic neurons and its role in OA synthesis has not been explored. Thus, there appears to be two potential mechanisms to deliver tyrosine to neurons. If tyrosine is in abundant supply, it would be transported directly; whereas, if tyrosine is in limited supply it could be synthesized from phenylalanine (Figure 1). It is unclear if the latter mechanism has a role in OAergic neurons.

Tyramine

The packing of classical and amino acid transmitters into secretory vesicles occurs through vesicular neurotransmitter transporters. Vesicular amine transporters (VATs) are proton-translocating antiporters that are well known to determine the quantity of vesicular content [88]. Vesicular transport has been observed for several classical transmitters including acetylcholine, monoamines, glutamate, GABA, and glycine. These transporters appear to share a similar membrane topology consisting of 12 putative transmembrane domains with a large hydrophilic loop between membrane domains 1 and 2. Substrate binding sites and the rate of transport are determined at sites near the transmembrane domains, whereas the cytoplasmic carboxy terminal is important for membrane sorting and localization [89]. Two vesicular monoamine transporters (VMAT) have been identified in mammals; VAMAT1 is expressed primarily in neuroendocrine cells and VMAT2 is expressed in all aminergic neurons in the mammalian CNS [90]. *Drosophila* VMAT (*DVMAT*) has been cloned and identified, and has two alternative splice variants that differ at the carboxy termini (*DVMAT-A* and *B*). *DVMAT-A* is expressed in all aminergic neurons of the fly (dopaminergic, serotoninergic and OAergic neurons) and binds and transports serotonin and dopamine with high affinity (K_m = 0.3-1.1 μM). Tyramine and octopamine inhibit 5-HT transport at equally low doses, IC_{50} = 0.71-1.1 μM, indicating that they have similar high affinities [91], and these transport affinities are similar to apparent Km values reported for other VAMAT2 transporters [92]. *DVMAT-B* is expressed in glia rather than neurons and appears to play a role in histamine transport and storage in the visual system [93]. Thus, *DVMAT-A* appears to be the vesicular tyramine transporter in octopaminergic neurons (Figure 1).

4.b. Co-substrates (Vitamin B6, Vitamin C, Copper)

Vitamin B6 Synthesis and Transport

Vitamin B6 plays a vital role as a cofactor for decarboxylases required for the synthesis of various neurotransmitters, including tyramine, serotonin, DOPA, and γ-aminobutyric acid (GABA) [57]. Vitamin B6 is comprised of three related compounds pyridoxine (PN), pyridoxal (PL) and pyridoxamine (PM) and their phosphorylated derivatives, pyridoxine 5'-phosphate (PNP), pyridoxal 5'-phosphate (PLP) and pyridoxamine 5'-phosphate (PMP). The movement of PN, PL, and PM through the blood brain barrier, blood-cerebral spinal fluid barrier, and brain cell membranes is a saturable, facilitated diffusion process, but the precise mechanism of transport remains unknown [94,95,96]. Once the non-phosphorylated forms are present in the celluar cytoplasm they are readily phosphorylated by pyridoxal kinase and PNP is further converted by PNP oxidase in liver and brain to PLP, the active cofactor. Recently, PNP oxidase has been cloned and characterized from the silkworm, *Bombyx mori* [97]. The insect oxidase has many features in common with other PNP oxidases. Similar structurally-conserved motifs, monomer structure, and substrate affinities ($K_{m,\ PNP}$ = 0.65 μM) all suggest that this enzyme is similar to other PNP oxidases. Little else is known, however, about the cellular distribution and functional role of this enzyme. Therefore, I suggest that a Vitamin B6 facilitative transporter is necessary for Tdc activity and should be present in OAergic neurons, however, its identification and precise localization is lacking (Figure 1).

Vitamin C Synthesis and Transport

Vitamin C is an essential nutrient required for all eukarotic organisms and is especially concentrated in the mammalian nervous system [98,99]. Vitamin C enters mammalian cells, including neurons, by two possible mechanisms. First, ascorbic acid (AA, oxidized Vitamin C) can be transported by a sodium-dependent Vitamin C transporter (SVCT), of which there are two known isoforms, SVCT1 and SVCT2 [100-102]. These mammalian symporters are membrane-associated glycoproteins with 12 transmembrane domains that are 66% identical, and mediate high affinity Na^+ dependent L-ascorbic acid transport that is necessary for the uptake of vitamin C in many tissues [103,104]. These transporters have different kinetic properties and cellular distributions; SVCT1 has a $K_{m,\ ascorbate}$ of ca. 100 μM and is present primarily in epithelial brush borders of kidney, intestine and liver [101,102,105], whereas SVCT2 has a $K_{m,\ ascorbate}$ of ca. 67 μM and is expressed in brain and eyes [100,106].

A second mechanism in which ascorbate can enter cells is via uptake of dehydroascorbic acid (DHA, reduced Vitamin C) through a widely expressed GLUT-type glucose transporter [107,108]. This facilitative glucose transporter is a member of a family of transporters that is diverse and composed of 14 isoforms, each of which are predicted to contain 12 transmembrane domains, and are classified into three subclasses according to their sequence and functional characteristics [109]. The Class I transporters are the most widely studied glucose-specific transporters and include the high affinity DHA transporter isoforms, GLUT 1, 2, and 4 ($K_{m,\ DHA}$ ca. 2 mM) which are widely distributed in many tissues, including the brain [110-112]. Once DHA is transported into the cytoplasm or the lumen of specific organelles (e.g. endoplasmic reticulum), it is recycled back to AA by several dehydroascorbic acid reductases (e.g. glutathioine, glutaredoxin, protein disfulfide isomerase) [105,113,114]. Thus, plasma membrane transport of DHA and AA and their interchangeable nature are important in the maintenance of cellular vitamin C levels.

The mechanism of ascorbate accumulation in synaptic vesicles is less well understood. The lumens of the endoplasmic reticulum (ER) and vesicles of the secretory pathway are characterized by high (millimolar) ascorbate concentrations, and ascorbate has multiple roles as an antioxidant and cofactor for intraluminal enzymes (e.g. copper type II, ascorbate-dependent monooxygenases). Dbh is also localized in neuroendocrine secretory vesicles (see above) where it uses ascorbic acid as a co-substrate, which generates semidehydroascorbate (SDA, mono-oxidized form of Vitamin C) [115]. Ascorbate and SDA do not cross vesicle membranes, however, the facilitated diffusion of DHA through a glucose transporter has been described in rat liver microsomes [116]. These observations suggest that ascorbate transport into the ER is dependent on its prior oxidation to DHA. The reduction of DHA (and SDA) to ascorbate then occurs inside the vesicles and may involve the transmembrane transfer of electrons by cytochrome b561, the second most abundant protein in chromaffin granule membranes [117]. The current structural models of cytochrome b561 suggests that it forms six transmembrane alpha-helices within the vesicle membrane and contains two hemes with differing redox potentials, each anchored by a pair of well-conserved histidyl residues, one near the cytosol and the other near the intravesicular face of the membrane [118,119]. This structural model supports the notion that cytochrome b561 can facilitate electron transfer across the vesicular membrane.

In sum, several transporters, enzymes, and redox reactions play a role in the transport of Vitamin C from the extracellular space to the lumen of secretory vesicles in mammalian endocrine cells. Little is known, however, about Vitamin C transport, redox, and storage in insects. Clearly, Vitamin C is critical for Tbh activity and genes encoding *Drosophila* AA transporters, GLUT-type transporters, and cytochrome b561 are expected to exist in the *Drosophila* genome, however, their presence and role in OA synthesis remains uninvestigated (Figure 1).

Copper Ion Transport

Copper is an ion essential for many enzymes associated with mitochondrial respiration, connective tissue formation, pigmentation, iron oxidation, neurotransmitter synthesis and processing, and antioxidant defense [120-122]. Copper's unique ability to accept and donate electrons is critical for these enzymatic reactions, however this same redox property of copper is responsible for its ability to generate deleterious reactive oxygen species. Thus, free copper concentration in cells is maintained at very low concentrations and copper-handling proteins have evolved to transport and deliver copper to specific sites of utilization [123,124]. Copper transport into cells is mediated Ctr1, a plasma membrane copper transporter that is a widely expressed homotrimeric transporter first identified in yeast [125,126]. Reduced copper appears to be the substrate of Ctr1-mediated copper uptake ($K_{m, copper}$ = 1 - 5 μM) and Ctr1 is thought to be the rate-limiting step of copper intake [125,127,128]. The detailed mechanisms of copper movement through CTR1 is unresolved, but likely involves a chain of copper exchange reactions between defined Cu(I)-binding sites and associated conformational switches in the transport protein [129].

Once in the cellular cytoplasm, copper chaperones bind and deliver copper to specific enzymes. For example, copper chaperones CCS and COX17 deliver copper to Cu/Zn superoxide dismutase in the cytoplasm and to cytochrome c oxidase in the mitochondria [130,131]. Other chaperones deliver copper to ATP7A, a copper transporter located in the Golgi complex of neurons and astrocytes [132,133]. ATP7A is a copper transporting protein that uses the energy of ATP hydrolysis to transport copper from the cytosol into the secretory pathway. [134-137]. Moreover, ATP7A plays a key role in providing copper for peptidylglycine-amidating monoxygenase and dopamine-monooxygenase [135,138].

Several studies have demonstrated a high conservation among copper homeostasis mechanisms in insects and mammals. First, copper appears to be concentrated in specific neurons in the insect CNS (including OAergic neurons) as revealed by a sulphide silver stain for heavy metals [139]. Second, three distinct *Drosophila* Ctr genes have been identified (CtrA, B, and C) and all three gene products have several structural features in common with other Ctr proteins, including three membrane-spanning domains, a hydrophilic methionine-rich amino terminus, and a Met-*X*3-Met motif found in the second transmembrane domain [140]. Recently, another putative symport divalent metal ion transporter has been identified in *Drosophila* that shares several functional features with Ctr1 proteins [141]. Interestingly, this transporter is expressed in many neurons of the larval nervous system, and appears to be localized subcellularly [142]. In addition, several copper chaperones and an ATP7 orthologue has been identified in *Drosophila* [143,144]. Thus, a number of *Drosophila* genes and gene products, similar to mammalian genes, are involved in copper uptake, transport and sequestration. These copper transport and homeostatic mechanisms would be expected in OAergic neurons and be vital for Tbh activity (Figure 1).

5. Summary and Future Directions

In summary, two synthetic enzymes (Tdc and Tbh), three plasma membrane active transport proteins (tyrosine, Vitamin C, and copper), two plasma membrane facilitative transporters (Vitamin B6 and DHA), one vesicular active transporter (tyramine), one vesicular facilitative transporter (DHA), and one vesicular ATPase pump (copper) are needed for octopamine synthesis (Figure 1). I argue here that the transport and availability of tyramine and Vitamin C (ascorbic acid and dehydroascorbic acid) into OAergic vesicles play key roles in the modulation of Tbh. First, tyrosine transport and metabolism is not likely to play a role in the regulation of OA synthesis. Cytoplasmic tyrosine concentrations in the insect nervous tissue are high due to the multiple mechanisms of transport and synthesis, and the apparent $K_{\text{m, tyrosine}}$ of Tdc is comparable to the apparent of the $E_{0.5,\ \text{phenylalanine}}$ of the Phe/Tyr transporter (0.050 mM and 0.194 mM, respectively) suggesting that as tyrosine is available to Tdc it is converted to tyramine. Moreover, Tdc covalently binds co-factor Vitamin B6 and Tdc does not appear to be subject to substrate or feedback inhibition. Thus, Tdc is unlikely to be influenced by fluctuating levels of co-factor. In contrast, Tbh is contained within secretory vesicles and its activity is influenced by relative amounts of tyramine and Vitamin C. Copper is unlikely to play a role in the modulation of Tbh because of its toxic nature, sequestration, and the high likelihood of it binding to Tbh early in the synthetic pathway. Conversely, vesicular tyramine transport affinity is low relative to the substrate affinities of Tdc and the tyrosine transporter (0.3 -11 μM, 87 μM, and 197 μM, respectively) suggesting that this vesicular transporter may limit substrate availability to Tbh. Little is known about Vitamin C concentrations and transport mechanisms in insects; however, the transport affinity of all Vitamin C transporters described to date are lower than the $\text{Km}_{\text{ascorbate}}$ of Tbh (ca. 16 mM), suggesting that the amount of Vitamin C available to Tbh is sub-saturating. Thus, the putative kinetics of substrate transport support the kinetic analysis of Tbh and strongly suggest that the availability of tyramine and Vitamin C affect OA synthesis. In summary, future studies focused on the identity and regulation of the vesicular Vitamin C transporter and the regulation of vesicular tyramine transport will likely reveal important control mechanisms of this vital biogenic amine, OA.

Acknowledgments

Research support from the National Science Foundation (IBN-0212554) and from the Dean of Faculty, Hamilton College, is gratefully acknowledged. The author also thanks T.

Elgren, A. Brodsky, S. Chaponis, and J. Chaponis for their constructive comments and proofreading on the text.

References

[1] Pfluger H-J, Stevenson P. Evolutionary aspects of octopaminergic systems with emphasis on arthropods. *Arthropod Structure & Development.* 2005;34:379-396.

[2] Erber J, Kloppenburg P. The modulatory effects of serotonin and octopamine in the visual system of the honey bee (*Apis mellifera*). I. Behavioral analysis of the motion-sensitive antennal reflex. *Journal of Comparative Physiology* 1995;176:111-118.

[3] Ramirez J-M, Orchard I. Octopaminergic modulation of the forewing stretch receptor in the locust *Locusta migratoria. Journal of Experimental Biology.* 1990;149:317-340.

[4] Linn CL, Roelofs WL. Modulatory effects of octopamine and serotonin on male sensitivity and periodicity of response to sex pheromone in the cabbage looper moth, *Trichuplusia ni. Archives of Insect Biochemistry and Physiology.* 1986;3: 161-171.

[5] Farooqui T. Octopamine-mediated neuromodulation of insect senses. *Neurochemical Research.* 2007; 32:1511–1529.

[6] Libersat F and Pflueger H-J. Monoamines and the orchestration of behavior. *Bioscience* 2004;54:17-25

[7] Roeder T. Tyramine and octopamine; ruling behavior and metabolism. *Annual Review of Entomology.* 2005; 50: 447-477.

[8] Wagener-Hulme C, Kuehn JC, Schulz DJ, Robinson GE. Biogenic amines and division of labor in honey bee colonies. *Journal of Comparative Physiology A.* 1999;184: 471-479.

[9] Lehman HK, Schultz DJ, Barron A, Wraight L, Hardison C, Whitney S, Takeuchi H, Robinson GE. Behavioral division of labor in honey bees: the role of tyramine beta-hydroxylase. *The Journal of Experimental Biology.* 2006;209: 2774-2784.

[10] Mons N, Geffard M. Specific antisera against the catecholamines: L-dihydroxyphenylalanine, dopamine, noradrenaline and octopamine tested by an enzyme-linked immunosorbant assay. *Journal of Neurochemistry.* 1987:48: 1826-1833.

[11] Eckert M, Rapus J, Nürnberger A, Penzlin H. A new specific antibody reveals octopamine -like immunoreactivity in the cockroach ventral nerve cord. *The Journal of Comparative Neurology.* 1992;322: 1-15.

[12] Spörhase-Eichmann U, Vullings HGB, Buijs RM, Hörner M, and Schürmann FW. Octopamine-immunoreactive neurons in the central nervous system of the cricket, *Gryllus bimaculatus. Cell and Tissue Research.* 1992;268: 287-304.

[13] Stevenson PA, Spörhase-Eichmann U. Localization of octopaminergic neurones in insects. *Comparative Biochemistry and Physiology.* 1995;110A: 203-215.

[14] Dacks AM, Christensen TA, Agricola HJ, Wollweber L, Hildebrand JG. Octopamine-immunoreactive neurons in the brain and subesophageal ganglion of the hawkmoth *Manduca sexta. Journal of Comparative Neurology.* 2005; 488: 255-68.

[15] Sinakevitch I, Strausfeld NJ. Comparison of octopamine-like immunoreactivity in the brains of the fruit fly and blow fly. *Journal of Comparative Neurology.* 2006;494: 460-475.

[16] Sinakevitch I, Niwa M, Strausfeld NJ. Octopamine-like immunoreactivity in the honey bee and cockroach: comparable organization in the brain and subesophageal ganglion. *Journal of Comparative Neurology.* 2005;488: 233-54.

[17] Vömel M, Wegener C. Neuroarchitecture of aminergic systems in the larval ventral ganglion of *Drosophila melanogaster*. PLoS ONE. 2008;3: e1848.

[18] Stevenson PA, Spörhase-Eichmann U. Localization of octopaminergic neurones in insects. *Comparative Biochemistry and Physiology.* 1995;110A: 203-215.

[19] Orchard I, Lange AB. Evidence for octopaminergic modulation of an insect visceral muscle. *Journal of Neurobiology.* 1984;16: 171-181.

[20] Allgäuer C, Honegger HW. The antennal motor system of crickets: modulation of muscle contractions by a common inhibitor DUM neurons, and proctolin. *Journal of Comparative Physiology A.* 1993;173: 485-494.

[21] Stevenson PA, Meuser S. Octopaminergic innervation and modulation of a locust flight steering muscle. *Journal of Experimental Biology.* 1997;200: 633-642.

[22] Bräunig P, Eder M. Locust dorsal unpaired median (DUM) neurones directly innervate and modulate hindleg proprioceptors. *The Journal of Experimental Biology.* 1998;201: 3333-3338.

[23] Mentel T, Duch C, Stypa H, Wegener G, Pflüger HJ. Central modulatory neurons control fuel selection in flight muscle of migratory locust. *Journal of Neuroscience.* 2003;23: 1109-1113.

[24] Schröter U, Malun D, Menzel R. 2007. Innervation pattern of suboesophageal ventral unpaired median neurones in the honeybee brain. Cell and Tissue Research. 2007;327:647–667.

[25] Bräunig P. The morphology of suboesophageal ganglion cells innervating the nervus corporis cardiaci III of the locust. *Cell and Tissue Research.* 1990;260: 95-108.

[26] Bräunig P. Suboesophageal DUM neurones innervate the principal neuropiles of the locust brain. *Philosophical Transactions of the Royal Society London B.* 1991;322: 221-240.

[27] Bräunig P, Burrows M. Projection patterns of posterior dorsal unpaired median (DUM) neurons of the locust suboesophageal ganglion. *Journal of Comparative Neurology.* 2004;478: 164-175.

[28] Raabe M. The neurosecretory–neurohemal system of insects: anatomical, structural and physiological data. *Advances in Insect Physiology.* 1983;17: 205-303.

[29] Fahrbach SE. Structure of the mushroom bodies of the insect brain. *Annual Review of Entomology.* 2006;51: 209-323.

[30] Heinze S, Homberg U. Linking the input to the output: new sets of neurons complement the polarization vision network in the locust central complex. *The Journal of Neuroscience.* 2009;29:4911-4921.

[31] Busch S, Selcho M, Ito K, Tanimoto H. A map of octopaminergic neurons in the *Drosophila* brain. *Journal of Comparative Neurology.* 2009;513:643-667.

[32] Hoyle, G. Generation of motor activity and control of behaviour: The role of the neuromodulator octopamine and the orchestration hypothesis. In: Kerkut GA, Gilbert L, eds. *Comparative Insect Physiology, Biochemistry and Pharmacology*. Toronto: Pergamon Press. 1985;5:607-621.

[33] Nagao T, Tanimura T. Distribution of biogenic amines in the cricket central nervous system. *Analytical Biochemistry*. 1988:171; 33-40.

[34] Nässel DR, Laxmyr L. Quantitative determination of biogenic amines and DOPA in the CNS of adult and larval blowflies, *Calliphora erythrocephala. Comparative Biochemistry and Physiology*. 1983;75C:259-265.

[35] Sloley BD, Okikaza S. Selective depletion of dopamine, octopamine and 5-hydroxytryptamine in the nervous tissue of the cockroach, *Periplaneta americana. Journal of Neurochemistry*. 1988;51:535-541.

[36] Bodnaryk RP. Changes in brain octopamine levels during metamorphosis of the moth *Mamestra configurata* Wlk. *Insect Biochemistry*. 1980;10:69-173.

[37] Linn CE, Roelofs WL. Levels of biogenic amines and peptides in individual corn earworm moths, *Helicoverpa zea*, using high performance liquid chromatography. Insect Biochemistry and Molecular Biology 1993;23:367-373.

[38] Davenport AP, Wright DJ. Octopamine distribution in the larvae and adults of two species of moth, *Spodoptera littoralis* and *Manduca sexta. Journal of Insect Physiology*. 1986;32:987-993.

[39] Evans PD. Octopamine. In: Kerkut GA, Gilbert L, eds. *Comparative Insect Physiology, Biochemistry and Pharmacology*. Toronto: Pergamon Press. 1985;11:499-530.

[40] Sparks TC, Geng C. Analysis of the biogenic amines in the central nervous system of the tobacco hornworm by high-performance liquid chromatography with 16 sensor electrochemical detection. *Analytical Biochemistry*. 1992;205:319-325.

[41] Linn CE, Poole DR, Roelofs WL. Studies on biogenic amines and their metabolites in nervous tissue and hemolymph of adult male cabbage looper moths – I. Quantitation of photoperiod changes. *Comparative Biochemistry and Physiology*. 1994;108C:73-85.

[42] Downer RGH, Hiripi L, Juhos S. Characterization of the tyraminergic system in the central nervous system of the locust, *Locusta migratoria migratoides. Journal of Neurochemistry*. 1993;18:1245-1248.

[43] Woodring JP, McBride LA, Fields P. The role of octopamine in handling and exercise-induces hyperglycaemia and hyperlipaemia in *Acheta domesticus. Journal of Insect Physiology*. 1989;41:613-617

[44] Lehman HE, Klukas KA, Gilchrist LS, Mesce KA. Steroid regulation of octopamine expression during metamorphic development of the moth *Manduca sexta. Journal of Comparative Neurology*. 2000;242:283-296.

[45] Livingstone MS, Temple BL. Genetic dissection of monoamine neurotransmitter synthesis in *Drosophila. Nature*. 1983;303:67-70.

[46] McClung C, Hirsh J. The trace amine tyramine is essential for sensitization to cocaine in *Drosophila. Current Biology*. 1999;9:853-857.

[47] Cole SH, Carney GE, McClung CA, Willard SS, Taylor BJ, Hirsh J. Two functional but noncomplementing *Drosophila* tyrosine decarboxylase genes: distinct roles for neural

tyramine and octopamine in female fertility. *Journal of Biological Chemistry.* 2005;280:14948-14955.

[48] Wallace BG. The biosynthesis of octopamine – characterization of lobster tyramine ß-hydroxylase. *Journal of Neurochemistry.* 1976;26:761-770.

[49] Lehman HK, Murgiuc CM, Hildebrand JG. Characterization and development regulation of tyramine-β-hydroxylase in the CNS of the moth *Manduca sexta. Insect Biochemisty and Molecular Biology.* 2000;30:377-386.

[50] Monastirioti M, Linn CE, White K. Characterization of *Drosophila* tyramine beta-hydroxylase gene and isolation of mutant flies lacking octopamine. *Journal of Neuroscience.* 1996;16:3900-3911.

[51] Scholz H. Influence of the biogenic amine tyramine on ethanol-induced behaviors in *Drosophila. Journal of Neurobiology.* 2005;63:199-214.

[52] Fox LE, Soll DR, Wu CF. Coordination and modulation of locomotion pattern generators in *Drosophila* larvae: effects of altered biogenic amine levels by the tyramine beta-hydroxlyase mutation. *Journal of Neuroscience.* 2006;25:1486-1498.

[53] Brembs B, Christiansen F, Pflüger HJ, Duch C. Flight initiation and maintenance deficits in flies with genetically altered biogenic amine levels. *Journal of Neuroscience.* 2007;27:11122-11131.

[54] Hardie SL, Zhang JX, Hirsh J. Trace amines differentially regulate adult locomotor activity, cocaine sensitivity, and female fertility in *Drosophila melanogaster. Developmental Neurobiology.* 2007;67:1396-1405.

[55] Zhou C, Rao Y, Rao Y. A subset of octopaminergic neurons are important for *Drosophila* aggression. *Nature Neuroscience.* 2008;9:1059-1067.

[56] Zhu MY, Juorio AV. Aromatic amino acid decarboxylase: biological characterization and functional role. *General Pharmacology.* 1995;26:681-696.

[57] Siegel G, Agranoff B, Albers R, Fisher S, Uhler M. *Basic neurochemistry: molecular, cellular, and medical aspects.* 6th ed. Philadelphila: Lippencott- Raven Publishers;1999.

[58] Restifo LL, White K. Molecular and genetic approaches to neurotransmitter and neuromodulator systems in *Drosophila. Advances in Insect Physiology.* 1990;22:116-219.

[59] Hare EE, Loer CM. Function and evolution of the serotonin-synthetic bas-1 gene and other aromatic amino acid decarboxylase genes in *Caenorhabditis. BCM Evolutionary Biology.* 2004;4:24.

[60] Nagy L, Hiripi L. Role of tyrosine, DOPA and decarboxylase enzymes in the synthesis of monoamines in the brain of the locust. *Neurochemistry International.* 2002;41:9-16.

[61] Sandmeier E, Hale T, Christen P. Multiple evolutionary origin of pyridoxal-5'-phosphate-dependent amino acid decarboxylases. *European Journal of Biochemistry.* 1994;221:997-1002.

[62] Facchini PJ, Huber-Allanach KL, Tari LW. Plant aromatic L-amino acid decarboxylases: evolution, biochemistry, regulation, and metabolic engineering applications. *Phytochemistry.* 2000;54:121-138.

[63] Southan C, Kruse LI. Sequence similarity between DBH and PAM: evidence for a conserved catalytic domain. *FEBS Letters.* 1989;255:116-120.

[64] Evans JP, Ahn K, Klinman JP. Evidence that dioxygen and substrate activation are tightly coupled in DβM: Implications for reactive oxygen species. *Journal of Biological Chemistry.* 2003;278:49691-49698.

[65] Klinman, J. The copper enzyme family of dopamine beta-monooxygenase and peptidylglycine alpha-hydroxylating monooxygenase: resolving the chemical pathway for substrate hydroxylation. *Journal of Biological Chemistry.* 2006;281:3013-3016.

[66] Robertson JG, Adams, GW, Medzihradszky KF, Burlingame, AL, Villafranca JJ. Complete assignment of disulfide bonds in bovine dopamine beta-hydroxylase. *Biochemistry.* 1994; 33:11563-11575.

[67] Prigge ST, Kolhekar AS, Eipper, BA, Mains RE, Amzel LM. Substrate-mediated electron transfer in peptidylglycine alpha-hydroxylating monooxygenase. *Nature Structural Biology* 1999;6:976–983.

[68] Hess C, McGuirl M, Klinman J. Mechanism of the insect enzyme, tyramine B-monooxygenase, reveals differences from the mammalian enzyme, dopamine B-monooxygenase. *Journal of Biological Chemistry.* 2008;283:3042-3049.

[69] Annis D, Miras-Portugal MT, Mandel P. Bovine adrenalmedullary dopamine b-hydroxylase: purification by affinity chromatography, kinetic studies and presence of essential histidyl residues. *Biochimica et Biophysica Acta.* 1973;327:313-327.

[70] Saxena A, Hensley P, Osborne JC, Fleming P. The pH dependent subunit dissociation and catalytic activity of bovine dopamine b-hydroxylase. *Journal of Biological Chemistry.* 1985;260:3386-3392.

[71] Dhawan S, Hensley P, Osborne JC., Fleming P. Adenosine 59-diphosphate-dependent subunit dissociation of bovine dopamine b-hydroxylase. *Journal of Biological Chemistry.* 1986;261:7680-7684.

[72] Stewart LC, Klinman JP. Dopamine beta-hydroxylase of adrenal chromaffin granules: structure and function. *Annual Review of Biochemistry.* 1988;57:551-592.

[73] Wimalasena K, Wimalasena DS. Continuous spectrophotometric assays for dopamine b-monooxygenase based on two novel electron donors: *N,N*-dimethyl-1,4-phenylenediamine and 2-aminoascorbic acid. *Analytical Biochemistry.* 1991;197:353-361.

[74] Fortin D, Coulon JF, Roberge AG. Comparative study of biochemical parameters and kinetic properties of dopamine b- hydroxylase activity from cat and rat adrenals. *Comparative Biochemistry and Physiology.* 1993;104 B:567-575.

[75] Hoyle G, Colquhoun W, Williams M. Fine structure of an octopaminergic neuron and its terminals. *Journal of Neurobiology.* 1980;11:103-126.

[76] Smith D. The organization and innervation of the lumninescent organ in a firefly, *Photuris pennsylvanica* (Coleoptera*). Journal of Cell Biology.* 1963;16:323-359.

[77] Lee HM, Wyse GA. Immunocytochemical localization of octopamine in the central nervous system of Limulus polyphemus: A light and electron microscopic study. *Journal of Comparative Neurology.* 1991;307:683-694.

[78] Bendtsen JD, Nielsen H, von Heijne G, Brunak S. Improved prediction of signal peptides: SignalP 3.0. *Journal of Molecular Biology.* 2004;340:783-795.

[79] Chen NH, Reith ME, Quick MW. Synaptic uptake and beyond: the sodium- and chloride-dependent neurotransmitter transporter family SLC6. *Pflügers Archiv European Journal of Physiology.* 2004;447:519-531.

[80] Boudko DY, Kohn AB, Meleshkevitch EA, Dasher MK, Seron TJ, Stevens BR, Harvey WR. Ancestry and progeny of nutrient amino acid transporters. *Proceedings of the National Academy of Sciences of the United States of America.* 2005;102:1360-1365.

[81] Thimgan MS, Berg JS, Stuart AE. Comparative sequence analysis and tissue localization of members of the SLC6 family of transporters in adult *Drosophila melanogaster. Journal of Experimental Biology,* 2006;209:3383-3404.

[82] Yamashita A, Singh SK, Kawate T, Jin Y, Gouaux E. Crystal structure of a bacterial homologue of Na+/Cl--dependent neurotransmitter transporters. *Nature.* 2005;437:215-223.

[83] Meleshkevitch EA, Assis-Nascimento P, Popova LB, Miller MM, Kohn AB, Phung EN, Mandal A, Harvey WR, Boudko DY. Molecular characterization of the first aromatic nutrient transporter from the sodium neurotransmitter symporter family. *The Journal of Experimental Biology.* 2006;209:3183-3198.

[84] Miller MM, Popova LB, Meleshkevitch EA, Tranc PV, Boudko DY. The invertebrate B0 system transporter, *D. melanogaster* NAT1, has unique d-amino acid affinity and mediates gut and brain functions. *Insect Biochemistry and Molecular Biology.* 2008;38:923-931.

[85] Fitzpatrick PF. Tetrahydropterin-dependent amino acid hydroxylases. *Annual Review of Biochemistry.* 1999;68:355-381.

[86] Neckameyer WS, White K. A single locus encodes both phenylalanine hydroxylase and tryptophan hydroxylase activities in *Drosophila. Journal of Biological Chemistry.* 1992;267:4199-4206.

[87] Coleman CM, Neckameyer WS. Substrate regulation of serotonin and dopamine synthesis in *Drosophila. Invertebrate Neuroscience.* 2004;5:85-96.

[88] Eiden LE. The vesicular neurotransmitter transporters: current perspectives and future prospects. *Journal of the Federation of American Societies for Experimental Biology.* 2000;14:2396-2400.

[89] Liu Y, Krantz DE, Waites C, Edwards RH. Membrane trafficking of neurotransmitter transporter in the regulation of synaptic transmission. *Trends in Cell Biology.* 1999;9:356-363.

[90] Peter D, Liu Y, Sternini C, de Giorgio R, Brecha N, Edwards RH. Differential expression of two vesicular monoamine transporters. *Journal of Neuroscience.* 1995;15:6179-6188.

[91] Greer CL, Grygoruk A, Patton DE, Ley B, Romero-Calderon R, Chang HY, Houshyar R, Bainton RJ, DiAntonio A, Krantz DE. A splice variant of the *Drosophila* vesicular monoamine transporter contains a conserved trafficking domain and functions in the storage of dopamine, serotonin, and octopamine. *Journal of Neurobiology.* 2005;64:239-258.

[92] Schuldiner S, Shirvan A, Linial M. Vesicular neurotransmitter transporters: from bacteria to humans. *Physiological Reviews.* 1995;75:369-392.

[93] Romero-Calderón R, Uhlenbrock G, Borycz J, Simon AF, Grygoruk A, A glial variant of the vesicular monoamine transporter is required to store histamine in the *Drosophila* visual system. *PLoS Genetics.* 2008;4:e1000245.
[94] Spector R. Vitamin B6 transport in the central nervous system. In vivo studies. *Journal of Neurochemistry.* 1978;30:881-887.
[95] Spector R. B6 transport in the central nervous system. In vitro studies. *Journal of Neurochemistry.* 1978;30:889-897.
[96] Spector R, Johanson CE. Vitamin transport and homeostasis in mammalian brain: focus on vitamins B and E. *Journal of Neurochemistry.* 2007;103:425-438.
[97] Huang S-H, Shi R-J, Zhang Y-J, Wang Z, Huang L-Q. Cloning and characterization of a pyridoxine 5'-phosphate oxidase from silkworm *Bombyx mori. Insect Molecular Biology.* 2009;18:365-371.
[98] Kratzing C, Kelly J, Oelrichs B. Ascorbic acid in neural tissue. *Journal of Neurochemistry.* 1982;44:1623-1624.
[99] Oke AF, May L, Adams RN. Ascorbic acid distribution patterns in human brain. A comparison with nonhuman mammalian species. *Annals of the New York Academy of Sciences.* 1987;498:1-12.
[100] Faaland CA, Race JE, Ricken G, Warner FJ, Williams WJ, Holtzman EJ. Molecular characterization of two novel transporters from human and mouse kidney and from LLC-PK1 cells reveals a novel conserved family that is homologous to bacterial and *Aspergillus* nucleobase transporters. *Biochimica et Biophysica Acta.* 1998;1442:353-360.
[101] Daruwala R, Song J, Koh WS, Rumsey SC, Levine M. Cloning and functional characterization of the human sodium-dependent vitamin C transporters hSVCT1 and hSVCT2. *FEBS Letters.* 1999;460:480-484.
[102] Tsukaguchi H, Tokui T, Mackenzie B, Berger UV, Chen X, Wang Y, Brubaker RF, Hediger MA. A family of mammalian Na+-dependent L-ascorbic acid transporters. *Nature.* 1999;399:70-75.
[103] Takanaga H, Mackenzie B, Hediger MA. Sodium-dependent ascorbic acid transporter family SLC23. *Pflügers Archiv European Journal of Physiology.* 2004;447:677-682.
[104] Savini I, Rossi A, Pierro C, et al. SVCT1 and SVCT2: key proteins for vitamin C uptake. *Amino Acids.* 2008;34:347-355.
[105] Rivas CI, Zúñiga FA, Salas-Burgos A, Mardones L, Ormazabal V, Vera JC. Vitamin C transporters. *Journal of Physiology and Biochemistry.* 2008;64:357-375.
[106] Rajana D, Huang W, Dutta B, Devoe LD, Leibach FH, Ganapathy V, Prasad PD. Human placental sodium-dependent Vitamin C transporter (SVCT2): molecular cloning and transport function. *Biochemical and Biophysical Research Communications.* 1999;262:762-768.
[107] Bianchi J, Rose RC. Glucose-independent transport of dehydroascorbic acid in human erythrocytes. *Proceedings of the Society for Experimental Biology and Medicine.* 1986;181:333-337.
[108] Vera JC, Rivas CI, Fischbarg J, Golde DW. Mammalian facilitative hexose transporters mediate the transport of dehydroascorbic acid. *Nature.* 1993;364:79-82.

[109] Mueckler M. Facilitative glucose transporters. *European Journal of Biochemistry.* 1994;219:713-725.

[110] Rumsey SC, Kwon O, Xu GW, Burant CF, Simpson I, Levine M. Glucose transporter isoforms GLUT1 and GLUT3 transport dehydroascorbic acid. *Journal of Biological Chemistry.* 1997;272:18982-18989

[111] Rumsey SC, Daruwala R, Al-Hasani H, Zarnowski MJ, Simpson IA, Levine M. Dehydroascorbic acid transport by GLUT4 in Xenopus oocytes and isolated rat adipocytes. *Journal of Biological Chemistry.* 2000;275:28246-28253.

[112] Caprile T, Salazar K, Astuya A, Cisternas P, Silva-Alvarez C, Montecinos H, Millán C, de Los Angeles García M, Nualart F. The Na+-dependent L-ascorbic acid transporter SVCT2 expressed in brainstem cells, neurons, and neuroblastoma cells is inhibited by flavonoids. *Journal of Neurochemistry.* 2009;108:563-577.

[113] Nualart FJ, Rivas CI, Montecinos VP, Godoy AS, Guaiquil VH, Golde DW, Vera JC. Recycling of Vitamin C by a bystander effect. *Journal of Biological Chemistry.* 2003;278:10128-10133.

[114] Linster CL, Van Schaftingen E. Vitamin C. Biosynthesis, recycling and degradation in mammals. *FEBS Journal.* 2007;274:1-22.

[115] Njus D, Kelley PM. The secretory-vesicle ascorbate-regenerating system: a chain of concerted H+/e(-) transfer reactions. *Biochimica et Biophysica Acta.* 1993;1144:235-248.

[116] Bánhegyi G, Marcolong P, Puskás F, Fulceri R, Mandl J, and Benedetti A. Dehydroascorbate and ascorbate transport in rat liver microsomal vesicles. *Journal of Biological Chemistry.* 1998;273:2758-2762.

[117] Njus D, Knoth J, Cook C, Kelly PM. Electron transfer across the chromaffin granule membrane. *Journal of Biological Chemistry.* 1983;258:27-30.

[118] Perin MS, Fried VA, Slaughter CA, Südhof TC. The structure of cytochrome b561, a secretory vesicle-specific electron transport protein. *European Molecular Biology Organization Journal.* 1988;7:2697-2703.

[119] Bashtovyy D, Bérczi A, Asard H, and Páli T. Structure prediction of the di-heme cytochrome b561 protein family. *Protoplasma.* 2003;221:31-40.

[120] La Fontaine S, Mercer J. Trafficking of the Copper-ATPases, ATP7A and ATP7B: role in copper homeostasis. *Archives of Biochemistry and Biophysics.* 2007;463:149-167.

[121] Madsen E, Gitlin JD. Copper and iron disorders of the brain. *Annual Review of Neuroscience.* 2007;30:317-337.

[122] Prohaska JR, Gybina AA. Intracellular copper transport in mammals. *Journal of Nutrition.* 2004;134:1003-1006.

[123] Changela A, Chen K, Xue Y, Holschen J, Outten CE, O'Halloran TV, Mondragon A. Molecular basis of metal-ion selectivity and zeptomolar sensitivity by CueR. *Science.* 2003; 301:1383- 1387.

[124] Finney LA, O'Halloran TV. Transition metal speciation in the cell: insights from the chemistry of metal ion receptors. *Science.* 2003;300:931-936.

[125] Dancis A, Haile D, Yuan DS, Klausner RD. The *Saccharomyces cerevisia* copper transport protein (Ctr1p). Biochemical characterization, regulation by copper, and

physiologic role in copper uptake. *Journal of Biological Chemistry.* 1994;269:25660-25667.

[126] Zhou B. Gitschier J. hCTR1: a human gene for copper uptake identified by complementation in yeast. *Proceedings of the National Academy of Sciences of the United States of America.* 1997;94:7481-7486.

[127] Lee J, Peña MMO, Nose Y, Thiele DJ. Biochemical characterization of the human copper transporter Ctr1. *Journal of Biological Chemistry.* 2002;277:4380-4387.

[128] Eisses JF, Kaplan JH. Molecular characterization of hCTR1, the human copper uptake protein. *Journal of Biological Chemistry.* 2002;277:29162-29171.

[129] De Feo CJ, Aller SG, Siluvai GS, Blackburn NJ, Unger VM. Three-dimensional structure of the human copper transporter hCTR1. *Proceedings of the National Academy of Sciences of the United States of America.* 2009;106:4237-4242.

[130] Amaravadi R, Glerum DM, Tzagoloff A. Isolation of a cDNA encoding the human homolog of COX17, a yeast gene essential for mitochondrial copper recruitment. *Human Genetics.* 1997;99:329-333.

[131] Casareno RLB, Waggoner D, Gitlin JD. The copper chaperone CCS directly interacts with copper/zinc superoxide dismutase. *Journal of Biological Chemistry.* 1998;273:23625-23628.

[132] Larin D, Mekios C, Das K, Ross B , Yang A-S , Gilliam TC. Characterization of the interaction between the Wilson and Menkes disease proteins and the cytoplasmic copper chaperone, HAH1p. *Journal of Biological Chemistry.* 1999;274:28497-28504.

[133] Yamaguchi Y, Heiny ME, Suzuki M, Gitlin JD. Biochemical characterization and intracellular localization of the Menkes disease protein. *Proceedings of the National Academy of Sciences of the United States of America.* 1996;93:14030-14035.

[134] Barnes N, Tsivkovskii R, Tsivkovskaia N, Lutsenko S. The copper-transporting ATPases, Menkes and Wilson disease proteins, have distinct roles in adult and developing cerebellum. *Journal of Biological Chemistry.* 2005;280:9640-9645.

[135] El Meskini R, Culotta VC, Mains RE, Eipper BA. Supplying copper to the cuproenzyme peptidylglycine alpha -amidating monooxygenase. *Journal of Biological Chemistry.* 2003;278:12278-12284.

[136] Petris MJ, Mercer JF, Culvenor JG, Lockhart P, Gleeson PA, Camakaris J. Ligand-regulated transport of the Menkes copper P-type ATPase efflux pump from the Golgi apparatus to the plasma membrane: a novel mechanism of regulated trafficking. *European Molecular Biology Organization Journal.* 1996;15:6084-6095.

[137] Linz R, Lutsenko S. Copper-transporting ATPases ATP7A and ATP7B: cousins, not twins. *Journal of Bioenergetics and Biomembranes.* 2007;39:403-407.

[138] Prohaska JR, Brokate B. Copper deficiency alters rat dopamine beta-monooxygenase mRNA and activity. *Journal of Nutrition.* 1999;129:2147-2153.

[139] Siegler M, Manley PE, Thompson KJ. Sulphide silber staining for endogenous heavy metals reveals subsets of dorsal unpaired median (DUM) neurons in insects. *Journal of Experimental Biology.* 1991;157:565-571.

[140] Zhou H, Cadigan KM, Thiele DJ. A copper-regulated transporter required for copper acquisition, pigmentation, and specific stages of development in *Drosophila* melanogaster. *Journal of Biological Chemistry.* 2003;278:48210-48218.

[141] Southon A, Farlow A, Norgate M, Burke R, Camakaris J. Malvolio is a copper transporter in *Drosophila melanogaster. Journal of Experimental Biology*. 2008; 211:709-716.

[142] Folwell JL, Baron CH, Shepherd D. Immunolocalisation of the *D. melanogaster Nramp* homologue *Malvolio* to gut and Malpighian tubules provides evidence that *Malvolio* and *Nramp2* are orthologous. *Journal of Experimental Biology*. 2006; 209:1988-1995.

[143] Southon A, Burke R, Norgate M, Batterham P, Camakaris J. Copper homoeostasis in *Drosophila melanogaster* S2 cells. *Biochemical Journal*. 2004;383:303-309.

[144] Norgate M, Southon A, Zou S, Zhan M, Sun Y, Batterham P, Camakaris J. Copper homeostasis gene discovery in *Drosophila melanogaster. Biometals*. 2007;20:683-697.

In: Biogenic Amines
Editors: T. Farooqui, A. A. Farooqui

ISBN: 978-1-60876-625-3

Chapter XVII

Perspective and Future Developments on Biogenic Amines in the CNS

Tahira Farooqui* and Akhlaq A. Farooqui
The Ohio State University, Columbus, Ohio 43210, USA.

Abstract

Biogenic amines and their metabolites are important signaling molecules that regulate neural cell functions in the peripheral nervous system and central nervous system (CNS) in vertebrates and invertebrates. Alterations in biogenic amines levels along with excitotoxicity, elevation in ROS production, induction of neuroinflammation and loss of synapses are observed in neurodegenerative process in Alzheimer disease, Parkinson disease, and Huntington disease. The neuronal population that degenerates in above neurodegenerative diseases modulates movements, learning and memory, processing sensory information, and decision making processes. Although, alterations in biogenic amine levels may not be the primary event in the pathogenesis of neurodegenerative diseases, but biogenic amine-mediated oxidative stress and synaptic loss may contribute to dementia and other behavioral problems. Thus, normalization of biogenic amine levels through medication may correct some abnormal behavioral and psychological symptoms of neurodegenerative diseases. Neuropsychiatric disorders involve abnormalities in cerebral cortex and limbic system, which shows changes in biogenic amine levels in a single microcircuit. The "biogenic amine hypothesis of depression and schizophrenia" states that depression is due to alteration functional activity of one or more brain biogenic amines (primarily norepinephrine, dopamine, and serotonin). Thus, drugs that inhibit the re-uptake and/or metabolism of these biogenic amines can be used as drugs for the treatment of depression and schizophrenia. Knowledge about pharmacokinetics and pharmacodynamics of drugs is important because drug dose-response is based on the principles of pharmacokinetics and

* Correspondence concerning this article should be addressed to: Tahira Farooqui, Department of Entomology/ Center for Molecular Neurobiology, The Ohio State University, Columbus, OH 43210-1220, Telephone: (614) 783-4369, Email: farooqui.2@osu.edu.

pharmacodynamics. The expanded use of pharmacokinetics-pharmacodynamic modelling for drug development may also be highly beneficial in improving the current state of therapeutics for neurodegenerative and neuropsychiatric diseases. In recent years, researchers are empowered with techniques of lipidomics, proteomics, and genomics. These procedures can detect minute amounts of biogenic amines and second messengers generated by biogenic amine in biological fluids. Establishment of automatic systems including databases and accurate analyses of biogenic amines and second messengers generated from enzymic and non-enzymic metabolism of neuronal membrane components will facilitate the identification of key biomarkers associated with neurodegenerative and neuropsychiatric diseases.

Introduction

Biogenic amines are biologically molecules that act as important messengers in the peripheral nervous system (PNS) and central nervous system (CNS) of both vertebrates and invertebrates. The five established biogenic amine neurotransmitters in vertebrates fall into three classes: (1) the catecholamines, such as norepinephrine, epinephrine and dopamine, (2) the indoleamine, such as serotonin, and (3) the imidazoleamine, such as histamine [1].

In invertebrates, norepinephrine and epinephrine are replaced by two phenolamines: tyramine and octopamine, which act as *bona fide* neurotransmitters/ neuromodulators/ neurohormones mediating diverse complex behaviors [1-2]. Tyramine and octopamine are present as trace amines in the mammalian brain. Trace amines, in general, act as neuromodulators rather than neurotransmitters in mammalian systems [3-5]. Therefore, they may modify the effect of primary neurotransmitters at a synapse, regulating synaptic plasticity [6]. Biogenic amines regulate many functions in the brain, including endocrine secretion, cognitive function, aggression, sleep and waking, emotional states, motivation, reward circuitry, decision making, and learning and memory [1, 2, 6].

In humans, the pathogenesis of various neurological diseases, such as Alzheimer disease (AD), Parkinson disease (PD), Huntington disease (HD), schizophrenia, or attention deficit hyperactivity disorder (ADHD), has been linked to abnormal levels of biogenic amines in the CNS. Therefore, development of specific drugs that can optimally and selectively influence the synthesis and release of specific biogenic amines, interaction with specific receptor subtypes, inhibition of uptake and reduction in the catabolic processes, with high efficacy and less side effects may be useful in safer treatment of neurodegenerative, neurological and psychiatric disorders.

Pathways for Synthesis and Degradation of Biogenic Amines

In brain catecholamines and phenolamines are synthesized from a non-essential amino acid, tyrosine. In catecholamines synthesis, tyrosine is first taken up from the blood stream to the CNS by active transport system. Tyrosine is hydroxylated and decarboxylated to dopamine by tyrosine hydroxylase (TH) and DOPA decarboxylase (DDC), respectively. The sequence stops at dopamine in dopamine-releasing neurons, but continues on to norepinephrine by dopamine β-hydroxylase (DBH) in norepinephrine-releasing neurons. Norepinephrine is converted to epinephrine by phenylethanolamine *N*-methyl transferase (PNMT) in a few nerve cells in the brain stem, and in epinephrine-releasing cells of the adrenal medulla. TH is the rate-limiting enzyme in the catecholamine biosynthesis. It requires tetrahydrobiopterin (BH_4) as a co-factor and is sensitive to oxygen concentration [7-8]. α-Methyl-*p*-tyrosine, a competitive inhibitor of TH, consequently inhibits the synthesis of all catecholamines, thus this knowledge is applied therapeutically to treat catecholamine excess.

In vertebrates, tyramine and octopamine are also synthesized from tyrosine in sequential reactions catalyzed by aromatic amino acid decarboxylase (AAAD) and dopamine β-hydroxylase (DBH), respectively. However, in invertebrates, synthesis of tyramine and octopamine from tyrosine occurs in reactions catalyzed by tyramine decarboxylase (TDC; a member of the aromatic amino acid family), and tyramine β-hydroxylase (TBH), respectively. Both DBH and TBH have similar properties [9]. Both enzymes recognize either tyramine or dopamine as substrates (but with different affinities), require oxygen and ascorbic acid as a cofactor, and bind with Cu^{2+}, suggesting that they may be evolutionarily related [9]. However, TDC has not been found in fruit fly *Drosophila* and in higher animals. Serotonin (5-hydroxytryptamine; 5-HT) is synthesized from an essential amino acid, L-tryptophan, in two reactions catalyzed by tryptophan hydroxylase (TPH) and monoamine oxidase (MAO). TPH uses the same pteridine cofactor as TH and also uses molecular oxygen. It catalyzes the hydroxylation of tryptophan into 5-hydroxytryptophan (5-HTP). Similar to TH, TPH is allosterically controlled through a phosphorylating action by Ca^{2+}-dependent or cAMP-dependent kinases. In the second step, 5-HTP is decarboxylated to 5-HT by 5-HTP decarboxylase. In brain, after conversion of 5-HT to 5-hydroxyindoleacetaldehyde by monoamine oxidase (MAO), it is dehydrogenated to 5-hydroxyindoleacetic acid (5-HIAA). In cerebrospinal fluid, the level of 5-HIAA is often used as an index of the rate of serotonin turnover. AAAD catalyzes the decarboxylation of L-dopa to dopamine and 5-HTP to serotonin. AAAD is required for the formation of catecholamines, indolamines and trace amines. Deficiency in AAAD impairs motor, cognitive, and several neurological behaviors in patients. Histamine is synthesized from an essential amino acid L-histidine in a decarboxylation reaction exclusively catalyzed by histidine decarboxylase (HDC), which like most decarboxylases, requires pyridoxal phosphate. Histamine breakdown involves its *N*-methylation by histamine *N*-methyltransferase (HMT) to 3-methylhistamine (3-MT), followed by its oxidative deamination by MAO to 3-methylimidazole-acetic acid. The key roles of HMT and MAO in histamine breakdown can be shown by rise in histamine or 3-MT after selective enzyme inhibition.

TRANSPORT SYSTEMS OF BIOGENIC AMINES

In neurons biogenic amines are transported through two systems. The first system transports the biogenic amines from the cytoplasm to the storage organelle, or 'synaptic vesicle' and the second transports biogenic amine from the extracellular space into the cytoplasm. Both systems transport biogenic amines uphill, against a concentration gradient, and are therefore coupled to the input of metabolic energy (ATP) [10,11]. Two different isoforms of vesicular monoamine transporters (VMATs) are involved in controlling the presynaptic vesicular packaging of biogenic amines such as serotonin, dopamine, norepinephrine, epinephrine, octopamine, tyramine and histamine in storage vesicles: VMAT1 and VMAT2 [12,13]. It was initially believed that VMAT1 is expressed exclusively in amine-containing endocrine/paracrine cells in the peripheral organs and PNS, whereas VMAT2 is predominantly found in CNS [12-14]. However, other studies have reported VMAT1 expression in the vertebrate brain [15]. The transport of positively charged amines utilizes an electrochemical gradient across the vesicular membrane established by proton pumping into the vesicle involving vacuolar ATPase and transporter.

Acetylcholine, a major excitatory neurotransmitter and a biogenic amine, is synthesized in the cytoplasm by choline acetyltransferase. It is transported by the vesicular acetylcholine transporter (VAChT) into synaptic vesicles of cholinergic neurons in the brain. Both VMAT and VAChT exchange one amine per two translocated protons with very similar initial velocity kinetics and pH dependencies [16]. VAChT transports acetylcholine and its synthetic analogs, but does not significantly transport choline and monoamines. However, VMAT transports monoamines and their synthetic analogs. The vesicular gradients of acetylcholine are much smaller than the driving force for uptake and vesicular gradients of monoamines, suggesting the regulatory mechanism of acetylcholine storage not found in monoamine storage [16].

VAChT gene (*VAChT*) is expressed in all known major cholinergic neurons in the PNS and CNS. Both VAChT and choline acetyltransferase are encoded by a single genetic locus, suggesting both genes are coregualted. *VAChT* is a specific marker for cholinergic neurons for studying cholinergic transmission in AD and other nervous system disorders [17-19]. Vesicular monoamine transporter1 gene (*VMAT1*), also known as SLC18A1, maps to bipolar disorder and Schizophrenia susceptibility locus, therefore, gene has been postulated to play a role in the etiology of these neuropsychiatric disorders [20-22]. The vesicular monoamine transporter 2 gene (*VMAT2*), also known as SLC18A2, controls loading of biogenic amines including catecholamines and indolamines into synaptic vesicles for exocytotic release [23]. Catecholamines when not transported to synaptic vesicles can be auto-oxidized in the cytosol and produce oxidative stress to the cell. For example, in dopaminergic neurons, VMAT2 is also a target for amphetamine and its analog (methamphetamine). The disruption of dopamine storage by methamphetamine in the striatum causes massive loss of dopaminergic neurons in mutant strain of mice lacking VMAT2, suggesting VMAT2 may be a regulator of methamphetamine-induced neurodegeneration [24]. Thus, vesicular capacity to store dopamine in neurons can determine the level of oxidative stress and cell death after methamphetamine abuse and idiopathic PD because in both cases one deals with mishandling of dopamine, degeneration of dopaminergic neurons, and use of dopamine rich regions in the brain. The storage of acetylcholine in synaptic vesicles plays a key role in maintaining cholinergic function.

The uptake of biogenic amines through monoamine transporters, located in the plasma membrane, following their release into the synaptic cleft from the synaptic vessels is the principal process for terminating monoaminergic transmission by clearing biogenic amines from the synaptic cleft [12,25,26]. Biogenic amine transporters are coupled with the fluxes of

Na^+, Cl^-, and in some systems K^+, to the reuptake of biogenic from the synaptic cleft to the nerve terminal. Dopamine transporter (DAT) requires two extracellular Na^+ and one Cl^- to cotransport with each dopamine molecule. Reuptake of monoamines into the nerve terminals and then into the synaptic vesicles allows recycling of same molecules. Biogenic amines interact with presynaptic and postsynaptic receptors after discharged from vesicles into the synaptic cleft by exocytosis following an action potential. Biogenic amines uptake transporters are targeted by major classes of antidepressant, psychostimulant and antihypertensive drugs [12,26,27]. DAT is the pharmacological target for widely abused psychostimulants such as cocaine, amphetamine and methamphetamine. Several classes of neurotoxins bind with these transporters and selectively poison specific neuronal groups [28-31]. Catecholaminergic toxins such as 6-hydroxydopamine (6-OH dopamine) and 1-methyl-4-phenyl-piperidinium ion (MPP^+) act as a substrate for DAT [32], and therefore have been used to induce selective dopaminergic cell death in various model systems of PD. One possible explanation for the selective loss of dopaminergic neurons in PD may be that patients have been exposed to such environmental dopaminergic neurotoxins, causing selective death of neurons. Collective evidence supports the view that biogenic amines uptake transporters represent a primary target for therapeutic intervention in the treatment of various psychiatric disorders, such as anxiety, depression, PD, schizophrenia as well as drug abuse.

Determination of Biogenic Amines

Several methods have been developed for the quantitation of biogenic amines. The most widely used method in the routine clinical laboratory for the measurement of catecholamines and their metabolites in biological fluids is HPLC. Original methods relied on fluorescence detection to gain analytical sensitivity, but the availability of electrochemical detection provides superior specificity and this mode of detection is the most common today [33]. Biogenic amines, including catecholamines and their metabolites, demonstrate different chemical stabilities in different biological fluids, stored under various conditions. A drawback with urine collections is the requirement to add concentrated acid to the collection container in order to stabilize catecholamines from oxidation, which introduces the possibility of deconjugation, artefactually elevating free catecholamine concentrations [34]. Other factors including recreational stimulants, such as caffeine and nicotine, and diet can also alter estimation results [34]. The combination of HPLC with tandem mass spectrometry (HPLC–MS/MS) is becoming more popular in clinical chemistry routine laboratories, particularly for detecting low levels of catecholamines and metabolites in tissue and biological fluids because it provides minimum interference with drugs and drug metabolites [35]. However, implementation of HPLC–MS/MS assays still requires substantial expertise and know-how.

Drug Dose and Response Relationship

Biogenic amine receptors agonists and antagonists are being used as drugs for treating neurodegenerative and neuropsychiatric diseases. Regardless of how these agonists work — through binding to the receptor as a partial agonist, full agonist and competitive or non-competitive antagonist, inhibiting or stimulating an enzyme reaction or chemically interacting with other drugs or endogenous neurotransmitters— its concentration in the vicinity of the receptor or at the site of action is important for determining the effect. Drug-drug interactions also exist with some of these agents (e.g. among antidepressants) and can usually be predicted from knowledge of their metabolism. However, when antidepressants are taken in overdose as the sole agents they are rarely capable of causing death but produce some common side effects, including seizures, nausea, vomiting, decreased level of consciousness, and tachycardia [36]. The serotonin syndrome, a potentially life-threatening adverse drug reaction, may occur with many drugs, following inadvertent interactions between drugs and overdose of drugs.

Drug dose-response is based on the principles of pharmacokinetics (absorption, distribution and excretion) and pharmacodynamics (effects of drugs on the body). Pharmacokinetics describes the drug concentration-time courses in body fluids resulting from administration of a certain drug dose, whereas pharmacodynamics describes the observed effect resulting from a certain drug concentration [37]. However, pharmacokinetics-pharmacodynamics links these two items providing a framework for modelling the time course of drug response [38]. It is quite complex to observe dose-response relationship clinically, because it depends on the duration time; weight, health, age, and stress on the individual. Drug-dose and response with the same dose may vary in a population and even in every individual. Indeed, a single individual may respond differently to the same drug at different times during the course of treatment. However, in carefully controlled *in vitro* systems, this relationship can be easily measured. Using dose response strategy, the required dose, frequency and the therapeutic index (ratio of the minimum toxic concentration to the median effective concentration; TD_{50}/ED_{50}) of a drug can be determined. The therapeutic index of a drug determines the "efficacy" and "safety" of a drug *in vivo*. The expanded use of pharmacokinetics-pharmacodynamic modelling for drug development is assumed to be highly beneficial in improving the current state of therapeutics for neurological and neuropsychiatric diseases.

Biogenic Amines in Neurodegenerative and Neuropsychiatric Diseases

Neurodegenerative and neuropsychiatric diseases are a group of neurological disorders with heterogeneous clinical and pathological expressions affecting specific subsets of neurons in specific functional anatomic regions of brain and spinal cord. Thus, in AD neurodegeneration mainly occurs in the nucleus basalis and hippocampal area, whereas in PD dopaminergic neurons in the substantia nigra undergo neurodegeneration. In HD neurodegeneration occurs in striatal medium spiny neurons and motor neurons located in the anterior part of spinal cord degenerate in ALS. Although, the exact cause and molecular mechanism of neurodegenerative diseases are fully understood, but it is becoming increasingly evident that multiple factors and mechanisms may contribute to the pathogenesis of disorders [39]. The most important risk factors for neurodegenerative diseases are old age, positive family history, unhealthy life style, endogenous factors, and exposure to toxic environment [39]. Other risk factors for neurodegenerative diseases include neuroinflammation, autoimmunity, cerebral blood flow, and blood-brain barrier dysfunction [40].

Despite the important differences in clinical manifestation, neurodegenerative diseases share some common characteristics such as their commencement late in life, the extensive neuronal death, and loss of synapses, elevation in ROS production, induction of neuroinflammation, excitotoxicity, alterations in biogenic amine levels, and the presence of cerebral deposits of misfolded protein aggregates [41,42]. These processes are closely associated with age-mediated decrease in cellular antioxidant defenses and resultant accumulation of lipid, protein and DNA damage in brain. These events play an important role in the etiology and pathogenesis of neurodegenerative diseases [40]. Multiple system dysfunction, including noradrenergic, serotonergic, dopaminergic, corticotropin-releasing factor, and somatostatin, may be responsible for alterations in biogenic amine levels, which in combination with glutamate and its receptors (through cross talk) may contribute to increase in oxidative stress and neuroinflammation in neurodegenerative diseases [43-46]. The neuronal population that degenerates in neurodegenerative diseases modulates movements, learning and memory, processing sensory information, and decision making processes [47]. It is well understood that biogenic amine dysfunction may not be the primary event in the pathogenesis of neurodegenerative diseases, but biogenic amine-mediated oxidative stress and synaptic loss may contribute to dementia and other behavioral problems. Thus, normalization of biogenic amine levels through medication may correct some abnormal behavioral and psychological symptoms of neurodegenerative diseases.

Neuropsychiatric disorders involve abnormalities in cerebral cortex and limbic system (thalamus, hypothalamus, hippocampus, and amygdale) [48]. Behavioral abnormalities are the hallmarks of many neuropsychiatric diseases, including schizophrenia, depression, and compulsive and bipolar disorders. Neuochemical studies indicate that in neuropsychiatric diseases several biogenic amine neurotransmitter systems (dopamine, serotonin and epinephrine) are simultaneously altered within a single microcircuit and each transmitter system shows circuitry changes in more than one region. Changes in microcircuits and neurotransmitters (synthesis and transport) may not only vary on a region-by-region basis, but also from one neuropsychiatric disease to another. Both macro- and microcircuitry within the specific brain system (such as limbic system) may serve as 'triggers' for the onset of neuropsychiatric condition [49,50]. It is also reported that alterations in cerebral blood flow and glucose utilization in the limbic system and prefrontal cortex of patients with major depression and other neuripsychiatric diseases [51,52]. The "biogenic amine hypothesis of depression and schizophrenia" postulates that depression is due to alteration functional activity of one or more brain biogenic amines (e.g. primarily norepinephrine, dopamine, and serotonin). Thus, compounds that inhibit the re-uptake and/or metabolism of these biogenic amines have been used to treat depression [53]. For example, the tricyclics antidepressants (TCA, named after their chemical structure composed of three rings of atoms), such as imipramine and amitriptyline, increase the levels of serotonin and norepinephrine in the brain by inhibiting their neuronal reuptake in the brain. TCAs antidepressants can be replaced by newer antidepressants in the selective serotonin reuptake inhibitor (SSRIs) class (such as citalopram and sertraline) or serotonin-norepinephrine reuptake inhibitors (SNRIs) class (such as mirtazapine and nefazodone). The overdose of TCA is especially fatal. The toxicity caused by TCAs antidepressants often becomes apparent after an overdose. In addition, the risk of drug interactions should always be considered when treating patients for depression. There is an urgent need for further research to determine relative efficacy of combination strategies (e.g TCAs antidepressants with MAO-Inhibitiors or TCAs with SSRIs) [54].

Although, the molecular mechanism associated with the pathogenesis of neurodegenerative and neuropsychiatric diseases remain unclear, but excitotoxicity, oxidative stress, inflammatory processes, and alterations in immunological parameters are closely associated with above neurological conditions [55]. Excitotoxicity, inflammation, and oxidative stress are interrelated processes that may induce neurodegeneration independently or synergistically. An upregulation of cross talk among excitotoxicity, oxidative stress, and neuroinflammation through neural membrane-derived lipid mediators may increase the vulnerability of neurons in acute neural trauma, neurodegenerative diseases, and neuropsychiatric disorders [55]. In neurodegenerative diseases, cross talk among excitotoxicity, oxidative stress, and neuroinflammation occurs at a slow rate. Oxygen, nutrients, and ATP are available to neurons so ionic homeostasis is maintained to a limited extent. These parameters result in a neurodegenerative process that takes many years to develop. In contrast, in nrurological condition, such as stroke, there is rapid depletion in ATP resulting in quick loss in ion homeostasis. These processes result in rapid neurodegeneration (in minutes) at the core of injury site through necrosis and slow neuronal death in the core surrounding area (hours to days) [55].

In neurodegenerative and neuropsychiatric diseases pain perception remains below the threshold of detection and immune system continues to attack brain tissue at the cellular and subcellular levels. This results in lingering of chronic inflammation for years causing continued insult to the brain tissue, ultimately reaching the threshold of detection many years after the onset of the neurodegenerative and neuropsychiatric diseases [56]. Many internal and external factors modulate the dynamic aspects of chronic inflammation and mild oxidative stress in neurodegenerative diseases [55]. The cross talk among excitotoxicity, oxidative stress, and neuroinflammation may be modulated by the diet, genetic, and environmental factors [57]. The onset of neurodegenerative and neuropsychiatric diseases is often subtle and usually occurs in mid to late life. Progression of neurodegenerative diseases depends not only on genetic and dietary factors, but also on environmental factors [58] leading to progressive cognitive and motor disabilities with devastating consequences. Perhaps in neurodegenerative and neuropsychiatric diseases, the intensity of glutamate and biogenic amine-mediated cross talk through second messengers generated during excitotoxicity, oxidative stress, and neuroinflammation may lead to unique manifestations that are characteristic features of AD, PD, depression and schizopherina. This cross talk may disturb neuronal lipid, protein, nucleic acid, and carbohydrate metabolism leading to irreversible neuronal damage [55,40]. It remains controversial whether excitotoxicity, oxidative stress, and neuroinflammation are the cause or consequence of neurodegeneration [59,60]. However, it is suggested that glutamate and biogenic amine-mediated signal transduction processes and cross talk may contribute to oxidative stress and neuroinflammation that ultimately kills neurons.

FUTURE PERSPECTIVES

Biogenic amines regulate mental functioning, blood pressure, body temperature, and many other bodily processes. Mental functioning, including mood, cognition, emotion, reward, learning and memory, is regulated by monoamines in the CNS. Therefore, any disturbance in monoaminergic activity in brain is associated with a variety of neurodegenerative and neuropsychiatric diseases. The physiological response of biogenic amines is mediated by G protein-coupled receptors (GPCRs) located in post-synaptic plasma membranes of target neurons. Biogenic amines-mediated signaling can be regulated by the interaction between different GPCRs forming homo- and hetero-oligomers, as well as by binding with receptor interacting proteins.

Studies on animal model systems indicate that the phenotype changes in neurodegenerative diseases caused by environmental factors, such as exposure to toxins and the pathogenetic mechanism of the disease are only loosely linked. Thus, neurotoxins (MPTP or 6-OH dopamine)-mediated animal and cell culture models indicate that injections of neurotoxins damage neurons in a non-specific manner [61-65]. Many investigators, who have assigned neurodegeneration and function to a particular brain area on the basis of neurotoxin-mediated injury, may not be justified in their contention. Alterations caused by MPTP or 6-OH dopamine are similar but not identical to neurochemical and pathological changes found in human PD and the extent and complexity of cross talk among various neurotransmitters in MPTP or 6-OH dopamine- injected mice brain may differ considerably from abnormal cross talk that occurs in human PD. This suggests that neurotoxins-mediated neurodegeneration and behavioral changes are similar but not identical to neurodegenerative and behavioral changes in PD and related diseases. In our opinion, assignment of function to a particular brain area on the basis of neurotoxin-mediated injury may not be justified because one can never be sure that other structures or fibers of passage have not been damaged during neurotoxin-mediated neurotoxicity. This may result in modification of numerous cell signaling steps during striatal deterioration. In genetic animal model, such as in insects and worms, transgenesis has opened new possibilities for screening genes/proteins and potential therapeutic molecules, but like neurotoxin model molecular events are not identical to human neurodegenerative diseases. For example, introducing a human mutated gene in insect, worm or animal induces neurodegeneration, but molecular mechanism and pathogenetic cascades may not be identical to human diseases [66-67]. Furthermore, in humans neurodegenerative diseases should be studied in parallel with their animal models to ensure that the model mimics most clinical manifestations of neurodegenerative diseases. This may not be possible due to a drastic difference in the life span of insects/worms/animals and humans.

In recent years, investigators are developing technologies related to lipidomics, proteomics, and genomics. These procedures can detect minute amounts of biogenic amines and second messengers generated by biogenic amine in biological fluids. Proteomics allows the separation and identification of proteins expressed by neural cells and subcellular organelles. Differential analysis of protein expression patterns in normal brain and brain from different neurological diseases (AD, PD, depression and schizophrenia) will allow the identification of diagnostic and/or prognostic biomarkers. Subsequently, validation of these markers may facilitate identification of new diagnostic tests and therapeutic targets. Using proteomics, more studies can be performed on determination of activities of enzymes associated with biogenic amine metabolism in biological fluids and biopsy and autopsy samples of brain tissue from not only from animal models but from patients with neurodegenerative and neuropsychiatric diseases that involve abnormal biogenic amine metabolism [68-70]. Recently, proteomic approach has been used to investigate the mechanisms that may dictate α-synuclein aggregation, Lewy body formation, and neurodegeneration in brains from PD patients [71,72]. Establishment of automatic systems including databases and accurate analyses of biogenic amines and second messengers generated from enzymic and non-enzymic metabolism of neuronal membrane components will facilitate the identification of key biomarkers associated with neurodegenerative diseases involving abnormal biogenic amine metabolism [73].

Microarray analysis of tissue samples from brain regions associated with AD, PD, depression, and schizophrenia can provide information on candidate genes that influence levels of biogenic amines, oxidative stress, and neuroinflammatory responses. Gene expression analysis of enzymes related to biogenic amine metabolism can also provide information on vulnerable brain regions in neurodegenerative and neuropsychiatric diseases [74]. Thus, gene expression profiles of susceptible neuronal populations may reveal mechanistic clues to the molecular mechanism underlying various neurodegenerative and neuropsychiatric diseases associated with abnormal biogenic amine metabolism. This information will facilitate molecular diagnostics and targets for drug therapy based on gene expression in body fluids such as CSF and blood. Levels second messengers generated by biogenic amine-mediated processes can then be used to monitors responses to drug therapy [68,74,75].

Neuroimaging techniques, namely positron emission tomography (PET), fluorodeoxyglucose-positron-emission tomography (FDG-PET), single photon emission computed tomography (SPECT), magnetic resonance imaging (MRI), functional MRI (fMRI) and diffusion tensor imaging (DTI) are increasingly used to obtain useful information not only on metabolism and rates of generation of second messengers, but also on abnormalities in brain structure and function in mild cognitive impairment (MCI) and its conversion to dementia, as well as early neurodegenerative and neuropsychiatric diseases (AD, PD, depression and schizophrenia) [76,77]. These procedures offer researchers and clinicians a new noninvasive window into the human brain and spinal cord.

PET utilizes incorporation of labeled radioligands to determine of its distribution in various brain regions as a function of time. For example, ^{11}C-RTI32 PET, a marker of noradrenergic and dopaminergic transporter binding, has been used to assess noradrenergic along with dopaminergic terminal function [78]. Serotonergic transporter binding can be monitored with ^{11}C-DASB PET and ^{123}I-β-CIT SPECT, whereas HT1A binding can be analyzed with ^{11}C-WAY100635 PET [78]. Based on the *in vivo* metabolism of various radioligands, PET can be used to image brain signaling and neuroplasticity in normal human brain and brain from patients with neurodegenerative diseases. [^{123}I]β-CIT (2β-carboxymethoxy-3-β-94-iodophenyl tropane), a dopamine transporter ligand/SPECT imaging has been used to study progression of early PD in human patients [79]. Based on [^{123}I]β-CIT/SPECT, it is reported that the rate of dopaminergic neuron loss in PD is significantly greater than that of healthy controls, and [^{123}I]β-CIT SPECT imaging provides a quantitative biomarker for the progressive nigrostriatal dopaminergic degeneration in PD. Molecular imaging with PET and SPECT has also been used to examine the relationship between cognitive processes and components of the dopaminergic system (pre-, intra-, and postsynaptic) in healthy subjects and patient with PD, schizophrenia, HD, and aging [80] It is demonstrated that alterations of DA within the fronto-striato-thalamic circuits might contribute to the cognitive impairments observed in PD, schizophrenia, and normal aging.

In future, β-amyloid, tau, parkin and synuclein imaging can be used as a non-invasive neuroimaging technique to visualize the accumulation of above mentioned proteins along with biogenic amine and biogenic amine generated second messengers in living human brain. Based on this procedure one can determine the accumulation of β-amyloid, tau, parkin and synuclein in the preclinical and clinical stages of AD and PD. Collective evidence suggests that above functional imaging methods can contribute inestimably to the understanding of physiological biogenic amine-mediated memory dysfunction in brains of normal subjects and patients with neurodegenerative diseases.

In conclusion, there is still a lot left to be learned about multiple biogenic amines-mediated signaling pathways, which can hopefully be unraveled by combining and studying different approaches, such as pharmacology, neurochemistry, and neurophysiology with molecular neurobiology and computational analysis. The precise contribution made by cross-talk between different biogenic amine receptors awaits elucidation.

References

[1] Roeder T. Octopamine in invertebrates. *Progress in Neurobiology.* 1999;59:533-561.

[2] Blenau W, Baumann A. Aminergic signal transduction in invertebrates: focus on tyramine and octopamine receptors. *Recent Res Devel Neurochem.* 2003;6:225-240.

[3] Berry MD. Mammalian central nervous system trace amines. Pharmacologic amphetamines, physiologic neuromodulators. *Journal of Neurochemistry.* 2004; 90:257-271.

[4] Paterson IA, Juorio AV, Boulton AA. 2-Phenylethylamine: a modulator of catecholamine transmission in the mammalian central nervous system? *J Neurochem* 1990; 55: 1827-1837.

[5] Berry MD. The Potential of Trace Amines and Their Receptors for Treating Neurological and Psychiatric Diseases. *Reviews on Recent Clinical Trials.*2007;2: 3-19.

[6] Farooqui T. Octopamine-mediated neuronal plasticity in honeybees: implications for olfactory dysfunction in humans. *Neuroscientist.* 2007; 13:304–322.

[7] Nagatsu T, Levitt M, Udenfriend S. Tyrosine hydroxylase. The initial step in norepinephrine biosynthesis. *Journal of Biological Chemistry*. 1964;239:2910–2917.

[8] Fitzpatrick PF. Tetrahydropterin-dependent amino acid hydroxylases. *Annual Review of Biochemistry*. 1999;68:355–381.

[9] Wallace BG. The biosynthesis of octopamine-characterization of a lobster tyramine β-hydroxylase. *Journal of Neurochemistry*. 1976;26:761-770.

[10] Rudnick G. Biogenic Amine Transporters of the Plasma Membrane. In: *Handbook of Biological Physics*. Volume 2, edited by W.N. Konings, H.R. Kaback and J.S. Lolkema. 1996, Elsevier Science B.V.

[11] Masson J, Sagné C, Hamon M, El Mestikawy S. Neurotransmitter transporters in the central nervous system. *Pharmacological Reviews*. 1999; 51: 439-464.

[12] Iversen LL. Uptake processes for biogenic amines. In: Iversen L, Iversen S, Snyder S (eds). *Handbook of Psychopharmacology*. Vol 3: Plenum Press: New York, 1975, pp381-432.

[13] Peter D, Liu Y, Sternini C, de Giorgio R, Brecha N, Edwards RH. Differential expression of two vesicular monoamine transporters. *Journal of Neuroscience*. 1995;15:6179–6188.

[14] Eiden LE, Schafer MK, Weihe E, Schutz B (2004). The vesicular amine transporter family (SLC18): amine/proton antiporters required for vesicular accumulation and regulated exocytotic secretion of monoamines and acetylcholine. *Pflugers Arch-European Journal of Physiology*. *447*: 636–640.

[15] Hansson SR, Hoffman BJ, Mezey E (1998). Ontogeny of vesicular monoamine transporter mRNAs VMAT1 and VMAT2. I. The developing rat central nervous system. *Brain Res Dev Brain Res 110*: 135–158.

[16] Parsons SM. Transport mechanisms in acetylcholine and monoamine storage. *FASEB Journal*. 2000;14:2423-2434.

[17] Arviddson U, Riedi M, Elde R, Meister B. Vesicular acetylcholine transporter (VAChT) protein: A novel and unique marker for cholinergic neurons in the central and peripheral nervous systems. *Journal of Comparative Neurology*. 1997;378;454-467.

[18] Gilmor ML, Nash NR, Roghani A, Edwards RH, Yi H, Hersch SM, Levey AI. Expression of the putative vesicular acetylcholine transporter in rat brain and localization in cholinergic synaptic vesicles. *Journal of Neuroscience*. 1996;16:2179-2190.

[19] de Castro BM, Pereira GS, Magalhães V, Rossato JI, De Jaeger X, Martins-Silva C, Leles B, Lima P, Gomez MV, Gainetdinov RR, Caron MG, Izquierdo I, Cammarota M, Prado VF, Prado MA. Reduced expression of the vesicular acetylcholine transporter causes learning deficits in mice. *Genes Brain and Behavior*. 2009;8:23-35.

[20] Tamminga CA, Holcomb HH. Phenotype of schizophrenia: a review and formulation. *Molecular Psychiatry*. 2005;10: 27–39.

[21] Lohoff FW, Weller AE, Bloch PJ, Buono RJ, Doyle GA, Ferraro TN, Berrettini WH. Association between polymorphisms in the vesicular monoamine transporter 1 gene (VMAT1/SLC18A1) on chromosome 8p and schizophrenia. *Neuropsychobiology*. 2008;57:55-60.

[22] Manji HK, Lenox RH. Signaling: cellular insights into the pathophysiology of bipolar disorder. *Biological Psychiatry. 2000;*48:518–530.

[23] Eiden LE. The vesicular neurotransmitter transporters: current perspectives and future prospects. *FASEB Journal*. 2000;14:2396-2400.

[24] Guillot TS, Shepherd KR, Richardson JR, Wang MZ, Li Y, Emson PC, Miller GW. Reduced vesicular storage of dopamine exacerbates methamphetamine-induced neurodegeneration and astrogliosis. *Journal of Neurochemistry*. 2008;106:2205–2217.

[25] Snyder SH, Kuhar MJ, Green AI, Coyle JT, Shakan EG. Uptake and subcellular localization of neurotransmitters in the brain. *International Review of Neurobiology*. 1970;13:127–158.

[26] Iversen L. Neurotransmitter transporters and their impact on the development of psychopharmacology. *British Journalof Pharmacology*. 2006;147;S82-S88.

[27] Iversen L. Neurotransmitter transporters: fruitful targets for CNS. *Molecular Psychiatry*. 2000;5:357-362.

[28] Javitch JA, D'amoto RJ, Strittmatter SM, Snyder SH. Parkinsonism-inducing neurotoxin. *N*-methyl-4-phenyl-1,2,3,6-tetrahydropyridine: uptake of the metabolite *N*-methyl-4-phenylpyridine by dopamine neurons explains selective toxicity. *Proceedings of the National Academy of Sciences of the United States of America*. 1985;82:2173–2177.

[29] Snyder SH, D'amato RJ. MPTP: a neurotoxin relevant to the pathophysiology of Parkinson's disease. *Neurology*. 1986; 36:250–258.

[30] Uretsky NJ, Iversen LL. Effects of 6-hydroxydopamine on catecholamine containing neurones in the rat brain. *Journal of Neurochemistry*. 1970; 17:269-278.

[31] Bell LJ, Iversen LL, Uretsky NJ. Time course of the effects of 6-hydroxydopamine on catecholamine-containing neurones in rat hypothalamus and striatum. *British Journal of Pharmacology*. 1970;40:790-799.

[32] Blum D, Torch S, Lambeng N, Nissou MF, Benabid AL, Sadoul R, Verna JM. Molecular pathways involved in the neurotoxicity of 6-OHDA, dopamine and MPTP: contribution to the apoptotic theory in Parkinson's disease. *Progress in Neurobiology*. 2001;64:135–172.

[33] Auger J, Boulay R, Jaillais B, Delion-Vancassel S. Analysis of biogenic amines by solid-phase microextraction and high-performance liquid chromatography with electrochemical detection. *Journal of Chromatography*. 2000;870:395–403.

[34] Whiting MJ, Doogue MP. Advances in Biochemical Screening for Phaeochromocytoma using Biogenic Amines. *Clinical Biochemist Reviews*. 2009;30:3-17.

[35] Peaston RT, Weinkove C. Measurement of catecholamines and their metabolites. *Annals of Clinical Biochemistry*. 2004;41:17-38

[36] Sarko J. Antidepressants,old and new. A review of their adverse effects and toxicity in overdose. *Emergency Medicine Clinics of North America*. 2000;18:637-654.

[37] Meibohm B, Derendorf H. Basic concepts of pharmacokinetic/pharmacodynamic (PK/PD) modeling. *International Journal of Clinical Pharmacology, Therapy and Toxicology*. 1997;35:401-413.

[38] Church MK, Gillard M, Sargentini-Maier ML, Poggesi I, Campbell A, Benedetti MS. From pharmacokinetics to therapeutics. *Drug Metabolism Reviews*. 2009;41:455-474.

[39] Farooqui T, Farooqui AA. Aging: an important factor for the pathogenesis of neurodegenerative diseases. *Mechanisms of Ageing and Development*. 2009; 130:203-215.

[40] Farooqui A.A. (2009) *In Beneficial effects of Fish Oil on Human Brain*, 2009, Springer, New York.

[41] Soto C. Unfolding the role of protein misfolding in neurodegenerative diseases. *Nature Reviews Neuroscience*. 2003;4:49-60.

[42] Ross CA, Poirier MA. Protein aggregation and neurodegenerative disease. *Nature Medicine*. 2004; 10 Suppl:S10-S17.

[43] Herrmann N, Lanctôt KL: From transmitters to treatment: the pharmacotherapy of behavioural disturbances in dementia. *Canadian Journal of Psychiatry*. 1997; 42(suppl 1):51S-64S

[44] Lawlor BA, Sunderland T, Mellow AM, Hill JL, Molchan SE, Murphy DL. Hyperresponsivity to the serotonin agonist m-chlorophenylpiperazine in Alzheimer's disease: a controlled study. *Archives of General Psychiatry*. 1989; 46:542–549.

[45] Bierer LM, Haroutunian V, Gabriel S, Knott PJ, Carlin LS, Purohit DP, Perl DP, Schmeidler J, Kanof P, Davis KL. Neurochemical correlates of dementia severity in Alzheimer's disease: relative importance of the cholinergic deficits. *Journal of Neurochemistry*. 1995; 64:749–760

[46] Herrmann N, Lanctôt KL, Khan LR. The role of norepinephrine in the behavioral and psychological symptoms of dementia. *Journal of Neuropsychiatry and Clinical Neurosciences*. 2004;16:261-276.

[47] Rao A.V. and Balachandran B. Role of oxidative stress and antioxidants in neurodegenerative diseases. *Nutritional Neuroscience*. 2002;5:291-309.

[48] Gallagher S. Neurocognitive models of schizophrenia: a neurophenomenological critique. *Psychopathology*. 2004;37:8-19.

[49] Benes FM. Emerging principles of altered neural circuitry in schizophrenia. Brain *Research Brain Research Reviews*. 2000;31:251–269.

[50] Harrison PJ. Neurochemical alterations in schizophrenia affecting the putative receptor targets of atypical antipsychotics. Focus on dopamine (D_1, D_3, D_4) and 5-HT_2a receptors. *Brain*. 1999;122:593–624.

[51] Ito H., Kawashima R., Awata S., Ono S., Sato K., Goto R., Koyama M., Sato M., and Fukuda H. (1996). Hypoperfusion in the limbic system and prefrontal cortex in depression: SPECT with anatomic standardization technique. *Journal of Nuclear Medicine*. 1996;37:410-414.

[52] Kimbrell TA, Ketter TA, George MS, Little JT, Benson BE, Willis MW, Herscovitch P, Post RM. Regional cerebral glucose utilization in patients with a range of severities of unipolar depression. *Biological Psychiatry*. 2002;51:237-252.

[53] Skolnick P, Lequtko P, Lequtko B, Li X, Bymaster FP. Current perspectives on the development of non-biogenic amine-based antidepressants. *Pharmacological Research*. 2001; 43:411-423.

[54] Schmauss M, Messer T. Combining antidepressants: a useful strategy for therapy resistant depression? *Fortschr Neurology Psychiatry*. 2009;77:316-325.

[55] Farooqui, A.A.; Horrocks, L.A. In *Glycerophospholipids in Brain: Phospholipases A_2 in Neurological Disorders* 2007, Springer, New York.

[56] Wood, P.L. (1998). *Neuroinflammation: Mechanism and Management*. 1998, Humana Press, Totowa, New Jersy.

[57] Kidd PM. Neurodegeneration from mitochondrial insufficiency: nutrients, stem cells, growth factors, and prospects for brain rebuilding using integrative management. *Alternative medicine review*. 2005;10:268-293.

[58] Graeber MB, Moran LB. Mechanisms of cell death in neurodegenerative diseases: fashion, fiction, and facts. *Brain Pathology*. 2002;12:385-390.

[59] Andersen JK. Oxidative stress in neurodegeneration: cause or consequence? *Nature Medicine*. 2004;10 Suppl:S18-S25.

[60] Juranek I, Bezek S. Controversy of free radical hypothesis: reactive oxygen species--cause or consequence of tissue injury? *General Physiology and Biophysics*. 2005;24:263-278.

[61] Uretsky NJ, Iversen LL. Effects of 6-hydroxydopamine on catecholamine containing neurones in the rat brain. *Journal of Neurochemistry*. 1970;17:269-278.

[62] Uretsky NJ, Schoenfeld RI. Effect of L-dopa on the locomotor activity of rats pretreated with 6-hydroxydopamine. *Nature New Biology*. 1971;234:157-159.

[63] Mandel RJ, Spratt SK, Synder RO, Leff SE. Midbrain injection of recombinant adeno-associated virus encoding rat glial cell line-derived neurotrophic factor protects nigral neurons in a progressive 6-hydroxydopamine-induced degeneration model of Parkinson's disease in rats. *Proceedings of the National Academy of Sciences of the United States of America*. 1997;94:14083-14088.

[64] Schmidt N, Ferger B. Neurochemical findings in the MPTP model of Parkinson's disease. *Journalof Neural Transmission*. 2001;108(11):1263-1282.

[65] Smeyne RJ, Jackson-Lewis V. The MPTP model of Parkinson's disease. *Brain Research Molecular Brain Research*. 2005;134:57-66.

[66] Langui D, Lachapelle F, Duyckaerts C. Animal models of neurodegenerative diseases. *Medical Science (Paris)*. 2007;23:180-186.

[67] Iijima K, Iijima-Ando K. *D*rosophila models of Alzheimer's amyloidosis: the challenge of dissecting the complex mechanisms of toxicity of amyloid-β 42. *Journal of Alzheimer's Disease*. 2008; 15:523-540.

[68] Jin J, Meredith GE, Chen L, Zhou Y, Xu J, Shie FS, Lockhart P, Zhang J. Quantitative proteomic analysis of mitochondrial proteins: relevance to Lewy body formation and Parkinson's disease. *Brain Research. Molecular Brain Research*. 2005;134:119-138.

[69] Hattoria N, Wanga M, Taka H, Fujimura T, Yoritaka A, Kubo S, Mochizuki H. Toxic effects of dopamine metabolism in Parkinson's disease. *Parkinsonism and Related Disorders*. 2009;15 Suppl 1: S35-S38.

[70] Van Laar VS, Dukes AA, Cascio M, Hastings TG. Proteomic analysis of rat brain mitochondria following exposure to dopamine quinone: implications for Parkinson disease. *Neurobiology of Disease*. 2008;29:477-489.

[71] Zhang J, Goodlett DR. Proteomic approach to studying Parkinson's disease. *Mol. Neurobiol.* 2004;29:271-288.

[72] Fasano M, Lopiano L. α-synuclein and Parkinson's disease: a proteomic view. *Expert Rev. Proteomics*. 2008;5:239-248.

[73] Chin MH, Qian WJ, Wang H, Petyuk VA, Bloom JS, Sforza DM, Laćan G, Liu D, Khan AH, Cantor RM, Bigelow DJ, Melega WP, Camp DG 2nd, Smith RD, Smith DJ. Mitochondrial dysfunction, oxidative stress, and apoptosis revealed by proteomic and transcriptomic analyses of the striata in two mouse models of Parkinson's disease. *Journal of Proteome Research*. 2008;7:666-677.

[74] Bogaerts V, Theuns J, van Broeckhoven C. (2008). Genetic findings in Parkinson's disease and translation into treatment: a leading role for mitochondria? *Genes, Brain and Behavior*. 2008;7:129-151.

[75] Rite I, Argüelles S, Venero JL, García-Rodriguez S, Ayala A, Cano J, Machado A. Proteomic identification of biomarkers in the cerebrospinal fluid in a rat model of nigrostriatal dopaminergic degeneration. *Journal of Neuroscience Research*. 2007;85:3607-3618.

[76] Thobois S, Jahanshahi M, Pinto S, Frackowiak R, Limousin-Dowsey P. PET and SPECT functional imaging studies in Parkinsonian syndromes: from the lesion to its consequences. *Neuroimage*. 2004;23:1-16.

[77] Weiner MW. Editorial: Imaging and Biomarkers Will be Used for Detection and Monitoring Progression of Early Alzheimer's Disease. *Journal of Nutrition, Health and Aging*. 2009;13:332.

[78] Brooks DJ, Piccini P. Imaging in Parkinson's disease: the role of monoamines in behavior. *Biological Psychiatry*. 2006;59:908-918.

[79] Marek K, Innis R, van Dyck C, Fussell B, Early M, Eberly S, Oakes D, Seibyl J. $^{[123I]}$beta-CIT SPECT imaging assessment of the rate of Parkinson's disease progression. *Neurology*. 2001;57:2089-2094.

[80] Cropley VL., Fujita M, Innis RB., Nathan PJ. Molecular imaging of the dopaminergic system and its association with human cognitive function. *Biological Psychiatry*. 2006;59:898-907.

INDEX

A

B

C

D

E

F

H

I

J

K

L

M

N

O

P

Q

R

S

T

U

V

W

X

Y

Z